Recent Progress in Spectroscopy

Recent Progress in Spectroscopy

Editor: Conrad Sinclair

NY RESEARCH
P R E S S

New York

Published by NY Research Press
118-35 Queens Blvd., Suite 400,
Forest Hills, NY 11375, USA
www.nyresearchpress.com

Recent Progress in Spectroscopy
Edited by Conrad Sinclair

© 2019 NY Research Press

International Standard Book Number: 978-1-63238-660-1 (Hardback)

Cataloging-in-Publication Data

Recent progress in spectroscopy / edited by Conrad Sinclair.
    p. cm.
Includes bibliographical references and index.
ISBN 978-1-63238-660-1
1. Spectrum analysis. 2. Spectroscope. I. Sinclair, Conrad.
QD272.S6 R43 2019
535.84--dc23

Contents

Preface

The world is advancing at a fast pace like never before. Therefore, the need is to keep up with the latest developments. This book was an idea that came to fruition when the specialists in the area realized the need to coordinate together and document essential themes in the subject. That's when I was requested to be the editor. Editing this book has been an honour as it brings together diverse authors researching on different streams of the field. The book collates essential materials contributed by veterans in the area which can be utilized by students and researchers alike.

Spectroscopy studies the interactions between matter and electromagnetic radiation. This is achieved by measuring radiation as a function of wavelength or frequency. Interactions of radiating energy with atoms, molecules and nuclei have led to the development of atomic, molecular and NMR spectroscopy. Other spectroscopic methods like auger spectroscopy, hadron spectroscopy, photoacoustic spectroscopy, spin noise spectroscopy, etc. are distinguished by their specific applications. Some of the practical applications of spectroscopy are measurement of different compounds in food, cure monitoring, toxic compounds measurement in blood samples, etc. This book provides comprehensive insights into the field of spectroscopy. It also traces the progress of this field and highlights some of its key concepts and applications. It is appropriate for students seeking detailed information in this area as well as for experts.

Each chapter is a sole-standing publication that reflects each author's interpretation. Thus, the book displays a multi-facetted picture of our current understanding of application, resources and aspects of the field. I would like to thank the contributors of this book and my family for their endless support.

Editor

Quantitative Estimating Salt Content of Saline Soil Using Laboratory Hyperspectral Data Treated by Fractional Derivative

Dong Zhang,[1,2] **Tashpolat Tiyip,**[1,2] **Jianli Ding,**[1,2] **Fei Zhang,**[1,2] **Ilyas Nurmemet,**[1,2] **Ardak Kelimu,**[1,2] **and Jingzhe Wang**[1,2]

[1]*College of Resources and Environment Science, Xinjiang University, Urumqi 830046, China*
[2]*Key Laboratory of Oasis Ecology, Xinjiang University, Urumqi 830046, China*

Correspondence should be addressed to Tashpolat Tiyip; tash@xju.edu.cn

Academic Editor: Petre Makreski

Most present researches on estimation of soil salinity by hyperspectral data have focused on the spectral reflectance or their integer derivatives but ignored the fractional derivative information of hyperspectral data. Motivated by this situation, the selected study area is the Ebinur Lake basin located in the southwest border in the Xinjiang Uygur Autonomous Region, China, with severe salinization. The field work was conducted from 15 to 25 October, 2014, and a total of 180 soil samples were collected from 45 sampling sites; after measuring the soil salt content and spectral reflectance in the laboratory, the range from 0 to 2 was divided into 11 orders (interval 0.2) and then the hyperspectral data were treated by 4 kinds of mathematical transformations and 11 orders of fractional derivatives. Combined with the soil salt content, partial least square regression method was applied for model calibrations and predictions and some indexes were used to evaluate the performance of models. The results showed that the retrieval model built up by 250 bands based on 1.2-order derivative of $1/\lg R$ had excellent capacity of estimating soil salt content in the study area ($RMSE_C = 14.685$ g/kg, $RMSE_P = 14.713$ g/kg, $R^2_C = 0.782$, $R^2_P = 0.768$, and RPD = 2.080). This study provides an application reference for quantitative estimations of other land surface parameters and some other applications on hyperspectral technology.

1. Introduction

Soil salinization is one of the most common but serious environmental problems worldwide and is considered as one of the main paths to land desertification [1]. Due to large evaporation and higher levels of groundwater table with relatively high soluble salt content [2], it often occurs in fragile arid and semiarid regions and causes the productivity loss of irrigated farmlands [3]. On the global scale, approximately 20% of irrigated lands are confronted with a severe threat of salinization and this figure will increase with great population pressure [4].

Faced with such large amounts of salt-affected land, timely detection and assessment of soil salinization become therefore extremely necessary and urgent for sustainable development [5]. However, conventional methods often require intensive field investigations restricted by limited funds and labor; thus, these could not meet the need of salinization monitoring for large areas [6]. Because of low-cost, rapid data acquisition, and large area coverage [7], remote sensing (especially hyperspectral remote sensing) shows as a promising tool to substitute or complement traditional methods and provides an overview of salinization on different spatial scales, and hyperspectral techniques have been successfully used for quantitative analysis of some indexes of the soil salinization [8–11].

Among the spectral analytical methods, derivative spectroscopy is a powerful mathematical tool and provides more useful information of spectral data than untreated data [12]. It is well-known that first derivative is the slope of the spectral curve and second derivative means the change in slope, that is, curvature of the spectral curve [13, 14]. Derivative analysis of hyperspectral data has been widely used to eliminate background noises, reduce the effects of baseline, solve overlapping problems, sharpen spectral features, capture subtle details of spectral curves, and increase the estimation accuracy of land surface parameters [15–20]. However, derivative spectroscopy has some disadvantages, such as spectral information loss and amplification of high-frequency noise [16].

As a more general case, fractional derivative is similar to integer derivative but extends the order of derivative to arbitrary [21]. And there are some successful applications on hyperspectral datasets treated by fractional derivative. Schmitt [22] introduced this method and applied it to laboratory measured near-infrared spectra of a liquid mixture of hemoglobin and milk. He found that fractional derivative affords more flexibility than integer derivative to adjust the order of the derivative to reduce baseline offsets and minimize high-frequency noise. Tong et al. [23] used fractional order Savitzky-Golay derivative (FOSGD) and stability competitive adaptive reweighted sampling (SCARS) in simulated, diesel, and Honghe tobacco spectral datasets to improve the performance of the multivariate calibration model. And they found that FOSGD has a better capacity to balance the contradiction of resolution and signal strength than integer derivative.

However, there are few studies on fractional derivative used in hyperspectral data of saline soil, and, to this regard and motivated by the previous works, this current research attempts to use laboratory hyperspectral data treated by fractional derivative combining with soil salt content data to build up a regression model for better accuracy in estimation. Specifically, the study aims to (1) analyze the relationship between soil salt content and hyperspectral data treated by fractional derivative, (2) develop a quantitative model for salt content estimation based on different fractional derivative, and (3) research the variation tendencies of some parameters of retrieval models with the order increasing.

2. Fractional Derivative Spectrometry Method

Fractional calculus is a theory branch of mathematics with a long history since the end of the 17th century and generalizes the classic integer derivative to arbitrary (noninteger) order [21, 24–26]. Fractional derivative has been widely used in some engineering fields because the models described by fractional derivative have better accuracy and higher efficiency than these built up based on integer derivative [27].

Although fractional derivative has a long history and many successful applications, its mathematical definition has still not been unified yet [28]. The most popular and often used definitions are Grünwald-Letnikov (G-L), Riemann-Liouville (R-L), and Caputo [26]. Due to being less complex than the others [21], G-L definition was employed in this study.

The v-order G-L fractional derivative of function $f(x)$ on the section $[a, b]$ is defined as

$$d^v f(x)$$
$$= \lim_{h \to 0} \frac{1}{h^v} \sum_{m=0}^{[(b-a)/h]} (-1)^m \frac{\Gamma(v+1)}{m!\Gamma(v-m+1)} f(x-mh), \tag{1}$$

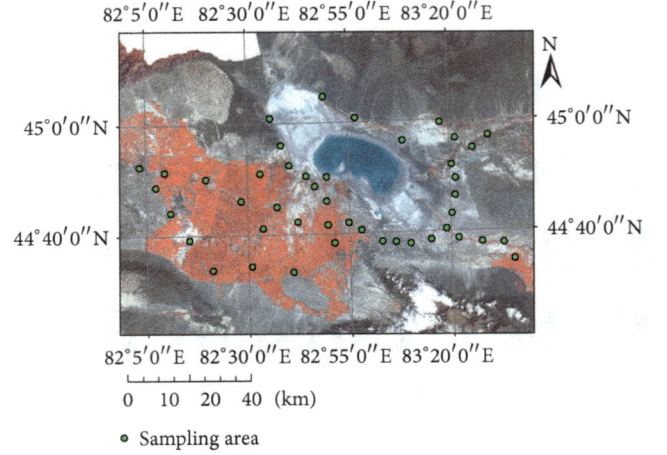

FIGURE 1: Study area and distribution of sampling sites.

where $[(b-a)/h]$ is the integer part of $(b-a)/h$ and the Gamma function is defined as follows [29]:

$$\Gamma(z) = \int_0^\infty \exp(-u) u^{z-1} du = (z-1)!. \tag{2}$$

Consider the fact that the instrument for hyperspectral measurement used in this study has a resampling spectral resolution of 1 nm, and, thus, set $h = 1$ and (1) becomes

$$\frac{d^v f(x)}{dx^v} \approx f(x) + (-v) f(x-1)$$
$$+ \frac{(-v)(-v+1)}{2} f(x-2) \tag{3}$$
$$+ \cdots \frac{\Gamma(-v+1)}{n!\Gamma(-v+n+1)} f(x-n).$$

Therefore, (3) could be regarded as the numerical algorithm for calculating the fractional derivative of hyperspectral data [30, 31], and zeroth order means the hyperspectral data are not treated by derivative algorithm.

3. Materials and Methods

3.1. Study Area. The Ebinur Lake ($44°44'$ ~$45°10'$N, $82°35'$ ~ $83°11'$E), situated in the southwest border area of China, is the biggest saltwater lake (about $542\,km^2$) in the Xinjiang Uygur Autonomous Region (Figure 1) [32, 33]. The lake is surrounded by the mountains on north, west, and south side [34]. This area is far from the oceans and belongs to temperate continental arid desert climate with little mean annual precipitation (100~200 mm), strong annual potential evaporation (1500~2000 mm), and strong wind (approximately 164 days per year with a wind speed above 8 m/s) [33, 35, 36]. Because of particular climate and position coupled with irrational human activities, the phenomenon of soil salinization and desertification in this area is very common but strong, and the ecological environment is extremely fragile.

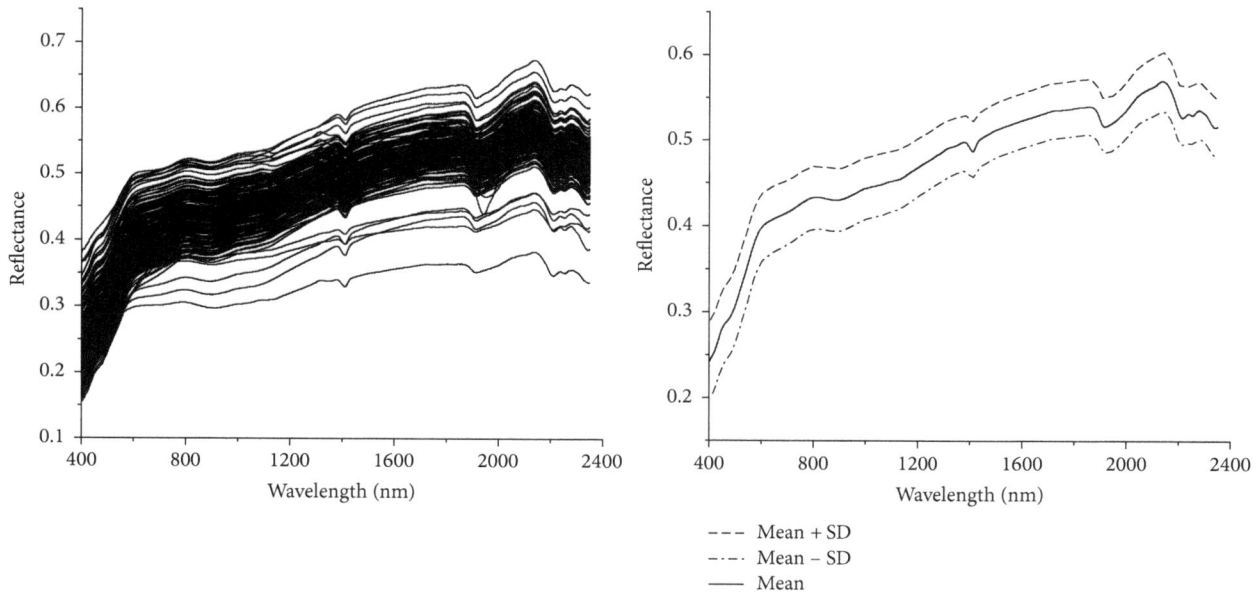

FIGURE 2: Smoothed spectral curves of 180 samples.

3.2. Field Campaign and Laboratory Experiment.

The field work of this study was conducted from 15 to 25 October, 2014, and a total of 180 soil samples (depth 0~20 cm) were collected from 45 sites (30 × 30 m square area, 4 samples in each site) (Figure 1). Before sampling, a handheld global positioning system equipment was used to record the coordinate of the sampling site and landscape photographs of the site were taken. Each soil sample (about 1 kg) was put into a plastic bag, sealed, labeled, and brought back to the laboratory. After being air dried sufficiently, all soil samples were crushed and passed through a 2 mm sieve for removing stones, weed roots, and other impurities. Each sample was divided into two equal parts for further analyses (soil properties analysis and laboratory spectroscopy measurement separately). The soil salt content and pH value of 180 soil samples were determined by a WTW inoLab® Multi 3420 Set B multiparameter measuring instrument (Wissenschaftlich-Technische Werkstätten GmbH, Germany) in 1 : 5 soil to distilled water extracts (20 g of soil sample added into 100 mL of distilled water) at 25°C.

3.3. Laboratory Reflectance Measurement.

For the purpose of controlling the light condition [37, 38], the reflectance measurement was conducted in the dark room with a widely used ASD FieldSpec®3 portable spectroradiometer (Analytical Spectral Device Inc., USA). The instrument covers the range from visible near-infrared to short wave infrared (350~2500 nm) with 2151 bands resampled to 1 nm [2, 39]. Petri dishes with a diameter of 15 cm and a depth of 2 cm were used to load sieved soil samples. After fully filling the dishes with the samples, the surfaces were scraped with a plastic ruler to ensure the same flat measurement surface [40]. The samples were illuminated with two 90 W tungsten halogen light sources placed on either side of the sample, and the light beams were set at 30° from vertical direction and the

distance between each of the lamps and the sample was set at 50 cm. The probe with an 8° viewing angle was fixed at a height of 10 cm perpendicular to the surface of the soil sample. The spectrometer was calibrated approximately every 10 minutes by measurements of dark current and a standard white spectralon reflectance panel (Spectralon Labsphere Inc., USA) [39, 41]. The spectral reflectance of each sample was collected 20 times.

3.4. Spectral Data Processing.

Before further analysis, spectral data pretreatment is a necessary and vital step to reduce the calculation errors. In order to minimize instrument noise [2], these 20 spectral curves of each sample were averaged after splice correction by ViewSpecPro software (version 6.0.11). Due to low signal-to-noise ratios [42], marginal ranges from 350 to 400 nm and from 2351 to 2500 nm were removed and not used in this study. Then, 180 spectral curves of sample were smoothed by Savitzky-Golay filter (polynomial order of 2 and frame size of 5, default settings in OriginPro 9.0.0) [43]. Smoothed spectral curves of soil samples are shown in Figure 2.

After preprocessing, in order to change nonlinear relations to linear and get more modelling results, the hyperspectral reflectance data (R) of 180 samples were transformed by some commonly nonlinear functions: root mean square ($\sqrt{R}$), inversion ($1/R$), logarithm ($\lg R$), and logarithm-inversion ($1/\lg R$). Particularly, $\lg(1/R)$ spectra was commonly used because absorbing components and their contributions often have near-linear relations with the $\lg(1/R)$ value [44], and because $\lg(1/R) = -\lg R$, so here $\lg R$ was applied in further modelling.

According to (3), their (R, $\sqrt{R}$, $1/R$, $\lg R$, and $1/\lg R$) 0~2nd fractional derivatives (interval 0.2) were calculated and the correlation coefficients between the soil salt content and each derivative treated data were computed under

TABLE 1: Statistical results of salt content (g/kg) and pH value.

Item	Min.	Max.	Mean	SD	CV/%	Skewness	Kurtosis
Salt content	0.0	196.0	14.739	15.610	105.909	3.730	15.202
pH value	7.9	9.718	8.546	0.415	4.851	1.028	0.537

SD: standard derivation; CV: coefficient of variation.

the Java programming integrated development platform Eclipse.

3.5. Estimation Model and Prediction Accuracy. In the aspect of quantitative research on hyperspectral data, partial least squares regression (PLSR) has been proved as a robust and reliable mathematical tool because of its advantage of solving colinearity problems [45–50]. Thus, in this research, PLSR was applied for model calibrations and predictions of soil salt content based on the hyperspectral data treated by G-L fractional derivative.

In order to evaluate the performances and accuracy of estimation models built up by PLSR, five indexes of models: the determinant coefficients of calibration (R^2_C) and prediction (R^2_P), root mean square errors of calibration ($RMSE_C$) and prediction ($RMSE_P$), and ratio of performance to deviation (RPD), were employed to perform assessment of calibrated models. Usually, a good and stable model should have high R^2_C, R^2_P, and RPD and low $RMSE_C$ and $RMSE_P$ [51–53]. In this step, all the calculations were carried out by MATLAB R2014b software (MathWorks Inc., USA).

4. Results, Analyses, and Discussion

4.1. Salinity Parameters. The descriptive statistics for the soil salt content and pH values of 180 samples collected in the study area are presented below in Table 1. The soil salt content exhibited a wide range from 0.0 to 196.0 g/kg with a mean value of 14.739 g/kg, a standard deviation (SD) of 15.610 g/kg, and a fairly high coefficient of variation (CV) of 105.909% (>100%). According to the soil salinity classification [54], the numbers of nonsaline, slightly, moderately, and heavily saline soil samples were 85, 33, 20, and 42 respectively. The pH value varied from 7.9 to 9.718 with a very low CV of 4.851% (<10%) [55]. Among 180 samples, there were 100 alkaline samples (pH between 7.5 and 8.5) and 80 strong alkaline samples (pH > 8.5) [56].

4.2. Spectral Features. On the basis of the different degrees of soil salinity mentioned above, 180 soil samples were classified into 4 categories and spectral curves of each category were averaged as a representative spectral curve of this degree (Figure 3). Four spectral curves followed similar basic shapes and there were three obvious absorption features located near 1400, 1900, and 2200 nm, respectively [38, 57]. Among 4 categories, nonsaline soil showed lowest reflectance and slightly saline soil displayed highest reflectance. It was easy to distinguish the differences among slightly, heavily, and nonsaline soil through the entire spectrum range (400~2350 nm), and, however, spectral curves of heavily and moderately

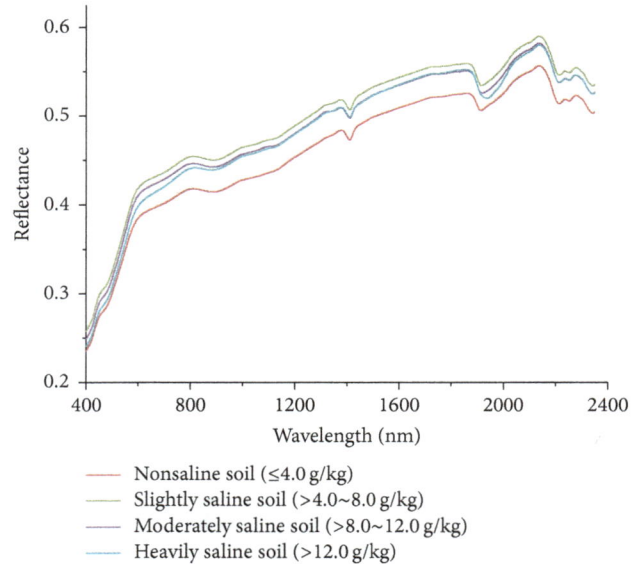

FIGURE 3: Spectral curves of soils with different degrees of salinization.

saline soils had some overlap sections but could be discriminated approximately from 400 to 900 nm and from 1900 to 2050 nm.

4.3. Correlations between Salt Content and Spectra. Band selection is an important process for constructing the regression model [58], and correlation coefficients between salt content and spectral reflectance are usually used to identify soil salinity sensitive bands [10]. All the correlation coefficients between soil salt content and fractional derivative values of raw reflectance data and mathematical transformations were tested with the significance level of 0.01 ($|r|$ = 0.192 or above). The curves of correlation coefficients of raw reflectance data are plotted in Figure 4. For raw reflectance data, no band passed the significance test at the level of 0.01, but with the order of derivative increasing, the correlation coefficients were raised beyond the 0.01 level in some wavelength ranges. In addition when the order increased from 0 to 0.6, variation tendency among the correlation coefficient curves of different orders detailed in the range from 600 to 1100 nm and from 2000 to 2200 nm and some other ranges, but when the order was greater than 0.6, the curves fluctuated greatly, and lacked regularity; thus more details could not be found in Figure 4.

In Figure 4, it is not clear how many bands passed the significance test at the level of 0.01, thus, the numbers of raw reflectance and 4 other transformations are counted and their

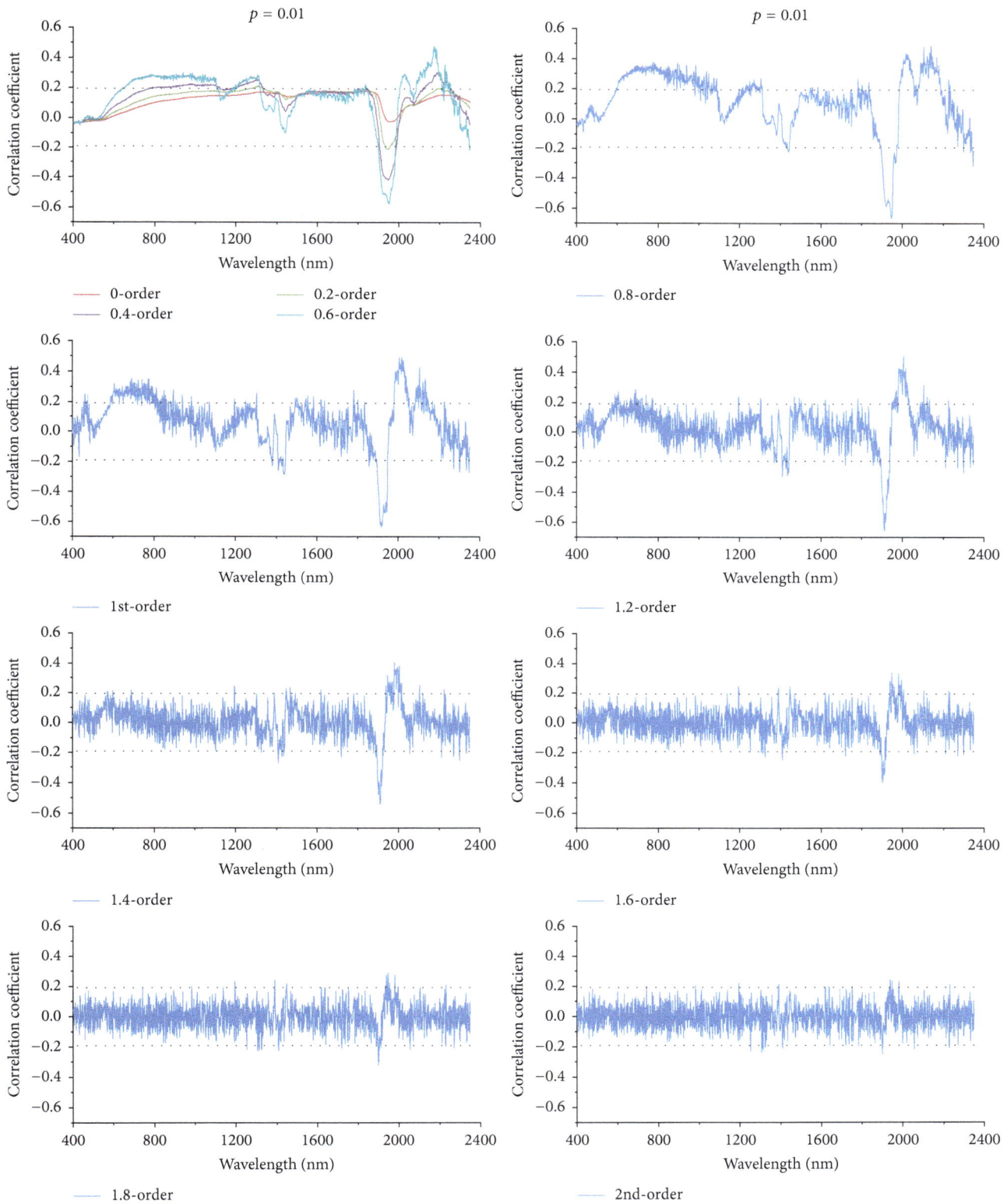

FIGURE 4: Correlation coefficients between salt content and raw reflectance data treated by fractional derivatives.

TABLE 2: The results of the models based on raw reflectance.

Order	Bands	Components	RMSE$_C$	R^2_C	RMSE$_P$	R^2_P	RPD
0	0						
0.2	112	3	26.644	0.283	25.004	0.332	1.224
0.4	736	3	25.165	0.361	22.733	0.448	1.346
0.6	907	4	19.260	0.626	16.552	0.707	1.849
0.8	845	4	17.642	0.686	16.484	0.710	1.856
1	510	4	16.143	0.737	16.938	0.694	1.807
1.2	234	4	15.394	0.761	16.599	0.706	1.844
1.4	123	3	16.924	0.711	16.743	0.701	1.828
1.6	75	3	17.919	0.676	19.965	0.575	1.533
1.8	43	3	21.438	0.536	23.999	0.385	1.275
2	29	2	26.596	0.286	30.427	0.011	1.006

TABLE 3: The results of the PLSR models based on $\sqrt{R}$.

Order	Bands	Components	RMSE$_C$	R^2_C	RMSE$_P$	R^2_P	RPD
0	0						
0.2	120	3	26.640	0.283	25.038	0.331	1.222
0.4	782	3	25.168	0.361	22.844	0.443	1.340
0.6	908	4	19.918	0.600	16.977	0.692	1.803
0.8	782	4	17.886	0.677	16.684	0.703	1.834
1	488	4	16.736	0.717	17.660	0.667	1.733
1.2	223	4	15.338	0.762	16.624	0.704	1.841
1.4	129	3	17.089	0.705	17.010	0.691	1.799
1.6	75	3	17.847	0.678	20.107	0.568	1.522
1.8	44	3	21.305	0.541	24.122	0.378	1.268
2	31	2	26.482	0.292	30.565	0.002	1.001

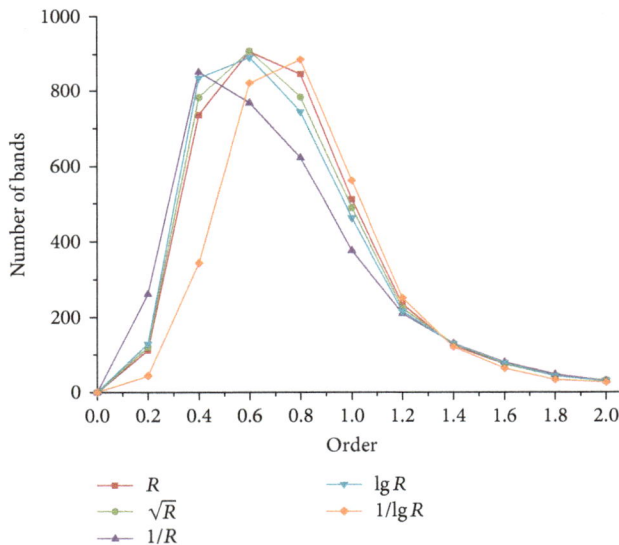

FIGURE 5: The numbers of bands passed the significance test and trend lines.

trend lines are shown in Figure 5. For these 5 mathematical forms of reflectance (R, $\sqrt{R}$, $1/R$, $\lg R$, and $1/\lg R$), no band passed the significance test, but with the increase of the derivative order, the numbers followed first increasing then decreasing trend, and all reached maximum at fractional order (R, $\sqrt{R}$, and $\lg R$ at 0.6 order, $1/\lg R$ at 0.8, and $1/R$ at 0.4 separately).

4.4. Model Calibration and Validation. The 180 samples were randomly divided into two parts: 144 (80%) for model calibration and 36 (20%) for model validation. In order to make full use of the hyperspectral data and take advantages of PLSR, all the bands whose correlation coefficient passed the significance test at the level of 0.01 were used as features to participate in the modelling process. The calibration and validation results of 55 models based on spectral data treated by mathematical transformations and different orders of fractional derivative are summarized in Tables 2–6.

As to the integer derivative, the models established on first derivative were much better than second derivative and the data untreated by derivative, because there was no band of spectral data without derivative treatment that passed the significance test and it had more obvious effect for first derivative than second derivative on raising the correlation coefficient. But, for fractional derivative, things had changed; the models based on the data treated by (3) had better results than the integer order models (lower RMSE$_C$ and RMSE$_P$ and higher R^2_C, R^2_P, and RPD).

TABLE 4: The results of the PLSR models based on $1/R$.

Order	Bands	Components	$RMSE_C$	R^2_C	$RMSE_P$	R^2_P	RPD
0	0						
0.2	261	3	27.107	0.258	26.925	0.226	1.136
0.4	851	3	25.111	0.363	23.346	0.418	1.311
0.6	767	3	22.481	0.489	20.527	0.550	1.491
0.8	622	4	19.510	0.615	17.151	0.686	1.784
1	376	4	18.012	0.672	18.780	0.623	1.629
1.2	209	4	16.672	0.719	18.360	0.640	1.667
1.4	130	3	18.366	0.659	17.858	0.659	1.713
1.6	80	3	18.121	0.668	20.744	0.540	1.475
1.8	47	3	21.771	0.521	23.194	0.425	1.319
2	29	2	26.389	0.297	29.987	0.040	1.020

TABLE 5: The results of the PLSR models based on $\lg R$.

Order	Bands	Components	$RMSE_C$	R^2_C	$RMSE_P$	R^2_P	RPD
0	0						
0.2	129	3	26.377	0.297	24.722	0.347	1.238
0.4	835	3	25.152	0.361	22.949	0.437	1.333
0.6	891	4	20.108	0.591	17.347	0.678	1.764
0.8	743	4	18.374	0.659	16.791	0.699	1.822
1	462	2	22.825	0.474	22.332	0.467	1.370
1.2	215	4	15.682	0.751	17.684	0.666	1.730
1.4	129	3	17.658	0.685	17.448	0.675	1.754
1.6	78	3	17.917	0.675	20.493	0.551	1.493
1.8	43	3	21.585	0.529	24.202	0.374	1.264
2	30	2	26.516	0.290	30.414	0.012	1.006

RPD is an important parameter to evaluate the performance of regression models and the ranges of <1.4, 1.4~2.0, and >2.0 correspondingly mean the model has a poor, receptible, and excellent capacity of predicting soil salinity [57, 59]. There were 30 models having acceptable results with RPD > 1.4, and among these 30 models there was only one best model which was built up by 250 bands based on 1.2-order derivative of $1/\lg R$ with 4 principal components, RPD = 2.080 (>2.0), lowest $RMSE_C$ (14.685 g/kg) and $RMSE_P$ (14.713 g/kg), highest R^2_C (0.782), and R^2_P (0.768). The scatter plot of measured and predicted soil salt content of the best model is shown in Figure 6. R^2 of measured and predicted values in calibration and validation set both reached 0.782 and all these figures meant that the calibrated model based on hyperspectral data treated by fractional derivative could be used to estimate the soil salinity in the study area.

5. Discussion

According to (3), when the order v = 1 or 2, the equation becomes the same as first- and second-derivative equation with derivative window that equals 1 [42, 60–62], and it can be seen from (3) that the integer derivative value of a band is related to the bands in the derivative window, while the fractional derivative value of a band has connections with the bands whose wavelength is less than this band. And that is

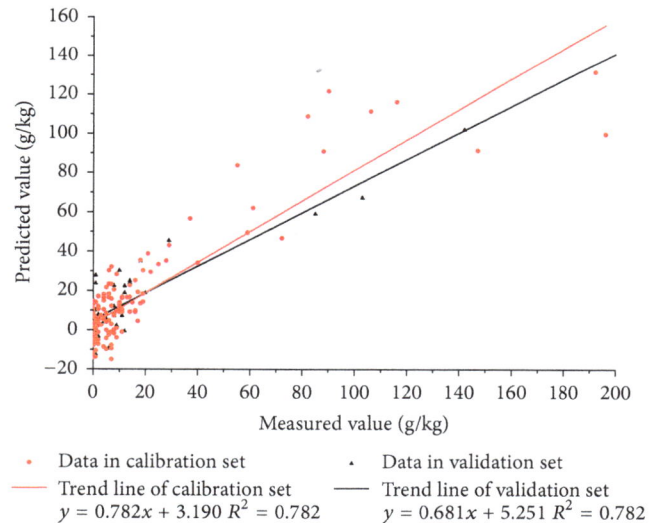

FIGURE 6: The relationship between measured and predicted soil salt content in calibration and validation set.

- Data in calibration set
- Data in validation set
- Trend line of calibration set $y = 0.782x + 3.190$ $R^2 = 0.782$
- Trend line of validation set $y = 0.681x + 5.251$ $R^2 = 0.782$

a big difference between fractional and integer derivative and the main cause of the results of this study, in which it is known that the integer derivative is unique and local, while fractional derivative is usually nonlocal and has memory [27, 63].

TABLE 6: The results of the PLSR models based on $1/\lg R$.

Order	Bands	Components	$RMSE_C$	R^2_C	$RMSE_P$	R^2_P	RPD
0	0						
0.2	44	2	26.712	0.279	24.910	0.337	1.228
0.4	344	2	25.786	0.328	24.180	0.375	1.265
0.6	820	3	21.687	0.525	17.868	0.659	1.712
0.8	885	4	16.670	0.719	15.541	0.742	1.969
1	560	4	15.685	0.751	16.339	0.715	1.873
1.2	250	4	14.685	0.782	14.713	0.768	2.080
1.4	120	3	16.689	0.718	17.844	0.660	1.715
1.6	63	3	18.855	0.641	21.079	0.525	1.452
1.8	33	2	25.618	0.337	27.881	0.170	1.097
2	26	2	27.024	0.262	29.384	0.078	1.041

Traditionally, there are big differences among the shapes of zeroth, first, and second derivatives, but fractional derivative could provide more useful information from hyperspectral data, because the order is extended to noninteger and it could add detail curves among integer derivative of spectral curves and, as a result, this effect could be directly manifested among the correlation coefficient curves of different orders of fractional derivative (Figure 4) [16, 63].

In the result section, the integer derivative indeed raised the correlation coefficients between reflectance data and soil salt content and also improved performances of models built up by PLSR to some extent, but, compared with fractional derivative, it truly lost information of some bands and decreased accuracy and performances of estimation models. Thus, fractional derivative could compensate for this disadvantage due to the flexibility in practice for conveniently choosing the suitable derivative order [64].

As is known to all, the first and second derivatives correspondingly mean the slope and curvature of spectral curves, and, however, the physical meaning of fractional derivative in spectroscopy has not been clarified yet. But it suggests that the order between 0 and 2 of fractional derivative could be described as the sensitivity to the slope and curvature of spectral curves; when the order increases from 0 to 1, the derivative value becomes more sensitive to the slope and less sensitive to reflectance, and while the order increases from 1 to 2, the derivative value turns out more sensitive to the curvature and less sensitive to the slope [22]. According to these suggestions, in this study, differences among correlation coefficient curves, the numbers of bands that passed the significance test, and the accuracy indexes of regression models ($RMSE_C$, $RMSE_P$, R^2_C, R^2_P, and RPD) were all manifestations of this sensitivity, and their tendencies did not directly increase or decrease but showed ups and downs to a certain degree, and some of them achieved optimal values at fractional orders (Figures 4 and 5 and Tables 2–6). The process of modelling in this study was a trying procedure to find a balance with suitable order, lowest $RMSE_C$, $RMSE_P$, and highest R^2_C, R^2_P, and RPD by PLSR; according to the performance evaluation indexes, the best model was finally discovered.

Indeed, there are some other studies with better performances than ours. Mashimbye et al. [65] used bagging PLSR with first-derivative reflectance data to estimate soil electrical conductivity in South Africa and validation R^2 reached 0.85. Peng et al. [66] combined visible near-infrared with mid-infrared hyperspectral data to predict total dissolved salts in Xinjiang by PLSR, and found $RMSE_C = 0.20$ g/kg, $RMSE_P = 0.43$ g/kg, $R^2_C = 0.96$, $R^2_P = 0.70$, and RPD = 2.14. According to the result of our study, they might grasp more details if fractional derivative was applied in their researches.

6. Conclusions

In this paper, the Ebinur Lake in the northwest border of Xinjiang, China, was chosen as the research area; combined with laboratory measured soil salt content and hyperspectral data of 180 samples, PLSR was employed to build up quantitative estimation models of soil salinity based on the hyperspectral data treated by mathematical transformations and fractional derivatives. The conclusions are as follows:

(1) The best retrieval model was built up by 250 bands based on 1.2-order derivative of $1/\lg R$ with 4 principal components, lowest $RMSE_C$ (14.685 g/kg) and $RMSE_P$ (14.713 g/kg), highest R^2_C (0.782), R^2_P (0.768), and RPD (2.080 > 2.0). This model had excellent capacity of estimating soil salt content in the study area.

(2) During the course of data processing and model calibration and validation, differences among correlation coefficient curves and the numbers of bands that passed the significance test, $RMSE_C$, $RMSE_P$, R^2_C, R^2_P, and RPD, showed some variation tendencies which were manifestations of the sensitivity to the slope and curvature of spectral curves.

(3) In the process of modelling, the integer derivative lost information and accuracy of quantitative estimation and, to some content, fractional derivative could compensate for this disadvantage because of the flexibility for choosing the suitable derivative order.

As an extension of integer derivative, and due to the flexibility of the order selection, fractional derivative could enrich the method of data preprocessing and dig for information lost by integer derivative from the spectral demission for making full use of hyperspectral data. Although this study is just an application of fractional derivative, it provides a reference for estimation of other parameters by using hyperspectral technology. Further researches should be focused on the physical meaning of fractional derivative in spectroscopy and promote for space-borne hyperspectral technology for precisely monitoring land surface parameters on large spatial scales.

Competing Interests

The authors declare that there are no competing interests regarding the publication of this paper.

Acknowledgments

This work was supported by National Natural Science Foundation of China (41130531, U1138303, U1503302, and 41561089) and the National Key Technology R&D Program (2014BAC15B01). The authors thank all the students in their team for their significant contribution to the fieldwork and laboratory experiments.

References

[1] M. Bouaziz, J. Matschullat, and R. Gloaguen, "Improved remote sensing detection of soil salinity from a semi-arid climate in Northeast Brazil," *Comptes Rendus—Geoscience*, vol. 343, no. 11-12, pp. 795–803, 2011.

[2] A. Sidike, S. Zhao, and Y. Wen, "Estimating soil salinity in Pingluo County of China using QuickBird data and soil reflectance spectra," *International Journal of Applied Earth Observation and Geoinformation*, vol. 26, no. 1, pp. 156–175, 2014.

[3] E. Scudiero, T. H. Skaggs, and D. L. Corwin, "Regional scale soil salinity evaluation using Landsat 7, Western San Joaquin Valley, California, USA," *Geoderma Regional*, vol. 2-3, pp. 82–90, 2014.

[4] G. I. Metternicht and J. A. Zinck, "Remote sensing of soil salinity: potentials and constraints," *Remote Sensing of Environment*, vol. 85, no. 1, pp. 1–20, 2003.

[5] Y. Weng, P. Gong, and Z. Zhu, "A spectral index for estimating soil salinity in the yellow river delta region of china using EO-1 hyperion data," *Pedosphere*, vol. 20, no. 3, pp. 378–388, 2010.

[6] J. Ding and D. Yu, "Monitoring and evaluating spatial variability of soil salinity in dry and wet seasons in the Werigan–Kuqa Oasis, China, using remote sensing and electromagnetic induction instruments," *Geoderma*, vol. 235-236, pp. 316–322, 2014.

[7] Y. Liu, W. Li, G. Wu, and X. Xu, "Feasibility of estimating heavy metal contaminations in floodplain soils using laboratory-based hyperspectral data—a case study along Le'an River, China," *Geo-Spatial Information Science*, vol. 14, no. 1, pp. 10–16, 2011.

[8] P. Jin, P. Li, Q. Wang, and Z. Pu, "Developing and applying novel spectral feature parameters for classifying soil salt types in arid land," *Ecological Indicators*, vol. 54, article 2325, pp. 116–123, 2015.

[9] J. Li, L. Pu, M. Zhu et al., "Monitoring soil salt content using HJ-1A hyperspectral data: a case study of coastal areas in Rudong County, Eastern China," *Chinese Geographical Science*, vol. 25, no. 2, pp. 213–223, 2015.

[10] S. Kumar, G. Gautam, and S. K. Saha, "Hyperspectral remote sensing data derived spectral indices in characterizing salt-affected soils: a case study of Indo-Gangetic plains of India," *Environmental Earth Sciences*, vol. 73, no. 7, pp. 3299–3308, 2015.

[11] J. Farifteh, F. Van der Meer, C. Atzberger, and E. J. M. Carranza, "Quantitative analysis of salt-affected soil reflectance spectra: a comparison of two adaptive methods (PLSR and ANN)," *Remote Sensing of Environment*, vol. 110, no. 1, pp. 59–78, 2007.

[12] K. Wiggins, R. Palmer, W. Hutchinson, and P. Drummond, "An investigation into the use of calculating the first derivative of absorbance spectra as a tool for forensic fibre analysis," *Science and Justice*, vol. 47, no. 1, pp. 9–18, 2007.

[13] H. Holden and E. LeDrew, "Accuracy assessment of hyperspectral classification of coral reef features," *Geocarto International*, vol. 15, no. 2, pp. 7–14, 2008.

[14] Y. Pu, W. Wang, J. Zhou, Y. Wang, and H. Jia, "Fractional differential approach to detecting textural features of digital image and its fractional differential filter implementation," *Science in China. Series F. Information Sciences*, vol. 51, no. 9, pp. 1319–1339, 2008.

[15] B. L. Becker, D. P. Lusch, and J. Qi, "Identifying optimal spectral bands from in situ measurements of Great Lakes coastal wetlands using second-derivative analysis," *Remote Sensing of Environment*, vol. 97, no. 2, pp. 238–248, 2005.

[16] S. S. Kharintsev and M. K. Salakhov, "A simple method to extract spectral parameters using fractional derivative spectrometry," *Spectrochimica Acta Part A: Molecular and Biomolecular Spectroscopy*, vol. 60, no. 8-9, pp. 2125–2133, 2004.

[17] F. Tsai and W. D. Philpot, "A derivative-aided hyperspectral image analysis system for land-cover classification," *IEEE Transactions on Geoscience and Remote Sensing*, vol. 40, no. 2, pp. 416–425, 2002.

[18] Z.-Y. Liu, J.-J. Shi, L.-W. Zhang, and J.-F. Huang, "Discrimination of rice panicles by hyperspectral reflectance data based on principal component analysis and support vector classification," *Journal of Zhejiang University SCIENCE B*, vol. 11, no. 1, pp. 71–78, 2010.

[19] C. Ruffin, R. L. King, and N. H. Younan, "A combined derivative spectroscopy and savitzky-golay filtering method for the analysis of hyperspectral data," *GIScience & Remote Sensing*, vol. 45, no. 1, pp. 1–15, 2013.

[20] S. T. Monteiro, Y. Minekawa, Y. Kosugi, T. Akazawa, and K. Oda, "Prediction of sweetness and amino acid content in soybean crops from hyperspectral imagery," *ISPRS Journal of Photogrammetry and Remote Sensing*, vol. 62, no. 1, pp. 2–12, 2007.

[21] B. Li and W. Xie, "Adaptive fractional differential approach and its application to medical image enhancement," *Computers & Electrical Engineering*, vol. 45, pp. 324–335, 2015.

[22] J. M. Schmitt, "Fractional derivative analysis of diffuse reflectance spectra," *Applied Spectroscopy*, vol. 52, no. 6, pp. 840–846, 1998.

[23] P. Tong, Y. Du, K. Zheng, T. Wu, and J. Wang, "Improvement of NIR model by fractional order Savitzky-Golay derivation (FOSGD) coupled with wavelength selection," *Chemometrics and Intelligent Laboratory Systems*, vol. 143, pp. 40–48, 2015.

[24] J. T. Machado, V. Kiryakova, and F. Mainardi, "Recent history of fractional calculus," *Communications in Nonlinear Science and Numerical Simulation*, vol. 16, no. 3, pp. 1140–1153, 2011.

[25] C. Li and G. Chen, "Chaos and hyperchaos in the fractional-order Rössler equations," *Physica A. Statistical Mechanics and Its Applications*, vol. 341, pp. 55–61, 2004.

[26] D. Sierociuk, T. Skovranek, M. Macias et al., "Diffusion process modeling by using fractional-order models," *Applied Mathematics and Computation*, vol. 257, pp. 2–11, 2015.

[27] J. Zhang and K. Chen, "Variational image registration by a total fractional-order variation model," *Journal of Computational Physics*, vol. 293, pp. 442–461, 2015.

[28] D. Tian, D. Xue, and D. Wang, "A fractional-order adaptive regularization primal-dual algorithm for image denoising," *Information Sciences*, vol. 296, pp. 147–159, 2015.

[29] A. Loverro, *Fractional Calculus: History, Definitions and Applications for the Engineer*, University of Notre Dame, 2004.

[30] N. He, J.-B. Wang, L.-L. Zhang, and K. Lu, "An improved fractional-order differentiation model for image denoising," *Signal Processing*, vol. 112, pp. 180–188, 2015.

[31] V. Garg and K. Singh, "An improved Grunwald-Letnikov fractional differential mask for image texture enhancement," *International Journal of Advanced Computer Science and Applications*, vol. 3, no. 3, pp. 130–135, 2012.

[32] J. Yao, Q. Zhao, and Z. Liu, "Effect of climate variability and human activities on runoff in the Jinghe River Basin, Northwest China," *Journal of Mountain Science*, vol. 12, no. 2, pp. 358–367, 2015.

[33] L. Ma, J. Wu, H. Yu, H. Zeng, and J. Abuduwaili, "The medieval warm period and the little ice age from a sediment record of Lake Ebinur, northwest China," *Boreas*, vol. 40, no. 3, pp. 518–524, 2011.

[34] D. Liu, J. Abuduwaili, J. Lei, G. Wu, and D. Gui, "Wind erosion of saline playa sediments and its ecological effects in Ebinur Lake, Xinjiang, China," *Environmental Earth Sciences*, vol. 63, no. 2, pp. 241–250, 2011.

[35] F. Zhang, T. Tiyip, V. C. Johnson et al., "The influence of natural and human factors in the shrinking of the Ebinur Lake, Xinjiang, China, during the 1972–2013 period," *Environmental Monitoring and Assessment*, vol. 187, no. 1, p. 4128, 2015.

[36] L. Ma, J. Wu, W. Liu, and J. Abuduwaili, "Distinguishing between anthropogenic and climatic impacts on lake size: a modeling approach using data from Ebinur Lake in arid northwest China," *Journal of Limnology*, vol. 73, no. 2, pp. 148–155, 2014.

[37] G. Zheng, D. Ryu, C. Jiao, and C. Hong, "Estimation of organic matter content in coastal soil using reflectance spectroscopy," *Pedosphere*, vol. 26, no. 1, pp. 130–136, 2016.

[38] H. Yang and J. Li, "Predictions of soil organic carbon using laboratory-based hyperspectral data in the northern Tianshan mountains, China," *Environmental Monitoring and Assessment*, vol. 185, no. 5, pp. 3897–3908, 2013.

[39] C. Gomez, Y. Le Bissonnais, M. Annabi, H. Bahri, and D. Raclot, "Laboratory Vis–NIR spectroscopy as an alternative method for estimating the soil aggregate stability indexes of Mediterranean soils," *Geoderma*, vol. 209-210, pp. 86–97, 2013.

[40] H. Liu, Y. Zhang, and B. Zhang, "Novel hyperspectral reflectance models for estimating black-soil organic matter in Northeast China," *Environmental Monitoring and Assessment*, vol. 154, no. 1–4, pp. 147–154, 2009.

[41] G. R. Mahajan, R. N. Sahoo, R. N. Pandey, V. K. Gupta, and D. Kumar, "Using hyperspectral remote sensing techniques to monitor nitrogen, phosphorus, sulphur and potassium in wheat (Triticum aestivum L.)," *Precision Agriculture*, vol. 15, no. 5, pp. 499–522, 2014.

[42] C. Lin, S. Zhou, and S. Wu, "Using hyperspectral reflectance to detect different soil erosion status in the Subtropical Hilly Region of Southern China: a case study of Changting, Fujian Province," *Environmental Earth Sciences*, vol. 70, no. 4, pp. 1661–1670, 2013.

[43] T. Shi, Y. Chen, Y. Liu, and G. Wu, "Visible and near-infrared reflectance spectroscopy-an alternative for monitoring soil contamination by heavy metals," *Journal of Hazardous Materials*, vol. 265, pp. 166–176, 2014.

[44] S. Schmidtlein and J. Sassin, "Mapping of continuous floristic gradients in grasslands using hyperspectral imagery," *Remote Sensing of Environment*, vol. 92, no. 1, pp. 126–138, 2004.

[45] K. Yu, Y. Zhao, X. Li, Y. Shao, F. Zhu, and Y. He, "Identification of crack features in fresh jujube using Vis/NIR hyperspectral imaging combined with image processing," *Computers and Electronics in Agriculture*, vol. 103, pp. 1–10, 2014.

[46] D. Wu, D. Sun, and Y. He, "Novel non-invasive distribution measurement of texture profile analysis (TPA) in salmon fillet by using visible and near infrared hyperspectral imaging," *Food Chemistry*, vol. 145, pp. 417–426, 2014.

[47] M. Kamruzzaman, Y. Makino, and S. Oshita, "Rapid and nondestructive detection of chicken adulteration in minced beef using visible near-infrared hyperspectral imaging and machine learning," *Journal of Food Engineering*, vol. 170, pp. 8–15, 2016.

[48] F. Khayamim, J. Wetterlind, H. Khademi, A. H. J. Robertson, A. Faz Cano, and B. Stenberg, "Using visible and near infrared spectroscopy to estimate carbonates and gypsum in soils in arid and subhumid regions of Isfahan, Iran," *Journal of Near Infrared Spectroscopy*, vol. 23, no. 3, pp. 155–165, 2015.

[49] C. Lu, B. Xiang, G. Hao, J. Xu, Z. Wang, and C. Chen, "Rapid detection of melamine in milk powder by near infrared spectroscopy," *Journal of Near Infrared Spectroscopy*, vol. 17, no. 2, pp. 59–67, 2009.

[50] A. M. Rady, D. E. Guyer, W. Kirk, and I. R. Donis-González, "The potential use of visible/near infrared spectroscopy and hyperspectral imaging to predict processing-related constituents of potatoes," *Journal of Food Engineering*, vol. 135, pp. 11–25, 2014.

[51] C. Gomez, P. Lagacherie, and G. Coulouma, "Regional predictions of eight common soil properties and their spatial structures from hyperspectral Vis–NIR data," *Geoderma*, vol. 189-190, pp. 176–185, 2012.

[52] R. Casa, F. Castaldi, S. Pascucci, A. Palombo, and S. Pignatti, "A comparison of sensor resolution and calibration strategies for soil texture estimation from hyperspectral remote sensing," *Geoderma*, vol. 197-198, pp. 17–26, 2013.

[53] D. Wu, H. Shi, S. Wang, Y. He, Y. Bao, and K. Liu, "Rapid prediction of moisture content of dehydrated prawns using online hyperspectral imaging system," *Analytica Chimica Acta*, vol. 726, pp. 57–66, 2012.

[54] F. Zhang, T. Tiyip, J. Ding et al., "Studies on the reflectance spectral features of saline soil along the middle reaches of Tarim River: a case study in Xinjiang Autonomous Region, China," *Environmental Earth Sciences*, vol. 69, no. 8, pp. 2743–2761, 2013.

[55] M. St. Luce, N. Ziadi, B. J. Zebarth, C. A. Grant, G. F. Tremblay, and E. G. Gregorich, "Rapid determination of soil organic matter quality indicators using visible near infrared reflectance spectroscopy," *Geoderma*, vol. 232–234, pp. 449–458, 2014.

[56] J. Liang, C. Chen, X. Song, Y. Han, and Z. Liang, "Assessment of heavy metal pollution in soil and plants from dunhua sewage irrigation area," *International Journal of Electrochemical Science*, vol. 6, no. 11, pp. 5314–5324, 2011.

[57] R. Srivastava, D. Sarkar, S. S. Mukhopadhayay et al., "Development of hyperspectral model for rapid monitoring of soil organic carbon under precision farming in the Indo-Gangetic Plains of Punjab, India," *Journal of the Indian Society of Remote Sensing*, vol. 43, no. 4, pp. 751–759, 2015.

[58] S. A. Medjahed, T. Ait Saadi, A. Benyettou, and M. Ouali, "Gray Wolf Optimizer for hyperspectral band selection," *Applied Soft Computing*, vol. 40, pp. 178–186, 2016.

[59] S. Nawar, H. Buddenbaum, and J. Hill, "Estimation of soil salinity using three quantitative methods based on visible and near-infrared reflectance spectroscopy: a case study from Egypt," *Arabian Journal of Geosciences*, vol. 8, no. 7, pp. 5127–5140, 2015.

[60] E. M. Abdel-Rahman, O. Mutanga, J. Odindi, E. Adam, A. Odindo, and R. Ismail, "A comparison of partial least squares (PLS) and sparse PLS regressions for predicting yield of Swiss chard grown under different irrigation water sources using hyperspectral data," *Computers and Electronics in Agriculture*, vol. 106, pp. 11–19, 2014.

[61] F. Yang, J. Li, X. Gan, Y. Qian, X. Wu, and Q. Yang, "Assessing nutritional status of Festuca arundinacea by monitoring photosynthetic pigments from hyperspectral data," *Computers and Electronics in Agriculture*, vol. 70, no. 1, pp. 52–59, 2010.

[62] N. Torbick and B. Becker, "Evaluating principal components analysis for identifying Optimal bands using wetland hyperspectral measurements from the Great Lakes, USA," *Remote Sensing*, vol. 1, no. 3, pp. 408–417, 2009.

[63] Y. Li, H. Tang, and H. Chen, "Fractional-order derivative spectroscopy for resolving simulated overlapped Lorenztian peaks," *Chemometrics and Intelligent Laboratory Systems*, vol. 107, no. 1, pp. 83–89, 2011.

[64] J. Zhang, Z. Wei, and L. Xiao, "A fast adaptive reweighted residual-feedback iterative algorithm for fractional-order total variation regularized multiplicative noise removal of partly-textured images," *Signal Processing*, vol. 98, pp. 381–395, 2014.

[65] Z. E. Mashimbye, M. A. Cho, J. P. Nell, W. P. De Clercq, A. Van niekerk, and D. P. Turner, "Model-based integrated methods for quantitative estimation of soil salinity from hyperspectral remote sensing data: A Case Study of Selected South African Soils," *Pedosphere*, vol. 22, no. 5, pp. 640–649, 2012.

[66] J. Peng, W. Ji, Z. Ma et al., "Predicting total dissolved salts and soluble ion concentrations in agricultural soils using portable visible near-infrared and mid-infrared spectrometers," *Biosystems Engineering*, 2016.

Selective Surface Sintering Using a Laser-Induced Breakdown Spectroscopy System

H. Jull,[1,2] **P. Ewart,**[1,3] **R. Künnemeyer,**[1,2] **and P. Schaare**[4]

[1]*School of Engineering, University of Waikato, Hamilton 3240, New Zealand*
[2]*The Dodd-Walls Centre for Photonic and Quantum Technologies, Hamilton 3240, New Zealand*
[3]*Waikato Institute of Technology, Hamilton 3200, New Zealand*
[4]*The New Zealand Institute for Plant & Food Research Ltd, Hamilton 3240, New Zealand*

Correspondence should be addressed to H. Jull; harrisson.jull@gmail.com

Academic Editor: Violeta Lazic

Titanium metal injection molding allows creation of complex metal parts that are lightweight and biocompatible with reduced cost in comparison with machining titanium. Laser-induced breakdown spectroscopy (LIBS) can be used to create plasma on the surface of a sample to analyze its elemental composition. Repetitive ablation on the same site has been shown to create differences from the original sample. This study investigates the potential of LIBS for selective surface sintering of injection-molded titanium metal. The temperature created throughout the LIBS process on the surface of the injection-molded titanium is high enough to fuse together the titanium particles. Using the ratio of the Ti II 282.81 nm and the C I 247.86 nm lines, the effectiveness of repetitive plasma formation to produce sintering can be monitored during the process. Energy-dispersive X-ray spectroscopy on the ablation craters confirms sintering through the reduction in carbon from 20.29 Wt.% to 2.13 Wt.%. Scanning electron microscope images confirm sintering. A conventional LIBS system, with a fixed distance, investigated laser parameters on injection-molded and injection-sintered titanium. To prove the feasibility of using this technique on a production line, a second LIBS system, with an autofocus and 3-axis translation stage, successfully sintered a sample with a nonplanar surface.

1. Introduction

Metal injection molding (MIM) is a fabrication method that uses the injection molding process and sintering to create metal parts with complex geometry (parts with varying surface heights). This process combines metal powders with a carrier system or binder, containing thermoplastic polymers and waxes to create a feedstock that can be injected into a mold to form a green part [1]. The binder is then gradually removed from the green part by a solvent debinding method, a thermal debinding method, or a combination of both. The part is then referred to as a brown part and is sintered in a furnace at high temperature whereby the particles coalesce or fuse together to form a fully consolidated metal part. Using titanium (Ti) metal in MIM allows parts to be created that are both light in weight and biocompatible [2]. The complex shapes formed offer a reduction in costs associated with titanium parts made by conventional fabrication processes

such as casting or machining and can even form parts not possible to produce by any other means. However, powder metallurgy (PM) is known to produce parts with mechanical properties inferior to those produced from wrought metals due to stress raisers and surface irregularities resulting from incomplete particle coalescence. Incomplete coalescence of titanium PM parts is known to occur due to low-quality starting powders, low sintering time, or temperature and also contaminated furnace atmosphere, all of which affect the outermost surface. These surface defects may be mitigated by optimizing the sinter cycle, secondary processes such as bead blasting and polishing or treatments such as electropolishing and coating.

Titanium MIM has the added problem of contamination through any residual binder components of which carbon and oxygen are the most prevalent and the most difficult to mitigate. The transformation of the titanium lattice structure during sintering is influenced by these impurities, and the formation of oxides/carbides also impinges on the coalescence. While the saturation levels of interstitial oxygen (wo $\leq$ 5%) are high, carbon compounds readily form on powder surfaces above 370°C. The presence of carbon on the titanium particles interferes with heat transfer, the oxide presence means the detection, and quantification of the elemental levels is difficult.

Sintering of metals can also be done with high-powered continuous wave (CW) lasers such as those used in direct metal laser sintering and selective laser sintering. This process requires a high-power density over a short time [3]. Laser-induced breakdown spectroscopy (LIBS) uses a high-power pulsed laser to ablate the surface of a sample and to create a plasma that contains the surface elements. The study presented here investigates whether LIBS can be used for selective surface sintering of injection-molded titanium while simultaneously measuring the amount of sintering that has taken place.

1.1. Laser-Induced Breakdown Spectroscopy. LIBS is a form of optical emission spectroscopy utilizing a high-powered, pulsed laser. The short pulse is focused onto a sample surface where it creates a high-power density in a localized area. A plasma is created from the ablated material [4]. Plasma formation is a complex process, which is still under investigation. The current models for ablation describing the change in thermal properties of the sample material are adequate for this study [5]. The models for nanosecond laser ablation state that when a laser hits a target, with enough high energy, the surface heats up, melts, vaporizes, and ionizes creating a plasma. If there is additional energy in the laser pulse, the plasma absorbs it, increasing the plasma temperature. Further heating of the surface of the target can occur if the plasma is hot enough [6, 7]. For a more detailed discussion on plasma ablation, refer to [8] and the references therein.

Substantial shot-to-shot variation is a problem for quantitative LIBS analysis. Plasmas created on identical samples can have different temperatures which affect emission line intensities [9]. Subtle changes in experimental conditions and surface composition can cause these temperature variations. To improve results, averages are generated by repeatedly measuring the same sample. The average spectra can then provide quantitative information on the composition of the sample.

The use of multiple LIBS pulses has been employed for different purposes. One use is removing surface layers off various materials. The LIBS laser is focused on the same surface position, and multiple pulses are used with spectroscopy feedback to determine when the surface layer has been successfully removed. Costela et al. [10] cleaned graffiti off urban buildings with minimal damage to the surface of the building. Li et al. [11] removed cobalt from tungsten carbide hardmetal. Diego-Vallejo et al. [12] used linear correlation and artificial neural networks to determine when the top layers of solar cells had been removed. Majewski et al. [13] cleaned contaminants from thermal barrier coatings, and Roberts et al. [14] removed rock-covering fossils. Maravelaki et al. [15] cleaned Pentelic marble covered in encrustations. Acquaviva et al. [16] cleaned the bust of St. Gregory the

Armenian of contaminants. Flores et al. [17] removed spines from prickled pears without visible damage to the pear, and Bengtsson et al. [18] cleaned salt off high-voltage insulators.

Another use of multiple LIBS pulses is depth profiling. Martelli et al. [19, 20] determined the different tissues in wheat by looking at the Mg II 279.55 nm to Mg I 285.22 nm intensity ratio as repetitive shots were taken at the same location. Abdelhamid et al. [21] investigated the effects of laser irradiance and working distance of a sample looking at how the intensity of Cu I 465.1 nm and Au I 431.5 nm changed after repetitive shots. Theodorakopoulos and Zafiropulos [22] looked at depth profiles of varnishes using the C_2 swan band. De Bonis et al. [23] observed the change in the ratio formed by Cu I 282.35 nm and Sn I 283.94 nm for an increasing number of shots on a corroded copper tin alloy. Most of the studies mentioned above used more than 100 pulses because of the small amount of surface material removed. The effects of heat propagation on the underlying surface were not addressed because they were irrelevant for the purposes employed. The heat distribution is determined by laser pulse parameters [8] and the optical coupling efficiency of the material. The ablation site can have differences compared to the original surface. These differences include melted and recondensed matter which no longer exhibits the original sample composition. Surface reflectivity and surface breakdown threshold change [7]. This is caused by high-power density pulses, heat, stress, and pressure during the plasma formation [24]. Inside the ablation crater, there is also recondensed material from the plasma [4]. The study presented in this paper utilizes the heat, stress, and pressure created by multiple plasma formations to change the surface properties to a desired state.

LIBS has previously been used as a process-monitoring technique. Boué-Bigne [25] investigated segregation and decarburization using LIBS on polished high-carbon steel. Elemental maps of the surfaces were produced using individual emission lines. One map created using a neutral manganese line was normalized with a neutral iron line removing the baseline gradient. Cabalín and Laserna [26] used pulsed laser deposition to coat silicon with manganin. They produced multielemental maps to monitor the coating constituents. Pedarnig et al. [27] utilized LIBS for in situ monitoring of plasma emission on thin-film deposition. Emission lines of interest were normalized by neutral and ionized yttrium lines and a neutral zinc line. Gruber et al. [28] designed a LIBS system for in situ analysis of solid and molten metals. Chromium, copper, manganese, and nickel were investigated using iron lines as an internal standard. Gruber et al. [29] also performed LIBS analysis on molten metal for chromium, nickel, and manganese but under a vacuum. Lopez-Moreno et al. [30] created a stand-off LIBS system to analyze molten metal as an in-process control. Quantitative analysis of chromium, nickel, and manganese was performed on steel samples at 1000°C. The present work investigates LIBS as a method for real-time monitoring of the amount of sintering taking place as a result of the LIBS process.

1.2. Internal Standards. In theory, taking a ratio of two emission lines can compensate for the fluctuations in the amplitude of spectra caused by experimental conditions. This is referred to as an internal reference, where all emission lines of interest are normalized using a single emission line. The emission line used as a reference is usually from an element with a constant concentration or of an element that makes up the majority of the plasma [6]. The sample can also be spiked with a foreign element of constant concentration, which can be used as a reference. An important consideration is to choose lines that have similar upper level electron energies (E_k). This will reduce the temperature effects, which cause large deviations in emission intensities. Temperature characteristics of each line need to be considered if variations are to be minimized. The magnitudes of the emission lines are to be comparable to reduce the statistical uncertainties produced by combining the percentage uncertainties of each signal. An example of this problem would be combining a strong line with a weak line. The percentage uncertainty, which is used to compute the resulting variance, would be larger for the weak line producing a high uncertainty.

This study investigates the use of multiple LIBS pulses, for postprocessing, to improve the outer surface of Ti MIM samples with complex geometry. This is achieved by the removal of residual carbon (compounds) from and below the surface thus enabling full coalescence. This process will be controlled by spectral feedback from the carbon and titanium lines. The hypothesis is that the heat caused by the LIBS process, from multiple pulses, is sufficiently high that the amount of surface sintered is relatively large compared to the ablated material. This would allow laser sintering and measurement of surface properties at the same time.

2. Experimental Setup

A commercial Spectrolaser 4000 unit (Laser Analysis Technologies, Australia) containing a Nd:YAG laser operating at 1064 nm with a pulse width of 7 ns and pulse energies of 50, 100, and 200 mJ was focused at a fixed distance perpendicular to the sample surface to generate the plasma. The spot diameter was about 650 μm, which results in an irradiance of about 8.6 GW cm^{-2} at 200 mJ per pulse. The wavelength range of 185.25–948.80 nm was captured by the four spectrometers inside the Spectrolaser unit. The spectrometers were set to integrate for 1 ms after a 1.27 μs delay from the onset of the laser pulse. This system can reach high laser pulse energies but does not allow changing of the lens-to-sample distance. The setup was used to investigate the effect that laser pulse energy has on thermally debound injection-molded titanium samples.

A second commercial system (LIBS-6, Applied Photonics, UK) containing a Nd:YAG laser (Big Sky Ultra, Quantel, France) with the same wavelength, pulse width, delay, and integration time as the Spectrolaser was set to 100 mJ per pulse. The built-in beam expander was focused to produce a spot size on the sample of 450 μm, roughly 80 mm from the beam expander producing an irradiance of 9.0 GW cm^{-2}. Each spectrum was acquired with six spectrometers (Avantes, The Netherlands) covering the range 182.26–908.07 nm. A 3-axis translation stage moved

the sample into the focal position of the LIBS beam. This system produces less laser pulse energy than the previous one but can change the lens-to-sample distance. The setup was used to demonstrate the selective sintering process at different heights using a simple autofocus system.

The Spectrolaser system was used with a fixed distance to the sample to acquire an average of 100 spectra from different locations on a single thermally debound sample and a single-sintered sample (see Section 2.2 for sample descriptions). Different locations were used to avoid the effects of previous pulses on the same site. Spectra were then acquired by sampling 10 different locations, on the thermally debound sample, 20 times at 50, 100, and 200 mJ per pulse. The second LIBS system was used with the autofocus function enabled to evaluate its use in removing surface carbon on a sample with varying surface height.

A commercial field emission scanning electron microscope (SEM) (Hitachi S-4700, Hitachi High Technologies, USA) with an accelerating voltage of 20 kV, emission current of 11,000 nA, and a working distance of 11,200 μm was used to investigate the morphology of the ablation craters. An energy-dispersive X-ray spectroscopy (EDS) attachment to the SEM (Noran System 6, Thermo Fisher Scientific, USA) was used to compare the composition of the craters to that obtained from the LIBS spectra.

2.1. Autofocus. A 635 nm, 4.5 mW, continuous wave laser directed on the ablation site was used in conjunction with a miniature CCD video camera, connected to the LIBS-6 module, to create an autofocus system based on the triangulation method. The laser was aligned so that the axis of the laser beam would intersect with the ablation site. The AForge.NET framework (a library of common computer vision algorithms [31]) was used to filter the image from the camera to only show the beam spot. The centroids of the resulting images (Figures 1(b), 1(d), and 1(f)) were calculated, and a model created to relate the centroid position of the laser spot with the distance from the ablation site. Figure 1(b) corresponds to the sample being too close to the LIBS-6 machine, Figure 1(d) too far from the LIBS-6 system, and Figure 1(f) at a distance where the focus of the LIBS beam is directly on the surface of the sample. This set-up was used in a feedback control system to move the vertical translation stage and hold the sample at the correct height.

2.2. Samples. The samples were made using a binder system comprised of polyethylene glycol 8000, polyethylene glycol 20,000, natural bees wax, linear low-density polyethylene, and polypropylene [32]. The metal powder was commercially pure titanium produced by the hydride-dehydride route with a particle size less than 325 μm (Specialty Metallurgical Products Co. Inc., USA). The binder system was combined with the titanium powder to produce a feedstock with a volume fraction of 30% binder to 70% metal powder. The feedstock was mixed using a corotating twin screw extruder (Thermo-Lab, Lab digital, USA) operating $D = 16$ mm screws, $L/D = 24$, $\omega = 0$–250 r/min, and $T = 30$–300°C. Injection molding was done using a fully hydraulic injection-molding machine

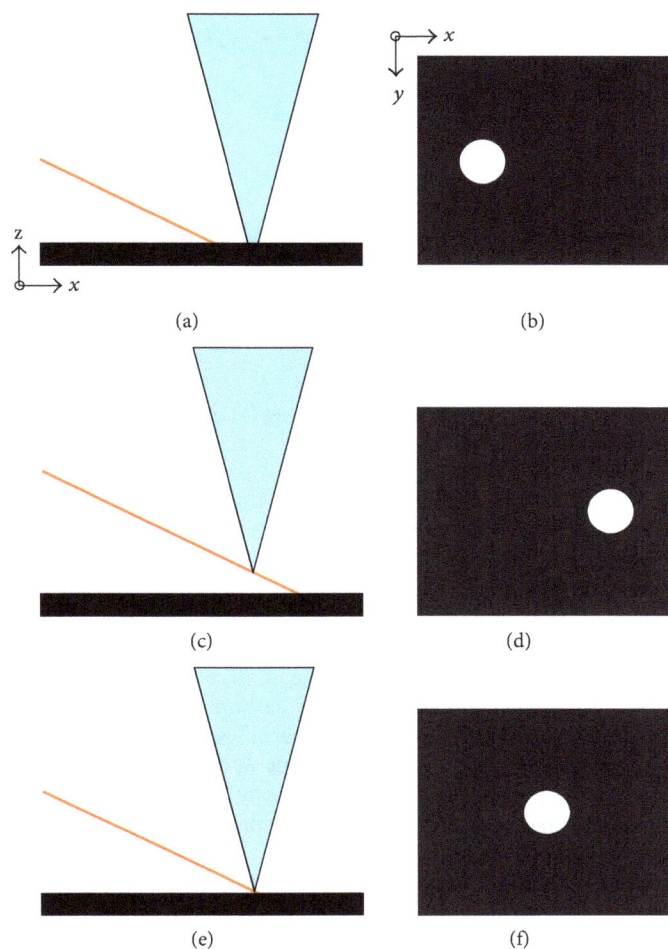

FIGURE 1: Triangulation method using a 635 nm CW laser (beam entering from the left). The vertical beam converging on the sample represents the LIBS beam. (a), (c), and (e) show where the autofocus laser strikes the sample surface. (b), (d), and (f) show the camera images of the filtered autofocus laser spots.

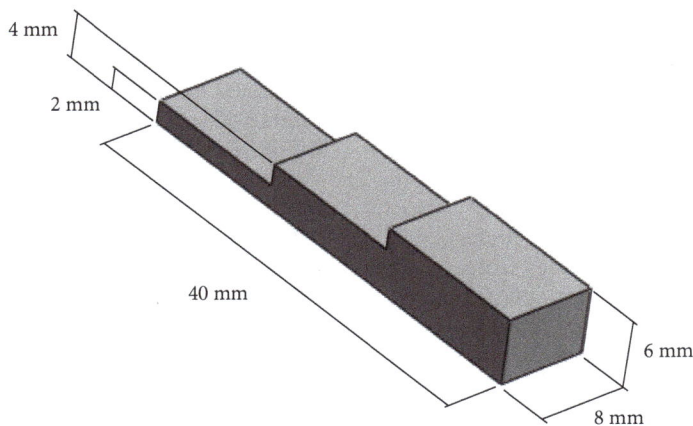

FIGURE 2: Geometry and dimensions for the titanium injection-molded part.

(35 M, Dr Boy, Germany) to create a part shape with complex geometry, that is, with multiple step heights. The dimensions of the part are shown in Figure 2.

Debinding was done using both solvent and thermal processes. A solvent debinding kit (SDK) assembled at the University of Texas-Pan American was used for the test

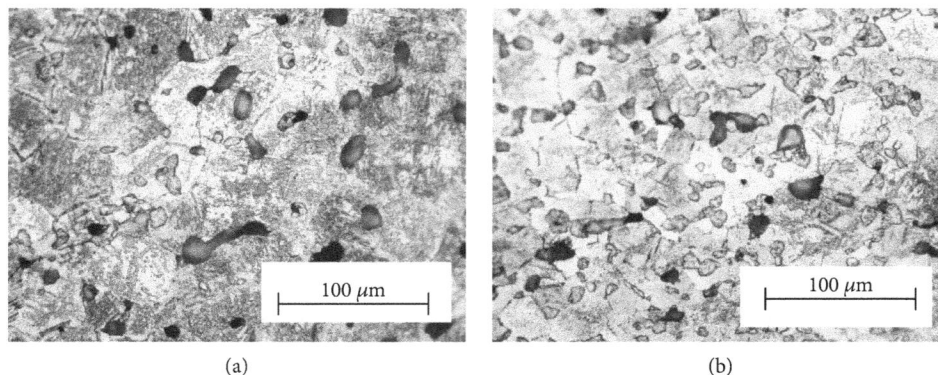

FIGURE 3: Microstructure of thermally sintered Ti part prior to laser sintering (a) polished external surface and (b) sectioned and polished internal microstructure.

samples [32]. Thermal debinding was done in a convective drying oven (FED53, Binder, USA) ($V = 53$ L, $T = 30–300°C$) with a gas tight chamber and programmable PID temperature control. After parts were solvent debound, the samples were identified using the designation "TD" for thermal debound (brown parts).

The brown parts were sintered in a furnace (ACME, China) by initial heating with an isothermal period at $470°C$ ($\beta = 5°C/min$, $t = 60$ min) followed by further heating to $800°C$, all under argon sweep gas ($\dot{V} = 10.0$ L/min). This was followed by a further heat increase and a final isothermal period at sinter temperature under vacuum ($5°C/min$, $t = 180$ min, $T = 1350°C$, $p = 2.0 \times 10 - 3$ mbar). The sintered parts were given the designation "S." Both TD and S samples were sanded to remove surface coating and finger print residue. Sectioned and polished S samples were analyzed by X-ray diffraction (XRD) (Empyrean, PANalytical, The Netherlands), with power settings 45 kV at 40 mA through a copper electrode, scanning from position 20 to $100°$ 2θ, using a step size of $0.026°$ 2θ, scan step time of 17.340 s in continuous scanning mode. Peak identification was done using the HighScore software (PANalytical, The Netherlands).

3. Results and Discussions

3.1. The TD and S Samples. The conventional means of determining binder level during MIM processing is by weighing each sample and comparing the mass to the theoretical value. This method is accurate for most solvent, and thermal debinding processes carried out under $250°C$. However, conventional thermal debinding temperatures are above $420°C$, at which point the thermoplastic binders are reduced to the constituent elements (carbon, oxygen, nitrogen, and hydrogen). Also, at these temperatures, titanium preferentially absorbs those elements with no indication from the component mass. Without destructive testing, it is difficult to determine whether binder residue levels are sufficiently low, and the desired mechanical properties are achieved. Higher elemental levels near the external surfaces can be seen (Figure 3(a)) compared to those internally (Figure 3(b)). The difference arises because the rate of elemental diffusion is slower than the rate of surface deposition

once the pores between the particles close during sintering and densities greater than about 94% are reached.

Even where levels of carbon and oxygen meet specified standards, they may still hinder coalescence of the particles during sintering and cause surface porosity. Surface porosity as seen in Figure 3 reduces the mechanical properties of the final part. The XRD analysis confirmed the presence of carbon and oxygen, as carbide formations, near the external surface [33]. The ability to reduce the elemental levels at the surface should therefore enable a decrease in porosity and subsequent increase in overall density leading to improvements in mechanical properties. Using LIBS to selectively sinter parts of the surface, postprocessing may have the ability to do this.

3.2. Difference in Carbon Content between the TD and S Samples. The average spectra of 100 shots on each of the TD and S samples are shown in Figure 4. These spectra were analyzed to find spectral lines that fulfilled the criteria discussed in Section 1.2. Each shot was aimed at different locations on the sample. The emission lines observed in the spectra of the S sample were weaker, and the background continuum was reduced. The Ti II lines were dominant, and the strong lines of Ti I, which are prevalent between 400 and 600 nm, were significantly reduced. These differences are shown in Figure 4. The UV range, where a number of well-identified Ti lines are present, is highlighted in Figure 5.

Lines that were identified as suitable for internal standardization were C I 247.86 nm ($E_k = 7.685$ eV) and Ti II 282.81 nm ($E_k = 8.132$ eV). Taking the ratio of these lines, over all 100 shots on the S sample are 2.45. The C I and Ti II line intensity depends on transition probabilities, electron temperature, and electron density of the plasma, which are governed by highly complex processes in a multicomponent plasma. Time profiles of the species differ as well. As the lifetime of the plasma is shorter than the minimum integration time of the spectrometers, the temporal behavior cannot be analyzed. However, the intensity ratio of the C I/Ti II lines can be calibrated against other measurements, like EDS, and should be useful to determine when enough carbon has been removed. A threshold value of 2.45 was used to indicate when the TD sample had been sintered. At this threshold, the

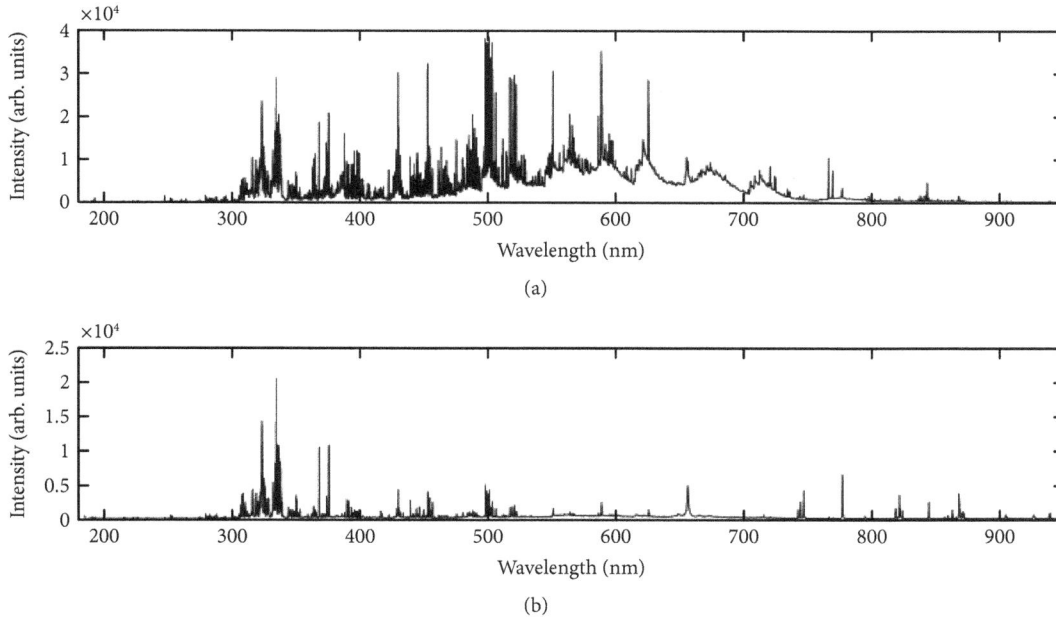

FIGURE 4: Average spectra acquired from 100 shots on the TD (a) and S (b) samples. Each shot was on a fresh surface which had not previously been sampled.

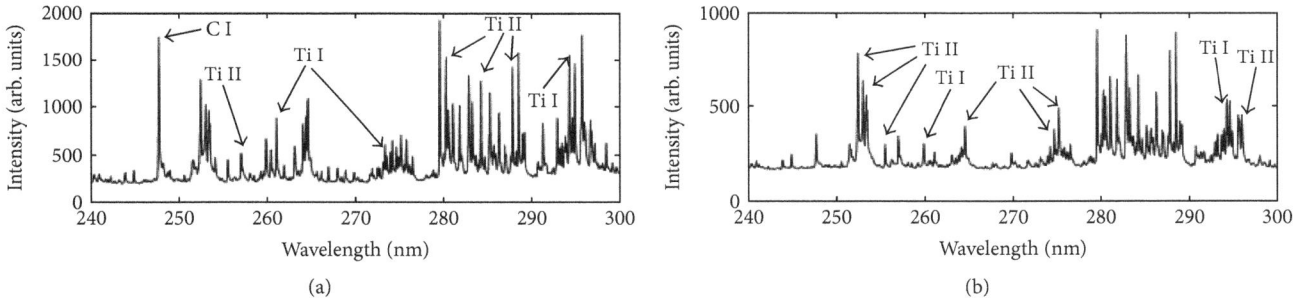

FIGURE 5: Enlarged section of Figure 4 showing some of the differences between the Ti I and Ti II emissions from the TD (a) and S (b) samples.

TABLE 1: Comparison of the Ti and C content.

	LIBS			EDS		
	Peak intensity ratio (Ti/C)	C (Wt.%)	Ti (Wt.%)	V (Wt.%)	O (Wt.%)	Si (Wt.%)
Thermally debound	0.762	20.29 ± 0.32	74.00 ± 0.68	—	5.71 ± 0.56	—
Sintered	2.45	1.79 ± 0.24	96.78 ± 0.69	0.24 ± 0.24	—	1.19 ± 0.08

amount of titanium and carbon in the TD sample surface would resemble that of the S sample.

EDS results show that there is a difference in carbon between the TD sample and the S sample of 18.5 Wt.%. Table 1 displays the results obtained from EDS. When the vanadium content is low, an incorrect concentration for vanadium is determined because of the overlap between the TiKβ (4.93 keV) and the VKα (4.95 keV) peaks in the EDS spectra [34]. The actual vanadium content was determined to be less than 0.01% V from the supplier (Specialty Metallurgical Products Co. Inc., USA). The silicon content is a residue from sanding the sample. The peak intensity ratio of the spectral lines is proportional to the amount of Ti II compared with C I present in the plasma. The amount of titanium in the plasma would be split between Ti I and Ti II. Therefore, the actual concentration ratio, which can be calculated from the EDS results, cannot easily be determined from the spectra without full knowledge of all plasma parameters, like electron temperature and electron density.

3.3. *Energy Effects on Multiple Shots.* Irradiating an ablation site multiple times with LIBS pulses and comparing the resulting spectra gives information on the amount of sintering taken place. The TD sample was irradiated in 10 different

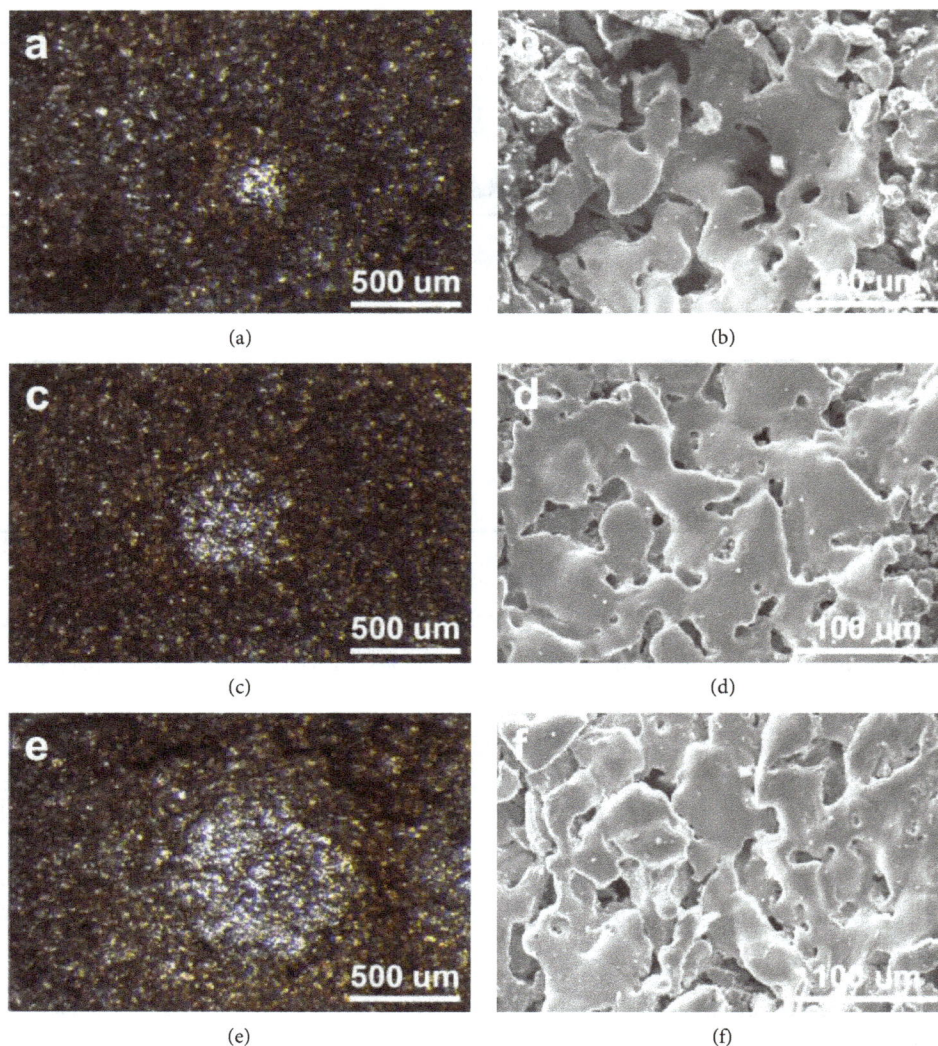

FIGURE 6: Ablation craters from 20 consecutive laser pulses using 50 mJ (a), (b); 100 mJ (c), (d); and 200 mJ (e), (f). (a), (c), and (e) are taken with a SEM.

locations by 20 successive pulses. This process was repeated at different laser pulse energies. The resulting spectra were investigated to determine the effect of laser pulse energy. Figure 6 shows the ablation sites at different laser energies. Figures 6(a), 6(c), and 6(e) are optical microscopy images and Figures 6(b), 6(d), and 6(f) are SEM images. The area being sintered increases as the laser pulse energy increases, as can be seen by the increasing spot in Figures 6(a), 6(c), and 6(e). SEM images of the centers of the ablation craters shown in Figures 6(a), 6(c), and 6(e) are displayed in Figures 6(b), 6(d), and 6(f). These images confirm that sintering has taken place. The edge of the sintered titanium particles are seen in Figure 6(b). The contrast between titanium particles that have and have not been sintered is displayed in Figure 7.

In addition to the line ratio mentioned above, other spectral lines were considered for their suitability to monitor sintering. Sixteen Ti I lines were investigated by calculating the ratio of the Ti I line intensity to the C I 247.86 nm

intensity. These ratios were normalized then scaled so that they could be compared to the above reference threshold value. Successive sampling using different energies is displayed in Figure 8 for the Ti I 398.18 nm/C I 247.86 nm intensity ratio. The error bars in Figure 8 correspond to the standard deviation of the 10 spectra that were used to find the average ratio for each point on the graph. All other emission line ratios exhibit a trend similar to that of Figure 8. The small initial value in the first shot may be due to surface contamination. The large fluctuations and error bars are produced by the large difference in ΔE_k (difference in E_k) between the Ti I lines and C I 247.86 nm which range from 3.26 eV to 5.29 eV. At these high-energy differences, the ratios cannot mitigate the differences in plasma temperature caused by shot-to-shot variations.

Using the same normalizing and scaling procedure as above, 47 Ti II to C I 247.86 nm intensity ratios were analyzed as well. The wavelengths of the Ti II lines were between 252.56 and 457.20 nm and had ΔE_k between −0.39 eV and

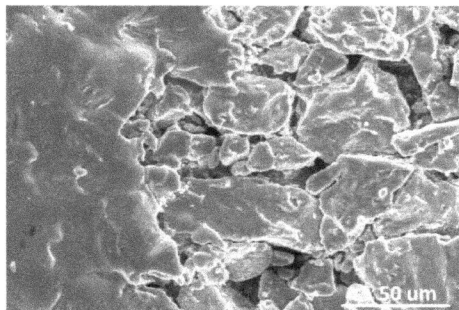

FIGURE 7: SEM showing the edge of sintered titanium. Unsintered particles to the right were not irradiated with the laser beam.

TABLE 2: Ionized titanium emission lines used in Figure 9.

λ (nm)	E_k (eV)	ΔE_k
253.13	5.03	2.65
264.59	8.57	−0.88
281.02	8.10	−0.41
283.22	4.95	2.73
351.08	5.42	2.26
352.03	5.57	2.12
353.54	5.57	2.12
362.48	4.64	3.04
416.36	5.57	2.12
433.79	3.94	3.75
450.13	3.87	3.82

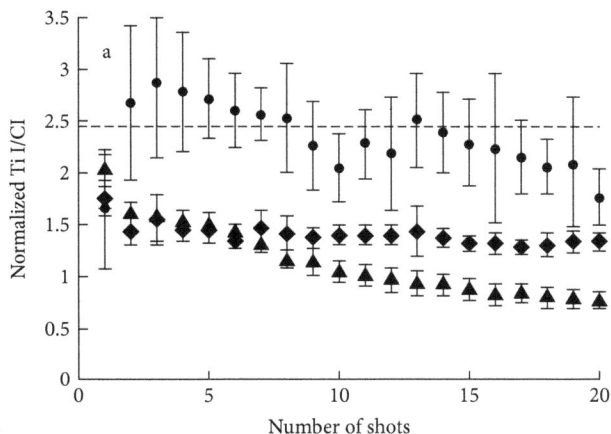

FIGURE 8: Evolution of the normalized Ti I 398.18 nm/C I 247.86 nm intensity ratio of the thermally debound sample. The dashed line shows the sintering threshold. The different energies used were 200 mJ (●), 100 mJ (♦), and 50 mJ (▲) per pulse.

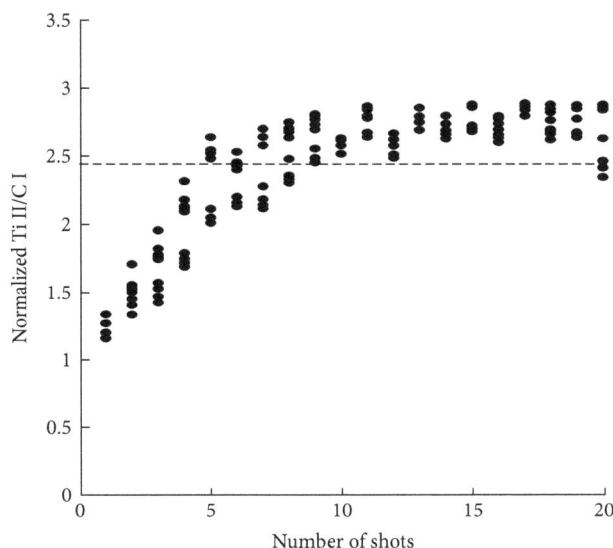

FIGURE 9: Evolution of multiple Ti II lines (Table 2) for a laser pulse energy of 200 mJ. Each point is an average of 10 shots. The dashed line shows the sintering threshold.

4.02 eV. A selection of these lines is shown in Table 2, and their variation with the number of 200 mJ laser shots in Figure 9. After 10 shots, all line ratios are above the sintering threshold.

Using Ti II rather than Ti I, substantial lines improved the sintering prediction. The Ti II lines that produced the best results were 264.59, 280.48, 281.02, 281.78, and 282.81 nm. Their upper energy level difference to the C I line, ΔE_k, is between −0.397 and −0.881 eV. All lines followed the trend shown in Figure 10 for the 282.81 nm line. However, C I 247.86 nm ($E_k = 7.685$ eV) and Ti II 282.81 nm ($E_k = 8.132$ eV) was identified as the most robust combination for internal standardization.

Figure 10 shows that 50 mJ successive shots bring no increase to the titanium to carbon intensity ratio but instead give a slow, steady decline. At 100 mJ, the ratio of peak intensities increases until the tenth shot where the ratio starts to decrease. It is not clear what causes this behavior. At 200 mJ, the intensity ratio increases until it surpasses the threshold of the S sample obtained from Section 3.2. After 10 shots, the threshold value is surpassed, and the TD surface composition is comparable to the S surface. The reason the

ratio continues to increase for 200 mJ is that the high-power density produced by the 200 mJ pulse raises the temperature in the sample through heating from the LIBS beam during plasma formation. The heat on the surface starts the sintering process, removing residual binder and fusing together the titanium particles. As the number of laser pulses increases, particle fusion increases. The particle fusion and necking in Figures 6, 7, and 11 are evidence that the surface has been sintered.

EDS results (Table 3) of the ablation crater created from the 200 mJ treatment (Figure 6(e)) show that the carbon content after 20 shots has reduced from 20.29 Wt.% to 2.13 Wt.% which is comparable to the S sample. The area examined by EDS is presented in Figure 6(f). This confirms the reduction in carbon seen in the titanium to carbon intensity ratio obtained from the LIBS spectra. These results show that there is enough heat penetrating into the sample

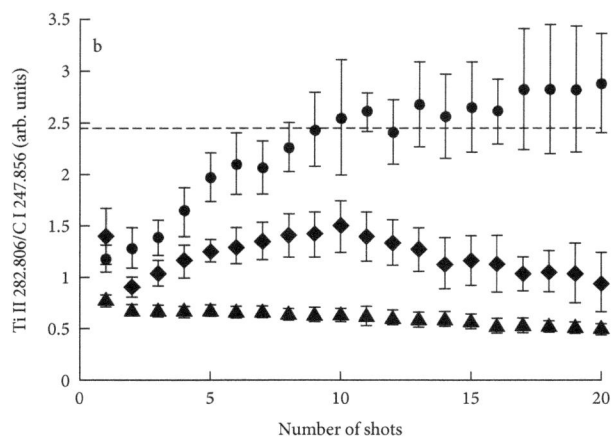

FIGURE 10: Evolution of the peak intensity ratio for Ti II 282.81 nm/ C I 247.86 nm on the thermally debound sample. The dashed line shows the sintering threshold. The different energies used were 200 mJ (●), 100 mJ (◆), and 50 mJ (▲).

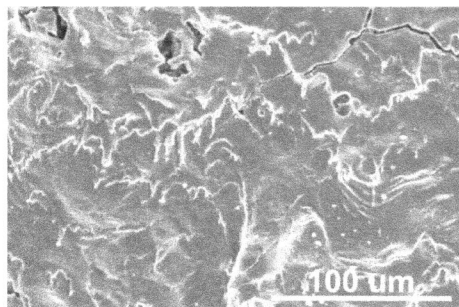

FIGURE 11: SEM image of sintering resulting from slowly moving the LIBS-6 beam after 10 shots.

creating sintering with very small removal of surface material due to plasma formation.

3.4. *Reduction of Carbon Signal in LIBS Spectra.* The focal position of the LIBS beam was positioned on the surface of the TD sample. The spot size of the laser, determined from the ablation crater, is approximately 650 μm in diameter. When successive shots are fired, sintering expands until it fills the entire spot size of the incident laser. The high-carbon content correlates to the unsintered part of the surface being sampled and is reduced as the area being sampled fuses together. When the entire area under the laser spot is sintered, the Ti/C ratio resembles that of sintered titanium.

It can be seen from the SEM images in Figure 6 that there has been sintering in all ablation craters from 50, 100, and 200 mJ laser pulses even though the LIBS intensity ratio does not reach the threshold value of the S sample. After 20 shots at 50 mJ, there is a sintered spot in the middle of the ablation crater which is about 250 μm in diameter. Because of the low laser energy, there is not enough heat generated on the surface of the sample to produce particle fusion over the

TABLE 3: Sample composition after repetitive shots compared to traditional sintering. Results obtained from EDS.

	C (Wt.%)	Ti (Wt.%)	V (Wt.%)	Si (Wt.%)
Ablation crater	2.13 ± 0.26	97.27 ± 0.84	0.59 ± 0.32	—
Sintered	1.79 ± 0.24	96.78 ± 0.69	0.24 ± 0.24	1.19 ± 0.08

entire spot size of the laser. Subsequent shots would still be sampling the carbon in the unsintered part of the ablation crater, and the Ti/C ratio does not reach the threshold value. The 100 mJ crater shows similar results with a slightly bigger sintered spot.

3.5. *Implementation.* The LIBS-6 system, coupled with the autofocus unit and 3-axis translation stage, was used to produce sintering on the surface of the sample steps at heights of 2, 4, and 6 mm (see Figure 12). The Spectrolaser system produced sintering at 200 mJ per pulse, with an irradiance of 8.6 GW cm^{-2}. Using the LIBS 6 system, which has a smaller spot size, increases irradiance on the sample causing higher temperatures and increased localized heating on the surface of the sample. With a pulse energy of 100 mJ, an irradiance of 9.0 GW cm^{-2} is produced which is comparable to the Spectrolaser system. Therefore, the LIBS-6 system will provide the same heat that the Spectrolaser system produces, entirely sintering the spot area after 10 shots.

The results of sintering a line on the TD sample using the LIBS-6 system and the ratio threshold of 2.45 to control the process are displayed in Figures 11 and 12. The LIBS-6 system had a spot diameter of 450 μm and a repetition rate of 20 Hz with the sample stage moving at a speed of roughly 0.4 mm/s^{-1}. It can be seen that the titanium particles have fused together forming a solid mass. After sintering had taken place, on an ablation site, the LIBS beam was moved to a new surface which overlapped the previous spot. Sintering was then repeated on the new surface. The liquid titanium formed on the new spot flowed over the previously sintered spot and has fused together. These regions are seen in Figure 11. This process was repeated until a sintered line was produced across the sample. Figure 12 shows the effectiveness of pulsed laser sintering a line across injected-molded titanium at step heights of 6, 4, and 2 mm using autofocus. The effect of heat propagation through the part can be seen by the titanium bluing.

4. Conclusions

LIBS has been proven useful for sintering-selected surfaces of injection-molded titanium. The LIBS beam, plasma, and residual heat of the sample surface all contribute to the heating of the injection-molded titanium causing sintering. By observing the C I 247.856 nm and the Ti II 282.806 nm lines and then comparing the intensity ratio to that of a conventionally sintered sample, the amount of sintering taking place can be determined. Repetitive shots continue heating the surface and provide spectral feedback to monitor whether the sintering process is complete. The effectiveness

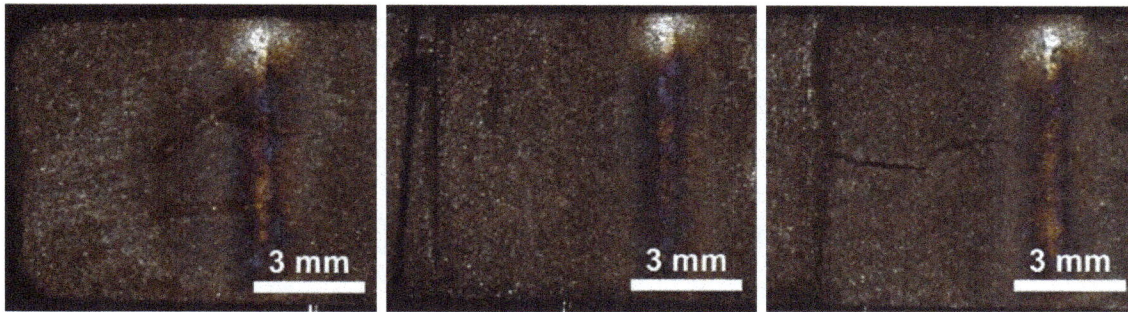

FIGURE 12: Optical micrograph of line sintering at different step heights. Left to right: 2, 4, and 6 mm step heights.

of using LIBS to determine when the sample has been sintered has been verified using EDS and SEM images. EDS results show a 18.5 Wt.% reduction in carbon content. It was found that 10 shots at an irradiance between 8.6 and 9.0 GW cm^{-2} is sufficient to sinter the spot produced by the LIBS beam.

Using the triangulation method described, an autofocus mechanism was produced and calibrated to the LIBS-6 system. This system was used on an injection-molded grey part with varying step heights to prove the viability of using LIBS in a production process on metal parts with complex geometry. Sintered lines were produced on different step heights of an injection-molded sample. The heat from repetitive plasma formations has colored the previously sintered titanium.

Additional Points

Highlights. The amount of sintering under a laser spot is evaluated using LIBS. Spectral measurements determine the relative titanium to carbon content. LIBS and Energy-dispersive X-ray Spectroscopy results display a reduction in carbon contents on the sintered surface.

Acknowledgments

The authors would like to acknowledge and offer continuing respect and admiration to deceased colleague Dr. Sadhana Talele, who was not able to see this work completed. This work was supported by New Zealand's Ministry of Business, Innovation and Employment under contract C11X1209.

References

[1] J. Beddoes and M. J. Bibby, "6 - powder metallurgy," in *Principles of Metal Manufacturing Processes*, J. Beddoes and M. J. Bibby, Eds., pp. 173–189, Butterworth-Heinemann, Oxford, 1999.

[2] T. Ebel, "17 - metal injection molding (MIM) of titanium and titanium alloys," in *Handbook of Metal Injection Molding*, D. F. Heaney, Ed., pp. 415–445, Woodhead Publishing, Cambridge, United Kingdom, 2012.

[3] A. Simchi, "Direct laser sintering of metal powders: mechanism, kinetics and microstructural features," *Materials Science and Engineering A*, vol. 428, pp. 148–158, 2006.

[4] D. Cremers and L. Radziemski, "Introduction," in *Handbook of Laser-Induced Breakdown Spectroscopy*, p. 3, John Wiley & Sons Ltd, West Sussex, UK, 2013.

[5] J. F. Ready, "Effects due to absorption of laser radiation," *Journal of Applied Physics*, vol. 36, pp. 462–468, 1965.

[6] J. P. Singh and S. N. Thakur, "Fundamentals of laser induced breakdown spectroscopy," in *Laser-Induced Breakdown Spectroscopy*, J. P. Singh and S. N. Thakur, Eds., p. 3, Elsevier, Oxford, UK, 2007.

[7] D. Cremers and L. Radziemski, "Basics of the LIBS plasma," in *Handbook of Laser-Induced Breakdown Spectroscopy*, D. Cremers and L. Radziemski, Eds., p. 29, John Wiley & Sons Ltd, West Sussex, UK, 2013.

[8] A. Bogaerts, Z. Chen, R. Gijbels, and A. Vertes, "Laser ablation for analytical sampling: what can we learn from modeling?" *Spectrochimica Acta Part B: Atomic Spectroscopy*, vol. 58, pp. 1867–1893, 2003.

[9] U. Panne, C. Haisch, M. Clara, and R. Niessner, "Analysis of glass and glass melts during the vitrification process of fly and bottom ashes by laser-induced plasma spectroscopy. Part I: normalization and plasma diagnostics," *Spectrochimica Acta Part B: Atomic Spectroscopy*, vol. 53, pp. 1957–1968, 1998.

[10] A. Costela, I. García-Moreno, C. Gómez, O. Caballero, and R. Sastre, "Cleaning graffitis on urban buildings by use of second and third harmonic wavelength of a Nd: YAG laser: a comparative study," *Applied Surface Science*, vol. 207, pp. 86–99, 2003.

[11] T. Li, Q. Lou, Y. Wei, F. Huang, J. Dong, and J. Liu, "Laser-induced breakdown spectroscopy for on-line control of selective removal of cobalt binder from tungsten carbide hardmetal by pulsed UV laser surface ablation," *Applied Surface Science*, vol. 181, pp. 225–233, 2001.

[12] D. Diego-Vallejo, D. Ashkenasi, A. Lemke, and H. J. Eichler, "Selective ablation of Copper-Indium-Diselenide solar cells monitored by laser-induced breakdown spectroscopy and classification methods," *Spectrochimica Acta Part B: Atomic Spectroscopy*, vol. 87, pp. 92–99, 2013.

[13] M. S. Majewski, C. Kelley, W. Hassan, W. Brindley, E. H. Jordan, and M. W. Renfro, "Laser induced breakdown spectroscopy for contamination removal on engine-run thermal barrier coatings," *Surface and Coatings Technology*, vol. 205, pp. 4614–4619, 2011.

[14] D. E. Roberts, A. du Plessis, J. Steyn, L. R. Botha, S. Pityana, and L. R. Berger, "An investigation of Laser Induced Breakdown Spectroscopy for use as a control in the laser removal of rock from fossils found at the Malapa hominin site, South Africa," *Spectrochimica Acta Part B: Atomic Spectroscopy*, vol. 73, pp. 48–54, 2012.

[15] P. V. Maravelaki, V. Zafiropulos, V. Kilikoglou, M. Kalaitzaki, and C. Fotakis, "Laser-induced breakdown spectroscopy as a diagnostic technique for the laser cleaning of marble," *Spectrochimica Acta Part B: Atomic Spectroscopy*, vol. 52, pp. 41–53, 1997.

[16] S. Acquaviva, M. L. De Giorgi, C. Marini, and R. Poso, "A support of restoration intervention of the bust of St. Gregory the Armenian: compositional investigations by laser induced breakdown spectroscopy," *Applied Surface Science*, vol. 248, pp. 218–223, 2005.

[17] T. Flores, L. Ponce, M. Arronte, and E. de Posada, "Free-running and Q: switched LIBS measurements during the laser ablation of Prickle Pears spines," *Optics and Lasers in Engineering*, vol. 47, pp. 578–583, 2009.

[18] M. Bengtsson, R. Grönlund, M. Lundqvist, A. Larsson, S. Krööll, and S. Svanberg, "Remote laser-induced breakdown spectroscopy for the detection and removal of salt on metal and polymeric surfaces," *Applied Spectroscopy*, vol. 60, pp. 1188–1191, 2006.

[19] M. R. Martelli, C. Barron, P. Delaporte, G. Viennois, X. Rouau, and A. Sadoudi, "Pulsed laser ablation: a new approach to reveal wheat outer layer properties," *Journal of Cereal Science*, vol. 49, pp. 354–362, 2009.

[20] M. R. Martelli, F. Brygo, P. Delaporte, X. Rouau, and C. Barron, "Estimation of wheat grain tissue cohesion via laser induced breakdown spectroscopy," *Food Biophysics*, vol. 6, pp. 433–439, 2011.

[21] M. Abdelhamid, S. Grassini, E. Angelini, G. M. Ingo, and M. A. Harith, "Depth profiling of coated metallic artifacts adopting laser-induced breakdown spectrometry," *Spectrochimica Acta Part B: Atomic Spectroscopy*, vol. 65, pp. 695–701, 2010.

[22] C. Theodorakopoulos and V. Zafiropulos, "Depth-profile investigations of triterpenoid varnishes by KrF excimer laser ablation and laser-induced breakdown spectroscopy," *Applied Surface Science*, vol. 255, pp. 8520–8526, 2009.

[23] A. De Bonis, B. De Filippo, A. Galasso, A. Santagata, A. Smaldone, and R. Teghil, "Comparison of the performances of nanosecond and femtosecond Laser Induced Breakdown Spectroscopy for depth profiling of an artificially corroded bronze," *Applied Surface Science*, vol. 302, pp. 275–279, 2014.

[24] F. O. Leme, Q. Godoi, P. H. M. Kiyataka, D. Santos, J. A. M. Agnelli, and F. J. Krug, "Effect of pulse repetition rate and number of pulses in the analysis of polypropylene and high density polyethylene by nanosecond infrared laser induced breakdown spectroscopy," *Applied Surface Science*, vol. 258, pp. 3598–3603, 2012.

[25] F. Boué-Bigne, "Laser-induced breakdown spectroscopy applications in the steel industry: rapid analysis of segregation and decarburization," *Spectrochimica Acta Part B: Atomic Spectroscopy*, vol. 63, pp. 1122–1129, 2008.

[26] L. M. Cabalín and J. J. Laserna, "Surface stoichiometry of manganin coatings prepared by pulsed laser deposition as described by laser-induced breakdown spectrometry," *Analytical Chemistry*, vol. 73, pp. 1120–1125, 2001.

[27] J. D. Pedarnig, J. Heitz, T. Stehrer et al., "Characterization of nano-composite oxide ceramics and monitoring of oxide thin film growth by laser-induced breakdown spectroscopy," *Spectrochimica Acta Part B: Atomic Spectroscopy*, vol. 63, pp. 1117–1121, 2008.

[28] J. Gruber, J. Heitz, H. Strasser, D. Bäuerle, and N. Ramaseder, "Rapid in-situ analysis of liquid steel by laser-induced breakdown spectroscopy," *Spectrochimica Acta - Part B Atomic Spectroscopy*, vol. 56, pp. 685–693, 2001.

[29] J. Gruber, J. Heitz, N. Arnold et al., "In situ analysis of metal melts in metallurgic vacuum devices by laser-induced breakdown spectroscopy," *Applied Spectroscopy*, vol. 58, pp. 457–462, 2004.

[30] C. Lopez-Moreno, S. Palanco, and J. J. Laserna, "Calibration transfer method for the quantitative analysis of high-temperature materials with stand-off laser-induced breakdown spectroscopy," *Journal of Analytical Atomic Spectrometry*, vol. 20, pp. 1275–1279, 2005.

[31] aforgenet.com, "AForge.NET framework," 2012, http://www.aforgenet.com/framework/.

[32] P. Ewart, S. Ahn, and D. Zhang, "Mixing titanium MIM feedstock: homogeneity, debinding and handling strength," *Powder Injection Moulding International*, vol. 5, pp. 54–59, 2011.

[33] P. Ewart, H. Jull, R. Künnemeyer, and P. N. Schaare, "Identification of contamination levels and the microstructure of metal injection moulded titanium," in *Key Engineering Materials*, T. Ebel and F. Pyczak, Eds., vol. 704, pp. 161–169, Trans Tech Publications Ltd, Zurich, Switzerland, 2016.

[34] J. C. Russ, "Chapter 3 - energy dispersive spectrometers," in *Fundamentals of Energy Dispersive X-ray Analysis*, J. C. Russ, Ed., pp. 17–41, Butterworth-Heinemann, Oxford, United Kingdom, 1984.

3

Detection of Water Contamination Events Using Fluorescence Spectroscopy and Alternating Trilinear Decomposition Algorithm

Jie Yu, Xiaoyan Zhang, Dibo Hou, Fang Chen, Tingting Mao, Pingjie Huang, and Guangxin Zhang

State Key Laboratory of Industrial Control Technology, College of Control Science and Engineering, Zhejiang University, Hangzhou 310027, China

Correspondence should be addressed to Dibo Hou; houdb@zju.edu.cn

Academic Editor: Jose M. Pedrosa

The method based on conventional index and UV-vision has been widely applied in the field of water quality abnormality detection. This paper presents a qualitative analysis approach to detect the water contamination events with unknown pollutants. Fluorescence spectra were used as water quality monitoring tools, and the detection method of unknown contaminants in water based on alternating trilinear decomposition (ATLD) is proposed to analyze the excitation and emission spectra of the samples. The Delaunay triangulation interpolation method was used to make the pretreatment of three-dimensional fluorescence spectra data, in order to estimate the effect of Rayleigh and Raman scattering; ATLD model was applied to establish the model of normal water sample, and the residual matrix was obtained by subtracting the measured matrix from the model matrix; the residual sum of squares obtained from the residual matrix and threshold was used to make qualitative discrimination of test samples and distinguish drinking water samples and organic pollutant samples. The results of the study indicate that ATLD modeling with three-dimensional fluorescence spectra can provide a tool for detecting unknown organic pollutants in water qualitatively. The method based on fluorescence spectra can be complementary to the method based on conventional index and UV-vision.

1. Introduction

Water pollution problem is attracting more and more attention, and monitoring of water quality is important in order to avoid health risk to residents. It is necessary to develop a technique for rapid water quality analysis in the case of unknown contaminants because of deterioration of water resources, indiscriminate discharge of wastewater, chemical leakage, and so on.

Currently, water pollution abnormality detection mainly depends on conventional water quality parameters. In the conventional detection, water anomaly detection methods are mainly based on traditional indexes of water quality. For example, Conde [1] developed regression models of artificial neural networks (ANNs) and relevance vector machines (RVMs) according to the normal water quality parameters and generated a discriminant classifier to discriminate the abnormal water from the normal water data online. However, the process of obtaining these water quality parameters has various problems, such as long analysis time, not sensitive enough, requiring reagents, and producing waste [2]. It cannot satisfy the online high-frequency water quality anomaly detection. Compared with the method of conventional detection, the method of water quality analysis based on spectra can measure samples directly by spectra without other operations such as extraction or separation, and its advantages are simple, rapid, and so on [3]. There have been many researches on water quality analysis based on UV-visible spectroscopy. Dürrenmatt and Gujer [4] analyzed the UV/Vis data with a two-staged clustering method consisting of the self-organizing map algorithm and the Ward clustering method to distinguish the industrial sewage and living sewage from the sewage treatment plant. Langergraber et al. [5, 6] continuously monitored the water quality through ultraviolet spectroscopy and judged the anomaly according to the three-dimensional spectrum and the historical spectrum of ultraviolet spectrum and time axis to obtain the good result. There are also some researches about distribution water quality anomaly detection by analyzing the ultraviolet spectroscopy [7, 8]. Hou et al., for example, integrated principal component analysis (PCA) with chi-square distribution to detect distribution water quality anomalies. However, detection limit of some organic matter based on UV-vision spectra is still not low enough to achieve the standard.

Three-dimensional fluorescence spectra have lower detection limit of organic matter than UV-visible spectra [9]; it can provide more complete spectral information; it has characteristics of high selectivity, high sensitivity, good reproducibility, requiring less sample, not damaging the structure of the sample, and so on [10]. So, three-dimensional fluorescence spectroscopy can be considered to be used as a tool of detecting water quality contaminant events and is complementary to conventional index methods and UV-vision spectra methods. It can detect some matters that the conventional index methods and UV-vision spectra methods cannot be used to detect, and the lower detection limit may be lower. Three-dimensional fluorescence spectra can be used to analyze water samples quantitatively. Three-dimensional fluorescence spectra are widely applied to measure the relative concentration of organic matter in water [11–13]. Three-dimensional fluorescence spectra are also applied to water qualitative analysis, mainly in the field of classification. The analysis of nature organic matter (NOM) in the sea, rivers, or lakes is used to classify the water samples from different sources. Baker [14] applied three-dimensional fluorescence spectroscopy to the water samples from 10 sample sites in six rivers and demonstrated that tryptophan- and fulvic-like fluorescence intensity is associated with whether the rivers accept the sewage treatment plant effluent or not. Pavelescu et al. [15] applied fluorescence EEM spectroscopy to the samples from 12 groundwater sources (wells) located in different surroundings, and fluorescence indices can be used to discriminate the water samples from different sources. Thus, studies about three-dimensional fluorescence spectra mainly highlight the quantification analysis and classification of specific known organic matter in water. However, there are few researches about unknown pollutant detection in water supply system.

The aim of this paper was to make qualitative detection of unknown pollutants for early warning when sudden pollution event happens in water supply system by using three-dimensional fluorescence spectra to make up for the lack of conventional method and UV-vision method in the water pollution qualitative detection. Fluorescence data were analyzed using matrix feature extraction method based on alternating trilinear decomposition (ATLD) and residual sum of squares, combining the threshold to judge the unknown samples and to detect the aqueous samples of organic contaminants.

2. Methodology

The basic idea of the paper is to establish a model of normal water sample. If a test sample does not conform to the model, it will be determined as an anomaly sample. Before the establishment of the model, the data need to be pretreated. Then, the ATLD algorithm is used to obtain the model. The model is applied to test samples to obtain the model data, and the model matrix is compared to the measured matrix to obtain the residual matrix as a basis of judging whether the water samples are abnormal or not. The main process of the method is shown in Figure 1.

2.1. Pretreatment. Three-dimensional fluorescence spectra could contain not only the fluorescence of the substance to be tested but also Rayleigh and Raman scattering. Since that scattering part does not meet the requirements of trilinear decomposition algorithm theory, it is not appropriate to analyze the scattering part with decomposition model. So it is essential to eliminate the scattering of the three-dimensional fluorescence spectra. Some researches eliminated the scattering by taking out the background of distilled water, which may still remain some scattering [16]. It is more likely to cause some problems when the concentration of the sample is relatively low. This paper adopts the Delaunay triangulation interpolation method [17] to preprocess the raw data in order to eliminate the impact of Rayleigh scattering. In the meanwhile, this paper eliminated the impact of Raman scattering by taking out the background of solvent.

When measuring the Raman spectra of ultrapure water on the excitation wavelength of 350 nm and the emission wavelength of 397 nm before the experiment starts up, it is discovered that there exists difference among the measurements. Consequently, in order to eliminate the difference, this paper did Raman normalization [18] to all of the three-dimensional fluorescence data, which equals to divide excitation-emission matrix (EEM) by the Raman peak value on the excitation/emission wavelength 350/397 nm of the ultrapure water.

2.2. Feature Extracting Based on ATLD and Residual Sum of Squares

2.2.1. ATLD Algorithm. Alternating trilinear decomposition (ATLD) algorithm, put forward by Wu et al. [19] in 1998, is an improvement to the conventional PARAFAC algorithm. Taking advantage of the alternating least squares theory, Moore-Penrose generalized inverse calculation based on singular value decomposition (SVD) as well as alternate iterations which are used to improve the performance of the trilinear decomposition. Compared to PARAFAC, ATLD, making use of generalized inverse and matrix diagonal elements extracting, is not sensitive to the number of components [20]; meanwhile, ATLD makes calculation by slice matrix, which accelerates the calculation speed and therefore improves the calculation efficiency.

Below is the trilinear model of a three-dimensional data matrix X, as Figure 2 shows:

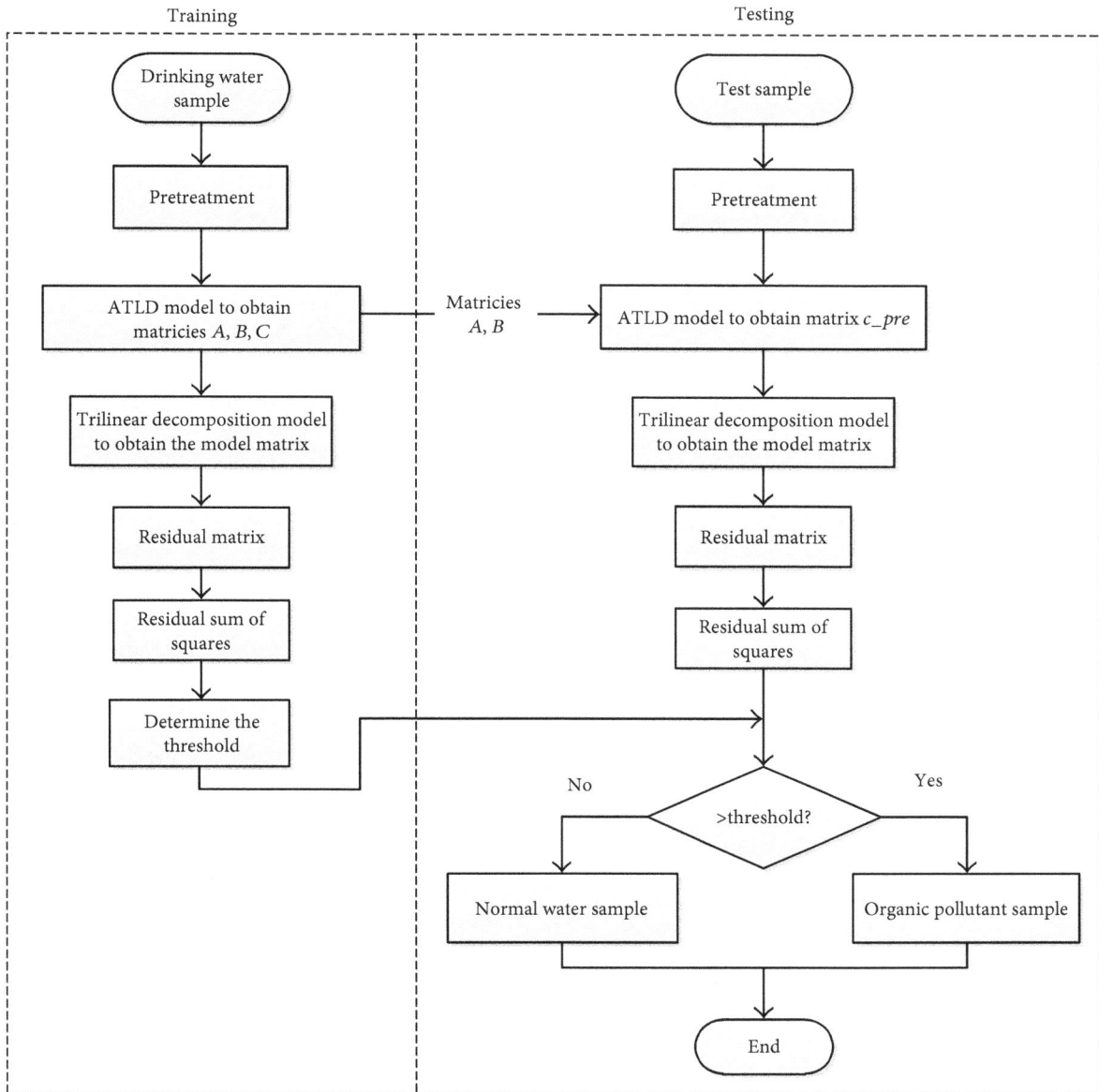

FIGURE 1: Qualitative discrimination based on ATLD.

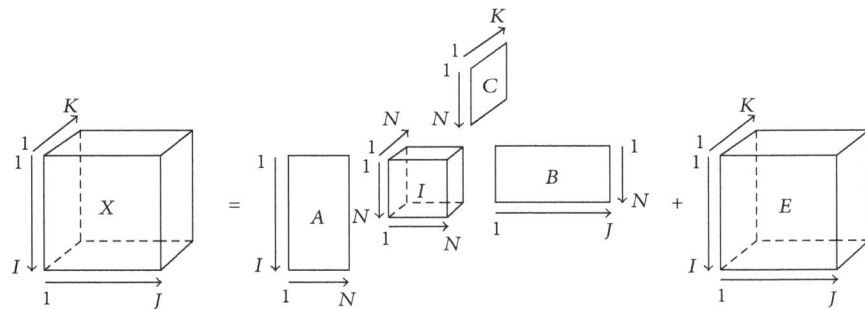

FIGURE 2: Graphical representation of trilinear model of three-way data array X. X is the original three-dimensional data matrix, and it can be decomposed to three matrices. A, B, and C are, respectively, the relative excitation matrix, relative emission matrix, and relative concentration matrix. Matrix E is the three-dimensional residual matrix.

$$x_{ijk} = \sum_{n=1}^{N} a_{in} b_{jn} c_{kn} + e_{ijk}, \tag{1}$$

where $i = 1, 2, ..., I; j = 1, 2, ..., J; k = 1, 2, ..., K;$ N represents the number of factors and factors are all the measurable components, including components which are useful for fluorescence data and components which are disturbance for fluorescence data; x_{ijk} represents the elements (i, j, k) of the three-dimensional data matrix X. In this paper, the three-dimensional data matrix X is just the three-dimensional fluorescence spectra matrix for k samples after pretreatment. a_{in} represents the element (i, n) of the relative excitation matrix A in size $I \times N$; b_{jn} represents the element (j, n) of the relative emission matrix B in size $J \times N$; c_{kn} represents the element (k, n) of the relative concentration matrix C in size $J \times N$; e_{ijk} represents the element (i, j, k) of the three-dimensional residual matrix E in size $I \times J \times K$.

First, ATLD gives matrices A and B randomly and alternately updates $A \times B \times C$ in the three separate directions of the three-dimensional data matrix, which are as follows:

$$a_{(i)}^{T} = \mathrm{diagm}\left(B^{+} X_{i..} \left(C^{T}\right)^{+}\right), \tag{2}$$

$$b_{(j)}^{T} = \mathrm{diagm}\left(C^{+} X_{.j.} \left(A^{T}\right)^{+}\right), \tag{3}$$

$$c_{(k)}^{T} = \mathrm{diagm}\left(A^{+} X_{..k} \left(B^{T}\right)^{+}\right), \tag{4}$$

where $i = 1, 2, ..., I;$ $j = 1, 2, ..., J;$ $k = 1, 2, ..., K;$ $\mathrm{diagm}(\cdot)$ represents taking the diagonal matrix elements as a column vector; A^{+}, B^{+}, C^{+}, respectively, represent the Moore-Penrose generalized reverse of matrices A, B, C; $a_{(i)}^{T}, b_{(j)}^{T}, c_{(k)}^{T}$, respectively, represent the transposition of the ith row vector of the relative excitation matrix A, the jth row vector of the relative emission matrix B, and the kth row vector of the relative concentration matrix C; matrices A and B normalized to unit length column by column every iterative loop.

Matrices A, B, C are solved by the alternative iteration of the iteration formulas (2), (3), and (4) until the objective function is converged. The rule of convergence is as follows:

$$\left| \frac{\mathrm{SSR}^{(m)} - \mathrm{SSR}^{(m-1)}}{\mathrm{SSR}^{(m-1)}} \right| < 1 \times 10^{-6}. \tag{5}$$

The residual sum of squares (SSR) is the loss function defined by ATLD, which is

$$\mathrm{SSR} = \sum_{i=1}^{I} \sum_{j=1}^{J} \sum_{k=1}^{K} \left(x_{ijk} - \sum_{n=1}^{N} a_{in} b_{jn} c_{kn} \right)^{2}. \tag{6}$$

As a result, ATLD could decompose the three-dimension fluorescence spectrum matrix $X(I \times J \times K)$ to three low-dimension matrices, which are known as relative excitation matrix $A(I \times N)$, relative emission matrix $B(J \times N)$, and relative concentration matrix $C(K \times N)$.

2.2.2. Feature Extracting Based on Residual Sum of Squares. Model parameters are obtained with ATLD algorithm, which is known as relative excitation matrix A, relative emission matrix B, and relative concentration matrix C of the background sample.

Known by formula (4), if the background sample matrix $X(I \times J)$, the relative excitation matrix A, and the relative emission matrix B of the background sample are substituted into formula (4), the relative concentration c_pre of each component of the sample X could be obtained based on the ATLD model. Moreover, if A, B, and c_pre are substituted into formula (1), the modeling value $\hat{X}$ of the sample could be given. As what formula (7) presents, the residual matrix E is obtained according to the difference of actual value X and the modeling value $\hat{X}$.

$$E_{I \times J} = X_{I \times J} - \hat{X}_{I \times J}. \tag{7}$$

The residual sum of squares is $\sum_{j=1}^{J} \sum_{i=1}^{I} e_{ij}$, where e_{ij} is the element (i, j) of the matrix E and can be obtained by calculating the sum of squares of the residual matrix E of all the elements. The residual sum of squares is defined as the basis of the qualitative discrimination.

2.3. Qualitative Discrimination Based on Threshold. The method based on threshold is often used for image segmentation because of its simple calculation. The grayscale of the image is usually divided into several parts through one or several thresholds and pixels belonging to the same part are considered as the same object. This paper separates test samples into two parts. One is normal drinking water samples, and the other is organic pollutant samples.

This paper aims at setting reasonable threshold of the object sequence for qualitative discrimination. In other words, through setting the threshold to the residual sum of squares, the aim that a new unknown sample is qualitatively identified can be reached. The setting of the threshold is a critical problem. If the threshold is too large, the polluted water sample cannot be detected; on the other hand, if the threshold is too small, the drinking water may be detected as the polluted water sample by mistake.

In the math analysis, the mean and standard deviation are often used to indicate the important characteristics of the data set. n times detection values are $x_1, x_2, ..., x_n$, then the mean is $\bar{x}$, and the standard deviation is σ.

Three times of the standard deviation of qualitative discrimination object sequence is usually used to detect the test sample qualitatively. Byer and Carlson [21] monitored 16,000 drinking water multiparameter samples online, and the data revealed that parameters were changed with the change of time and flow. Suppose that the data follow Gaussian distribution, the samples within the scope of three times of the standard deviation (3σ) occupied 99.96%. In the research of Byer and Carlson, three times of the standard deviation (3σ) is just the threshold of organic matter sample detection. For test samples, if the detection value x_i is in the scope of $\bar{x} \pm 3\sigma$, it is determined to be drinking water samples, otherwise determined to be organic pollutant samples.

In this paper, the standard for the setting of threshold is that false positive rate of drinking water samples is under 10%, so two times of the standard deviation of the residual

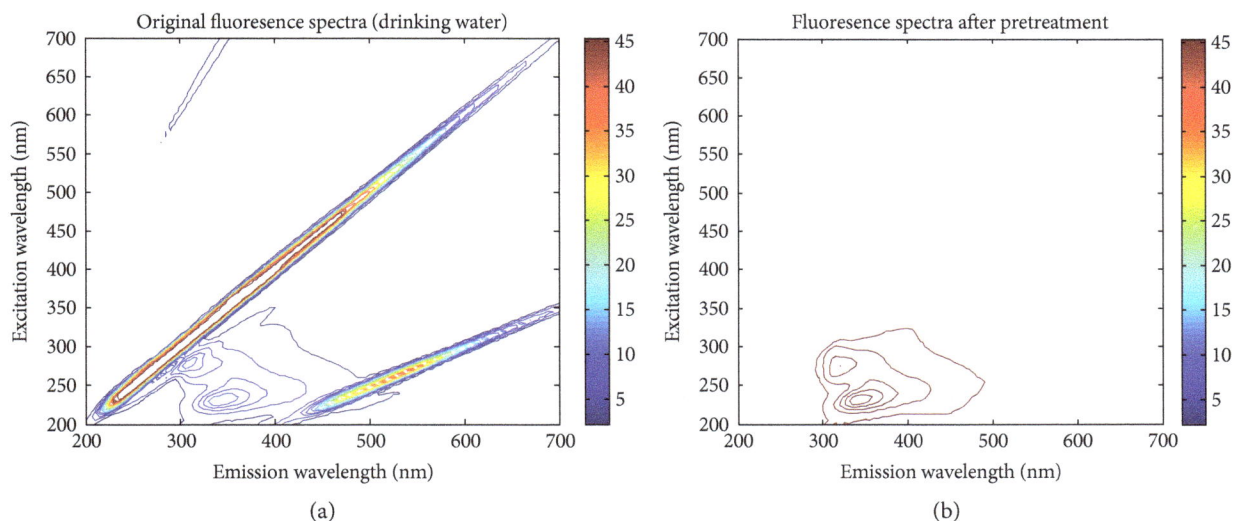

FIGURE 3: Three-dimensional fluorescence spectra of background drinking water samples. (a) is the spectrum before pretreatment. (b) is the spectrum after pretreatment.

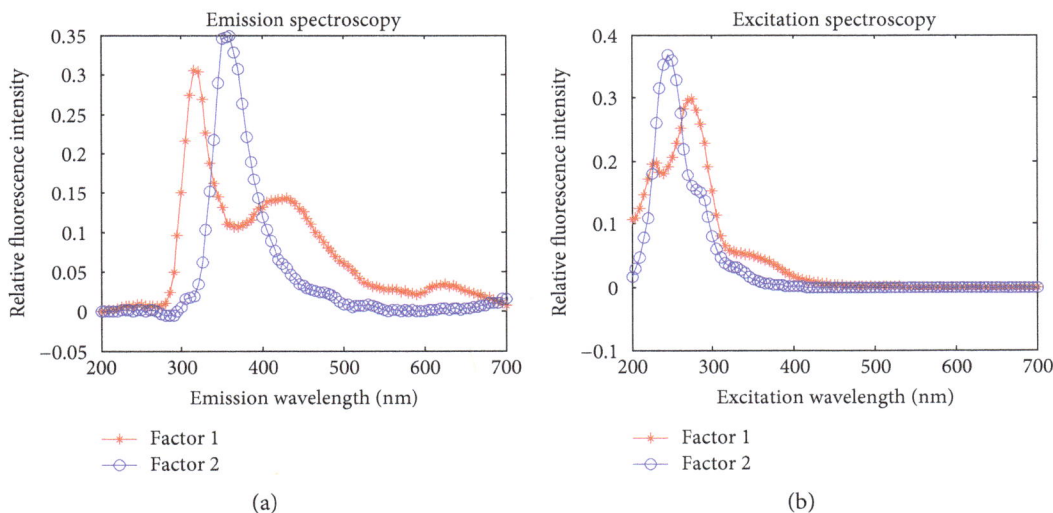

FIGURE 4: Relative excitation matrix and relative emission matrix of drinking water samples after ATLD. (a) is relative excitation matrix. (b) is relative emission matrix.

sum of squares of 10 background drinking water samples is used as the threshold for qualitative discrimination. Explanatorily, the residual sums of squares of 10 drinking water samples $x_1, x_2, \ldots, x_{10}$ are used to get the mean $\bar{x}$ first; then, $x_1, x_2, \ldots, x_{10}$ and $\bar{x}$ are put together to get the standard deviation σ, and 2σ is defined as the threshold of qualitative discrimination.

3. Results and Discussion

3.1. Sampling. One group of samples was the three kinds of organic solution with fluorescent characteristics in the concentrations of $2\,\mu/L$, $4\,\mu/L$, $6\,\mu/L$, $8\,\mu/L$, $10\,\mu/L$, $20\,\mu/L$, $30\,\mu/L$, and $40\,\mu/L$, and the experiment was repeated three times. The other group of samples was in the concentrations of $1\,\mu/L$, $3\,\mu/L$, $5\,\mu/L$, $7\,\mu/L$, $15\,\mu/L$, $25\,\mu/L$, $35\,\mu/L$, and $45\,\mu/L$,

and the experiment was repeated two times. In each experiment, ten drinking water samples were collected at the interval of half an hour from the laboratory tap as the background samples. Fluorescence spectra, in the form of excitation-emission matrices (EEMs), were recorded in the excitation wavelength range from 200 to 700 nm and emission wavelength range from 200 to 700 nm. Scanning intervals for excitation and emission were both 5 nm.

3.2. Pretreatment of Spectra. Three-dimensional fluorescence spectra measured by the spectrometer contain Rayleigh and Raman scattering, as shown in Figure 3(a). The Delaunay triangulation interpolation method was used to make the scattering pretreatment of the raw data. In order to remove the noise of instrument, EEM was subjected to Raman normalization after removing scattering, that is, divided by Raman

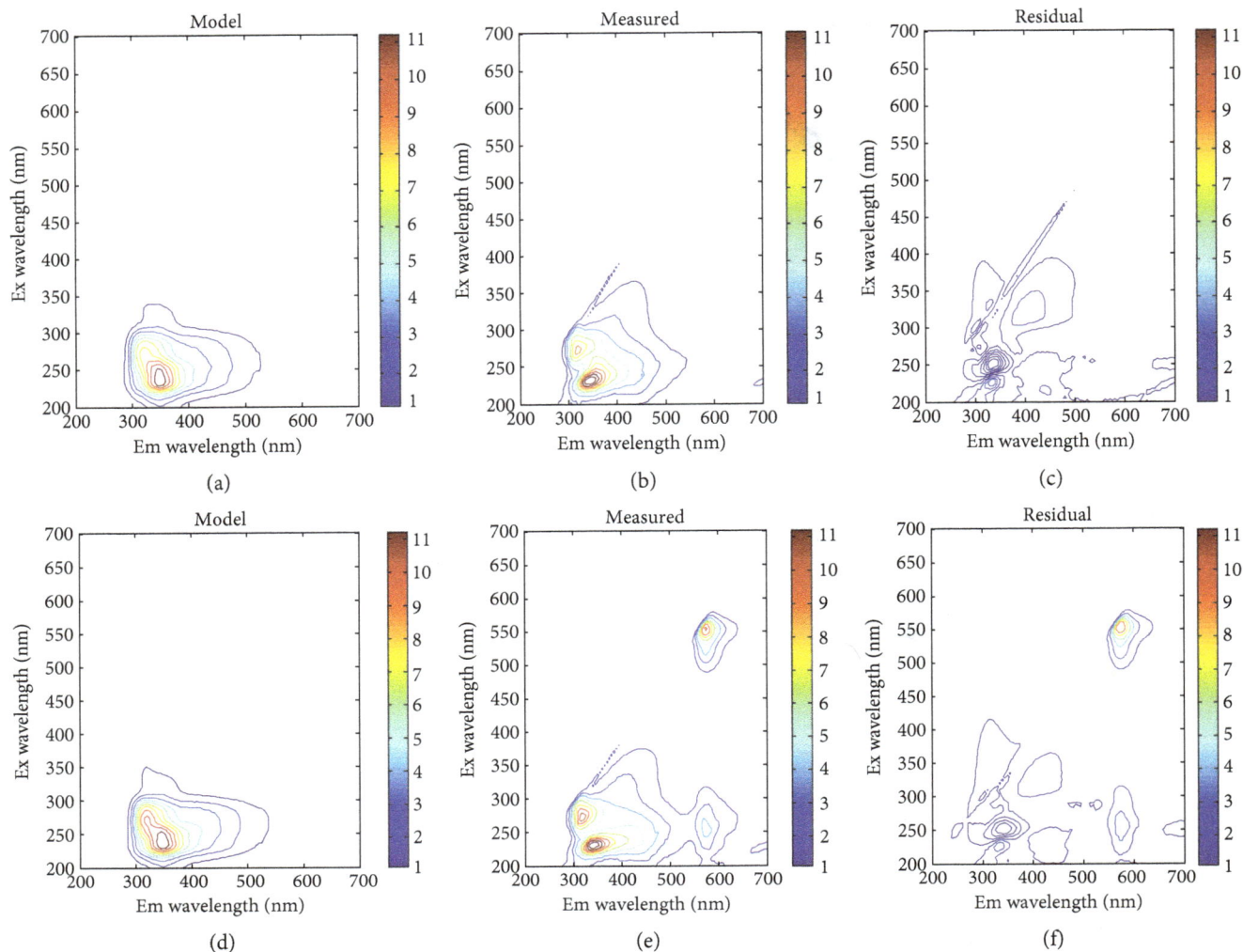

FIGURE 5: Fluorescence characteristics of background drinking water and rhodamine B solution samples. (a), (b), and (c) are, respectively, modeling value, measured value, and residual value of drinking water. (d), (e), and (f) are, respectively, modeling value, measured value, and residual value of 30 ug/L rhodamine B solution.

scattering value of water at $\lambda_{ex} = 350$ nm, $\lambda_{em} = 397$ nm. The charts before and after the pretreatment are shown in Figure 3. It is clear that the three-dimensional fluorescence spectra before pretreatment had significant scattering, and spectral pretreatment removed scattering and retained effective fluorescence spectra part.

3.3. Qualitative Determination Based on ATLD and Threshold.
ATLD was applied to background drinking water samples to establish the model. Because of the complexity of the composition of water samples, the number of factors in the ATLD model is hard to identify. However, there are two peaks in the spectra of drinking water after pretreatment, as in Figure 3(b); according to the common dissolved organic matter with fluorescent characteristic in the water, the two peaks may correspond to tyrosine-like protein organic matter and tryptophan-like protein organic matter, so it can be seen as a mixture of two kinds of organic matters

without other fluorescent organic matters, and the number of factors is identified as $N = 2$. Relative excitation matrix A, relative emission matrix B of water samples (Figure 4), and relative concentration matrix C of each component can be obtained. In Figure 4, factor 1 and factor 2 correspond to the two peaks in original fluorescence spectra of water samples, which may represent tyrosine-like protein organic matter and tryptophan-like protein organic matter.

Relative excitation matrix A, relative emission matrix B, and the test sample matrix are used to estimate relative concentration matrix c_pre of each component of test samples according to formula (4). According to the trilinear decomposition principle, combining relative excitation matrix A and relative emission matrix B with relative concentration matrix c_pre can obtain spectra data modeling value of unknown test samples, as is shown in Figure 5(a). Compared with Figures 5(a) and 5(b), it can be found in Figure 5(b) that the spectra of water samples had two fluorescence peaks; the

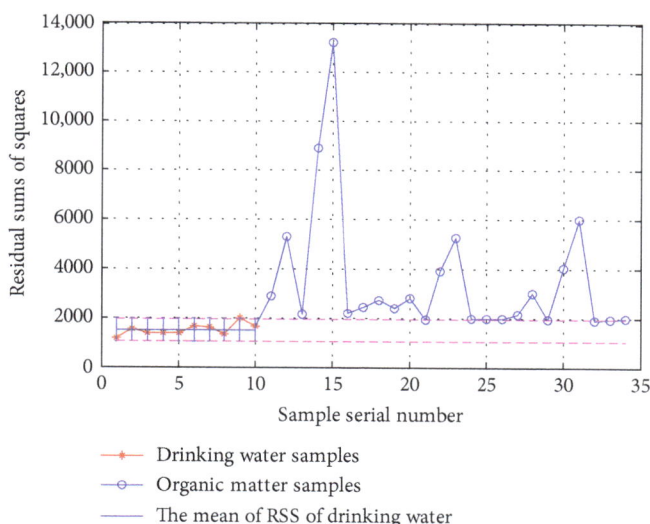

FIGURE 6: The residual sums of squares. "∗" is drinking water samples, and "○" is organic matter samples. The range between the two dashed lines represents the samples judged as the normal samples.

model established by ATLD as Figure 5(a) can characterize the fluorescence signals caused by main organic matters in the drinking water.

The measured value subtracting the modeling value of spectra matrix can obtain residual matrix (Figure 5(c)), which can indicate the fluorescence signal caused by the noise of instrument and the remaining scattering part that cannot be represented by the model. The ATLD model established with the drinking water can be used to explain its main organic matter composition and content.

For organic matter samples, three-dimensional fluorescence spectra can be considered as superposition of drinking water spectra and organic matter spectra. Combined with Figure 5(d), it can be found that the ATLD model based on drinking water background can represent the part of drinking water spectra in the organic matter solution. However, according to Figure 5(f), rhodamine B not belonging to the background water sample cannot be explained by the model. It can be used as the basis of distinguishing the background drinking water samples and organic pollutant samples.

According to the residual matrix, the residual sums of squares of background drinking water samples and test samples are, respectively, calculated, as a detecting target sequence of qualitative determination. The residual sums of squares of one group experiment data are shown in Figure 6.

The mean and standard deviation of residual sums of squares of drinking water sample were calculated, and then the residual sum of squares of each unknown test sample subtracted the mean of background water samples. If the difference is larger than the threshold, the sample is judged to be organic pollutant sample; otherwise, drinking water sample. The result of qualitative discrimination based on the threshold depends largely on the threshold selection. As is mentioned in the second part of this paper, the standard of selecting the threshold is that false-positive rate of background drinking water samples is less than 10%. So two times

TABLE 1: The detecting result of the samples.

Group	Mean	Standard deviation	Total number of OM samples	Total number of detected	Detection rate
1	1844.39	301.99	24	22	91.67%
2	2017.03	310.34	24	20	83.33%
3	1510.16	223.26	24	20	83.33%
4	1617.39	303.89	27	20	77.78%
5	1679.49	325.21	27	25	92.57%

of standard deviation of drinking water samples is selected as the threshold. The experiment result is shown in Table 1.

It can be seen from the result of qualitative discrimination that the detection rates of organic matter samples are all more than 77.78%. 107 organic matter samples are detected from the total 126, so the detection rate of all five groups of experiments is 84.92%. Some contaminants having similar spectra with drinking water were not detected, because the spectral peaks of contaminants overlap with those of drinking water. The result of detection rate in group 4 is not good. The possible reason is that the scattering part was not removed completely in many water samples of group 4, which resulted in the model established in group 4, and was not accurate enough because the scattering part cannot be explained by the trilinear model.

4. Conclusion

This paper studied a problem of unknown contaminant qualitative detection in water but not a qualitative problem of known contaminant detection. A method based on ATLD and threshold was applied to analyze three-dimensional fluorescence spectra in order to detect the unknown organic pollutants with fluorescent characteristics in the water. ATLD algorithm was used to establish the model of normal water sample, and the residual matrix was obtained through the difference of the model matrix and the measured matrix. The residual sum of squares was calculated according to the residual matrix and compared with the threshold to judge the test sample which was an organic pollutant sample or a normal water sample. In order to verify the theory, the experiments of analyzing the spectra of water samples and organic contaminant samples were launched.

The result shows that ATLD model extracting feature can be used to qualitatively discriminate whether test samples are polluted if the pollutants are with fluorescent characteristics. However, the detection rate of qualitative discrimination method based on the ATLD model fluctuated. This may be related that the method of extracting feature based on ATLD and residual sum of squares is a method depending on the background samples, that is, the result of the method is related to the quality and quantity of the samples.

In general, qualitative discrimination method based on ATLD and residual sum of squares, combining the threshold, can be used to detect the unknown organic pollutants with fluorescent characteristics in drinking water, but the detection rate is low which can be a problem worthy of further study. Besides, the method proposed in the paper does not apply to the situation of overlapped peaks well, and this will also be the work of further research.

Acknowledgments

This work was funded by the National Natural Science Foundation of China (no. 61573313) "Online water-quality anomaly detection, classification, and identification based on multi-source information fusion" and (no. U1509208) "Research on big data analysis and cloud service of urban drinking water network safety," and the Key Technology Research and Development Program of Zhejiang Province (no. 2015C03G2010034) "Research on intelligent management and long-effective mechanism for river regulation and maintenance."

References

[1] E. F. Conde, *Environmental Sensor Anomaly Detection Using Learning Machines*, Dissertations & Theses-Gradworks, Utah State University, 2011.

[2] J. Dahlen, S. Karlsson, M. Backstrom, J. Hagberg, and H. Pettersson, "Determination of nitrate and other water quality parameters in groundwater from UV/Vis spectra employing partial least squares regression," *Chemosphere*, vol. 40, no. 1, pp. 71–77, 2000.

[3] W. Bourgeois, J. E. Burgess, and R. M. Stuetz, "On-line monitoring of wastewater quality: a review," *Journal of Chemical Technology & Biotechnology*, vol. 76, no. 4, pp. 337–348, 2001.

[4] D. J. Dürrenmatt and W. Gujer, "Identification of industrial wastewater by clustering wastewater treatment plant influent ultraviolet visible spectra," *Water Science and Technology*, vol. 63, no. 6, pp. 1153–1159, 2011.

[5] G. Langergraber, J. K. Gupta, A. Pressl et al., "On-line monitoring for control of a pilot-scale sequencing batch reactor using a submersible UV/VIS spectrometer," *Water Science and Technology*, vol. 50, no. 10, pp. 73–80, 2004.

[6] G. Langergraber, J. V. D. Broeke, W. Lettl, and A. Weingartner, "Real-time detection of possible harmful events using UV/vis spectrometry," *Spectroscopy Europe*, vol. 18, no. 4, pp. 19–22, 2006.

[7] D. Hou, S. Liu, J. Zhang, F. Chen, P. Huang, and G. Zhang, "Online monitoring of water-quality anomaly in water distribution systems based on probabilistic principal component analysis by UV-Vis absorption spectroscopy," *Journal of Spectroscopy*, vol. 2014, no. 20, pp. 1–9, 2014.

[8] D. Hou, J. Zhang, Z. Yang, S. Liu, P. Huang, and G. Zhang, "Distribution water quality anomaly detection from UV optical sensor monitoring data by integrating principal component analysis with chi-square distribution," *Optics Express*, vol. 23, no. 13, pp. 17487–17510, 2015.

[9] J. G. Xu, *Fluorophotometric Analysis*, Science Press, Beijing, 2006.

[10] W. Yuanqing, *Three-Dimensional Fluorescence Spectrum Detection Method for Measuring the Concentration of Organic Pollutants in Water*, Zhejiang University, 2011.

[11] K. M. G. Mostofa, T. Yoshioka, E. Konohira, E. Tanoue, K. Hayakawa, and M. Takahashi, "Three-dimensional fluorescence as a tool for investigating the dynamics of dissolved organic maker in the Lake Biwa watershed," *Limnology*, vol. 6, no. 2, pp. 101–115, 2005.

[12] M. V. Bosco, M. P. Callao, and M. S. Larrechi, "Simultaneous analysis of the photocatalytic degradation of polycyclic aromatic hydrocarbons using three-dimensional excitation–emission matrix fluorescence and parallel factor analysis," *Analytica Chimica Acta*, vol. 576, no. 2, pp. 184–191, 2006.

[13] Y. Shutova, A. Baker, J. Bridgeman, and R. K. Henderson, "Spectroscopic characterisation of dissolved organic matter changes in drinking water treatment: from PARAFAC analysis to online monitoring wavelengths," *Water Research*, vol. 54, no. 4, pp. 159–169, 2014.

[14] A. Baker, "Fluorescence excitation-emission matrix characterization of some sewage-impacted rivers," *Environmental Science & Technology*, vol. 35, no. 5, pp. 948–953, 2001.

[15] G. Pavelescu, L. Ghervase, C. Ioja, S. Dontu, and R. Spiridon, "Spectral fingerprints of groundwater organic matter in rural areas," *Romanian Reports in Physics*, vol. 65, no. 3, pp. 1105–1113, 2013.

[16] R. G. Zepp, W. M. Sheldon, and M. A. Moran, "Dissolved organic fluorophores in southeastern US coastal waters: correction method for eliminating Rayleigh and Raman scattering peaks in excitation–emission matrices," *Marine Chemistry*, vol. 89, no. 1, pp. 15–36, 2004.

[17] R. F. Chen, P. Bissett, P. Coble et al., "Chromophoric dissolved organic matter (CDOM) source characterization in the Louisiana Bight," *Marine Chemistry*, vol. 89, no. 1, pp. 257–272, 2004.

[18] A. J. Lawaetz and C. A. Stedmon, "Fluorescence intensity calibration using the Raman scatter peak of water," *Applied Spectroscopy*, vol. 63, no. 8, pp. 936–940, 2009.

[19] H. L. Wu, M. Shibukawa, and K. Oguma, "An alternating trilinear decomposition algorithm with application to calibration of HPLC–DAD for simultaneous determination of overlapped chlorinated aromatic hydrocarbons," *Journal of Chemometrics*, vol. 12, no. 1, pp. 1–26, 1998.

[20] Y. Yongjie, *A Research on Three-Way Second-Order Calibration Algorithms and Its Applications*, Hunan University, 2009.

4

Solid-State ^{13}C NMR Spectroscopy Applied to the Study of Carbon Blacks and Carbon Deposits Obtained by Plasma Pyrolysis of Natural Gas

Jair C. C. Freitas, Daniel F. Cipriano, Carlos G. Zucolotto, Alfredo G. Cunha, and Francisco G. Emmerich

Laboratory of Carbon and Ceramic Materials, Department of Physics, Federal University of Espírito Santo, Av. Fernando Ferrari 514, 29075-910 Vitória, ES, Brazil

Correspondence should be addressed to Jair C. C. Freitas; jairccfreitas@yahoo.com.br

Academic Editor: Nikša Krstulović

Solid-state ^{13}C nuclear magnetic resonance (NMR) spectroscopy was used in this work to analyze the physical and chemical properties of plasma blacks and carbon deposits produced by thermal cracking of natural gas using different types of plasma reactors. In a typical configuration with a double-chamber reactor, N_2 or Ar was injected as plasma working gas in the first chamber and natural gas was injected in the second chamber, inside the arc column. The solid residue was collected at different points throughout the plasma apparatus and analyzed by ^{13}C solid-state NMR spectroscopy, using either cross polarization (CP) or direct polarization (DP), combined with magic angle spinning (MAS). The ^{13}C CP/MAS NMR spectra of a number of plasma blacks produced in the N_2 plasma reactor showed two resonance bands, broadly identified as associated with aromatic and aliphatic groups, with indication of the presence of oxygen- and nitrogen-containing groups in the aliphatic region of the spectrum. In contrast to DP experiments, only a small fraction of ^{13}C nuclei in the plasma blacks are effectively cross-polarized from nearby ^{1}H nuclei and are thus observed in spectra recorded with CP. ^{13}C NMR spectra are thus useful to distinguish between different types of carbon species in plasma blacks and allow a selective study of groups spatially close to hydrogen in the material.

1. Introduction

Plasma pyrolysis of natural gas is a promising way of producing high-purity carbon blacks without generation of environmentally harmful products. Upon plasma pyrolysis, methane decomposes to produce hydrogen and a solid, carbon-rich residue commonly designed as "plasma black" [1, 2]. These materials are promising for several applications, including the production of inks, electrodes, and catalyst supports [3–5], which means that their profitable use would aggregate value to the process of plasma conversion of natural gas. The detailed understanding of the structure of plasma blacks in the atomic scale is thus of interest aiming their possible practical applications and the optimization of their production methods.

Solid-state ^{13}C nuclear magnetic resonance (NMR) is a spectroscopic method that yields results sensitive to the local atomic environment and which has been used in studies of carbon materials of diverse types, including peat, humic substances, coal, coke, and chars, among others [6]. In the ^{13}C NMR spectra of these materials, contributions due to aromatic and aliphatic groups are readily separated based on the ^{13}C isotropic chemical shifts, which fall typically between 0 and 90 ppm for aliphatic and in the range 110–160 ppm for aromatic groups. A widespread experimental approach used in solid-state ^{13}C NMR studies of carbon materials involves the use of cross polarization (CP), in order to enhance the polarization of rare ^{13}C nuclei through their interaction with abundant ^{1}H nuclei, combined to magic angle spinning (MAS). In general, CP experiments are less time-consuming in comparison with experiments based on the direct polarization (DP) of ^{13}C nuclei, since in CP the relevant longitudinal relaxation time is the one associated with the abundant ^{1}H nuclei (T_{1H}), which is usually much shorter than the ^{13}C spin-lattice relaxation time (T_{1C}). However, the

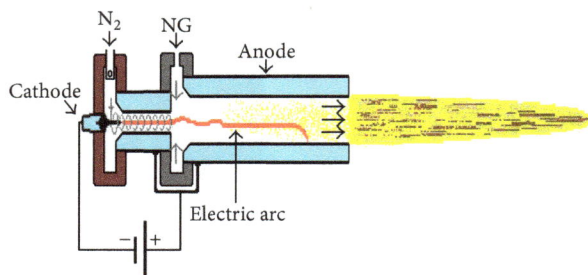

FIGURE 1: Illustration of the plasma reactor used for thermal cracking of natural gas (NG), composed of a double chamber with N_2 as the working gas.

efficiency of CP is strongly dependent on the magnitude of ^{1}H-^{13}C dipolar coupling, which means that the spectral intensities in a ^{13}C NMR spectrum recorded with CP will be affected by differences in the rate of polarization transfer from ^{1}H to chemically distinct ^{13}C nuclei [6–8]. Also, there are many technologically important types of carbon materials with low (or even zero) hydrogen content, including carbon blacks, carbon nanotubes, nanodiamonds, nanographites, graphene, and amorphous carbon films. For such materials, it is certainly more advantageous to record the ^{13}C NMR spectra using DP, even if these experiments are more time-consuming, allowing a larger portion of carbon atoms to be observed in DP experiments as compared to the CP method. There are plenty of examples in the literature of the successful use of ^{13}C DP/MAS experiments for the analysis of carbon materials, making possible the determination of the fractions of carbon atoms with sp^2 or sp^3 hybridization in carbon films [9, 10], the monitoring of the chemical reduction of graphene oxide [11, 12], the study of the correlation between chemical shifts and structural parameters in heat-treated carbons [13, 14], the study of relaxation processes of ^{13}C nuclei interacting with paramagnetic centers in nanocarbons [15, 16], and many others.

In this work, solid-state ^{13}C NMR spectroscopy was used to study plasma blacks and carbon deposits produced by thermal cracking of natural gas using different types of plasma reactors. The obtained spectra provided a detailed characterization of these materials from the point of view of the local chemical environment of carbon atoms in the material. Both DP and CP approaches were employed, allowing, on the one hand, a thorough characterization of the chemically distinct carbon atomic sites and, on the other hand, the study of minor contributions due to hydrogen-containing groups eventually present in the material.

2. Experimental

The plasma blacks studied in this work were produced as byproducts of the plasma pyrolysis of natural gas (NG). In a typical configuration with a double-chamber reactor (shown in Figure 1), N_2 was injected as plasma working gas in the first chamber and NG was injected in the second chamber, inside the arc column. The double-chamber plasma torches were composed of a tungsten cathode and a copper step nozzle as the anode. Other plasma gases (e.g., Ar or H_2) were also used in different configurations. The solid residue was collected at different points throughout the reactor. These products were first characterized by thermogravimetry (TG), using a Shimadzu TGA-50H instrument, with constant heating-rate of 20°C/min up to 1000°C under O_2 flow (50 mL/min); X-ray diffraction (XRD), using a Shimadzu XRD-6000 powder diffractometer operating with a Cu-Kα X-ray source (wavelength λ = 1.5418 Å); scanning electron microscopy (SEM), using a Shimadzu SSX-550 electron microscope; and elemental analysis, performed with a Leco CHNS932 instrument.

Solid-state NMR spectra were recorded at room temperature in a Varian/Agilent spectrometer operating at the NMR frequencies of 399.73 and 100.52 MHz for ^{1}H and ^{13}C nuclei, respectively (magnetic field of 9.4 T), using ^{1}H →^{13}C CP or ^{13}C DP experiments. The carbon-rich powders were mixed with kaolin to avoid problems related to the radiofrequency (RF) penetration in the material (skin depth effect) and to allow a proper tuning of the RF probe [6]. Next, the samples were packed into 4 mm diameter rotors for carrying out MAS experiments, with a spinning rate of 14 kHz. In CP experiments, the duration of the ^{1}H $\pi/2$ pulse was 3.6 μs, the contact time was 500 μs, and the recycle delay was 5.0 s. In the case of DP experiments, a pulse sequence is composed of a ^{13}C $\pi/2$ pulse (with 3.3 μs duration) immediately followed by a pair of π pulses and the subsequent detection of the free induction decay (FID) was employed, in order to suppress spurious background NMR signals coming from some carbon-containing parts of the NMR probe [17]; the recycle delay was 15.0 s in the DP experiments. All spectra were obtained by Fourier transform of the FIDs and the chemical shifts were referenced to tetramethylsilane (TMS), using hexamethylbenzene (HMB) as secondary reference.

3. Results and Discussion

TG and elemental analysis results showed the plasma blacks were primarily composed by carbon, with small amounts of hydrogen (1-2 wt.%) and nitrogen (< 1 wt.%) and reduced ash content (~2 wt.%). XRD patterns revealed no crystalline phase, with the detection of only the broad maxima associated with the turbostratic structure typical of disordered carbon materials [18]. SEM images revealed the presence of spherules with varied sizes, typically < 1 μm, depending on the point of collection of the residue inside the plasma apparatus (see discussion below).

Some typical ^{13}C CP/MAS NMR spectra recorded for plasma blacks produced in N_2 and in Ar plasma reactors are shown in Figure 2. Two resonance bands were clearly observed in the case of the sample produced in the N_2 plasma reactor (Figure 2(a)), broadly identified as coming from aromatic and aliphatic groups. The presence of oxygen- and nitrogen-containing groups is suggested by the chemical shifts observed in the aliphatic region of the spectrum [6]. In the case of the sample produced using the Ar plasma

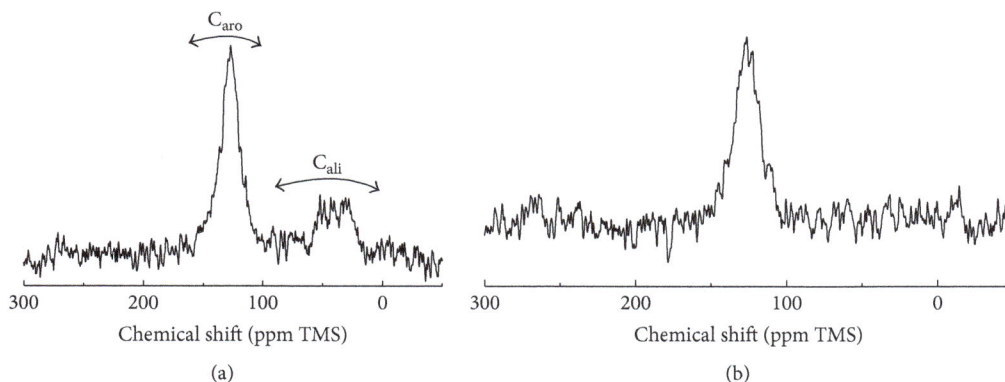

FIGURE 2: ^{13}C CP/MAS NMR spectra of plasma blacks produced by pyrolysis of natural gas in plasma reactors using (a) N$_2$ or (b) Ar.

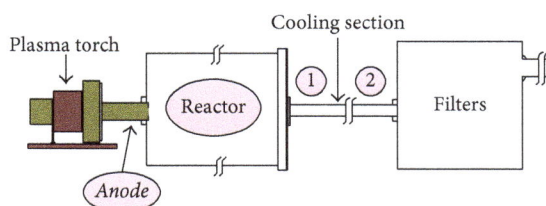

FIGURE 3: Illustration of the apparatus used for Ar plasma reactions; solid residues were collected at four points inside the apparatus: anode, reactor, cooler 1, and cooler 2.

reactor (Figure 2(b)), only an intense aromatic contribution was observed.

As mentioned before, it is worth emphasizing that spectra recorded with CP do not reveal the whole carbon content in the material, being representative only of regions close to hydrogen-containing groups [6, 9, 10, 13]. This is in contrast to what is expected for DP spectra, which contain contributions from essentially all carbon-containing groups. In plasma blacks, with a quite reduced hydrogen content, it is expected that only a small fraction of ^{13}C nuclei are effectively cross-polarized from nearby ^{1}H nuclei; that is the main reason why the spectra exhibited in Figure 2 present poor signal-to-noise ratio.

In order to establish a detailed comparison between CP and DP NMR spectra of plasma blacks exhibiting different structural features, a set of samples were collected at different points inside the plasma apparatus in an experiment involving the use of Ar plasma for achieving the thermal cracking of natural gas. Figure 3 shows an illustration of the apparatus, with indication of the four points where the solid residues were collected (anode, reactor, cooler 1, and cooler 2). The materials collected at each of these points remained in contact with the hot zone of the plasma torch for different times, so the structural and chemical characteristics of these materials are expected to change from one collection point to another.

Figure 4 shows the SEM images of the materials collected at each of these points. It is clear that the materials collected in the coolers present the smallest particle sizes, with a

morphology similar to what is typically observed for carbon blacks [19, 20]. On the other hand, the materials collected in the main section of the reactor and, especially, inside the anode are composed of larger particles (mostly with diameter > 1 μm), forming agglomerates, which is a consequence of the exposition of these materials to high temperatures for long periods inside the plasma apparatus. The XRD patterns recorded for these samples (shown in Figure 5) reveal the same trends: the widths of the Bragg peaks, particularly the ones corresponding to the (002) peak at $2\theta \cong 26°$, are much smaller for the samples collected in the anode and in the reactor as compared to samples collected in the coolers. This is a further evidence of the better structural order and larger crystallite sizes of the materials that were deposited in the hottest zones of the plasma apparatus.

The ^{13}C CP/MAS and DP/MAS NMR spectra of the samples collected at different points in the Ar plasma apparatus are shown in Figures 6 and 7, respectively. In the case of CP spectra (Figure 6), no NMR signal was obtained for the sample collected in the anode, which is easily explained since the hydrogen content of this sample (also indicated in Figure 6) was below the detection limit of the elemental analysis instrument used in this work. For the other samples, with hydrogen contents around 1 wt.%, CP spectra were successfully recorded, showing the presence of a well-defined resonance associated with aromatic carbons. It is likely that the resonances observed in CP-derived spectra are due to ^{13}C nuclei located at or close to the edges of the graphene-like planes. Being near hydrogen-containing groups, these nuclei are effectively cross-polarized and give rise to the relatively narrow resonances observed with CP.

On the other hand, the DP spectra were effectively observed for all collected samples, including the sample collected in the anode, as shown in Figure 7. The comparison between DP and CP spectra shows that a larger chemical shift range is covered in the DP case, with the resonances observed with the DP method being generally broader in comparison to the corresponding CP spectra. This is a consequence of the fact that in DP experiments the overall carbon content in the material is observed and not only those ^{13}C nuclei close to ^{1}H nuclei. The resonance peaks in the DP spectra

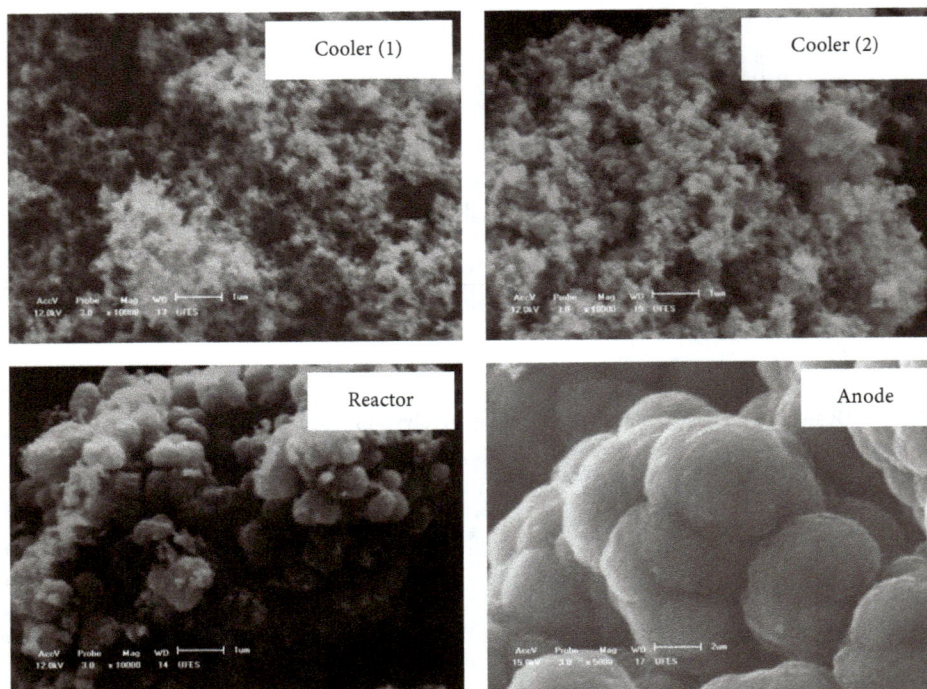

FIGURE 4: SEM images of the carbon residues collected at different points in the Ar plasma apparatus.

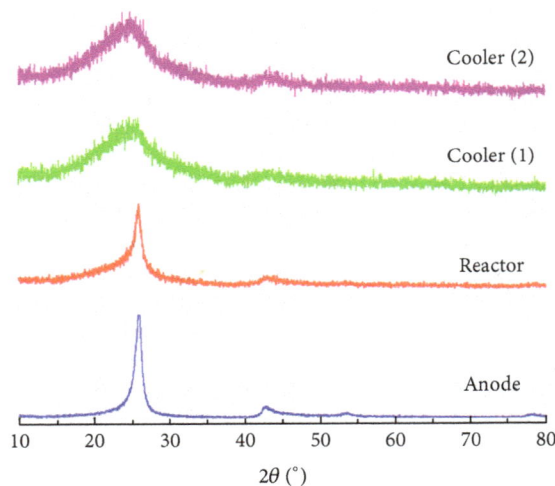

FIGURE 5: XRD patterns recorded for the carbon residues collected at different points in the Ar plasma apparatus.

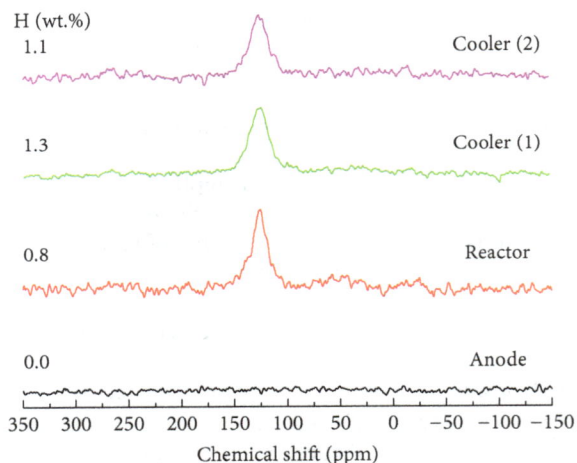

FIGURE 6: ^{13}C CP/MAS NMR spectra of the carbon residues collected at different points in the Ar plasma apparatus. The hydrogen contents of each sample (obtained by elemental analysis) are also informed.

of the samples collected in the reactor and in the coolers were detected close to 126 ppm, which is a value consistent with the predominantly aromatic character of the material; the corresponding line shapes were found to be similar to the ones observed in carbon materials obtained by the carbonization of organic precursors at high temperatures [13] and also in amorphous carbon films [9, 10].

The large broadening observed in the DP spectra of the carbon residues even when using MAS is attributed to the occurrence of a wide distribution of shifts associated with the locally anisotropic magnetic susceptibility of graphene-like planes [14]. Thus, the extent of this broadening and also the peak position are expected to reveal details about structural aspects of the material. The existence of graphene-like planes arranged in the turbostratic structure within the spherules observed in the SEM images is thus related to the large broadening observed in the ^{13}C NMR spectra of the analyzed samples. The observation of a large broadening and a sizeable upfield shift for the resonance peak in the

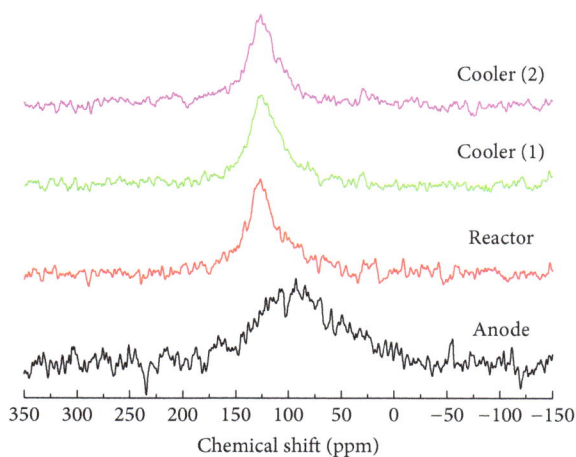

FIGURE 7: ^{13}C DP/MAS NMR spectra of the carbon residues collected at different points in the Ar plasma apparatus.

^{13}C DP/MAS NMR spectrum of the sample collected inside the anode is particularly remarkable (Figure 7). The peak position was found close to 90 ppm, which is a value quite low in comparison to the typical chemical shift range of sp^2 carbons [6], but is similar to what has been observed in graphite samples [14, 21]. This large upfield shift and the severe broadening of this resonance are effects associated with the improved structural organization of this material, as also revealed by the SEM and XRD results previously discussed, which is a consequence of the long residence time of this material in the high-temperature region inside the anode. Thus, this sample exhibits structural features very distinct from the other ones, leading to the resemblance of the corresponding NMR spectrum to the spectra commonly observed for graphitic materials [14, 21].

A scenario similar to the one described here has been reported in the NMR analyses of hydrogenated amorphous carbon films, where the comparison between CP and DP spectra has allowed the distinction between sp^2- and sp^3-like carbon atoms [9, 10]. In these materials, the ^{13}C CP/MAS NMR spectra are representative of just a part of the carbon network, but, given the improved sensitivity of CP experiments compared to DP ones, it is possible to detect even minor contributions associated with hydrogen-containing chemical groups. As an example, Pan et al. [9] estimated that the CP/MAS spectrum recorded for an amorphous carbon film represented only 1.5% of the total amount of carbon atoms in the material, with the spectrum indicating the presence of both sp^2- and sp^3-like contributions. In a related work, Cho et al. [10] used a combination of DP/MAS and CP/MAS spectra to study amorphous hydrogenated carbon films prepared by sputtering. The DP/MAS spectra were employed to achieve information on the total contents of sp^2- and sp^3-like groups; on the other hand, the CP/MAS spectra recorded with spectral editing methods allowed the assessment of the chemical changes occurring in films with different hydrogen contents. The analysis of the ^{13}C NMR spectra of the plasma blacks and carbon deposits described

here allows a similar assessment of contributions from carbon atoms located far within the graphene-like planes or those located close to the plane edges and thus more susceptible to the polarization transfer from nearby ^{1}H nuclei present in functional groups.

4. Conclusion

This work shows how solid-state ^{13}C NMR spectroscopy methods can be useful to distinguish between different types of carbon-containing groups and to allow a selective study of groups spatially close to hydrogen in carbon blacks and carbon deposits obtained by plasma pyrolysis of natural gas. For samples collected at different points along the plasma apparatus, the results evidenced a good correlation between the structural features (revealed by XRD and SEM) and the spectral characteristics observed in the ^{13}C DP/MAS NMR spectra. Even with little hydrogen content in the plasma blacks and in the deposited carbon residues, the combination of CP and DP methods allowed the detection of signals coming from carbon atoms close to hydrogen-containing groups at the plane edges and also from carbon atoms located far within the graphene-like planes.

Competing Interests

The authors declare that there is no conflict of interests regarding the publication of this paper.

Acknowledgments

The authors are grateful to Petrobras and to the Brazilian funding agencies CNPq, FAPES, CAPES, and FINEP.

References

[1] L. Fulcheri and Y. Schwob, "From methane to hydrogen, carbon black and water," *International Journal of Hydrogen Energy*, vol. 20, no. 3, pp. 197–202, 1995.

[2] J. R. Fincke, R. P. Anderson, T. A. Hyde, and B. A. Detering, "Plasma pyrolysis of methane to hydrogen and carbon black," *Industrial and Engineering Chemistry Research*, vol. 41, no. 6, pp. 1425–1435, 2002.

[3] W. Cho, S.-H. Lee, W.-S. Ju, Y. Baek, and J. K. Lee, "Conversion of natural gas to hydrogen and carbon black by plasma and application of plasma carbon black," *Catalysis Today*, vol. 98, no. 4, pp. 633–638, 2004.

[4] H.-M. Yang, W.-K. Nam, and D.-W. Park, "Production of nano-sized carbon black from hydrocarbon by a thermal plasma," *Journal of Nanoscience and Nanotechnology*, vol. 7, no. 11, pp. 3744–3749, 2007.

[5] T. C. S. Evangelista, G. T. Paganoto, M. C. C. Guimarães, and J. Ribeiro, "Raman spectroscopy and electrochemical investigations of Pt electrocatalyst supported on carbon prepared through plasma pyrolysis of natural gas," *Journal of Spectroscopy*, vol. 2015, Article ID 329730, 7 pages, 2015.

[6] J. C. C. Freitas, A. G. Cunha, and F. G. Emmerich, "Solid-state NMR methods applied to the study of carbon materials," in

Chemistry and Physics of Carbon, L. R. Radovic, Ed., vol. 31, pp. 85–170, CRC Press, Boca Raton, Fla, USA, 2012.

[7] R. J. Smernik, J. A. Baldock, and J. M. Oades, "Impact of remote protonation on ^{13}C CPMAS NMR quantitation of charred and uncharred wood," *Solid State Nuclear Magnetic Resonance*, vol. 22, no. 1, pp. 71–82, 2002.

[8] P. Conte, R. Spaccini, and A. Piccolo, "State of the art of CPMAS ^{13}C-NMR spectroscopy applied to natural organic matter," *Progress in Nuclear Magnetic Resonance Spectroscopy*, vol. 44, no. 3-4, pp. 215–223, 2004.

[9] H. Pan, M. Pruski, B. C. Gerstein, F. Li, and J. S. Lannin, "Local coordination of carbon atoms in amorphous carbon," *Physical Review B*, vol. 44, no. 13, pp. 6741–6745, 1991.

[10] G. Cho, B. K. Yen, and C. A. Klug, "Structural characterization of sputtered hydrogenated amorphous carbon films by solid state nuclear magnetic resonance," *Journal of Applied Physics*, vol. 104, no. 1, Article ID 013531, 2008.

[11] S. Stankovich, D. A. Dikin, R. D. Piner et al., "Synthesis of graphene-based nanosheets via chemical reduction of exfoliated graphite oxide," *Carbon*, vol. 45, no. 7, pp. 1558–1565, 2007.

[12] Y. Si and E. T. Samulski, "Synthesis of water soluble graphene," *Nano Letters*, vol. 8, no. 6, pp. 1679–1682, 2008.

[13] J. C. C. Freitas, T. J. Bonagamba, and F. G. Emmerich, "Investigation of biomass- and polymer-based carbon materials using ^{13}C high-resolution solid-state NMR," *Carbon*, vol. 39, no. 4, pp. 535–545, 2001.

[14] J. C. C. Freitas, F. G. Emmerich, G. R. C. Cernicchiaro, L. C. Sampaio, and T. J. Bonagamba, "Magnetic susceptibility effects on ^{13}C MAS NMR spectra of carbon materials and graphite," *Solid State Nuclear Magnetic Resonance*, vol. 20, no. 1-2, pp. 61–73, 2001.

[15] A. M. Panich, "Solid state nuclear magnetic resonance studies of nanocarbons," *Diamond and Related Materials*, vol. 16, no. 12, pp. 2044–2049, 2007.

[16] A. M. Panich and A. I. Shames, "Nuclear spin-lattice relaxation in nanocarbon compounds caused by adsorbed oxygen," *Diamond and Related Materials*, vol. 20, no. 2, pp. 201–204, 2011.

[17] D. G. Cory and W. M. Ritchey, "Suppression of signals from the probe in bloch decay spectra," *Journal of Magnetic Resonance*, vol. 80, no. 1, pp. 128–132, 1988.

[18] Z. Q. Li, C. J. Lu, Z. P. Xia, Y. Zhou, and Z. Luo, "X-ray diffraction patterns of graphite and turbostratic carbon," *Carbon*, vol. 45, no. 8, pp. 1686–1695, 2007.

[19] J. B. Donnet, R. C. Bansal, and M. J. Wang, *Carbon Black*, Marcel Dekker, New York, NY, USA, 2nd edition, 1993.

[20] K. S. Kim, S. H. Hong, K.-S. Lee, and W. T. Ju, "Continuous synthesis of nanostructured sheetlike carbons by thermal plasma decomposition of methane," *IEEE Transactions on Plasma Science*, vol. 35, no. 2, pp. 434–443, 2007.

[21] M. A. Vieira, G. R. Gonçalves, D. F. Cipriano et al., "Synthesis of graphite oxide from milled graphite studied by solid-state ^{13}C nuclear magnetic resonance," *Carbon*, vol. 98, pp. 496–503, 2016.

5

Discrimination of Nasopharyngeal Carcinoma from Noncancerous *Ex Vivo* Tissue Using Reflectance Spectroscopy

Zhihong Xu,[1] Wei Huang,[1,2] Duo Lin,[1,3] Shanshan Wu,[1] Maowen Chen,[1] Xiaosong Ge,[1] Xueliang Lin,[1] and Liqing Sun[4]

[1] *Key Laboratory of Optoelectronic Science and Technology for Medicine, Ministry of Education, Fujian Normal University, Fuzhou 350007, China*
[2] *Fujian Metrology Institute, Fuzhou 350003, China*
[3] *College of Integrated Traditional Chinese and Western Medicine, Fujian University of Traditional Chinese Medicine, Fuzhou, Fujian 350122, China*
[4] *Affiliated Fuzhou First Hospital of Fujian Medical University, Fuzhou 350009, China*

Correspondence should be addressed to Wei Huang; huang84wei@163.com

Academic Editor: Rickson C. Mesquita

Reflectance spectroscopy is a low-cost, nondestructive, and noninvasive method for detection of neoplastic lesions of mucosal tissue. This study aims to evaluate the capability of reflectance spectroscopy system under white light (400–700 nm) with a multivariate statistical analysis for distinguishing nasopharyngeal carcinoma (NPC) from nasopharyngeal benign *ex vivo* tissues. High quality reflectance spectra were acquired from nasopharyngeal *ex vivo* tissues belonging to 18 noncancerous and 19 cancerous subjects, and the combination of principal component analysis-linear discriminant analysis (PCA-LDA) along with leave-one-spectrum-out cross-validation (LOOCV) diagnostic algorithm was subsequently employed to classify different types of tissue group, achieving a diagnostic sensitivity of 73.7% and a specificity of 72.2%. Furthermore, in order to distinguish NPC from nasopharyngeal benign *ex vivo* tissues based on reflectance spectra simply, spectral intensity ratios of oxyhemoglobin ($R540/R576$) were used as an indicator of the carcinogenesis associated transformation in the hemoglobin oxygenation. This tentative work demonstrated the potential of reflectance spectroscopy for NPC detection using *ex vivo* tissue and has significant experimental and clinical value for further *in vivo* NPC detection in the future.

1. Introduction

Nasopharyngeal carcinoma (NPC) is a prominent malignant tumor that occurs in the top and the side wall of nasopharyngeal cavity with no obvious early symptom, which directly causes harm to human health [1]. The incidence of NPC has remained high in southeast Asia, northern Africa, parts of the Middle East, and so forth. We should be concerned about the fact that approximately 80% of NPC cases were reported from China, especially in southern provinces, such as Guangdong and Fujian [2–4]. NPC was defined as a special disease with a remarkable racial and geographical distribution worldwide [5]. According to the epidemiological investigation, NPC has various factors to its etiology, and the main relevant risk factors include inherited genetic factors,

the dormant infection of Epstein-Barr virus (EBV), and natural environment factor [6]. The clinical reports indicated that NPC patients treated in Stages I and Stages II have 5-year survival rates of approximately 90% and 75%, respectively. However, early stage of NPC has no obvious symptoms until the middle and advanced stages of disease [2, 6, 7]. Therefore, early diagnosis is the effective way to improve the survival rates of patient and promote the quality of their life.

Currently common detection and screening method for NPC is dependent on clinicians' experience, which easily causes possible diagnostic errors owing to lack of the biochemical information of suspicious lesions. Although the histopathological examination of biopsy is still the golden standard for NPC detection [8], it involves the destruction and invasion of tissue from suspected lesions. Diagnostic techniques based on optical spectroscopy, in contrast, can provide quantitative information in real time from tissues, which are excellent in the field of precancerous tumor localization and pathology analysis.

Reflectance spectroscopy has attracted wide attention in human tissue study including oral, breast, lung, bladder, cervical, esophagus, kidney, and prostate tissue [9–14], due to its simple operation and low investment cost of equipment. In reflectance, broadband beam interaction with the biological tissue surface and two forms of cases can occur, that is, multiple elastic scattering and absorption phenomenon [15]. In this process, part of broadband beam returns as reflectance spectral signals, which are able to provide the optical properties of the tissue [16]. The reflectance spectral signatures are dependent on biological tissue morphology, surface characteristics, and structural components [13, 15]. Hence, reflectance spectra have certain advantages for explaining major tissue features within early cancerous transform, for example, the cellular metabolism, capillary distribution, and hemoglobin oxygenation [15, 17–19].

Zonios et al. reported a diffuse reflectance spectra study on human adenomatous colon polyps, showing that differences can be ascribed to hemoglobin concentration and effective scatterer size between normal and adenomatous tissue sites [15]. Utzinger and coworkers have investigated an exploratory pilot study to measure reflectance spectra at different source-detector separations from normal and neoplastic ovarian tissue and found the changes of intensities of the reflectance spectrum between 550 and 580 nm, which are dramatically attributed to the absorption of blood oxygenation [20]. Subhash and coworkers have characterized reflectance spectral intensity ratio of oxygenated hemoglobin absorption bands ($R540/R570$) associated with different types of lesions and histopathological conditions, and results showed that the $R540/R570$ ratio for malignant lesions was always lower than that for normal areas [13]. Li et al. used the reflectance ratio $R540/R575$ method to distinguish malignant from nonmalignant sites in human ex vivo gastric epithelial tissues and found some pronounced differences in various pathological types [21]. Zhao et al. reported a study about oxygenated hemoglobin absorption band ratios of esophageal epithelial tissues at different heat treatment temperature, showing the capability of employing oxygenated hemoglobin band ratios $R540/R570$ as indicators of malignancy [22]. Recent results have shown that tissue absorption in the short wavelength region and visible region is assigned to hemoglobin, with the oxygenation forms exhibiting different absorption spectral characteristic [23]. These reports demonstrated that reflectance spectroscopy could obtain spectral information about the pathological changes of tissue, which can serve as important supplement information for clinical diagnosis. Nevertheless, as far as we are concerned, still few reports have been found using reflectance spectroscopy for NPC detection.

In this preliminary study, we investigated reflectance spectral signals from human nasopharyngeal ex vivo tissue with different pathological types. Reflectance spectra were subjected to multivariate statistical tools, including principal component analysis-linear discriminant analysis (PCA-LDA)

algorithm together with the leave-one-spectrum-out cross-validation (LOOCV). To validate the feasibility of reflectance spectroscopy as a tool to differentiate noncancerous ex vivo tissues from NPC, the receiver operating characteristic (ROC) curve was employed. Empirical method using reflectance spectral intensity ratios of oxyhemoglobin bands ($R540/R576$) was used to distinguish cancer from benign ex vivo tissue.

2. Materials and Methods

2.1. Patients and Sample Preparations. Nasopharyngeal ex vivo tissues for this study were collected from 37 patients who were recruited for biopsy from nasopharyngeal cavity lesions with written informed consent. This procedure was authorized by the Ethics Committee of Affiliated Fuzhou First Hospital of Fujian Medical University. All nasopharyngeal tissues were divided into two groups with clinical diagnosis (19 NPC samples and 18 benign tissue samples), as shown in Table 1. The pathological types of NPC patients contain 17 undifferentiated nonkeratinizing carcinomas and 2 differentiated nonkeratinizing carcinomas. The pathological types of benign tissues consist of 15 chronic inflammations, 2 lymphomas, and 1 lymphoid tissues hyperplasia. These fresh tissue samples were immediately stored at 4°C until being sent to the laboratory moments later, where they were stored in a minus 80°C refrigerator until experimental measurements. These targets were placed on an aluminum plate and thawed at room temperature immediately before spectral measurement (Figure 1).

2.2. Laboratory Instruments. All the reflectance spectra of nasopharyngeal ex vivo tissues were recorded using a special measurement system (Verisante Technology, Inc., Vancouver, Canada); more details were presented in previous report [24]. In brief, this detection system included a light source (a xenon arc lamp), a spectrometer (USB2000, Ocean Optics), and a PC. A band-pass filter (400–700 nm) was installed at the reflectance spectrograph entrance to obstruct the reflected unwanted light to improve reflectance signals detection. The exposure time of the spectrometer is 200 ms. An optical fiber is used to collect reflectance spectra, of which the entrance corresponds to a spot size of 1 mm diameter on the sample. The dimensions of the samples are generally between 1.8 and 2.5 mm. Three reflectance spectra were obtained from different positions for each ex vivo tissue and averaged to a mean spectrum so that we could acquire representative spectra for each sample. This detection system was calibrated by a certified reflectance standard (WS-1-SL, Labsphere) before each measurement to eliminate interference by the status of the system.

2.3. Data Processing and Analysis. To eliminate the interference of slow-moving backgrounds, the raw spectra were smoothed by using Savitzky-Golay method with points of window of 50 and polynomial order of 5. Reflectance spectra in the 400–700 nm range were subjected to principal component analysis (PCA) and linear discriminant analysis (LDA) combined with LOOCV by using SPSS statistical software

TABLE 1: Sample characteristic.

Sample type	Pathologic types	Number
Nasopharyngeal cancerous (NPC) patients ($n = 19$)	Undifferentiated nonkeratinizing carcinomas	17
	Differentiated nonkeratinizing carcinomas	2
Nasopharyngeal noncancerous patients ($n = 18$)	Chronic inflammations	15
	Lymphoma	2
	Lymphoid tissues hyperplasia	1

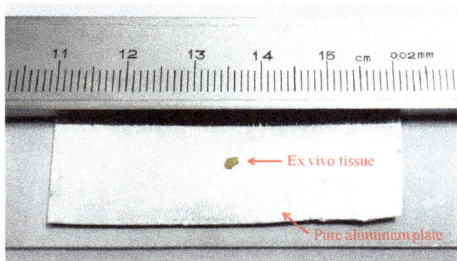

FIGURE 1: Nasopharyngeal *ex vivo* tissue placed on the pure aluminum plate. Pure aluminum and sample are pointed out by the black arrow.

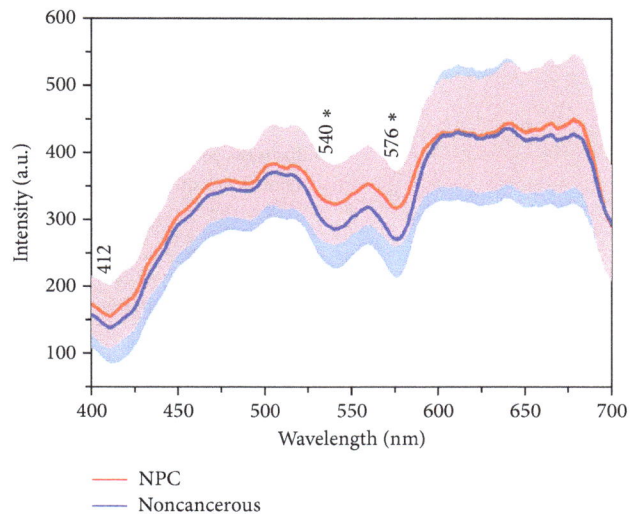

— NPC
— Noncancerous

FIGURE 2: Mean reflectance spectra from NPC (red line, $n = 19$) and noncancerous *ex vivo* tissue (blue line, $n = 18$). Asterisks indicated significant difference (p value < 0.05) after t-test, in which the p values are 0.006 and 1.6×10^{-4} for 540 nm and 576 nm, respectively. Areas (shaded) illustrate the respective standard deviations of the means spectra.

(SPSS Inc., Chicago). PCA-LDA was employed to achieve a diagnostic algorithm for classification between different groups within the dataset. Briefly, PCA is a statistical method that is used on the dataset to reduce the dimensions of the data, while finding a series of the linear combination of the original variables, so-called principal components (PCs) [25, 26]. An independent-samples t-test was used to determine significant PCs (p value < 0.05) [25, 27]. The findings, which represent most of the significant variance in the dataset, are selected as input data for the development of LDA model for tissue classification [28]. LDA is a supervision algorithm that maximizes the covariance in the data between different groups and minimizes the variances within group [27]. The classification efficiency of the PCA-LDA diagnostic models for correctly predicting the sample groups (i.e., normal versus malignancy) was evaluated by LOOCV method so that to eliminate the risk of overfitting [29, 30]. The receiver operating characteristic (ROC) curves were plotted to evaluate the performance of the above algorithm [26].

3. Results

3.1. Analysis of Reflectance Spectra.

Figure 2 shows the comparison of mean *ex vivo* reflectance spectra with standard deviations (SD) of the biopsy lesions from malignant sites and benign mucosa sites with the same detection system setup. Prominent valleys of reflectance spectra were observed in 412 nm, 540 nm, and 576 nm which can be attributed to absorption from hemoglobin bands [31, 32]. For all the spectra measured, the absorption peak intensity from NPC group was slightly weaker than that of noncancerous group in these absorption valleys. Notable absorption variations in reflectance intensity are more significant at green light region

(around 540 nm and 576 nm bands), indicating an increase of microvasculature and hemoglobin [12, 13]. Significant differences were found in the spectral valleys at 540 nm (p value $= 0.006 < 0.05$) and 576 nm (p value $= 1.6 \times 10^{-4} < 0.05$) between the two groups by using the analysis of t-test.

3.2. Multivariate Statistical Analysis.

Multivariate analyses including PCA and LDA were explored to evaluate the feasibility of classifying cancerous samples from noncancerous ones by employing the recorded reflectance dataset. Figure 3 shows this multivariate statistical analysis in a flowchart. In the first step the dataset underwent dimension reduction using unsupervised PCA to extract the principal components (PCs) which represent the measured spectra by a weighted linear combination. The PCs were tested by independent-samples t-test. PC3 and PC6 have the significant variance in the dataset (p value < 0.05) and were selected to discriminate the two pathological groups, in which the p values of are 0.022 and 0.007 for PC3 and PC6, respectively. The selected PCs (PC3 and PC6) were subsequently imported into LDA model with LOOCV method to evaluate classification efficiency.

The scatter plot of the diagnostically significant PC3 versus PC6 scores is displayed in Figure 4. Also displayed is the diagnosis equation (0.738 PC3 + 0.836 PC6 + 0.017 = 0) that separates the two groups of *ex vivo* tissue according to the result of statistical analysis. The diagnosis equation is defined as $Z = aX + bY$, where a and b are the standardized canonical discriminant function coefficients of PC3 and PC6 (0.738 and 0.836, resp.). This decision line is controlled by choosing the coordinate parameters of X and Y. Figure 5 shows the posterior probability generated from PCA-LDA algorithm belonging to noncancerous and NPC subjects, with a diagnostic sensitivity of 73.7% (14/19) and a specificity of

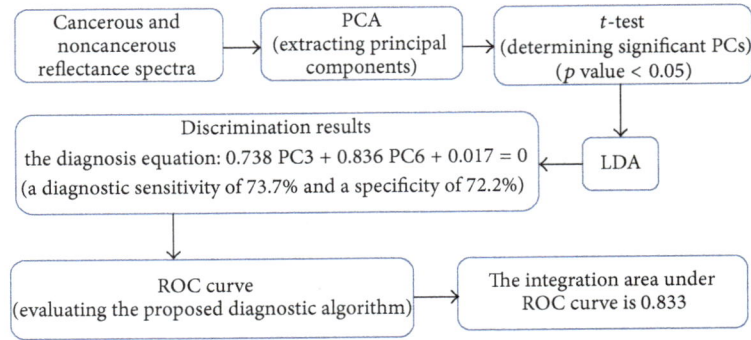

FIGURE 3: The work flowchart of multivariate statistical analysis.

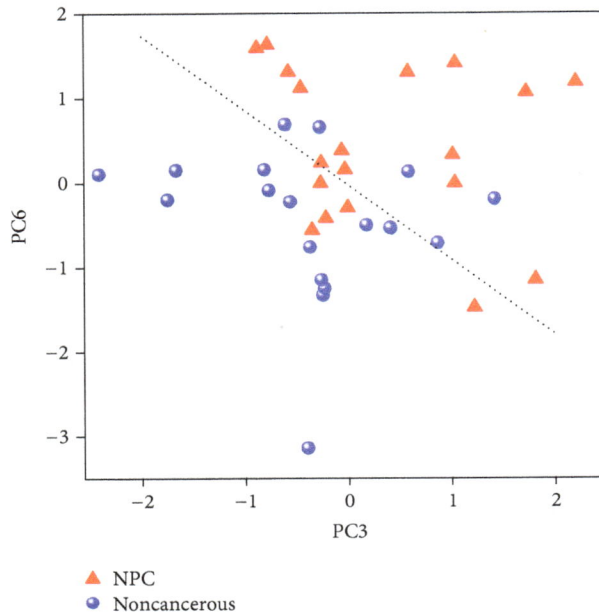

FIGURE 4: A two-dimension scattered point diagram of the PCA result for NPC group (red triangles ▲) and noncancerous group (blue circle ●). The dotted lines (0.738 PC3 + 0.836 PC6 + 0.017 = 0) served as a diagnostic method for NPC sample and noncancerous sample differentiation.

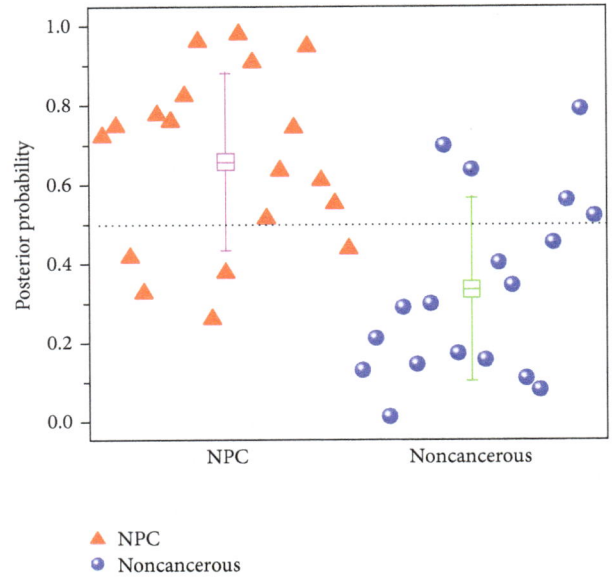

FIGURE 5: Scatter plots of the posterior probability belonging to NPC and nasopharyngeal noncancerous subjects (▲—NPC group; ●—noncancerous group), generated from PCA-LDA diagnosis algorithms along with LOOCV method. Error bars represent the mean and standard deviation of NPC (magenta) and nasopharyngeal noncancerous (green) *ex vivo* tissue posterior probability. The dotted lines represented the sensitivities and specificities for separating NPC from nasopharyngeal noncancerous tissue.

72.2% (13/18). The dotted lines were set to 0.5, where the total number of misclassified tissues was minimized [33], which represented the sensitivities and specificities for separating NPC from nasopharyngeal noncancerous *ex vivo* tissue. The sensitivity and specificity were defined as [34]

$$\text{Sensitivity} = \frac{\text{True Positives}}{\text{True Positives} + \text{False Negatives}},$$
$$\text{Specificity} = \frac{\text{True Negatives}}{\text{True Negatives} + \text{False Positives}}. \tag{1}$$

Figure 6 is the ROC curve, which was generated from the posterior probability plot at different threshold levels to evaluate the proposed diagnostic algorithm. The integration area under ROC curve (AUC) is 0.833, which further demonstrates that using reflectance spectra with multivariate statistical algorithms may be viable for discriminating NPC group from benign *ex vivo* tissue group.

3.3. Analysis of Hemoglobin Absorption Ratio R540/R576. Significant differences were found in the spectral valleys at 540 nm (p value = 0.006 < 0.05) and 576 nm (p value = 1.6×10^{-4} < 0.05) between the two groups by using the analysis of t-test. The ratios of intensity of 540 nm and 576 nm were calculated for each original reflectance spectrum. Then all the ratios were examined by t-test, and the p value is 1.89×10^{-7}. The ratios of $R540/R576$ were calculated by the related original data, and then these original ratios were averaged.

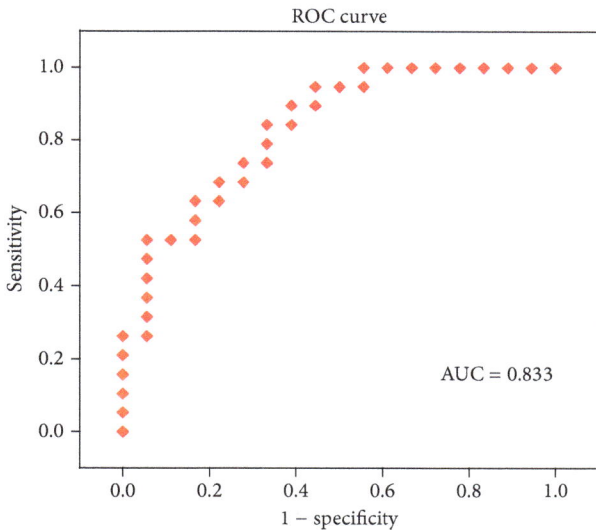

FIGURE 6: ROC curve for discriminating NPC from nasopharyngeal noncancerous *ex vivo* tissue for the training dataset by employing PCA-LDA diagnosis algorithms combined with LOOCV method. The AUC is 0.833.

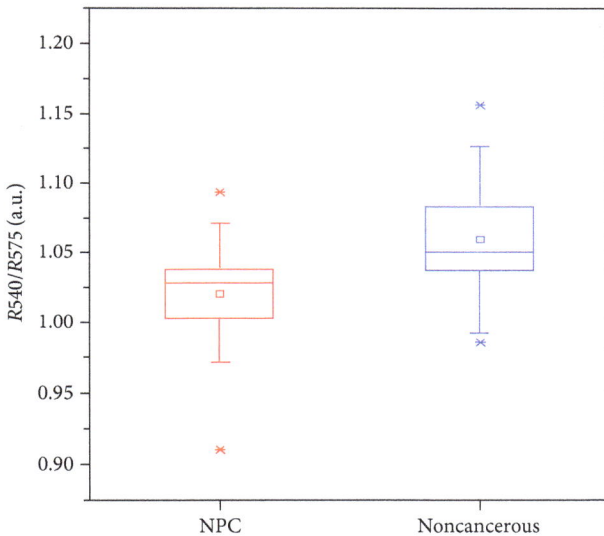

FIGURE 7: Mean reflectance ratios ($R540/R576$) of NPC (red box) versus those of noncancerous group (blue box).

Finally, the mean ratios were used to draw the boxplots (Figure 7). Figure 7 shows the comparison of intensities and standard deviations of the mean $R540/R576$ ratios between nasopharyngeal noncancerous (blue box) and NPC (red box) *ex vivo* tissues. In the case of noncancerous tissue, the mean $R540/R576$ ratio was found to be higher than that of nasopharyngeal malignancy group (1.06 versus 1.02).

4. Discussion

In this work, we investigated reflectance spectra obtained from benign and malignant *ex vivo* tissues in the human nasopharynx. The results of this tentative study show that there were some differences in reflectance spectra of *ex vivo* tissue between NPC and noncancerous subjects, demonstrating potential of reflectance spectroscopy for NPC detection and preliminary precancer screening. Our spectral measurement (Figure 2) shows that reflectance intensity curves of both *ex vivo* benign and malignant mucosa tissues are similar at 400–700 nm. However, the differences of reflectance intensity exist in curve valleys at around 412, 540, and 576 nm region owing to the absorption from oxyhemoglobin bands [31, 32]. These spectral absorption valleys are much more significant in the malignant tumor spectrum at 500–600 nm bands. The spectral absorption characteristics at 540 nm and 576 nm in *ex vivo* nasopharyngeal mucosa may arise from the absorption from oxyhemoglobin associated with high demand of oxygen by cancerous tissues, leading to a decrease in the content of oxygen levels in the blood [13, 23]. Therefore, the spectral absorption of cancerous group is weaker than that of noncancerous group at 540 nm and 576 nm (Figure 2). In the case of malignant lesion tissues, reflectance spectral intensities at 412 nm, 540 nm, and 576 nm bands were slightly stronger than that of noncancerous tissues. One possible explanation for the special spectral intensity differences is that cancerous tissues are often characterized by abnormally histological structure and microvascular volume [12, 13, 15]. In addition, collagen is more densely distributed in the submucosa [15], which is usually able to affect reflectance property of tissue surface.

These preliminary results may indicate the carcinogenesis process of nasopharyngeal carcinoma, since the cancer cells' growth and proliferation relate intimately on the capillary proliferation and nutrient requirements, which caused the metabolic disorders during this period. The tissular capillary will continue to spread, which may lead to the anomalies of histological structure. Furthermore, the cancerous tissue clonal expansion and evolution cause a significant decrease in oxygenated hemoglobin in tumor tissue during the long process of malignant cancer [13, 15, 35]. So, malignant tissues are of low blood-oxygen content owing to metabolic disorders [15]; that is, oxygenated hemoglobin content will be greatly reduced in cancerous tissues.

To further evaluate the capability of reflectance spectroscopy for differentiating nasopharyngeal malignant tumor from benign *ex vivo* tissue, multivariate statistical analysis (PCA-LDA) was performed on the spectral analysis. PCA-LDA yields a sensitivity of 73.7% (14/19) and a specificity of 72.2% (13/18), which demonstrates the availability of the diagnostic algorithm for distinguishing NPC from noncancerous tissue samples by using reflectance spectra.

At the same time, we used a simple empirical method; that is, reflectance spectral intensity ratios ($R540/R576$) were used as an indicator of neoplastic associated changes in the hemoglobin oxygenation. The mean intensity of $R540/R576$ ratio of oxyhemoglobin for malignant and benign tissue was 1.02 and 1.06, respectively. This computation coincides with the results of Subhash et al. [13]. The mean reflectance spectral intensity ratio ($R540/R576$) of oxyhemoglobin was weaker for NPC compared to noncancerous *ex vivo* tissue, which is probably for the reason that malignant tissues are much hungry for oxygen.

Finally, we have compared the diagnostic result of the $R540/R576$ method with that of the PCA-LDA algorithm. The $R540/R576$ method generated a sensitivity of 78.9% and specificity of 69.8% for differentiation between *ex vivo* malignant and noncancerous tissues. The PCA-LDA diagnostic algorithm has a sensitivity of 73.7% and a specificity of 72.2%. As PCA-LDA diagnostic algorithm used entire *ex vivo* range of reflectance spectra for classification by the algorithm, the PCA-LDA method gives a more comprehensive analysis for the dataset. Therefore, the diagnostic specificity of the PCA-LDA algorithm is slightly better than that of the $R540/R576$ method. So, the $R540/R576$ method is used as a kind of simple empirical methods. The discrimination results of PCA-LDA are more comprehensive and reliable. Therefore, reflectance spectroscopy could be one of the screening methods for nasopharyngeal carcinoma detection.

It is worth mentioning that *ex vivo* reflectance spectral intensities and absorption bands intensities of oxyhemoglobin in both normal and malignant tissues decreased over time [13]. Subhash et al. observed that there are some spectral differences between *ex vivo* and *in vivo* normal tissues [13]. To be a more effective diagnosis method, the features of *in vivo* nasopharyngeal carcinoma from that of noncancerous one need further investigation with noninvasive *in vivo* detection in the future.

5. Conclusions

In summary, we studied the characteristics of reflectance spectra from benign and malignant *ex vivo* tissues in the human nasopharynx for the first time. Mean reflectance spectra of benign and malignant *ex vivo* tissue showed significant differences in the absorption bands from oxyhemoglobin bands. Using PCA-LDA diagnostic algorithms, we got a sensitivity of 73.7% and a specificity of 72.2% for NPC detection, indicating the potential of reflectance spectroscopy for *ex vivo* diagnosis of nasopharyngeal malignant tumor. In order to differentiate NPC from noncancerous *ex vivo* tissue based on reflectance spectra simply, mean spectral intensity ratios of oxyhemoglobin ($R540/R576$) were employed as an indicator of cancerization associated transformation in the hemoglobin oxygenation. These results demonstrated that reflectance spectroscopy combined with multivariate statistical tools can be used to indicate the inherent information like blood hemoglobin content, which has the potential to detect nasopharyngeal cancerous tissues quickly in real time, presenting a significant experimental and clinical value for further nasopharyngeal cancer detection *in vivo* in the future.

Competing Interests

The authors declare that they have no competing interests.

Acknowledgments

This work was supported by the Program for Changjiang Scholars and Innovative Research Team in University (no. IRT15R10) and the National Natural Science Foundation of China (nos. 61210016, 61178090, and 61405036) and the Project of Science Foundation of Ministry of Health and United Fujian Provincial Health and Education Project for Tackling the Key Research (no. WKJ-FJ-01). The authors thank Professor Haishan Zeng (British Columbia Cancer Research Centre, Vancouver, BC, Canada) for technical support and helpful advice and thank Verisante Technology, Inc. (Vancouver, Canada), for instrumental support.

References

[1] J. H. C. Ho, "An epidemiologic and clinical study of nasopharyngeal carcinoma," *International Journal of Radiation Oncology, Biology, Physics*, vol. 4, no. 3-4, pp. 183–198, 1978.

[2] T. Liu, "Issues in the management of nasopharyngeal carcinoma," *Critical Reviews in Oncology/Hematology*, vol. 31, no. 1, pp. 55–69, 1999.

[3] K. W. Lo, K. F. To, and D. P. Huang, "Focus on nasopharyngeal carcinoma," *Cancer Cell*, vol. 5, no. 5, pp. 423–428, 2004.

[4] W. I. Wei and J. S. T. Sham, "Nasopharyngeal carcinoma," *The Lancet*, vol. 365, no. 9476, pp. 2041–2054, 2005.

[5] W.-H. Jia, Q.-H. Huang, J. Liao et al., "Trends in incidence and mortality of nasopharyngeal carcinoma over a 20–25 year period (1978/1983–2002) in Sihui and Cangwu counties in southern China," *BMC Cancer*, vol. 6, article 178, 2006.

[6] D. T. T. Chua, J. S. T. Sham, D. L. W. Kwong, D. T. K. Choy, G. K. H. Au, and P. M. Wu, "Prognostic value of paranasopharyngeal extension of nasopharyngeal carcinoma: a significant factor in local control and distant metastasis," *Cancer*, vol. 78, no. 2, pp. 202–210, 1996.

[7] L. Xu, J. Pan, J. Wu et al., "Factors associated with overall survival in 1706 patients with nasopharyngeal carcinoma: significance of intensive neoadjuvant chemotherapy and radiation break," *Radiotherapy and Oncology*, vol. 96, no. 1, pp. 94–99, 2010.

[8] N. Stone, C. Kendall, N. Shepherd, P. Crow, and H. Barr, "Near-infrared Raman spectroscopy for the classification of epithelial pre-cancers and cancers," *Journal of Raman Spectroscopy*, vol. 33, no. 7, pp. 564–573, 2002.

[9] F. Koenig, R. Larne, H. Enquist et al., "Spectroscopic measurement of diffuse reflectance for enhanced detection of bladder carcinoma," *Urology*, vol. 51, no. 2, pp. 342–345, 1998.

[10] I. Georgakoudi, B. C. Jacobson, J. Van Dam et al., "Fluorescence, reflectance, and light-scattering spectroscopy for evaluating dysplasia in patients with Barrett's esophagus," *Gastroenterology*, vol. 120, no. 7, pp. 1620–1629, 2001.

[11] S. K. Chang, Y. N. Mirabal, E. N. Atkinson et al., "Combined reflectance and fluorescence spectroscopy for *in vivo* detection of cervical pre-cancer," *Journal of Biomedical Optics*, vol. 10, no. 2, Article ID 024031, 2005.

[12] Y. S. Fawzy, M. Petek, M. Tercelj, and H. Zeng, "*In vivo* assessment and evaluation of lung tissue morphologic and physiological changes from non-contact endoscopic reflectance spectroscopy for improving lung cancer detection," *Journal of Biomedical Optics*, vol. 11, no. 4, Article ID 044003, 2006.

[13] N. Subhash, J. R. Mallia, S. S. Thomas, A. Mathews, P. Sebastian, and J. Madhavan, "Oral cancer detection using diffuse reflectance spectral ratio R540/R575 of oxygenated hemoglobin bands," *Journal of Biomedical Optics*, vol. 11, no. 1, Article ID 014018, 2006.

[14] Z. Volynskaya, A. S. Haka, K. L. Bechtel et al., "Diagnosing breast cancer using diffuse reflectance spectroscopy and intrinsic fluorescence spectroscopy," *Journal of Biomedical Optics*, vol. 13, no. 2, Article ID 024012, 2008.

[15] G. Zonios, L. T. Perelman, V. Backman et al., "Diffuse reflectance spectroscopy of human adenomatous colon polyps *in vivo*," *Applied Optics*, vol. 38, no. 31, pp. 6628–6637, 1999.

[16] H. Zeng, A. McWilliams, and S. Lam, "Optical spectroscopy and imaging for early lung cancer detection: a review," *Photodiagnosis and Photodynamic Therapy*, vol. 1, no. 2, pp. 111–122, 2004.

[17] R. K. Jain, "Determinants of tumor blood flow: a review," *Cancer Research*, vol. 48, no. 10, pp. 2641–2658, 1988.

[18] L. T. Perelman, V. Backman, M. Wallace et al., "Observation of periodic fine structure in reflectance from biological tissue: a new technique for measuring nuclear size distribution," *Physical Review Letters*, vol. 80, article 627, 1998.

[19] K. Bensalah, D. Peswani, A. Tuncel et al., "Optical reflectance spectroscopy to differentiate benign from malignant renal tumors at surgery," *Urology*, vol. 73, no. 1, pp. 178–181, 2009.

[20] U. Utzinger, M. Brewer, E. Silva et al., "Reflectance spectroscopy for *in vivo* characterization of ovarian tissue," *Lasers in Surgery and Medicine*, vol. 28, no. 1, pp. 56–66, 2001.

[21] L. Q. Li, H. J. Wei, Z. Y. Guo et al., "Oxygenated hemoglobin diffuse reflectance ratio for in vitro detection of human gastric pre-cancer," *Laser Physics*, vol. 20, no. 7, pp. 1667–1672, 2010.

[22] Q. L. Zhao, Z. Y. Guo, J. L. Si et al., "Heat treatment of human esophageal tissues: effect on esophageal cancer detection using oxygenated hemoglobin diffuse reflectance ratio," *Laser Physics*, vol. 21, no. 3, pp. 559–565, 2011.

[23] S.-H. Tseng, C.-K. Hsu, J. Y.-Y. Lee, S.-Y. Tzeng, W.-R. Chen, and Y.-K. Liaw, "Noninvasive evaluation of collagen and hemoglobin contents and scattering property of *in vivo* keloid scars and normal skin using diffuse reflectance spectroscopy: pilot study," *Journal of Biomedical Optics*, vol. 17, no. 7, Article ID 077005, 2012.

[24] H. Zeng, M. Petek, M. T. Zorman, A. McWilliams, B. Palcic, and S. Lam, "Integrated endoscopy system for simultaneous imaging and spectroscopy for early lung cancer detection," *Optics Letters*, vol. 29, no. 6, pp. 587–589, 2004.

[25] R. Shaikh, T. K. Dora, S. Chopra et al., "In vivo Raman spectroscopy of human uterine cervix: exploring the utility of vagina as an internal control," *Journal of Biomedical Optics*, vol. 19, no. 8, Article ID 087001, 2014.

[26] S. Duraipandian, W. Zheng, J. Ng, J. J. H. Low, A. Ilancheran, and Z. Huang, "Simultaneous fingerprint and high-wavenumber confocal Raman spectroscopy enhances early detection of cervical precancer *in vivo*," *Analytical Chemistry*, vol. 84, no. 14, pp. 5913–5919, 2012.

[27] J. Mo, W. Zheng, J. J. H. Low, J. Ng, A. Ilancheran, and Z. Huang, "High wavenumber raman spectroscopy for *in vivo* detection of cervical dysplasia," *Analytical Chemistry*, vol. 81, no. 21, pp. 8908–8915, 2009.

[28] Z. Huang, M. S. Bergholt, W. Zheng et al., "In vivo early diagnosis of gastric dysplasia using narrow-band image-guided Raman endoscopy," *Journal of Biomedical Optics*, vol. 15, no. 3, Article ID 037017, 2010.

[29] D. Lin, G. Chen, S. Feng et al., "Development of a rapid macro-Raman spectroscopy system for nasopharyngeal cancer detection based on surface-enhanced Raman spectroscopy," *Applied Physics Letters*, vol. 106, no. 1, Article ID 013701, 2015.

[30] M. Jermyn, K. Mok, J. Mercier et al., "Intraoperative brain cancer detection with Raman spectroscopy in humans," *Science Translational Medicine*, vol. 7, no. 274, Article ID 274ra19, 2015.

[31] P. Chikezie, A. Akuwudike, and F. Chilaka, "Absorption spectra of normal adults and patients with sickle cell anaemia haemoglobins treated with hydrogen peroxide at two pH values," *Iranian Journal of Blood & Cancer*, vol. 5, no. 4, pp. 129–135, 2013.

[32] W. G. Zijlstra and A. Buursma, "Spectrophotometry of hemoglobin: absorption spectra of bovine oxyhemoglobin, deoxyhemoglobin, carboxyhemoglobin, and methemoglobin," *Comparative Biochemistry and Physiology Part B: Biochemistry and Molecular Biology*, vol. 118, no. 4, pp. 743–749, 1997.

[33] J. Y. Qu, H. Chang, and S. Xiong, "Fluorescence spectral imaging for characterization of tissue based on multivariate statistical analysis," *Journal of the Optical Society of America A*, vol. 19, no. 9, pp. 1823–1831, 2002.

[34] J. Y. Qu, P. Wing, Z. Huang et al., "Preliminary study of *in vivo* autofluorescence of nasopharyngeal carcinoma and normal tissue," *Lasers in Surgery and Medicine*, vol. 26, no. 5, pp. 432–440, 2000.

[35] R. Mallia, S. S. Thomas, A. Mathews et al., "Oxygenated hemoglobin diffuse reflectance ratio for *in vivo* detection of oral pre-cancer," *Journal of Biomedical Optics*, vol. 13, no. 4, Article ID 041306, 2008.

6

Raman Spectroscopy in Colorectal Cancer Diagnostics: Comparison of PCA-LDA and PLS-DA Models

Wenjing Liu,[1,2] **Zhaotian Sun,**[3] **Jinyu Chen,**[4] **and Chuanbo Jing**[3]

[1]*State Key Laboratory of Environmental Chemistry and Ecotoxicology, Research Center for Eco-Environmental Sciences, Chinese Academy of Sciences, Beijing 100085, China*
[2]*University of Chinese Academy of Sciences, Beijing 100049, China*
[3]*No. 4 Hospital, Jinan 250031, China*
[4]*National Center for Mathematics and Interdisciplinary Sciences, Academy of Mathematics and Systems Science, Chinese Academy of Sciences, Beijing 100190, China*

Correspondence should be addressed to Chuanbo Jing; chbjing@eyou.com

Academic Editor: Feride Severcan

Raman spectra of human colorectal tissue samples were employed to diagnose colorectal cancer. High-quality Raman spectra were acquired from normal and cancerous colorectal tissues from 81 patients. Subtle Raman variations, such as for peaks at $1134 \, cm^{-1}$ (protein, C-C/C-N stretching) and $1297 \, cm^{-1}$ (lipid, C-H$_2$ twisting), were observed between normal and cancerous colorectal tissues. The average peak intensity at 1134 and $1297 \, cm^{-1}$ was increased from approximately 235 and 72 in the normal group, respectively, to 315 and 273 in the cancer group. The variations of Raman spectra reflected the changes of cell molecules during canceration. The multivariate statistical methods of principal component analysis-linear discriminant analysis (PCA-LDA) and partial least-squares-discriminant analysis (PLS-DA), together with leave-one-patient-out cross-validation, were employed to build the discrimination model. PCA-LDA was used to evaluate the capability of this approach for classifying colorectal cancer, resulting in a diagnostic accuracy of 79.2%. Further PLS-DA modeling yielded a diagnostic accuracy of 84.3% for colorectal cancer detection. Thus, the PLS-DA model is preferable between the two to discriminate cancerous from normal tissues. Our results demonstrate that Raman spectroscopy can be used with an optimized multivariate data analysis model as a sensitive diagnostic alternative to identify pathological changes in the colon at the molecular level.

1. Introduction

Colorectal cancer has high morbidity and mortality rates and is the third most commonly diagnosed cancer as well as the third leading cause of cancer death for both males and females in the United States [1]. Accurately detecting this cancer is a crucial and foremost step toward improving the survival rate of patients with colorectal cancer. Currently, colonoscopy and histopathology are standard screening and diagnostic techniques for colorectal tissues. Though colonoscopic screening has significantly increased the survival rate of patients with colorectal cancer, it remains a challenge to distinguish adenomas and early adenocarcinomas from benign hyperplastic polyps using colonoscopy [2]. This difficulty is due mainly to the fact that conventional white light reflectance colonoscopy deeply relies on subjective visual assessment of colorectal polyps [3]. The gold standard for cancer diagnostics is histopathology, which is based on the visual investigation of tissue biopsies. A pathologist can diagnose a sample using specific staining, for example, with hematoxylin and eosin, to highlight the focus. Disadvantages of histopathology include time-consuming sample preparation and the subjectivity of pathologists [4]. It is of great necessity to develop an objective and sensitive technique that can assist clinicians in the differential diagnosis of benign and malignant cysts.

Raman spectroscopy, a vibrational analysis technique, is gaining popularity in cancer diagnostics. This technology investigates molecular vibrations that can be used for functional group identification and compositional analysis. Extensive research has demonstrated that Raman spectroscopy can support gold-standard techniques and substantially improve clinical diagnostics [5–9]. Raman measurements on biopsies can help pathologists identify the tumor margins in a fast and precise way.

Raman spectroscopy has been employed to study human colorectal tissues *in vivo* or *ex vivo* to collect spectral information for cancer diagnosis. Since biochemical changes only lead to subtle changes in the Raman spectra, statistical methods are necessary to extract diagnostic information [10, 11]. Typical statistical methods can be categorized into supervised and unsupervised approaches. The unsupervised approach relies only on the Raman spectra to make a decision, whereas the supervised method uses additional information acquired by the gold-standard method. Principal component analysis (PCA), a frequently used unsupervised approach, reduces the number of variables and assesses the data as a first step. Following PCA, a supervised approach such as linear discriminant analysis (LDA), which takes advantage of PCA and the histopathological results, can classify tissues or cells [12–14]. A study on mice with colon cancer showed that the PCA-LDA model can correctly discriminate tumors from healthy tissues with an accuracy of 86.8% [15].

Another commonly used supervised approach, partial least-squares-discriminant analysis (PLS-DA), can provide additional group affinity information by classifying memberships as zeros and ones and thus can maximize the variations between groups of samples. PLS-DA rotates the latent variables (LVs) to achieve maximum group separation [16, 17]. Thus, the LVs consider the diagnostically relevant variations rather than the significant differences in the dataset. PLS-DA model has been employed to analyze colon tissues [3, 18]. In the previous study of Bergholt et al., the PLS-DA model was performed to diagnose the colorectal cancer with an accuracy of 88.8% [3].

Different models can result in different diagnostic performance when employed to analyze the same dataset. Thus, the use of a proper statistical model plays an important role in achieving diagnostic accuracy. The diagnostic performances obtained from different statistical models in terms of sensitivity and specificity were compared to find the optimal model. Based on this optimal model, the suitable diagnostic method was identified in the target tissue system. However, the relevant study on model comparison in the colorectal cancer diagnosis is lacking. This study evaluated and compared two statistical models for colorectal tissue classification. Our aim is to bridge the knowledge gap in identifying the appropriate model for Raman spectroscopy in cancer diagnosis.

Two multivariate statistical methods, PCA-LDA and PLS-DA, were used in combination with leave-one-patient-out cross-validation to establish the discrimination model. This work demonstrates that Raman spectroscopy is a prospective tool in the diagnosis of colorectal cancer during clinical examinations and that the PLS-DA model is superior in detecting spectral differences between normal and cancerous colorectal tissues.

2. Experimental Methods

2.1. Sample Preparation. The formalin-fixed, paraffin-embedded colorectal tissues were retrieved from the Jinan No. 4 Hospital in accordance with the regulations of its ethics committee. The Jinan No. 4 Hospital has approved this study. Normal regions were outside of the tumor areas in the tissue that was obtained during surgery. The paraffin-embedded tissues were sectioned into 10 μm thick sections. Each section was put on a glass microscope slide and stained with hematoxylin and eosin for histopathological diagnosis of the suspected area. The adjacent section was placed on a glass slide without being stained for Raman spectroscopy analysis [12, 19]. The histological analysis was conducted by professional medical doctors who are board certified pathologists.

2.2. Raman Spectroscopy. Raman spectra were acquired with a 10 s integration time in the spectral range of 400~4000 cm^{-1} using a Raman system (Horiba JY HR evolution, France) equipped with an Olympus BXFM open space optical microscope and a charge-coupled device (CCD) detector. A 532 nm laser was focused through a 100x objective (NA = 0.9, WD = 0.21 mm) to excite the samples. The laser power on the sample was about 1.33 mW. A 520.7 cm^{-1} band of silicon wafer was used for calibration. The spectral resolution was about 0.65 cm^{-1}, and the wavenumber accuracy was ±0.03 cm^{-1}. The normal spectra were acquired from healthy regions outside the tumor areas in tissue.

2.3. Data Processing and Multivariate Data Analysis. A linear baseline correction was applied to the Raman spectra using Labspec6 software (Horiba JY). About 10 spectra were collected for each tissue and then averaged. Mean-centering was carried out prior to multivariate statistical analysis to remove common variance from the colorectal tissue Raman spectra dataset.

PCA-LDA and PLS-DA methods were applied for discriminant analysis. Leave-one-patient-out cross-validation was used to validate and optimize the PLS-DA model. Distinct molecular features of the colorectal tissues were extracted and visualized through loadings and scores. The statistical significance among the PCA/PLS scores for normal and cancerous tissues was calculated using a p value less than 0.05.

The PCA-LDA statistical analysis was performed using in-house written scripts. The PLS-DA statistical analysis was carried out with PLS toolbox (Eigenvector Research, Wenatchee, US). All statistical analyses were carried out in the Matlab programming environment (Mathworks Inc., Natick, US).

3. Results and Discussion

3.1. Raman Spectra of Colorectal Tissues. Figure 1(a) shows the averaged Raman spectra of normal ($n = 78$) and cancerous ($n = 81$) colorectal tissues, and the peak assignments are listed in Table S1 in Supplementary Material available online at http://dx.doi.org/10.1155/2016/1603609 [20–26].

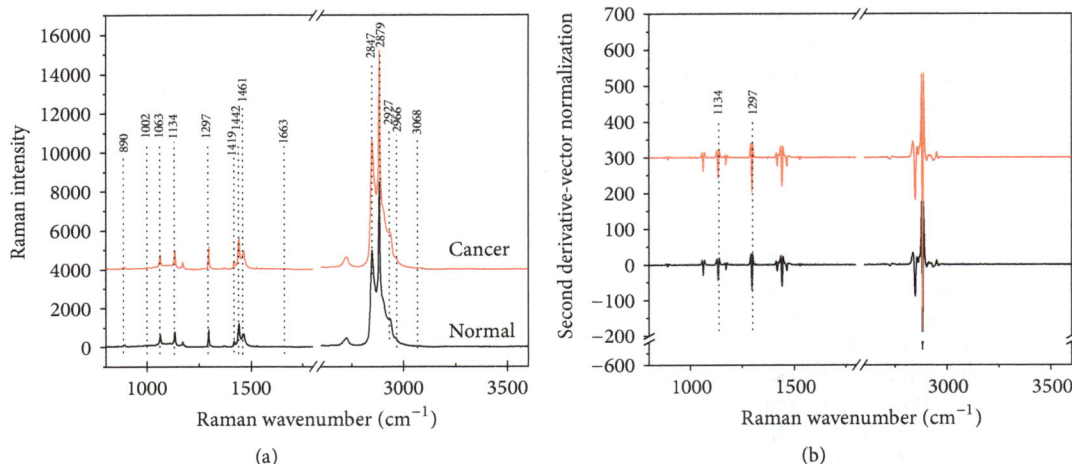

FIGURE 1: (a) Averaged Raman spectra for normal ($n = 78$) and cancerous ($n = 81$) colorectal tissues. (b) Second derivative-vector normalized spectra of Raman spectra from normal and cancerous tissues. The broken interval (-//-) indicates the region of 1800–2600 cm^{-1}, which does not contain tissue-related biochemical information.

General Raman-active tissue components were comparable among colorectal tissues, and subtle variations, even though highly molecule-specific, were observed including peak position and intensity. Prominent Raman bands were observed for normal and cancerous colorectal tissue at about 1063 (lipids/collagen), 1134 (fatty acids and proteins), 1174 (L-tryptophan), 1297 (lipids and phospholipids), 1414 (lipids), 1442 (fatty acids and triglycerides), 1461 (lipids/proteins), 2847 (fatty acids and triglycerides), 2879 (lipids), and 2927 cm^{-1} (proteins and lipids). The difference between normal and cancerous tissues reflects the molecular changes in the tissue associated with the dysplastic progression (Figure 1(b)). For instance, the peak intensities at 1134 and 1297 cm^{-1} increased significantly in cancerous tissue, relative to the normal tissue, suggesting a higher amount of lipid material compared with the normal tissue (Figure S1). But the subtle variations were also hard to differentiate two types of tissues. This motivated further studies of PCA-LDA and PLS-DA to analyze the suitability of each in colorectal cancer diagnosis.

3.2. PCA-LDA Analysis of Raman Spectra. To reduce the dimension and complexity of the biological dataset, we performed PCA-LDA on normal and cancerous colorectal tissues in the spectral range of 400–4000 cm^{-1}. PCA modeling is able to extract most fundamental features, resolving highly specific biomolecular information. Figure 2 shows that the first two PC components accounted for 98% (PC1, 82.7%; LV2, 15.3%) of the total Raman variations around the major Raman peak positions. These two PC components alone contributed to the most characteristic vibrational frequencies. They are dominated by the vibrational features of fatty acids, lipids, proteins, and nucleic acids from the colorectal tissues (Figure 3). The PC1 loading contained Raman peaks for fatty acids (1134 cm^{-1}); proteins (1174 cm^{-1} from L-tryptophan and 1461 cm^{-1} from C-H wagging); and lipids and fatty acids (1442 cm^{-1} from CH$_2$ or CH$_3$ deformations, 2847 cm^{-1} from

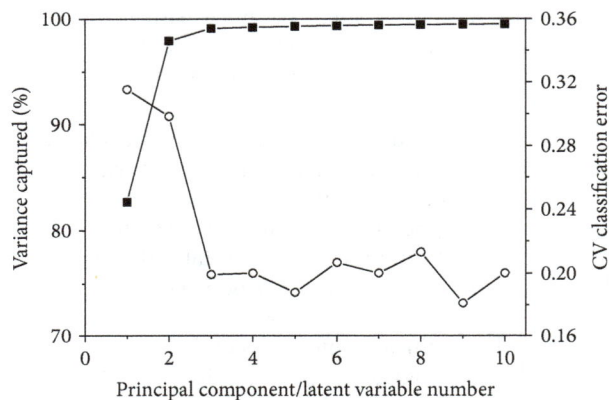

FIGURE 2: Variance captured percent and classification error as function of model complexity (i.e., retained number of PCs and LVs) using PCA-LDA and PLS-DA together with leave-one-patient-out cross-validation.

symmetric CH$_2$ stretching, and 2879 cm^{-1} from asymmetric CH$_2$ stretching). The loading on PC2 captured Raman peaks similar to those in PC1 loading, reflecting the main components of Raman spectra. To cross-validate the classification, we used the LDA model with the leave-one-patient-out approach for cross-validation. The PCA-LDA model resulted in a sensitivity of 72.8% and a specificity of 85.9%, which finally yielded a diagnostic accuracy of 79.2%.

3.3. PLS-DA Model for Predicting Cancer. The PLS-DA diagnostic model was performed with leave-one-patient-out cross-validation, to achieve a Raman spectral dataset using an optimum number of components. The optimum number of components was estimated using the local minimum of cross-validation classification error values and was determined to be 3 LVs (Figure 2). Figure 3 shows three significant LV loadings for the Raman spectral dataset. The loading

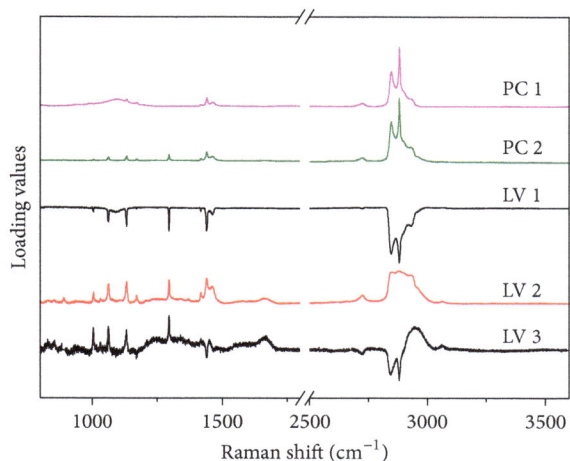

FIGURE 3: Loading plot of PCA components (PC) and PLS components (LV) in the each model calculated from the Raman spectra of colorectal tissues. Each loading is shifted vertically for better visualization. The broken interval (-//-) indicates the region of ~2000–2500 cm^{-1}, which does not contain much tissue-related biochemical information.

on LV1 contained the following specific Raman peaks from lipids: 1063 cm^{-1} (C-C stretching), 1297 cm^{-1} (CH$_2$ twisting), 1461 cm^{-1} (C-H wagging), and 2879 cm^{-1} (asymmetric CH$_2$ stretching). It also contained fatty acids, as evidenced by these peaks: 1442 cm^{-1} (CH$_2$ or CH$_3$ deformation), 2847 cm^{-1} (symmetric CH$_2$ stretching), and 2927 cm^{-1} (symmetric CH$_3$ stretching). The loading on LV2-captured Raman peaks, aside from lipids and fatty acids, was mainly associated with proteins, as evidenced by the peaks at 1002 cm^{-1} (C-C stretching, phenylalanine), 1174 cm^{-1} (L-Tryptophan), 1663 cm^{-1} (amide I α-helix, C=O stretching), 2927 cm^{-1} (symmetric CH$_2$ stretching), and 3068 cm^{-1} (nucleic acids/proteins, C-H aromatic vibration). Thus, Raman spectroscopy associated with PLS-DA modeling using 3 LVs provides highly specific signatures of various biomolecules, rendering a sensitivity of 77.7%, a specificity of 91.0%, and collectively a diagnostic accuracy of 84.3% (Table S2).

3.4. Comparison of PCA-LDA and PLS-DA. Figure 4 shows the box chart of significant PC and LV scores to visualize different degrees of diagnostic efficiency. The PCA scores show the classification comparisons between normal and cancerous tissues through PC1 and PC2. Compared with PCs and the other LVs, LV2 ($p = 1.16e - 7$) shows the greatest efficacy in distinguishing colorectal cancer. Analysis of the LV2 scores showed that increased protein and nuclear contents occurred during the neoplastic progression in the colon tissue, indicating the elevated number of cells associated with cancerous development. Channelling this increased biomolecule biosynthesis is absolutely required for tumorigenic transformation [27]. The cancerization could induce lipid and nucleic acid changes, which are reflected in the Raman spectra and further analysis. Human cancer cells express high levels of lipogenic enzymes to meet the great

demand for lipid synthesis [28]. Meanwhile, changes in the levels of nucleic acids were associated with tumor burden and malignant progression [29]. Receiver operating characteristic (ROC) curves (Figure 5) were also generated from spectral datasets to further evaluate the separation. The integrated area under the ROC curve of the PLS-DA model was 0.856, while the integrated area of the PCA-LDA model was 0.696, substantiating the efficiency of using the Raman technique with PLS-DA for diagnosing cancerous colorectal tissues.

Table S2 shows that the PLS-DA model combined with Raman spectroscopy had better diagnostic performance, compared with the PCA-LDA model. Previous studies reported that the PLS-DA model provided a diagnostic sensitivity of 90.9% and specificity of 83.3% for differentiating adenomas from hyperplastic polyps [18]. The PCA-LDA model could distinguish cancer from normal colon tissues in mice with diagnostic accuracy of 86.8% [15]. The diagnostic results from the literature indicated that the PLS-DA model resulted in better diagnostic accuracy than the PCA-LDA model, even in different tissue systems.

In the PLS-DA model, the diagnostic specificity (91.0%) in our study was higher than that (83.3%) in previous study, while the sensitivity (77.7%) from our study was lower than that (90.9%) from previous study. Compared with previous studies, the diagnostic accuracy (79.2%) from the PCA-LDA model in our study was lower than that (86.8%) in the mice colon tissue study. The different statistics may be attributed to the difference in Raman spectra collection and sample preparation. The diagnostic results of our study and previous studies can be acceptable for the clinical application.

PCA is a classical technique for dimensionality reduction. This method identifies several principal directions with a high variance, especially for high-dimensional data X with small number of samples and large number of features, such as Raman spectra. By projecting the original data of X onto these directions, much of the information of X will be maintained by just a small number of these new projected variables (i.e., LVs) [30]. However, PCA only involves one set of data. On the other hand, PLS realizes dimensionality reduction by considering the relations between two data blocks (X and Y) across the same samples. PLS maximizes the covariance between X and Y, which balances the requirement to explain as much variance as possible by considering the correlated relationships between X and Y [31]. Thus, the three LVs identified using PLS captured not only the high variance in the spectral dataset, but also the relationship between the spectral dataset and sample class. In this experiment, the accuracy of PLS-DA modeling (84.3%) was much higher than that of PCA-LDA (79.2%). Thus, out of the two, the PLS-DA model is preferable for discriminating cancerous from normal tissues following the strategies in this study.

4. Conclusions

In summary, Raman spectroscopy was applied as a sensitive diagnostic alternative for identifying pathologic changes (e.g., dysplasia) in colon tissue at the molecular level, using an optimized multivariate data analysis model. In a side-by-side comparison of PCA-LDA and PLS-DA with respect

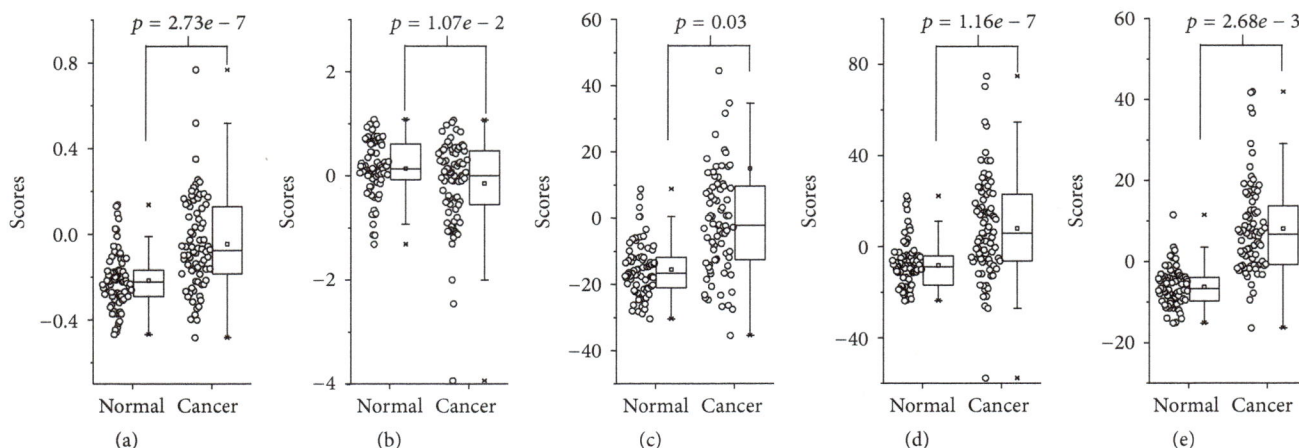

FIGURE 4: Box charts of the two PCA component (PC) scores and three significant PLS component (LV) scores calculated from the Raman dataset for normal and cancerous tissue types: (a) PC1 score, (b) PC2 score, (c) LV1 score, (d) LV2 score, and (e) LV3 score. The line within each box represents the median, while the lower and upper boundaries of the box indicate the first (25th percentile) and the third (75th percentile) quartiles, respectively. Error bars (whiskers) represent the 1.5-fold interquartile range.

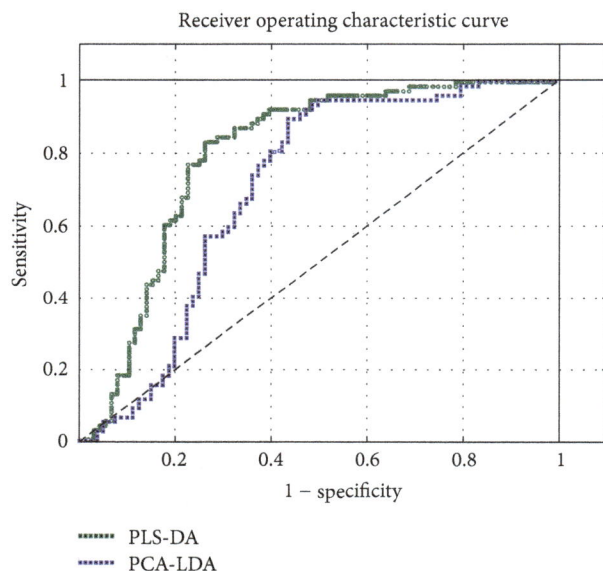

FIGURE 5: Receiver operating characteristic (ROC) curves of discrimination results from Raman spectra for colorectal tissue using Raman spectroscopy and PCA-LDA (blue) and PLS-DA (green) together with the leave-one-patient-out cross-validation method.

to the characterization of molecular profiles (e.g., proteins, lipids, and nucleic acids) of normal and cancerous colorectal tissues, the PLS-DA model was found to be a superior choice. The subtle Raman variations among normal and cancerous colorectal tissues are associated with cancerous tissue transformation. Confocal Raman spectroscopy is a promising technique in the diagnosis and characterization of colorectal cancer.

Competing Interests

The authors declare that they have no competing interests.

Acknowledgments

The authors acknowledge the financial support of the Strategic Priority Research Program of the Chinese Academy of Sciences (XDB14020201).

References

[1] R. Siegel, J. Ma, Z. Zou, and A. Jemal, "Cancer statistics, 2014," *CA Cancer Journal for Clinicians*, vol. 64, no. 1, pp. 9–29, 2014.

[2] M. B. Wallace and R. Kiesslich, "Advances in endoscopic imaging of colorectal neoplasia," *Gastroenterology*, vol. 138, no. 6, pp. 2140–2150, 2010.

[3] M. S. Bergholt, W. Zheng, K. Lin et al., "Characterizing variability of in vivo Raman spectroscopic properties of different anatomical sites of normal colorectal tissue towards cancer diagnosis at colonoscopy," *Analytical Chemistry*, vol. 87, no. 2, pp. 960–966, 2015.

[4] K. Eberhardt, C. Stiebing, C. Matthäus, M. Schmitt, and J. Popp, "Advantages and limitations of Raman spectroscopy for molecular diagnostics: an update," *Expert Review of Molecular Diagnostics*, vol. 15, no. 6, pp. 773–787, 2015.

[5] M. S. Bergholt, W. Zheng, K. Y. Ho et al., "Fiberoptic confocal raman spectroscopy for real-time in vivo diagnosis of dysplasia in Barrett's esophagus," *Gastroenterology*, vol. 146, no. 1, pp. 27–32, 2014.

[6] I. Taleb, G. Thiéfin, C. Gobinet et al., "Diagnosis of hepatocellular carcinoma in cirrhotic patients: a proof-of-concept study using serum micro-Raman spectroscopy," *Analyst*, vol. 138, no. 14, pp. 4006–4014, 2013.

[7] S. Borel, E. A. Prikryl, N. H. Vuong et al., "Discrimination of normal and malignant mouse ovarian surface epithelial cells *in vitro* using Raman microspectroscopy," *Analytical Methods*, vol. 7, no. 22, pp. 9520–9528, 2015.

[8] O. J. Old, L. M. Fullwood, R. Scott et al., "Vibrational spectroscopy for cancer diagnostics," *Analytical Methods*, vol. 6, no. 12, pp. 3901–3917, 2014.

[9] A. Sahu, S. Tawde, V. Pai et al., "Raman spectroscopy and cyto-pathology of oral exfoliated cells for oral cancer diagnosis," *Analytical Methods*, vol. 7, no. 18, pp. 7548–7559, 2015.

[10] C. Beleites, A. Bonifacio, D. Codrich, C. Krafft, and V. Sergo, "Raman spectroscopy and imaging: promising optical diagnostic tools in pediatrics," *Current Medicinal Chemistry*, vol. 20, no. 17, pp. 2176–2187, 2013.

[11] K. H. Esbensen and P. Geladi, "Principles of proper validation: use and abuse of re-sampling for validation," *Journal of Chemometrics*, vol. 24, no. 3-4, pp. 168–187, 2010.

[12] A. Salman, G. Sebbag, S. Argov, S. Mordechai, and R. K. Sahu, "Early detection of colorectal cancer relapse by infrared spectroscopy in 'normal' anastomosis tissue," *Journal of Biomedical Optics*, vol. 20, no. 7, Article ID 075007, 2015.

[13] A. Synytsya, M. Judexova, D. Hoskovec, M. Miskovicova, and L. Petruzelka, "Raman spectroscopy at different excitation wavelengths (1064, 785 and 532 nm) as a tool for diagnosis of colon cancer," *Journal of Raman Spectroscopy*, vol. 45, no. 10, pp. 903–911, 2014.

[14] A. Molckovsky, L.-M. Wong Kee Song, M. G. Shim, N. E. Marcon, and B. C. Wilson, "Diagnostic potential of near-infrared Raman spectroscopy in the colon: differentiating adenomatous from hyperplastic polyps," *Gastrointestinal Endoscopy*, vol. 57, no. 3, pp. 396–402, 2003.

[15] A. Taketani, R. Hariyani, M. Ishigaki, B. B. Andriana, and H. Sato, "Raman endoscopy for the in situ investigation of advancing colorectal tumors in live model mice," *Analyst*, vol. 138, no. 14, pp. 4183–4190, 2013.

[16] M. Hedegaard, C. Krafft, H. J. Ditzel, L. E. Johansen, S. Hassing, and J. Popp, "Discriminating isogenic cancer cells and identifying altered unsaturated fatty acid content as associated with metastasis status, using K-means clustering and partial least squares-discriminant analysis of raman maps," *Analytical Chemistry*, vol. 82, no. 7, pp. 2797–2802, 2010.

[17] F. C. De Lucia Jr., J. L. Gottfried, C. A. Munson, and A. W. Miziolek, "Multivariate analysis of standoff laser-induced breakdown spectroscopy spectra for classification of explosive-containing residues," *Applied Optics*, vol. 47, no. 31, pp. G112–G122, 2008.

[18] M. S. Bergholt, K. Lin, J. Wang et al., "Simultaneous fingerprint and high-wavenumber fiber-optic Raman spectroscopy enhances real-time *in vivo* diagnosis of adenomatous polyps during colonoscopy," *Journal of Biophotonics*, vol. 9, no. 4, pp. 333–342, 2016.

[19] S. Argov, J. Ramesh, A. Salman et al., "Diagnostic potential of Fourier-transform infrared microspectroscopy and advanced computational methods in colon cancer patients," *Journal of Biomedical Optics*, vol. 7, no. 2, pp. 248–254, 2002.

[20] Q. Matthews, A. Jirasek, J. Lum, X. Duan, and A. G. Brolo, "Variability in raman spectra of single human tumor cells cultured *in vitro*: correlation with cell cycle and culture confluency," *Applied Spectroscopy*, vol. 64, no. 8, pp. 871–887, 2010.

[21] L. Pantanowitz, J. H. Sinard, W. H. Henricks et al., "Validating whole slide imaging for diagnostic purposes in Pathology: guideline from the College of American pathologists Pathology and Laboratory Quality Center," *Archives of Pathology and Laboratory Medicine*, vol. 137, no. 12, pp. 1710–1722, 2013.

[22] J. De Gelder, K. De Gussem, P. Vandenabeele, and L. Moens, "Reference database of Raman spectra of biological molecules," *Journal of Raman Spectroscopy*, vol. 38, no. 9, pp. 1133–1147, 2007.

[23] N. Stone, C. Kendall, N. Shepherd, P. Crow, and H. Barr, "Near-infrared Raman spectroscopy for the classification of epithelial pre-cancers and cancers," *Journal of Raman Spectroscopy*, vol. 33, no. 7, pp. 564–573, 2002.

[24] I. Notingher, "Raman spectroscopy cell-based biosensors," *Sensors*, vol. 7, no. 8, pp. 1343–1358, 2007.

[25] J. Kneipp, T. Bakker Schut, M. Kliffen, M. Menke-Pluijmers, and G. Puppels, "Characterization of breast duct epithelia: A Raman Spectroscopic Study," *Vibrational Spectroscopy*, vol. 32, no. 1, pp. 67–74, 2003.

[26] C. Krafft, L. Neudert, T. Simat, and R. Salzer, "Near infrared Raman spectra of human brain lipids," *Spectrochimica Acta Part A: Molecular and Biomolecular Spectroscopy*, vol. 61, no. 7, pp. 1529–1535, 2005.

[27] R. A. Cairns, I. S. Harris, and T. W. Mak, "Regulation of cancer cell metabolism," *Nature Reviews Cancer*, vol. 11, no. 2, pp. 85–95, 2011.

[28] J. A. Menendez and R. Lupu, "Fatty acid synthase and the lipogenic phenotype in cancer pathogenesis," *Nature Reviews Cancer*, vol. 7, no. 10, pp. 763–777, 2007.

[29] H. Schwarzenbach, D. S. B. Hoon, and K. Pantel, "Cell-free nucleic acids as biomarkers in cancer patients," *Nature Reviews Cancer*, vol. 11, no. 6, pp. 426–437, 2011.

[30] E. B. Corrochano, *Handbook of Geometric Computing: Applications in Pattern Recognition, Computer Vision, Neuralcomputing, and Robotics*, Springer Science & Business Media, Berlin, Germany, 2005.

[31] R. Rosipal and N. Krämer, "Overview and recent advances in partial least squares," in *Subspace, Latent Structure and Feature Selection*, vol. 3940 of *Lecture Notes in Computer Science*, pp. 34–51, Springer, Berlin, Germany, 2006.

7

Quantitative Classification of Quartz by Laser Induced Breakdown Spectroscopy in Conjunction with Discriminant Function Analysis

A. Ali,[1,2] **M. Z. Khan,**[1] **I. Rehan,**[1] **K. Rehan,**[1,3] **and R. Muhammad**[1]

[1]*Department of Applied Physics, Federal Urdu University of Arts, Science and Technology, Islamabad 44,000, Pakistan*
[2]*National Center for Physics, Quaid-i-Azam University Campus, Islamabad 44,000, Pakistan*
[3]*International College, UCAS, Beijing 100190, China*

Correspondence should be addressed to I. Rehan; irehanyousafzai@gmail.com

Academic Editor: Jau-Wern Chiou

A responsive laser induced breakdown spectroscopic system was developed and improved for utilizing it as a sensor for the classification of quartz samples on the basis of trace elements present in the acquired samples. Laser induced breakdown spectroscopy (LIBS) in conjunction with discriminant function analysis (DFA) was applied for the classification of five different types of quartz samples. The quartz plasmas were produced at ambient pressure using Nd:YAG laser at fundamental harmonic mode (1064 nm). We optimized the detection system by finding the suitable delay time of the laser excitation. This is the first study, where the developed technique (LIBS+DFA) was successfully employed to probe and confirm the elemental composition of quartz samples.

1. Introduction

Quartz is one of the most abundant minerals found on the earth after feldspar. The crystal structure of quartz is a continuous arrangement of SiO4 silicon-oxygen tetrahedra with every oxygen particle being shared between two tetrahedra, thereby giving a general chemical formula of SiO2. Quartz has many types according to impurities present in it. Basically, it occurs in all environments and is an essential constituent of many rocks. Previously quartz was classified on the basis of its physical appearance and the possibility of determination of its structural essentials using an optical microscope. A literature survey showed that different experiments have been performed to study the quartz crystals. Laser induced breakdown spectroscopy was used as a rapid tool for the discrimination and analysis of minerals and geological samples [1]. For the quantitative analysis of quartz minerals, a digital analytical system was developed and applied [2]. Laser spectroscopy in conjunction with the Raman spectroscopy was also used for identifying the composition of different minerals. Due to its high power of laser energy, LIBS was also applied for the investigation and interpretation of lithium in pegmatite minerals [3]. This technique was also utilized for the identification of polymer based samples [4, 5]. To improve the capability of LIBS technique, it has been employed in combination with discriminant function analysis to identify and differentiate between different kinds of polymers [6]. Similarly, the combination of LIBS and DFA was also applied to analyze the *Pseudomonas aeruginosa* bacteria's colony in blood [7]. The calibration-free LIBS (CF-LIBS) method can be considered an important tool but it still requires more experimental efforts for getting recognized as a useful tool for practical applications [8].

Laser induced breakdown spectroscopy (LIBS) is a multipurpose method fit for analyzing the sample in its original structure and thereby catching up from their point of limitations of other methods in terms of long sample preparation that could lead to the amendment of the chemical composition of the original test sample. The emission spectrum is a finger-print of any material and it can provide a complete elemental composition of any substance [9].

We applied laser induced breakdown spectroscopy in coincidence with the discriminant function analysis for the quantitative investigation of five different types of quartz

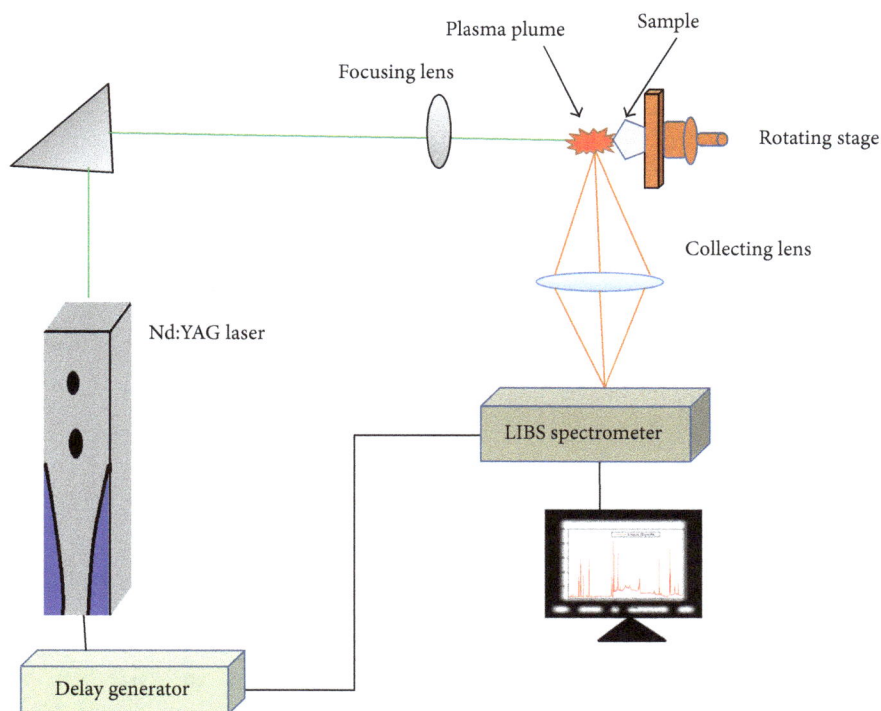

FIGURE 1: Schematic diagram of LIBS setup.

samples, namely, crystal quartz, lapis quartz, pink quartz, tourmaline quartz, and amethyst quartz. For this purpose the samples were collected from Waziristan (North, South) (crystal quartz), Mohmand Agency (pink quartz), and Gilgit-Baltistan (lapis and tourmaline) areas of Khyber Pakhtunkhwa, Pakistan, and one sample (Amethyst) was collected from Minas Gerais, Brazil, South America (SA). The mountains are rich with precious minerals in Pakistan but lack facilities and scientific methods of exploration of these treasures are not proper. Apparently, all types of quartz are quite similar. They can be classified on the basis of their colors, but the difference in colors can be due to different impurities present in quartz as well as different weather conditions in that area.

In this work, plasmas from quartz were produced utilizing Q-switched Nd:YAG laser at fundamental harmonic mode (1064 nm) in air. The emission spectra of quartz samples were registered by LIBS setup and major constituents were identified with the help of NIST database [10]. Due to similar elemental composition (Na, Li, Mn, Si, etc.) of the acquired quartz samples, all the samples were similar qualitatively. Discriminant function analysis (DFA) was employed for calculating the percentage profiles of the observed species in the collected quartz samples which is a strong mathematical tool, and has never been used for the classification of quartz. The developed technique (LIBS+DFA) for the classification of quartz has potential applications in the field of core mining.

2. Experimental Setup

The experimental setup used in the present work was illustrated in Figure 1. Briefly, it consists of a Nd:YAG laser

(Brilliant, Quantel, France), having a pulse width of 5 ns and repetition rate of 10 Hz. The laser beam was focused through a 10 cm quartz lens on the surface of quartz target. The sample was mounted on a rotating stage, which was rotated continuously using a step motor to provide a fresh target surface after each laser pulse. The emitted radiation from the plume was collected by an optical fiber (high-OH, core diameter of 600 μm) having collimating lens (0–45°) field view placed at right angle to the direction of the plume. The optical fiber was connected to the spectrometer (LIBS2000+, Ocean Optics, USA) of focal length 0.4 m and spectral resolution $\lambda/\Delta\lambda \cong 5000$. The LIBS detection system was equipped with five HR2000 high-resolution miniature spectrometers each having the slit width of 5 μm with 2400 lines/mm gratings thereby covering a spectral range of 200–720 nm. Every spectrometer has 2048-element linear silicon charge coupled device (CCD) array with optical resolution $\approx$0.05 nm. The system was calibrated by recording the well-known lines of neon, argon, and mercury with $\approx$0.02 nm measurement uncertainty. All the five spectrometers installed in the LIBS2000 were manufacturer calibrated in efficiency using the DH-2000-CAL standard light source. The pulse energy was varied through OOILIBS software. The laser pulse energy was measured using a dedicated energy meter (NOVA-QTL, P/N 1Z01507, Sr. number 56461). The time delay between the laser firing and the opening of the detection system was optimized and set ~3 μm.

In order to get well resolved emission spectrum at the early stage of plasma generation, the data acquisition time was optimized and found to be 850 ns. Also the optimal laser energy of 100 mJ/pulse was chosen to avoid the absorption of laser photon by the plasma plume at the higher end of the

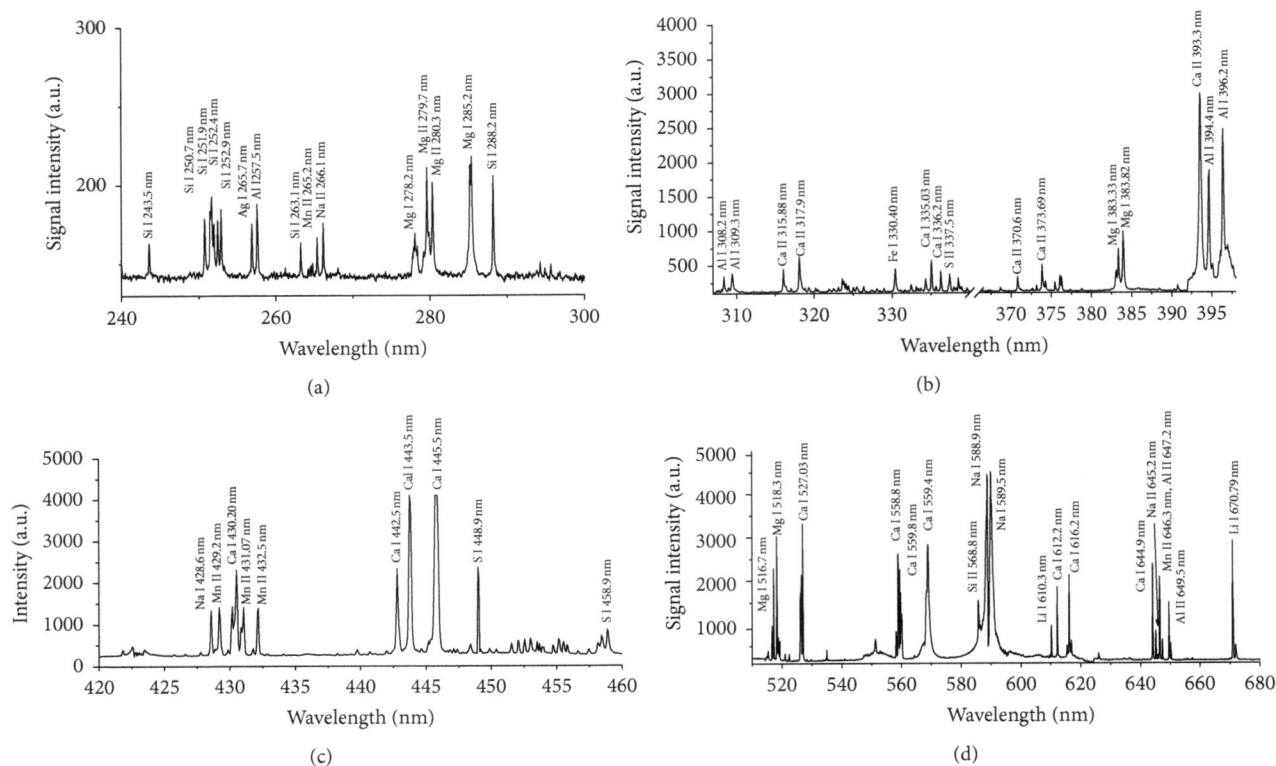

FIGURE 2: The representative emission spectrum of lapis quartz along with the relevant atomic emission identified, covering a spectral range of 240–680 nm.

laser energy with no compromising on the signal to noise ratio. We used an exposure time of 0.2 s and a slit width of 50 μm. To minimize the statistical error, the output data were averaged for 20 laser shots. From the spot diameter of ~103 μm, a laser fluence of about 44 J cm^{-2} was approximated.

3. Result and Discussion

3.1. Study of Emission Spectra. A representative emission spectrum of lapis quartz was shown in Figure 2. For time and space saving we did not show all the emission spectra. To analyze each of the emission spectra we divided each of them into four parts thereby covering a spectral region of 240–680 nm.

The dominant spectral emission lines were atomic and singly ionized lines mainly coming from Al, Ca, Li, and Si and some lines of Fe have been labeled on the graph. Only those strong spectral emissions were taken in calculations which were common in all the emission spectra. The major transitions of different element observed in the emission spectra of quartz were Si I [(243.51), (250.69), (252.41), (252.85), (263.12), (288.15), (298.76)] nm, Li I (670.79 nm), Na I [(588.995), (589.594), (285.28)] nm, Al I [(394.70), (396.42)] nm, Ca I [(393.62), (397.13), (422.67)] nm, and so forth. The elemental impurities present within quartz such as copper, lithium, silver, and calcium were reported in the literature [11]. The optical emission spectrums of all five different types of quartz were registered and presented in

Figure 3. Due to these similarities, the classification of the collected quartz samples was quite difficult by qualitative analysis of LIBS spectrum. The longest wavelength detectable range of our experimental setup was 720 nm; that is why we were unable to detect oxygen in the collected quartz samples.

A detailed qualitative analysis of the emission spectra of all types of quartz was performed and a small wavelength region of the emission spectrums was presented in Figure 4.

The emission spectrum reveals that, Ca I, II transitions are common in all spectrums, but Al I is not present in the emission spectrums of amethyst and crystal quartz. Therefore, on the basis of detailed spectroscopic analysis of emission spectrum, the collected quartz can be classified into two classes of class I (amethyst, crystal quartz) and class II (lapis, pink, and tourmaline). From the relative intensities of Ca I and Al I emission lines, it can be inferred that the Ca and Al concentration could be different in all five types of quartz. However, the complete classification of the quartz samples requires a model for quantitative analysis.

3.2. Discriminant Function Analysis. In last few years, the statistical methods are effectively used in a joint venture with the experiments, especially in the field of laser spectroscopy. Discriminant function analysis (DFA) is one of those statistical techniques which are used to quantify and predict the group membership of more than one group. In this work, we used this method to classify all the established set of observations of the different types of quartz on the basis

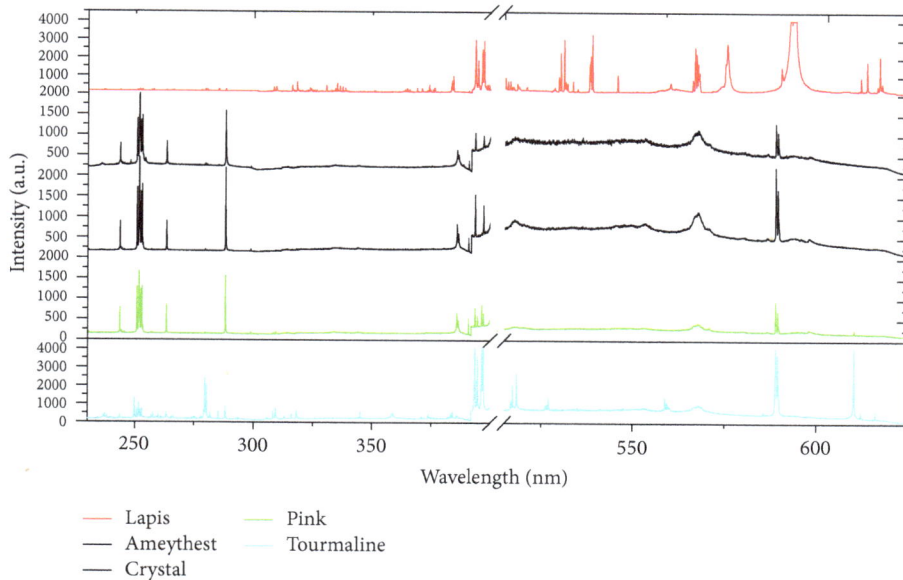

FIGURE 3: Plot showing a comparison between optical emission spectra of all the collected quartz samples.

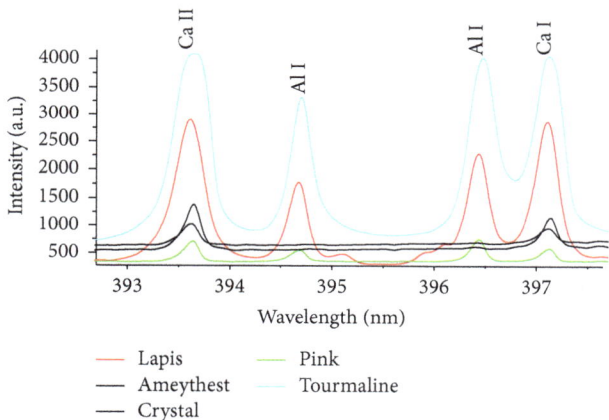

FIGURE 4: Optical emission spectrum showing Al I and Ca I, II transition in all five types of quartz.

of independent variable (emission intensities) from each spectrum (individually each spectrum was treated as single input data point) [12].

The mathematical formula for each set of n variables $X_1, X_2, \ldots, X_n$ is given as

$$D = b_1 X_1 + b_2 X_2 + \cdots + b_n X_n + c, \qquad (1)$$

where D is a dependent variable called discriminant function, X's are the predictive variable or original variable, b's are the discriminant coefficient or weight for that variable, and c is a constant.

From (1), it is clear that discriminant function is the linear combination of the predictive variables which is responsible for group discrimination. The fundamental motivation behind DFA was to deliver a model to predict the group

memberships to figure out the dimensions in which the groups were emphatically differentiated [6, 13]. In discriminant function, each predictive variable demonstrates its contribution to classify the groups by weighting b's on that direction. The possible number of discriminant functions (dimension) is less than the number of groups and very nearly equivalent to the predictor variables. DFA point out the established set of b's to maximize the variance between groups and variance within groups ratios. These functions do not depend upon each other (orthogonal), so it is inferred that discriminations between groups because of these functions do not overlap [6, 12]. The discriminant function analysis was handled in three stages. In the first stage, canonical correlation analysis generates a set of canonical discriminant functions (shown in Table 2) which are basically the eigenvectors of the observed information and gave a base to maximize the difference between the groups. If there were N groups to be distinguished, then $N - 1$ discriminant functions were constructed. The primary functions established a strong differentiation between the groups of all combinations of the variables. The auxiliary function again maximizes the discrimination between groups but at this time secondary function controls for primary factor and so on. In the second stage, a test (F test) was characterized for the significance in the set of discriminant functions, based on the variances in the groups means, used to ascertain the Wilks' lambda (also called u statistics) whose range is from 0 to 1. Zero implies that all grouping means are different and one implies that all grouping variables were not distinctive. The last step was the classification, if the result was significant then reject the null hypothesis and the occurrence of classification of the dependent variable appeared by canonical discriminant function analysis of each single data point spectrum [6]. There were five groups of different types of quartz; therefore

TABLE 1: Variables that are used[a,b] in the analysis at each step.

Step	Entered	Statistic	df1	df2	df3	Wilks' lambda Exact F Statistic	df1	df2	Sig.	Approximate F Statistic	df1	df2	Sig.
1	Si I (243.51)/Li I (670.79)	0.025	1	4	148.000	1419.00	4	148.000	0.000				
2	Si I (250.69)/Li I (670.79)	0.002	2	4	148.000	818.586	8	294.000	0.000				
3	Si I (243.52)/Na I (588.99 + 589.59)	0.000	3	4	148.000					590.966	12	386.571	0.000
4	Si I (288.15)/Li I (670.79)	0.000	4	4	148.000					496.521	16	443.620	0.000
5	Si I (250.69) Na I (588.99 + 589.59)	0.000	5	4	148.000					408.230	20	478.544	0.000
6	Si I (263.12)/Li I (670.79)	0.000	6	4	148.000					341.600	24	500.077	0.000
7	Si I (288.15)/Na I (588.99 + 589.59)	0.000	7	4	148.000					293.669	28	513.410	0.000
8	Si I (263.12)/Li I (670.79)]	0.000	8	4	148.000					256.773	32	521.577	0.000

TABLE 2: The first 4 canonical discriminant functions used in the analysis.

Function	Eigenvalue	% of variance	Cumulative %	Canonical correlation
1	72.923	56.8	56.8	0.993
2	47.804	37.2	94.0	0.990
3	7.641	6.0	100.0	0.940
4	0.048	0.0	100.0	0.215

four discriminant function scores as well as eight prediction variables were figured out by DFA procedure as tabulated in Table 1. A stepwise DFA was performed to choose a small number of predicators that were responsible for the classification among all the five types of quartz. To know the behavior of the classification a Wilks' lambda was calculated at each step. Wilks' lambda is the ratio of within group sums of square to the sum of square and proportion of total variance in discriminant function scores. We conclude from Table 1 that eight steps were taken and each step included new variable and all were significant as $p < 0.000$.

The total discriminant function numbers acquired by DFA process were less than the total groups as $N - 1$, where N is the total numbers of groups which were five; therefore, discriminant capacities numbers were four which was significant ($p < 0.000$). For statistical analysis, eight indicators were utilized. The discriminants with their corresponding power of discrimination were tabulated in Table 2. In the first column the eigenvalue is a characteristic background (root) of every function, representing how well the corresponding function distinguishes the groups in the analysis, in such a way that the greater the eigenvalue is the better the discrimination will be. In the % variance column, we see that the initial three functions have the highest percentage of variance. As inserted in Table 2, the DF1 were contributing 56.8% of variance between the groups and 37.2% were contributed by DF2; just 94% of the variance is shared by the DF1 and DF2. Accumulation of the initial three variances was 100%. The last column of canonical correlation shows the correlation between the group members (dependent variables)

and discriminant scores, which represents that these three functions were highly correlated (it ranges from 0 to 1; a high correlation tells us about a function that discriminates well). Therefore the first function was the most imperative function in explanatory power on the grounds that a high correlated predictor contributes more information in the analysis, the second was the next more important, and so on.

Standardized canonical discriminant function coefficients were presented in Table 3, which were playing very important role in discrimination. The plus and minus sign show the direction of relationships. The unique contributions to the discriminant functions were evaluated as can be observed Table 3. The strongest predictors in the primary function (dimension) were Si I (252.41)/Li I (670.79), Si I (252.41)/Na I (588.99 + 589.59), and Si I (252.85)/Li I (670.79) nm scores. The other strongest predictors in secondary function were Si I (288.15)/Na I (588.986 + 589.58), Si I (288.15)/Li I (670.79), and Si I (243.51)/Na I (588.986 + 589.58) and in the third function column, the next strongest predictors were Si I (252.85)/Li I (670.891), Si I (288.15)/Li I (670.79), Si I (252.85)/Li I (670.79), and so on. The ratios were taken because Li (at 670.79 nm) and Na (at 588.9 nm, 589.5 nm) have a major contribution in the collected samples. The Si I (263.12)/Li I (670.79) and others were the weakest predictors as in Table 3. Hence the powerful factor or strongest predictor which plays an important role in the classification of quartz was [Si I (243.51)/Li I (670.79)] ratio. The classification results were shown in Table 4, which were based on the DFA. In Table 4, the actual group membership against the predicted group membership was presented. By performing DFA, the quartz was classified in ~99.3% of the original group cases successfully. To find out the expected accuracy for group membership prediction, we used a leave-one-out method which is also known as jackknife method [14, 15]. In this method, the identification of each data point was calculated. The process was repeated for all of the observations, leaving out one at each period, and the fact that all of the observations have misclassified case ratios informs us about the actual prediction rate error. In the last, the predicted accuracy was calculated. Therefore using leave-one-out method, all of the correct classifications were done

TABLE 3: Standardized canonical discriminant function coefficients.

	Function			
	1	2	3	4
[Si I (243.51)/Li I (670.79)]	−2.391	1.704	−0.864	−3.105
[Si I (252.41)/Li I (670.79)]	7.641	0.661	0.531	3.279
[Si I (252.85)/Li I (670.79)]	−4.620	−4.877	2.149	−4.737
[Si I (263.12)/Li I (670.79)]	0.609	0.161	−0.497	0.483
[Si I (288.15)/Li I (670.79)]	−0.768	3.392	−1.300	4.151
[Si I (243.51)/Na I (588.99 + 589.59)]	−3.079	2.997	6.506	0.615
[Si I (252.41)/Na I (588.99 + 589.59)]	5.787	−8.814	−2.952	−2.072
[Si I (288.15)/Na I (588.99 + 589.59)]	−2.749	5.610	−2.784	1.913

TABLE 4: Predicted group membership of quartz and its classification.

			Predicted group membership of quartz					
		ID	Lapis	Crystal	Pink	Tourmaline	Amethyst	Total
Original	Count	Lapis	29	0	0	1	0	30
		Crystal	0	30	0	0	0	30
		Pink	0	0	30	0	0	30
		Tourmaline	0	0	0	30	0	30
		Amethyst	0	0	0	0	30	30
	%	Lapis	96.8	0.0	0.0	3.2	0.0	100.0
		Crystal	0.0	100.0	0.0	0.0	0.0	100.0
		Pink	0.0	0.0	100.0	0.0	0.0	100.0
		Tourmaline	0.0	0.0	0.0	100.0	0.0	100.0
		Amethyst	0.0	0.0	0.0	0.0	100.0	
Cross-validated[a]	Count	Lapis	29	0	0	1	0	30
		Crystal	0	30	0	0	0	30
		Pink	0	0	30	0	0	30
		Tourmaline	0	0	0	30	0	30
		Amethyst	0	0	0	0	30	30
	%	Lapis	96.8	0.0	0.0	3.2	0.0	100.0
		Crystal	0.0	100.0	0.0	0.0	0.0	100.0
		Pink	0.0	0.0	100.0	0.0	0.0	100.0
		Tourmaline	0.0	0.0	0.0	100.0	0.0	100.0

[a]Cross-validation is done only for those cases in the analysis. In cross-validation, each case is classified by the functions derived from all cases other than that case.
[b]99.3% of original grouped cases correctly classified.
[c]99.3% of cross-validated grouped cases correctly classified.

99.3%, which clearly indicates that the conjunction of DFA with LIBS provides a powerful tool for classification of quartz.

The discriminant function scores were calculated by multiplication of the discriminant coefficients and their corresponding predicted variables as given in (1). In Figure 5, the scores of discriminant function 1 and discriminant function 2 were plotted. The same group member's data were scattered about the mean and differences between groups were large as compared to the variance of within groups which are the measure of that scatter (see Figure 5). The amethyst quartz and crystal quartz are somewhat congested relatively at the two neared focuses as shown and characterized from the other three examples. In other words, the complete characterization of every one of the five different types of the collected quartz is impractical by plotting just two function scores. Therefore, a three-dimensional plot of function 1, function 2, and function 3 scores was exhibited in Figure 6.

Figure 6 demonstrates that the amethyst quartz and crystal quartz were put in the two opposite surfaces of the 3D plot and the other three are practically at the same line due to the missing of the Al I component. The crystal quartz, pink quartz, and amethyst quartz which were having the weak emission lines of Ca I are clearly separated compared to the other two (tourmaline and lapis quartz).

4. Conclusion

The LIBS detection system was improved through finding the suitable delay time between the laser excitation and data acquirement system. LIBS technique in combination

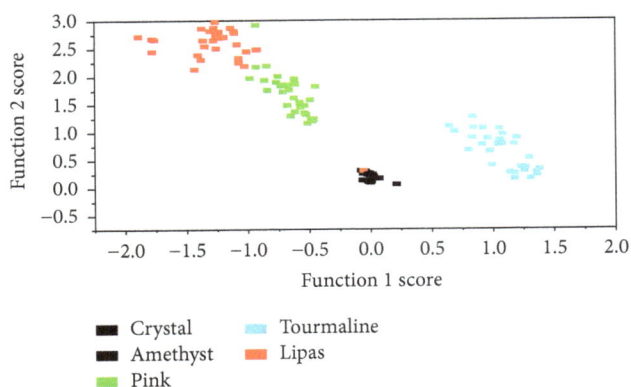

FIGURE 5: Plot showing the comparison between function score 1 and function score 2.

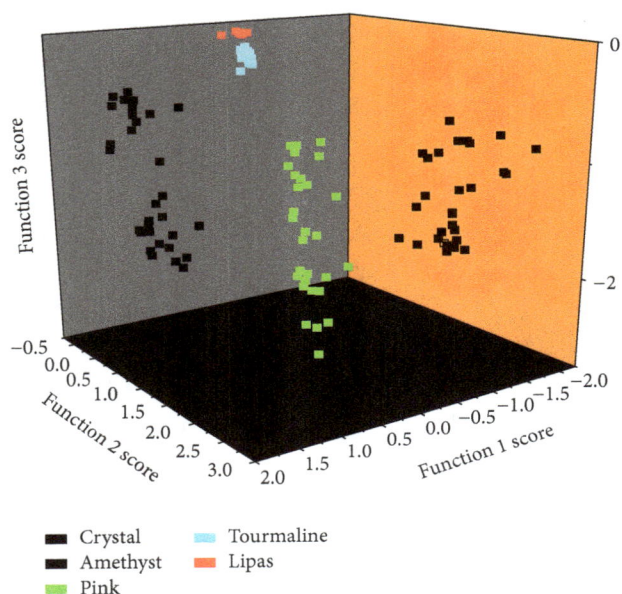

FIGURE 6: 3D plot of all five types of quartz.

References

[1] R. S. Harmon, J. Remus, N. J. McMillan et al., "LIBS analysis of geomaterials: geochemical fingerprinting for the rapid analysis and discrimination of minerals," *Applied Geochemistry*, vol. 24, no. 6, pp. 1125–1141, 2009.

[2] I. D. Delbem, R. Galéry, P. R. G. Brandão, and A. E. C. Peres, "Semi-automated iron ore characterisation based on optical microscope analysis: quartz/resin classification," *Minerals Engineering*, vol. 82, pp. 2–13, 2015.

[3] M. T. Sweetapple and S. Tassios, "Laser-induced breakdown spectroscopy (LIBS) as a tool for in situ mapping and textural interpretation of lithium in pegmatite minerals," *American Mineralogist*, vol. 100, no. 10, pp. 2141–2151, 2015.

[4] C. A. Munson, J. L. Gottfried, E. G. Snyder, F. C. De Lucia Jr., B. Gullett, and A. W. Miziolek, "Detection of indoor biological hazards using the man-portable laser induced breakdown spectrometer," *Applied Optics*, vol. 47, no. 31, pp. G48–G57, 2008.

[5] J. L. Gottfried, F. C. De Lucia Jr., C. A. Munson, and A. W. Miziolek, "Standoff detection of chemical and biological threats using laser-induced breakdown spectroscopy," *Applied Spectroscopy*, vol. 62, no. 4, pp. 353–363, 2008.

[6] M. Banaee and S. H. Tavassoli, "Discrimination of polymers by laser induced breakdown spectroscopy together with the DFA method," *Polymer Testing*, vol. 31, no. 6, pp. 759–764, 2012.

[7] S. J. Rehse, J. Diedrich, and S. Palchaudhuri, "Identification and discrimination of Pseudomonas aeruginosa bacteria grown in blood and bile by laser-induced breakdown spectroscopy," *Spectrochimica Acta—Part B Atomic Spectroscopy*, vol. 62, no. 10, pp. 1169–1176, 2007.

[8] I. Rehan, K. Rehan, S. Sultana, M. O. ul Haq, M. Z. K. Niazi, and R. Muhammad, "Spatial characterization of red and white skin potatoes using nano-second laser induced breakdown in air," *The European Physical Journal Applied Physics*, vol. 73, no. 1, Article ID 10701, 2016.

[9] R. W. Campbell, *A Conceptual Analysis of Introductory Physic Textbooks*, ProQuest, 2008.

[10] Y. Ralchenko, *Memorie della Società Astronomica Italiana*, vol. 3–8, 2005.

[11] H. H. Koehn, *New Mexico Institute of Mining and Technology*, 1977.

[12] S. J. Rehse and Q. I. Mohaidat, "The effect of sequential dual-gas testing on laser-induced breakdown spectroscopy-based discrimination: application to brass samples and bacterial strains," *Spectrochimica Acta Part B: Atomic Spectroscopy*, vol. 64, no. 10, pp. 1020–1027, 2009.

[13] R. A. Fisher, "The use of multiple measurements in taxonomic problems," *Annals of Eugenics*, vol. 7, no. 2, pp. 179–188, 1936.

[14] J. T. Bromberger, L. L. Schott, K. A. Matthews et al., "Association of past and recent major depression and menstrual characteristics in midlife: study of Women's Health Across the Nation," *Menopause*, vol. 19, no. 9, pp. 959–966, 2012.

[15] C. J. Huberty, *Applied Discriminant Analysis*, Wiley, New York, NY, USA, 1994.

with discriminant function analysis (DFA) was employed for the qualitative analysis of five different types of quartz. The strongest predictor in discriminating of quartz was the intensities ratios of Si I (252.41)/Li I (670.79), Si I (252.41)/Na I (588.99 + 589.59), and Si I (252.85)/Li I (670.79) spectral lines. With the conjunction of DFA and LIBS, the group membership of unknown quartz was predicted 99% precisely with 1% predicted error rate. Besides the precise results of LIBS+DFA, it was extremely straightforward, cost effective, and very quick. Therefore, it can be applied for the classification and identification of quartz. The developed technique (LIBS+FDA) was found reliable to investigate the elemental composition of quartz samples.

Competing Interests

The authors have no competing interests.

8

Rapid Detection of Tetracycline Residues in Duck Meat Using Surface Enhanced Raman Spectroscopy

Jinhui Zhao,[1] Ping Liu,[2] Haichao Yuan,[1] Yijie Peng,[1] Qian Hong,[1] and Muhua Liu[1]

[1]*Optics-Electrics Application of Biomaterials Lab, College of Engineering, Jiangxi Agricultural University, Nanchang 330045, China*
[2]*College of Animal Science and Technology, Jiangxi Agricultural University, Nanchang 330045, China*

Correspondence should be addressed to Muhua Liu; suikelmh@sohu.com

Academic Editor: Christoph Krafft

A rapid detection method based on surface enhanced Raman spectroscopy (SERS) was proposed in this paper in order to realize the detection of tetracycline residues in duck meat. Firstly, surface enhanced Raman spectra characteristics of tetracycline aqueous solution, duck meat extract, and duck meat extract containing tetracycline were analyzed. Secondly, the effect of the addition amount of duck meat extract containing tetracycline on SERS intensity and the effect of the adsorption time on SERS intensity were discussed, respectively. Thirdly, SERS intensity ratio at 1272 and 1558 cm^{-1} (I_{1272}/I_{1558}) was used to establish the SERS calibration curve. A good linearity relationship between the tetracycline concentration in duck meat extract and I_{1272}/I_{1558} was obtained, and the linear regression equation and the correlation coefficient (r) were $y = 0.0177x + 0.1213$ and 0.950, respectively. The average recovery of tetracycline in duck meat extract was 101~108% with relative standard deviation (RSD) of 2.4~4.6%. The experimental results showed that the method proposed in this paper was a good detection scheme for the rapid detection of tetracycline residues in duck meat.

1. Introduction

Tetracycline ($C_{22}H_{24}N_2O_8$), which is a kind of broad-spectrum tetracycline antibiotics, is often used to prevent and treat the duck diseases or is added to the duck feedstuff as the additive. However, tetracycline residues content in duck meat easily exceeds the standard within legal limits due to the unsuitable usage of tetracycline, and the excessive tetracycline residues in duck meat may further cause some potential harm to human health, such as kidney toxicity and fetal malformation [1, 2]. Therefore, the regulations for the Maximum Residue Limits (MRL) in meat have been established in China, the EU, and the USA [1, 3]. For instance, tetracycline residues content in duck meat in China and USA cannot exceed 0.1 and 0.2 ppm, respectively [1, 3].

It is highly imperative to detect all the duck products in advance in order to prevent the products containing the tetracycline residues from entering the markets. The conventional detection methods for the antibiotic residues in meat include enzyme-like immunosorbent assay [4–6], microbiological method [7, 8], and physical and chemical

detection method [9]. Although these methods have high detection precision, they cannot meet the rapid, mass, and field detection requirements of tetracycline residues in duck meat owing to the tedious pretreatment process or the time-consuming detection process and so forth. Hence, it is highly needed to investigate a rapid detection method of tetracycline residues in duck meat.

Surface enhanced Raman spectroscopy (SERS) is a highly specific and sensitive method for the detection of chemical and biochemical compounds in chemistry, microbiology, pharmacology, biochemistry, and environmental science [10, 11]. Zhang et al. utilized SERS to analyze prohibited aquaculture drugs including enrofloxacin, furazolidone, and malachite green in fish products [11]. Li et al. applied SERS to detect the residuals of the prohibited and restricted drugs including malachite green and crystal violet in fish muscle, and the lowest concentrations detected were 1.0 and 20 $\mu g/kg$, respectively [12]. Ma et al. used SERS to detect sulfamerazine and sulfamethazine in fish, and the detection limits were 0.16 and 0.59 ppm, respectively [13]. Li et al. applied improved surface enhanced Raman scattering on microscale Au hollow spheres to detect different concentrations of tetracycline aqueous solution, and a good linear response was obtained [14]. Compared to the SERS detection of tetracycline aqueous solution, the SERS detection condition of tetracycline residues in duck meat is more complicated. For instance, the composition of duck meat can have effect on intensities of tetracycline SERS characteristic peaks; even some tetracycline SERS characteristic peaks disappear. Moreover, some tetracycline SERS characteristic peaks can overlap with SERS peaks of the composition of duck meat. So far, the relevant literature about the detection of tetracycline residues in duck meat using SERS was not reported.

The objective of this paper was to investigate the rapid detection method of tetracycline residues in duck meat using SERS. Firstly, SERS spectra characteristics of tetracycline aqueous solution, duck meat extract, and duck meat extract containing tetracycline were analyzed. Secondly, the effect of the addition amount of duck meat extract containing tetracycline on SERS intensity (peak height) and the effect of the adsorption time on SERS intensity were explored, respectively. Lastly, the linear regression equation between the tetracycline concentration in duck meat extract and SERS intensity ratio at 1272 and 1558 cm^{-1} (I_{1272}/I_{1558}) was established to predict the tetracycline residues in duck meat.

2. Materials and Methods

2.1. Materials and Reagents. Sheldrakes were purchased from the vegetable market of Jiangxi Agricultural University. Tetracycline (98.0%) was purchased from Standard Substances Network of China. OTR202 (gold nanoparticles), and OTR103 (gold colloid enhancement reagent) were purchased as SERS enhancement substrate from Opto Trace Technologies, Inc. Ethyl acetate was analytical reagent grade. Ultrapure water was obtained using T10 laboratory water purifier (Hunan Kertone Water Treatment Co., Ltd.) for the preparation of all aqueous solutions. All chemicals are of analytical reagent grade and used without further purification.

2.2. Instruments. RamTracer®-200 portable Raman spectrometer (Opto Trace Technologies, Inc.), T6 series UV-Vis spectrophotometer (Beijing Purkinje General Instrument Co., Ltd.), JK-50B ultrasonic cleaner (Hefei Jinnike Machinery Co., Ltd.), FA1004B electronic weigher (Shanghai Shangping instrument Co., Ltd.), T10 laboratory water purifier, JW-1024 low speed centrifuge (Anhui Jiaven Equipment Industry Co., Ltd.), JJ-2B tissue disintegrator (Jintan Jinnan Instrument Factory), and VORTEX-5 whirlpool mixer (Haimen Kylin-Bell Lab Instruments Co., Ltd.) were used in this study.

2.3. Experimental Methods

(1) Preparation of duck meat extract: firstly, the breast meats were removed from the sheldrakes and the membranes were eliminated from the breast meats for the following experiment. Secondly, 200 g of duck breast meat was crushed in meat emulsion by using tissue disintegrator, and then the meat emulsion was cryopreserved below −18°C. Thirdly, 5 g of meat

emulsion and 20 mL of ethyl acetate were added into 50 mL centrifuge tube every time. Fourthly, the supernatant was taken out after whirlpool mixing for 2 minutes, oscillating for 10 minutes, and centrifuging (4500 r/min) for 15 minutes. Fifthly, the residues were extracted for two times using the same method, and then two extracted supernatants were mixed together. Lastly, the supernatant was filtered with rapid filter papers.

(2) Preparation of tetracycline standard solution: 5 mg of tetracycline was added into 50 mL brown volumetric flask, and then 100 mg/L tetracycline standard solution was obtained after ultrasonic dissolving with ultrapure water. Subsequently, 100 mg/L tetracycline standard solution was diluted to different concentrations with ultrapure water during the course of the experiment.

(3) Preparation of spiked samples: 5 mg of tetracycline was added into 50 mL brown volumetric flask, and then the spiked sample containing 100 mg/L tetracycline was obtained after ultrasonic dissolving with duck meat extract. Subsequently, the spiked sample containing 100 mg/L tetracycline was diluted to different tetracycline concentrations by using duck meat extract in the following experiment.

(4) 500 μL of OTR202, 20 μL of the analyzed solution, and 100 μL of OTR103 were added to 2 mL sample bottle in turn, and then SERS spectra were collected after this system was well blended. SERS spectra were collected 5 times for each sample, and their average intensities were used as original SERS spectrum intensities of each sample. Spectral range of 400~1800 cm^{-1} was used as the following analysis of SERS spectra.

(5) The parameters of portable Raman spectrometer were set as follows: laser power was 200 mW, laser wavelength was 785 nm, the scanning range of Raman wavelength was 100~3300 cm^{-1}, spectral resolution was 6 cm^{-1}, the range of Raman intensity was 0~60000 a.u., and the integration time was 10 sec.

3. Results and Discussion

3.1. Pretreatment of SERS Spectra. The original SERS spectrum of duck meat extract containing 12.0 mg/L tetracycline was showed in Figure 1(a). As seen from Figure 1(a), the original SERS spectrum included not only Raman signal of the sample detected but also fluorescence and all other kinds of background signals. Thus, it was indispensable to reduce the effect of the fluorescence and all other kinds of background signals on SERS analysis in order to improve the reliability of SERS qualitative, semiquantitative, or quantitative analysis. Adaptive iteratively reweighted penalized least squares (air-PLS), which need not any user intervention and initial information, is a highly effective background subtraction method [15, 16]. So, air-PLS was employed to remove the fluorescence and all other kinds of background signals in

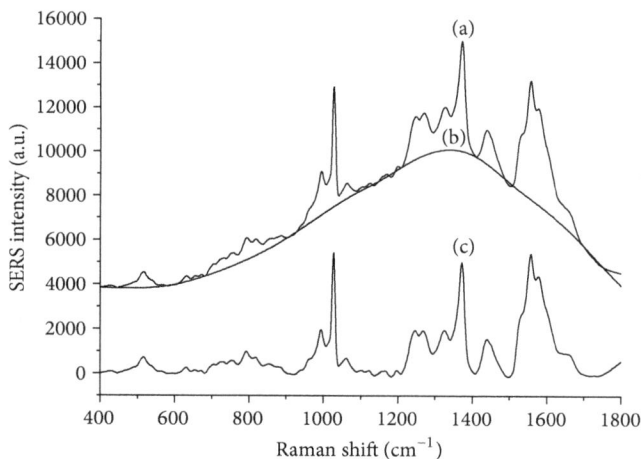

FIGURE 1: Result of background subtraction using air-PLS. (a) Original SERS spectrum; (b) background signals fitted; (c) SERS spectrum subtracted by background signals.

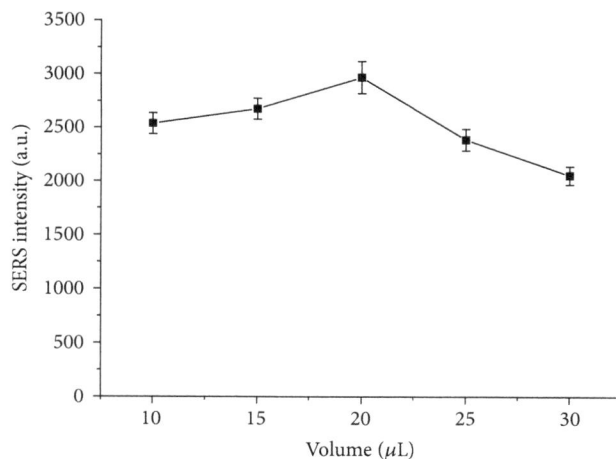

FIGURE 2: SERS spectra of (a) tetracycline aqueous solution, (b) duck meat extract, and (c) duck meat extract containing tetracycline. The inset shows that tetracycline is a derivative of naphthacene according to the chemical structure of tetracycline.

this paper. As showed from Figure 1(c), the fluorescence and all other kinds of background signals on curve (a) were subtracted perfectly by using air-PLS under the condition of keeping the peak shapes of SERS spectra.

3.2. SERS Spectra Characteristics of Samples.

The OTR202 has a strong absorption peak at 544 nm according to UV-Vis absorption spectra of OTR202, which is the typical surface plasmon resonance absorption of gold nanoparticles. Moreover, its half-peak width is approximately 79 nm.

SERS spectra of tetracycline aqueous solution, duck meat extract, and duck meat extract containing tetracycline were showed in Figure 2. The major SERS characteristic peaks of tetracycline aqueous solution could be observed at 520, 792,

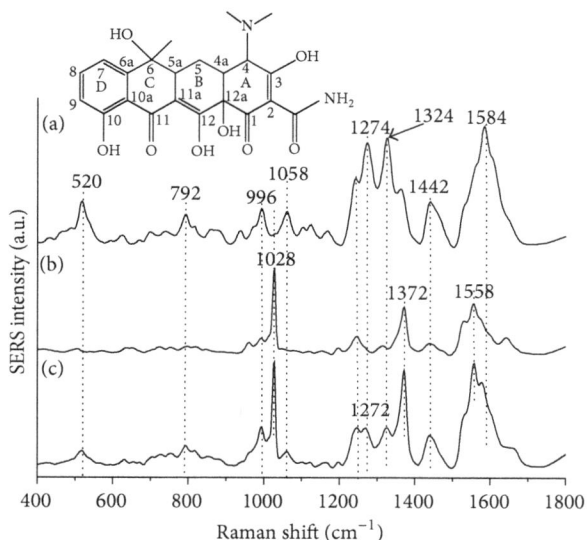

FIGURE 3: Effect of the addition amount of duck meat extract containing tetracycline on SERS intensity.

996, 1058, 1274, and 1584 cm^{-1} (± 3 cm^{-1}) in Figure 2(a). SERS characteristic peak of tetracycline aqueous solution at 1274 cm^{-1} has blue-shifted to 1272 cm^{-1} on SERS spectrum of duck meat extract containing tetracycline, which was probably caused by the charge transfer that took place between gold nanoparticles and tetracycline [13]. The main SERS characteristic peaks of duck meat extract could be seen at 1028, 1372, and 1558 cm^{-1} (± 3 cm^{-1}) in Figure 2(b). As showed in Figure 2(c), SERS characteristic peaks of tetracycline aqueous solution at 520, 1058, and 1274 cm^{-1}, which did not overlap with SERS characteristic peaks of duck meat extract, appeared on SERS spectrum of duck meat extract containing tetracycline. So, SERS characteristic peaks at 520, 1058, and 1272 cm^{-1} could be selected as SERS characteristic peaks for the detection of tetracycline residues in duck meat extract. Tetracycline is a derivative of naphthacene according to the chemical structure of tetracycline in Figure 2(a). The characteristic peak at 1058 cm^{-1} was attributed to stretching vibration of CO3 [17, 18]. The characteristic peak at 1272 cm^{-1} was ascribed to bending vibration of CH4,4a,5,5a, OH12, and amid-NH and stretching vibration of CO10, CO3, CH7,8,9, amid-NC, C4aC5, and benzene ring D [17, 18]. In conclusion, it was entirely feasible, at least in theory, to detect tetracycline residues in duck meat using SERS according the above analysis.

3.3. Effect of Addition Amount of Samples on SERS Intensity.

To analyze the effect of the addition of the samples on SERS intensity, SERS intensity at 1272 cm^{-1} was investigated under the condition of the different addition amounts (10, 15, 20, 25, and 30 μL) of duck meat extract containing tetracycline and the fixed volumes of OTR202 (500 μL) and OTR103 (100 μL). As showed from Figure 3, SERS intensities at 1272 cm^{-1} were different when different volumes of duck meat extract containing tetracycline were mixed with gold nanoparticles. The possible reason was that their mixture ratio had the significant impact on adsorption effect and then affected SERS intensity. SERS intensity at 1272 cm^{-1} was strongest

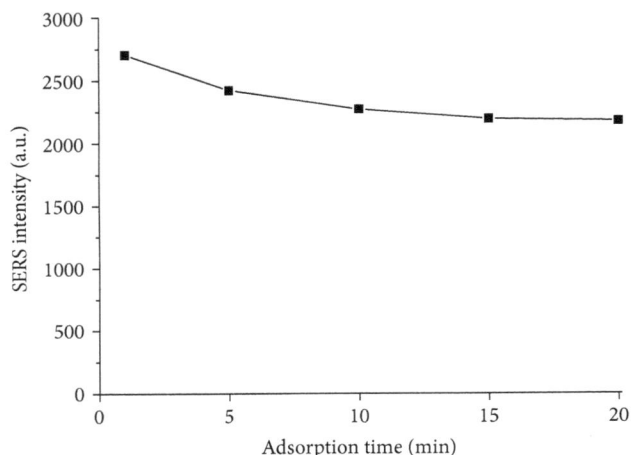

FIGURE 4: Effect of adsorption time on SERS intensity.

TABLE 1: Predicted results of tetracycline in duck meat ($n = 5$).

Spiked ($mg \cdot L^{-1}$)	Detected ($mg \cdot L^{-1}$)	Recovery (%)	Relative standard deviation (%)
10	10.639 ± 0.321	106.39	4.597
15	16.201 ± 0.431	108.01	3.510
20	20.114 ± 0.354	100.57	2.376

while the addition amount of duck meat extract containing tetracycline was 20 μL. This showed that adsorption effect on the detected sample was best when the volume of duck meat extract containing tetracycline was 20 μL. Therefore, 20 μL was determined as the addition amount of duck meat extract containing tetracycline.

3.4. Effect of Adsorption Time on SERS Intensity. In order to study the effect of adsorption time on SERS intensity, the change of SERS intensity at 1272 cm^{-1} with adsorption time was explored on the condition of 500 μL of OTR202, 20 μL of the same concentration of the sample detected, and 100 μL of OTR103. With the increases of adsorption time between duck meat extract containing tetracycline and SERS active substrate, SERS intensities of the mixture solution at 1272 cm^{-1} gradually reduced (Figure 4). This might be because the aggregation was produced to some extent after gold nanoparticles were mixed with duck meat extract containing tetracycline, and the enhancement effects, which were produced by combining active hotspots determining SERS enhancement effects with duck meat extract containing tetracycline, decreased gradually along with the increases of adsorption time. SERS intensity at 1272 cm^{-1} was strongest when adsorption time was 1 min. This indicated that SERS enhancement effects, which were produced by combining active hotspots with duck meat extract containing tetracycline, were best when adsorption time was 1 min. Therefore, 1 min was ascertained as the best adsorption time.

3.5. SERS Calibration Curve and Predicted Results. SERS spectra of duck meat extract containing different tetracycline concentrations were showed in Figure 5. SERS characteristic peak of tetracycline aqueous solution at 1272 cm^{-1} could be observed on SERS spectrum of duck meat extract containing tetracycline (4 mg/L), and the peak position at 1272 cm^{-1} was stable. This illustrated that OTR202 and OTR103 could be used as SERS active substrate for the detection of tetracycline residues in duck meat. Thus, 1272 cm^{-1} could be selected as the characteristic peak of semiquantitative or quantitative analysis of tetracycline residues in duck meat.

In order to decrease the effects of external factors on SERS intensity at 1272 cm^{-1} and take full advantage of ratio effect, SERS intensity ratio at 1272 and 1558 cm^{-1} (I_{1272}/I_{1558}) was used to establish the SERS calibration curve between the tetracycline concentration in duck meat extract and I_{1272}/I_{1558}. SERS characteristic peak at 1558 cm^{-1} was caused by duck meat extract. As showed from the inset in Figure 5, the linear regression equation was $y = 0.0177x + 0.1213$, and the correlation coefficient (r) was 0.950, in which x and y represented the tetracycline concentration in duck meat extract and I_{1272}/I_{1558}, respectively. This illustrated that a good linearity relationship between the tetracycline concentration (4~25 mg/L) in duck meat extract and I_{1272}/I_{1558} was obtained.

The SERS calibration curve obtained from above was applied to predict the tetracycline concentration in duck meat extract, and the predicted results were showed in Table 1. The average recovery of tetracycline in duck meat extract was 101~108% with the relative standard deviation (RSD) of 2.4~4.6%, which indicated that this method had a good precision and satisfactory reproducibility. The detection limit of tetracycline in duck meat extract could reach 1.120 mg/L ($S/N = 3$). The experimental results showed that the pretreatment method adopted in this paper was more simple and rapid than the conventional detection methods such as enzyme-like immunosorbent assay, microbiological method, and physical and chemical detection method. However, this method still needs to be improved owing to the disadvantage of its high detection limit, which is possibly because surface enhancement factor of SERS substrate is affected by many factors, such as particle size, shape, and uniformity of the distribution of nanoparticles [19], and the composition of duck meat can decrease intensities of tetracycline SERS characteristic peaks in duck meat extract. So, it will be a good selection, in future studies, to explore a pretreatment method with less effect on intensities of tetracycline SERS characteristic peaks in duck meat extract or a nanoparticle with the better enhancement effect to decrease the detection limit. Also, the method proposed in this paper could provide the important technical support for achieving the rapid detection of tetracycline residues in duck meat.

4. Conclusions

A simple and rapid method using OTR202 and OTR103 as SERS enhancement substrate was developed for the detection of tetracycline residues in duck meat. The effects of addition amount of samples and adsorption time on SERS intensity were analyzed, respectively. A good linearity relationship between the tetracycline concentration in duck meat extract

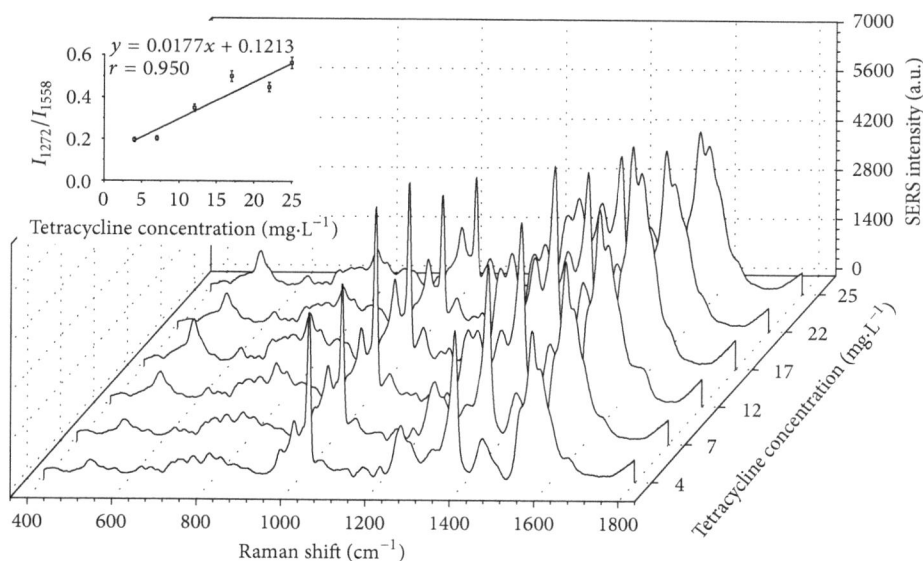

FIGURE 5: SERS spectra of duck meat extract containing different tetracycline concentrations. The inset shows the linear relationship between the tetracycline concentration and SERS intensity ratio at 1272 and 1558 cm^{-1} (I_{1272}/I_{1558}).

and I_{1272}/I_{1558} was obtained, and the linear regression equation and r were $y = 0.0177x + 0.1213$ and 0.950, respectively. The pretreatment method in this paper was simple and rapid, and the detection limit of tetracycline in duck meat extract could reach 1.120 mg/L. Therefore, the method proposed in this paper was a good detection scheme for the rapid detection of tetracycline residues in duck meat.

Competing Interests

The authors declare that they have no competing interests.

Authors' Contributions

Jinhui Zhao and Ping Liu contributed equally to this work and should be considered co-first authors.

Acknowledgments

This research was supported by the National Natural Science Foundation of China (Grant no. 31660485), Science and Technology Support Project of Jiangxi Province, China (Grant no. 2012BBG70058), and External Science and Technology Cooperation Plan of Jiangxi Province, China (Grant no. 20132BDH80005).

References

[1] X. D. Gao, S. E. Chen, Y. L. Ye, X. J. Wang, J. K. Deng, and S. C. Wei, "Determination of terramycin, minocycline and aureomycin in livestock, poultry meat and salmon by high performance liquid chromatography," *Journal of Food Safety and Quality*, vol. 5, no. 2, pp. 369–376, 2014.

[2] J.-H. Zhao, H.-C. Yuan, M.-H. Liu, H.-B. Xiao, Q. Hong, and J. Xu, "Rapid determination of tetracycline content in duck meat

using particle swarm optimization algorithm and synchronous fluorescence spectrum," *Spectroscopy and Spectral Analysis*, vol. 33, no. 11, pp. 3050–3054, 2013.

[3] General Adminstration of Quality Supervision, Inspection and Quarantine of the People's of China and Standardization Administration of the People's Republic of China, 'Fresh and frozen poultry product', GB16869-2005, 2005.

[4] Y. Zhou, C.-Y. Li, Y.-S. Li et al., "Monoclonal antibody based inhibition ELISA as a new tool for the analysis of melamine in milk and pet food samples," *Food Chemistry*, vol. 135, no. 4, pp. 2681–2686, 2012.

[5] X. D. Yan, X. Z. Hu, H. C. Zhang, J. Liu, and J. P. Wang, "Direct determination of furaltadone metabolite, 3-amino-5-morpholinomethyl-2-oxazolidinone, in meats by a simple immunoassay," *Food and Agricultural Immunology*, vol. 23, no. 3, pp. 203–215, 2012.

[6] J. Li, J. Liu, H.-C. Zhang, H. Li, and J.-P. Wang, "Broad specificity indirect competitive immunoassay for determination of nitrofurans in animal feeds," *Analytica Chimica Acta*, vol. 678, no. 1, pp. 1–6, 2010.

[7] O. G. Nagel, M. C. Beltrán, M. P. Molina, and R. L. Althaus, "Novel microbiological system for antibiotic detection in ovine milk," *Small Ruminant Research*, vol. 102, no. 1, pp. 26–31, 2012.

[8] P. K. Dang, G. Degand, S. Danyi et al., "Validation of a two-plate microbiological method for screening antibiotic residues in shrimp tissue," *Analytica Chimica Acta*, vol. 672, no. 1-2, pp. 30–39, 2010.

[9] G. F. Pang, *Compilation of Official Methods of Analysis for Pesticide Residues and Veterinary Drug Residues*, Standards Press of China, Beijing, China, 2009.

[10] Y. Zhang, W. Yu, L. Pei, K. Lai, B. A. Rasco, and Y. Huang, "Rapid analysis of malachite green and leucomalachite green in fish muscles with surface-enhanced resonance Raman scattering," *Food Chemistry*, vol. 169, pp. 80–84, 2015.

[11] Y. Y. Zhang, Y. Q. Huang, F. L. Zhai, R. Du, Y. D. Liu, and K. Q. Lai, "Analyses of enrofloxacin, furazolidone and

malachite green in fish products with surface-enhanced Raman spectroscopy," *Food Chemistry*, vol. 135, no. 2, pp. 845–850, 2012.

[12] C. Li, K. Lai, Y. Zhang, L. Pei, and Y. Huang, "Use of surface-enhanced Raman spectroscopy for the test of residuals of prohibited and restricted drugs in fish muscle," *Acta Chimica Sinica*, vol. 71, no. 2, pp. 221–226, 2013.

[13] H. K. Ma, X. H. Han, C. H. Zhang, X. Zhang, X. F. Shi, and J. Ma, "The study of sulfonamide antibiotics in fish based on surface-enhanced Raman spectroscopy technology," *Acta Laser Biology Sinica*, vol. 23, no. 6, pp. 560–565, 2014.

[14] R. Li, H. Zhang, Q.-W. Chen, N. Yan, and H. Wang, "Improved surface-enhanced Raman scattering on micro-scale Au hollow spheres: synthesis and application in detecting tetracycline," *Analyst*, vol. 136, no. 12, pp. 2527–2532, 2011.

[15] C. X. Fang, J. H. Li, and Y. Z. Liang, "Determination of MTBE in gasoline by raman spectroscopy combined with baseline correction method," *Journal of Instrumental Analysis*, vol. 31, no. 5, pp. 541–545, 2011.

[16] X. Li and Y. Lü, "Background subtraction in raman measurement of ethanol concentration," *Journal of Beijing Information Science and Technology University*, vol. 28, no. 2, pp. 27–30, 2013.

[17] X. Y. Chen, *Study on Antibiotics Raman Spectroscopy Based on Partial Least Squares*, Tianjin University, Tianjin, China, 2012.

[18] C. F. Leypold, M. Reiher, G. Brehm et al., "Tetracycline and derivatives—assignment of IR and Raman spectra via DFT calculations," *Physical Chemistry Chemical Physics*, vol. 5, no. 6, pp. 1149–1157, 2003.

[19] F.-L. Zhai, Y.-Q. Huang, X.-C. Wang, and K.-Q. Lai, "Surface-enhanced Raman spectroscopy for rapid determination of β-agonists in swine urine," *Chinese Journal of Analytical Chemistry*, vol. 40, no. 5, pp. 718–723, 2012.

9

Fourier Transform Infrared Spectroscopy of "Bisphenol A"

Ramzan Ullah,[1,2] **Ishaq Ahmad,**[2] **and Yuxiang Zheng**[1]

[1]*Shanghai Ultra-Precision Optical Manufacturing Engineering Center, Department of Optical Science and Engineering, Fudan University, Shanghai 200433, China*
[2]*Department of Physics, COMSATS Institute of Information Technology, Islamabad 45550, Pakistan*

Correspondence should be addressed to Ramzan Ullah; ramzanullah@gmail.com and Yuxiang Zheng; yxzheng@fudan.edu.cn

Academic Editor: Eugen Culea

FTIR (400–$4000 \, \text{cm}^{-1}$) spectra of "Bisphenol A" are presented. Absorption peaks (400–$4000 \, \text{cm}^{-1}$) are assigned on the basis of Density Functional Theory (DFT) with configuration as B3LYP 6-311G++ (3df 3pd). Calculated absorption peaks are in reasonable reconciliation with experimental absorption peaks after scaling with scale factor of 0.9679 except C-H and O-H stretching vibrations.

1. Introduction

"Bisphenol A" is a carbon based synthetic compound used primarily in the plastic industry and epoxy resins. BPA mimics hormone-like behavior which questions upon its use in food containers and baby bottles and so forth. BPA has been controversially associated with a number of adverse health effects including neurotoxicity, genotoxicity, and carcinogenicity [1]. BPA ($C_{15}H_{16}O_2$) molecule is shown in Figure 1. BPA has been used tremendously in many areas like optical media, food containers, medical devices, electronics, protective coatings, and automotive industry are very few to mention. Besides that, the debate over the use of BPA for public health safety continues where US Food and Drug Administration (FDA) considers BPA totally safe for food containers and packaging [2] but some other researches oppose this idea [3]. In this state of perplexity, banning the use of BPA in some applications mostly related to foods and human contact is on the rise [4]. Situation further gets worse when some researches show that "Bisphenol S" is also toxic like BPA which is a common replacement for BPA to bypass the law [5]. Keeping in view its widespread use and significance, its optical properties are very important. Some environmental hormones have already been studied by THz, Raman, and FTIR spectroscopy [6]. BPA has also been studied by THz spectroscopy where one of its vibrational modes was observed but in FTIR spectroscopy, main focus was on

CH and OH stretching [7]. Various molecular vibrations of BPA have been identified through Raman spectroscopy by our research group [8]. However, Raman spectroscopy and FTIR spectroscopy are complementary to each other. To get a complete picture of the molecular dynamics, both Raman and FTIR spectroscopic studies are essential. So, here we present a complementary FTIR spectroscopic study of BPA which will help a lot to understand the behavior of this molecule to a wide extent. The goal of this study is to find molecular vibrations of BPA in the Infrared regime which will pave the way to a better understanding of the interaction of this molecule and in finding the origin of its toxicity. This study may also help researchers and health agencies to devise better ways for the characterization and detection of BPA.

2. Experimental Set-Up

Spectrum 100 FTIR [9] from PerkinElmer is used to record the spectra from 400 to $4000 \, \text{cm}^{-1}$. BPA, CAS number (80-5-7), is purchased from "Sinopharm Chemical Reagent Co. Ltd." [10]. Sample as well as Potassium Bromide (KBr) is ground to very fine powder and then a tablet is formed under high pressure after mixing them. This tablet is then put on the optical path of Infrared radiation to take the measurements. All the measurements were taken at room temperature and atmospheric pressure.

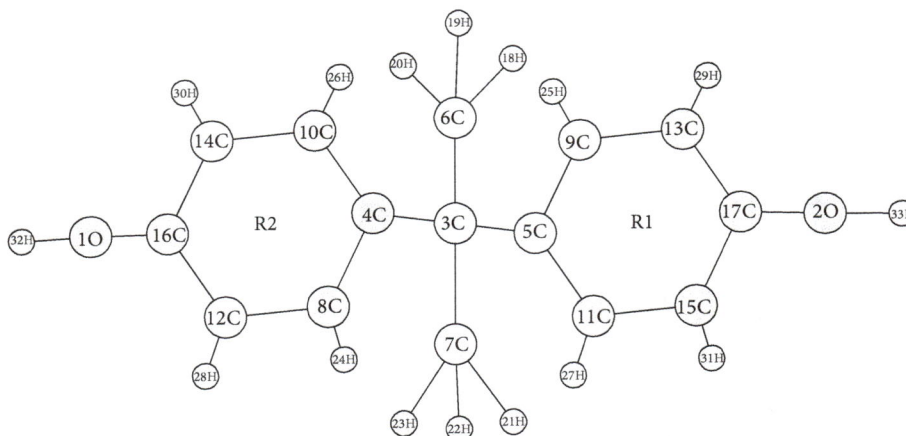

FIGURE 1: Bisphenol A molecule.

3. DFT Calculations

Gaussian 09 package [11] is utilized to carry out DFT related calculations which are optimized to minimum and Ground State method B3LYP with 6-311G basis set [12] is used together with (3df, 3pd) and ++ diffuse functions [13]. MOLVIB [14, 15] is used to calculate Potential Energy Distributions (PEDs) given in Table 1. Another table given in the Supplementary Material contains more information about simulation and values of important parameters obtained like energy, dipole moment and so forth (see Supplementary Material available online at http://dx.doi.org/10.1155/2016/2073613).

4. Results and Discussion

Figure 2 shows experimental spectra of "Bisphenol A" in the range 400–1700 cm^{-1} along with calculated DFT spectra in absorbance mode for easy comparison. Absorption peaks are reasonably in good agreement in the region 400 to 1700 cm^{-1} after scaling with scale factor of 0.9679 for vibrational frequencies calculated by Andersson and Uvdal [16]. However, the large deviation of experimental C-H and O-H stretching vibrations with the DFT peaks as shown in Figure 3 has been addressed by Ullah et al. [7].

Experimental peak of 531 cm^{-1} is assigned to 543 cm^{-1} B3LYP which is a combination of different motions including CO, CC out of plane bending, and CCC interactions in both the rings and otherwise. 531 cm^{-1} experimental absorption peak is not clearly visible in Figure 2 which lies behind 552 cm^{-1}. It is clear from Table 1 that 552 cm^{-1} and 564 cm^{-1} experimental peaks are mainly due to CO, CC, and CH out of plane bending along with little amount of torsions in both the rings and hence are assigned to 563 cm^{-1} and 572 cm^{-1} (B3LYP), respectively, as shown in Figure 2. FTIR absorption peak of 721 cm^{-1}, which, according to Table 1, is assigned to simulated value of 734 cm^{-1} and corresponding PED analysis shows that the major contribution in this vibrational mode comes from the torsions in both the rings and CC out of plane bending. Absorption peak of 734 cm^{-1} is assigned to 768 cm^{-1} which is caused by CC stretching. Next absorption

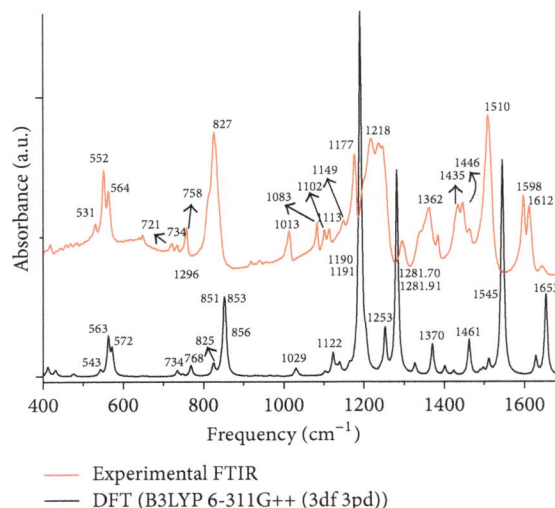

FIGURE 2: FTIR (400–1700 cm^{-1}) spectra of "Bisphenol A" along with DFT spectra. Values of some peaks are given near the tip of the respective peak.

peak, 758 cm^{-1} which is clearly marked with its value in Figure 2, is really interesting to see that it is due to the CH out of plane bending motion and we can clearly see that there is only one experimental peak with this value but B3LYP shows two distinct peaks of 825 cm^{-1} and 826 cm^{-1} in Table 1. Hence, we assign 758 cm^{-1} to both of these values. Moreover, this CH out of plane bending does not come from single carbon and hydrogen atoms. It is actually the combination of all the CH out of plane bending motions in both the rings. A relatively high absorption peak of 827 cm^{-1} (Figure 2) is assigned to 4 simulated vibrational modes of 844, 851, 853, and 856 cm^{-1} due to their closeness. The major contribution in 844 cm^{-1} mode is from CO stretching and CC stretching in both the rings. The other three come mainly from CH out of plane bending movements. So the experimental peak 827 cm^{-1} might belong to any of them. In a similar fashion, 1013 cm^{-1} is assigned to 3 simulated values of 1026, 1029, and 1033 cm^{-1}. Interestingly, the first and

TABLE 1: Experimental as well as DFT absorption peaks along with PEDs.

Sr. number	Experimental FTIR (cm^{-1})	Simulated (B3LYP) cm^{-1}	PED (%)
1	531	543	ν CCC4 (36), γ COob (12), ν CCCR2 (9), ν CCCR1 (9), γ CCob (5)
2	552	563	γ COob (28), γ CCob (15), ν CCC4 (14), γ CHob (12), τ R1t (7), τ R2t (6)
3	564	572	γ COob (34), γ CHob (15), γ CCob (13), ν CCC4 (12), τ R2t (9), τ R1t (8)
4	721	734	τ R1t (23), τ R2t (23), γ COob (20), γ CCob (19), γ CHob (7)
5	734	768	ν CC4 (26), ν CCC4 (20), ν CO (8), δ CCHb (6), ν CCHa (6), ν CCCR1 (5)
6	758	825	γ CHob (95)
7	758*	826	γ CHob (95)
8	827	844	ν CO (23), ν CCR2 (14), ν CCR1 (14), ν CCCR2 (12), ν CCCR1 (12), ν CC4 (6)
9	827*	851	γ CHob (43), ν CO (9), ν CCR2 (9), ν CCR1 (8), γ COob (6)
10	827*	853	γ CHob (70), γ COob (11)
11	827*	856	γ CHob (52), γ COob (8), ν CCR1 (5), ν CO (5), ν CCR2 (5)
12	1013	1026	ν CCHa (28), ν CCHb (28), ν CCCR1 (8), ν CCCR2 (7), ν CCR1 (7), ν CCR2 (6)
13	1013*	1029	ν CCCR1 (21), ν CCCR2 (21), ν CCR1 (15), ν CCR2 (15), ν CCHR1 (12), ν CCHR2 (11)
14	1013*	1033	ν CCHa (16), ν CCHb (16), ν CCCR2 (13), ν CCCR1 (11), ν CCR2 (11), ν CCR1 (9)
15	1083	1122	ν CC4 (26), ν CCR1 (12), ν CCR2 (12), ν CCHR1 (9), ν CCHR2 (9), ν CCC4 (7)
16	1102	1132	ν CCR2 (15), ν CCR1 (15), ν CCHR2 (15), ν CCHR1 (15), ν CC4 (10), ν CCHb (7)
17	1113	1139	ν CCHR2 (16), ν CCHR1 (16), ν CCR2 (14), ν CCR1 (14), ν CCHa (9), ν CCHb (9)
18	1149	1164	ν CC4 (29), ν CCC4 (17), ν CCHR2 (12), ν CCHR1 (12), ν CCR2 (7), ν CCR1 (7)
19	1177	1190	ν COH (55), ν CCR1 (12), ν CCR2 (8), ν CO (7), ν CCHR1 (7)
20	1177*	1191	ν COH (56), ν CCR2 (13), ν CCR1 (10), ν CCHR2 (7), ν CO (6), ν CCHR1 (5)
21	1218	1253 (Tentative)	ν CC4 (45), ν CCC4 (13), ν CCHa (8), ν CCHb (8), ν CCR1 (6), ν CCR2 (6)
22	1218*	1281.70 (Tentative)	ν CO (55), ν CCR2 (10), ν CCR1 (9), ν CCHR2 (8), ν CCHR1 (7)
23	1218*	1281.91 (Tentative)	ν CO (54), ν CCR1 (9), ν CCHR1 (9), ν CCR2 (8), ν CCHR2 (8)
24	1296	1326	ν CCR1 (31), ν CCR2 (29), ν CCC4 (12), ν CCHR1 (9), ν CCHR2 (8)
25	1362	1370.15	ν CCHR2 (32), ν CCHR1 (20), ν COH (16), ν CCR2 (16), ν CCR1 (9)
26	1362*	1370.41	ν CCHR1 (33), ν CCHR2 (20), ν COH (17), ν CCR1 (16), ν CCR2 (10)
27	1384	1401	δ HCHb (23), δ CCHb (23), ν HCHa (22), ν CCHa (22), ν CC4 (5)
28	1435	1461	ν CCR2 (26), ν CCHR2 (24), ν CCR1 (11), ν CCHR1 (11), ν CCC4 (10), ν COH (9)
29	1446	1462	ν CCR1 (26), ν CCHR1 (25), ν CCR2 (11), ν CCHR2 (11), ν COH (9), ν CCC4 (8)
30	1463	1510.8	ν HCHa (44), ν HCHb (44)
31	1463*	1511.55	ν HCHb (46), ν HCHa (45)
32	1510	1545	ν CCHR2 (26), ν CCHR1 (24), ν CCR2 (16), ν CCR1 (14), ν CO (8)
33	1510*	1548	ν CCHR1 (26), ν CCHR2 (24), ν CCR1 (16), ν CCR2 (15), ν CO (8)
34	1598	1628	ν CCR1 (33), ν CCR2 (33), ν CCHR1 (7), ν CCHR2 (7), ν COH (6)
35	1612	1653	ν CCR2 (39), ν CCR1 (22), ν CCHR2 (14), ν CCHR1 (8), ν CCCR2 (6)
36	1612*	1654	ν CCR1 (39), ν CCR2 (22), ν CCHR1 (14), ν CCHR2 (8), ν CCCR1 (6)
37	2870	3029	ν CH3a (59), ν CH3b (41)
38	2870*	3034	ν CH3b (59), ν CH3a (41)
39	2933	3093	ν CH3a (57), ν CH3b (42)
40	2964	3098	ν CH3b (56), ν CH3a (44)
41	2975	3104	ν CH3a (77), ν CH3b (22)
42	2975*	3105	ν CH3b (79), ν CH3a (20)
43	3337	3833.87	ν OH (100)
44	3337*	3834.41	ν OH (100)

*Multiple assignments.

Ob: out of plane bending; t and τ: torsion. R1: Ring 1, R2: Ring 2, a: Label a, b: Label b, ν: stretching, δ: in-plane deformation, γ: out-of-plane deformation, and R: ring.

Scale factor for DFT B3LYP 6-311++G (3df, 3pd) is 0.9679.

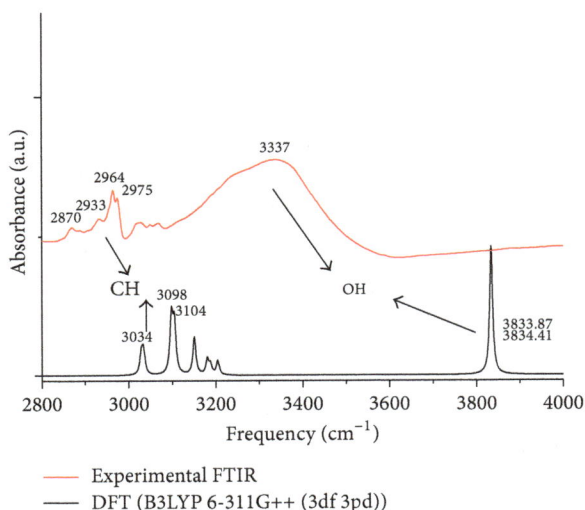

FIGURE 3: FTIR (2800–4000 cm^{-1}) spectra of "Bisphenol A" along with DFT spectra. Values of some peaks are given near the tip of the respective peak.

last peaks (1026 and 1033 cm^{-1}) have similar contributions as given by PED analysis in Table 1 but the middle one 1029 cm^{-1} has different motions. However, they are too close to each other that DFT also suggests a single experimental peak in such circumstance. So 1013 cm^{-1} has once again multiple assignments. 1083 cm^{-1} and 1102 cm^{-1} are related to B3LYP values of 1122 cm^{-1} and 1132 cm^{-1}, respectively, which are contributed mostly by CC stretching in both the rings. 1113 cm^{-1} and 1149 cm^{-1} are hereby assigned to 1139 cm^{-1} and 1164 cm^{-1}, respectively, according to Table 1 and relative PED analysis shows that these modes are mainly because of CCH and CC interactions in the molecule. Multiple assignment of the experimental peak of 1177 cm^{-1} to both 1190 cm^{-1} and 1191 cm^{-1} is due to their nearness. The respective PED analysis shows that the interactions are similar in such a way that for one DFT peak 1190 cm^{-1}, it is the motion in one ring and for the other 1191 cm^{-1}, it is the same motion but in the second ring. Since their frequencies are bit different, DFT gives two peaks but experimentally, it shows only one as shown in Figure 2. Immediately next to the 1177 cm^{-1} peak, there is rather a broadband peak with small fluctuations in Figure 2. Its value is given in Table 1 as 1218 cm^{-1} which is the highest point along the broad peak. If we compare it with DFT spectra, a small peak of 1253 cm^{-1} might be assigned to it. However, due to broadband nature of this curve, it is hard to assign it with certain degree of confidence. Moreover, DFT spectra also show a rather high peak adjacent to 1253 cm^{-1}, as given in Figure 2, which is actually two peaks (1281.70 cm^{-1} and 1281.91 cm^{-1}) according to Table 1. These two peaks can also be assigned to 1218 cm^{-1}. However, the shape of the experimental curve does not show a good behavior which might be due to some defect and hence tentative assignment is given here. PED analysis shows the former as a result of CC stretching and the last two from CO stretching. 1296 cm^{-1} is assigned to 1326 cm^{-1} which is

from CC stretching in two rings. 1362 cm^{-1} is assigned to two absorption peaks of 1370.15 cm^{-1} and 1370.41 cm^{-1} resulting from CCH interactions in the rings but with different configurations. In a similar fashion, 1384 cm^{-1} can be assigned to 1401 cm^{-1} originating from HCH and CCH interactions. 1435 cm^{-1} and 1446 cm^{-1} are attributed to 1461 cm^{-1} and 1462 cm^{-1}, respectively, resulting from CC and CCH in the rings. 1463 cm^{-1} which is not very clear in Figure 2 is assigned to two DFT peaks, 1510.8 cm^{-1} and 1511.55 cm^{-1}, both of which are resulting from HCH interactions (not occurring in the rings) in different ways as shown in PED analysis. A sharp absorption peak of 1510 cm^{-1} in Figure 2 goes to 1545 cm^{-1} and 1548 cm^{-1} where CCH and CC modes are prevalent. 1598 cm^{-1} is assigned to calculated frequency of 1628 cm^{-1} and similarly experimental peak of 1612 cm^{-1} to 1653 cm^{-1} and 1654 cm^{-1} (multiple). Spectra end here and after a long pause, peaks start appearing again near 3000 cm^{-1} where C-H and then O-H stretching interactions exist which are clearly marked in Figure 3. The detailed analysis and assignment of these modes can be found in [6]. Spectra in the range 1700 to 2800 cm^{-1} are not shown due to nonexistence of any peak.

Simulated values of B3LYP given in Table 1 are raw values. These values can be corrected by using a scale factor of 0.9679 as discussed earlier in this section. DFT values are rounded off to nearest whole number. Only major peaks are labeled in Figures 2 and 3. The detail of the absorption peaks both experimental as well as calculated is given in Table 1. Assignment is done on the basis of comparison of the relative intensities of the experimental and calculated spectra.

According to DFT calculations, there are 14 C-H stretching vibrations in different combinations between 3000 and 3300 cm^{-1}. Only few of them are visible in Figure 2. As "Bisphenol A" is relatively large molecule and many peaks are very near to each other, so it is hard to assign the individual peaks distinctly without multiple assignments as indicated at the end of Table 1.

5. Conclusion

We have presented FTIR (400–4000 cm^{-1}) spectra of "Bisphenol A." Peaks (400–4000 cm^{-1}) have been assigned on the basis of Density Functional Theory (DFT) with configuration as B3LYP 6-311G++ (3df 3pd). Calculated absorption peaks are in reasonable agreement with experimental peaks after scaling with scale factor of 0.9679 except C-H and O-H stretching vibrations.

Competing Interests

The authors declare that they have no competing interests.

Acknowledgments

This work has been partially supported by the National Natural Science Foundation of China (no. 61275160). The authors would like to thank Professor S. Y. Wang for providing computer resources.

References

[1] http://www.who.int/foodsafety/publications/fs_management/No_05_Bisphenol_A_Nov09_en.pdf.

[2] http://www.fda.gov/Food/IngredientsPackagingLabeling/Food-AdditivesIngredients/ucm355155.htm.

[3] A. M. Hormann, F. S. Vom Saal, S. C. Nagel et al., "Holding thermal receipt paper and eating food after using hand sanitizer results in high serum bioactive and urine total levels of bisphenol A (BPA)," *PLoS ONE*, vol. 9, no. 10, Article ID e110509, 2014.

[4] http://www.bisphenol-a-europe.org/factsmyths/8/24/Certain-countries-have-banned-BPA.

[5] R. Viñas and C. S. Watson, "Bisphenol S disrupts estradiol-induced nongenomic signaling in a rat pituitary cell line: effects on cell functions," *Environmental Health Perspectives*, vol. 121, no. 3, pp. 352–358, 2013.

[6] R. Ullah, S. U.-D. Khan, M. Aamir, and R. Ullah, "Terahertz time domain, Raman and fourier transform infrared spectroscopy of acrylamide, and the application of density functional theory," *Journal of Spectroscopy*, vol. 2013, Article ID 148903, 17 pages, 2013.

[7] R. Ullah, H. Li, and Y. Zhu, "Terahertz and FTIR spectroscopy of 'Bisphenol A'," *Journal of Molecular Structure*, vol. 1059, no. 1, pp. 255–259, 2014.

[8] R. Ullah and Y. Zheng, "Raman spectroscopy of 'Bisphenol A'," *Journal of Molecular Structure*, vol. 1108, pp. 649–653, 2016.

[9] http://www.research.usf.edu/rf/docs/perkin-elmer-spectrum-100-ftir-users-guide.pdf.

[10] http://en.reagent.com.cn/.

[11] M. J. Frisch, G. W. Trucks, H. B. Schlegel et al., *Gaussian 09, Revision A.01*, Gaussian, Inc, Wallingford, Conn, USA, 2009.

[12] V. A. Rassolov, J. A. Pople, M. Ratner, P. C. Redfern, and L. A. Curtiss, "6-31G* basis set for third-row atoms," *Journal of Computational Chemistry*, vol. 22, no. 9, pp. 976–984, 2001.

[13] T. Clark, J. Chandrasekhar, G. W. Spitznagel, and P. V. Schleyer, "Efficient diffuse function-augmented basis sets for anion calculations. III. The 3-21+G basis set for first-row elements, Li-F," *Journal of Computational Chemistry*, vol. 4, no. 3, pp. 294–301, 1983.

[14] T. Sundius, "Molvib—a flexible program for force field calculations," *Journal of Molecular Structure*, vol. 218, pp. 321–326, 1990.

[15] T. Sundius, "A new damped least-squares method for the calculation of molecular force fields," *Journal of Molecular Spectroscopy*, vol. 82, no. 1, pp. 138–151, 1980.

[16] M. P. Andersson and P. Uvdal, "New scale factors for harmonic vibrational frequencies using the B3LYP density functional method with the triple-ζ basis Set 6-311+G(d,p)," *Journal of Physical Chemistry A*, vol. 109, no. 12, pp. 2937–2941, 2005.

10

Terahertz Spectroscopy and Brewster Angle Reflection Imaging of Acoustic Tiles

Patrick Kilcullen,[1] **Mark Shegelski,**[1] **MengXing Na,**[2] **David Purschke,**[2] **Frank Hegmann,**[2] **and Matthew Reid**[1]

[1]*Department of Physics, University of Northern British Columbia, Prince George, BC, Canada V2N 4Z9*
[2]*Department of Physics, University of Alberta, Edmonton, AB, Canada T6G 2E1*

Correspondence should be addressed to Patrick Kilcullen; kilcull@unbc.ca

Academic Editor: Arnaud Cuisset

A Brewster angle reflection imaging apparatus is demonstrated which is capable of detecting hidden water-filled voids in a rubber tile sample. This imaging application simulates a real-world hull inspection problem for Royal Canadian Navy Victoria-class submarines. The tile samples represent a challenging imaging application due to their large refractive index and absorption coefficient. With a rubber transmission window at approximately 80 GHz, terahertz (THz) sensing methods have shown promise for probing these structures in the laboratory. Operating at Brewster's angle allows for the typically strong front surface reflection to be minimized while also conveniently making the method insensitive to air-filled voids. Using a broadband THz time-domain waveform imaging system (THz-TDS), we demonstrate satisfactory imaging and detection of water-filled voids without complicated signal processing. Optical properties of the tile samples at low THz frequencies are also reported.

1. Introduction

The transparency of many dry, nonconductive materials to radiation at THz frequencies has allowed new applications of broadband THz pulses to nondestructive evaluation (NDE) and the imaging of concealed objects and surfaces [1–3]. THz radiation is electromagnetic radiation found in roughly the 0.1 THz to 10 THz range, between the microwave and infrared portions of the electromagnetic spectrum. THz radiation is nonionizing and has submillimeter wavelength, making it a competitive imaging technology to X-rays without having the associated health risks [4]. For example, it has proven suitable for imaging defects in foam insulation within the fuel tanks of the NASA space shuttle [5]. THz reflection tomography [6] has been demonstrated using coherently detected THz pulses and has been used for applications ranging from the imaging of delaminations [7] and under paint corrosion [8, 9] to art conservation [10]. THz radiation is also extremely sensitive to water which

absorbs it strongly in the far infrared. This has been exploited to image water content and diffusion in cork [11] and to identify concealed water-filled voids [7].

The motivation for this project was to investigate the feasibility of a reflection imaging system for the hull inspection of Royal Canadian Navy Victoria-class submarines (Figure 1). In order to operate with enhanced stealth, these vessels are fitted with an acoustic rubber tiling for the purposes of absorbing emanating noise and providing sonar cloaking. The design of the acoustic tiles generally features internal voids for enhanced acoustic absorption. These tiles, as well as a connective grouting compound, thus form an opaque covering over virtually the entire hull surface.

In evaluating the integrity of the pressurized hull, it is of paramount importance to discern states of corrosion of the underlying steel in addition to delamination and seawater ingress within the tile voids and grouting interfaces. A nondestructive means of on-site inspection with THz radiation has therefore been sought after. Such nondestructive

FIGURE 1: The HMCS Victoria, one of Canada's four Victoria-class submarines, shown in dry dock (Department of National Defence photo).

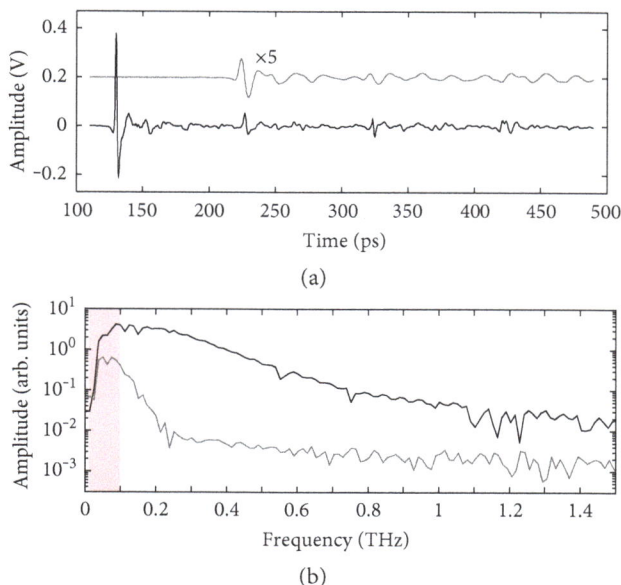

FIGURE 2: Example THz signals produced by the THz-TDS apparatus. (a) Time-domain comparison of a thru-air THz waveform (lower, black) and a thru-tile waveform (upper, gray). The thru-tile signal has been offset and magnified ×5 for clarity. (b) Log scale frequency domain comparisons corresponding to the thru-air (black) and thru-tile (gray) waveforms. The 0.01 to 0.10 THz frequency band used for the transmission image has been highlighted in red.

evaluation would be done in practice from the outside surface of the hull, with the radiation reflecting off the metal subsurface or intermediate tile layers.

It has been observed that, as a prelude to further corrosion, sites of water ingress are typically characterized by the internal tile voids becoming flooded with seawater. The ability to detect hidden water-filled voids could therefore constitute an early warning for corrosion, and an easy signal for tile replacement and regrouting.

Due to security reasons that precluded the use of actual tile samples, representative rubber tile samples or "analogue tiles" were used in the laboratory for this study. Each of the tile samples was approximately 4 inches square by 1 inch thick and consisted of three laminated layers (see Figure 5, oval inset). Engineered voids were simulated by holes drilled into the middle layer.

2. THz Apparatus

The THz-TDS system used in this work was a high speed Picometrix T-Ray 4000 system, generating linearly polarized, pulsed THz radiation detected in an 80 ps window, with a bandwidth of approximately 2 THz, modified to operate at a 1000 waveforms/sec acquisition rate. The emitter and detector are fibre-coupled photoconductive switches, allowing the system to be easily mounted in a raster-scanning configuration for the imaging studies.

Example THz signals from the THz-TDS system are presented in Figure 2 for transmission through both air and analogue tile. The time-domain waveforms presented in Figure 2(a) were stitched from several waveform observations (80 ps duration) obtained from the device. After transmission through the tile sample, the initial THz pulse is delayed and strongly attenuated. A window of THz transmission for the analogue rubber can be seen in Figure 2(b) centered at approximately 80 GHz.

3. Imaging Experiments

The large refractive index and absorption of the analogue tile samples was extracted from transmission and reference time-domain measurements using the method presented in [12, 13]. The extracted absorption coefficient and refractive index are shown in Figure 3.

Transmission imaging scans were initially taken of the analogue sample tiles. A direct transmission image taken of the internal air-filled void structure of tile sample "A1-8" is shown in Figure 4 which was generated from waveform data captured at each position. The color map corresponds to the amplitude of the transmitted spectral content in the 0.01 THz to 0.10 THz range (highlighted in Figure 2(b)) computed from the Fourier transform of the time-domain data. Variation in transmission is primarily attributed to a nonuniform bonding glue application between layers. The two internal voids are clearly observable from transmission minima at their edges due to the deflection and subsequent trapping of the penetrating radiation.

The large refractive index of the analogue tiles leads to a large front surface reflection at normal incidence through air at THz frequencies. In addition for reflection imaging, the large absorption of the material diminishes the amplitude of collected signals which must pass through any absorptive layers *twice* before detection. For example, at 80 GHz and at normal incidence, the expected reflected intensity coefficient from the top surface of a water-filled void within our sample would be 0.0146,

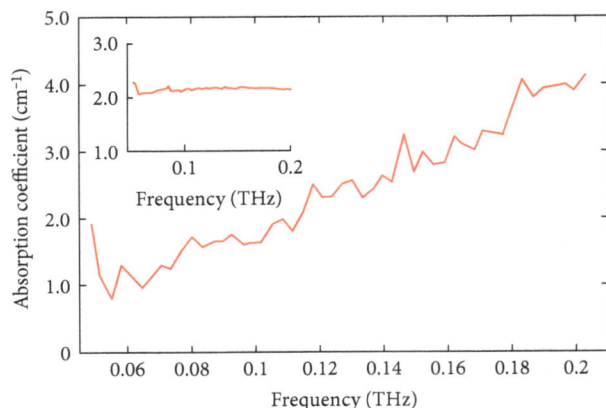

FIGURE 3: Propagation properties of the analogue tiles. The absorption coefficient is plotted as a function of frequency, shown in units of inverse centimeters. Inset: index of refraction, plotted as a function of frequency. These measurements were computed using the method presented in [12, 13].

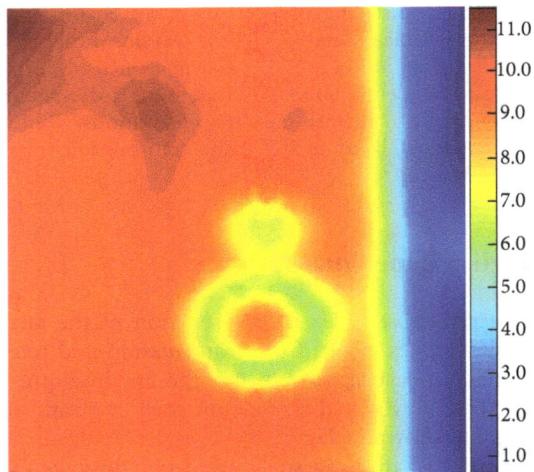

FIGURE 4: Spectral content transmission image (0.01 THz to 0.10 THz) of tile sample "A1-8" with air-filled voids taken at a grid resolution of 2 mm. Transmitted energy is shown colored from blue (low transmission) to red (high transmission) with the scale shown in arbitrary units. Both circular voids are discernible. The large and small cylindrical voids of this sample are 2.2 cm and 0.75 cm in diameter, respectively, with a separation of 0.8 cm. The transmission cut-off on the right hand side of the image was caused by metal apparatus used to hold the tile sample.

whereas the expected reflected intensity from the front surface would be greater by a factor of 9.5. A value of $n_{water} = 3.98 + i2.13$ at 80 GHz [14] was used to estimate these values, along with the data of Figure 3.

Due to the large refractive index and strong absorption of the analogue tiles, we attempted reflection imaging at 80 GHz from the top rubber-rubber tile interface, located at a depth of approximately 0.25 inches below the top tile surface (see Figure 5, oval inset). Operating at the Brewster angle with TM-polarization minimized the front surface reflection and maximized signal penetration.

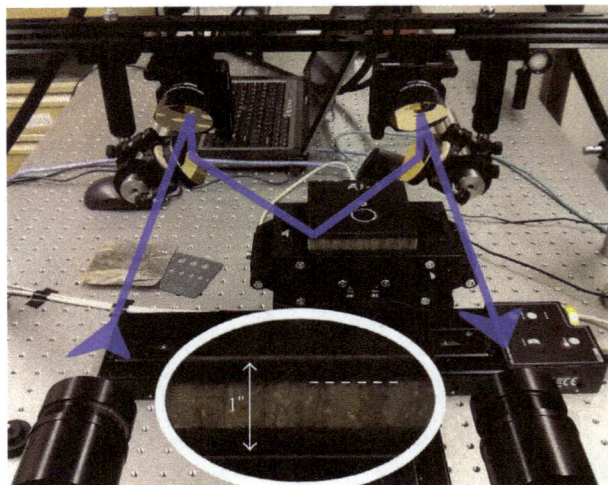

FIGURE 5: Sample tile "A1-8" positioned ready for scanning. The two black cylinders on the left and right are the Teflon-collimating lens enclosures of the THz emitter and detector, respectively. The path of the THz beam is highlighted in blue. Upon exiting the transmitter the THz radiation is linearly polarized horizontally across the optical table. The rectangular tile sample is positioned on top of a precision lab jack and two X-Y positioning stages. Oval inset: an edge closeup of the three-layered tile sample. The dashed horizontal line indicates the target depth of the reflection imaging at the first rubber-rubber interface.

Brewster angle microscopy and imaging was developed to exploit dielectric contrast to probe surface interfaces [15], where it has been exceedingly successful at investigating even monolayers at surface interfaces [15, 16]. We employ the Brewster angle here to make our imaging set-up as insensitive as possible to the surface interface, while exploiting dielectric contrast to probe subsurface interfaces, which would otherwise be obscured by strong front-surface back reflections.

Our Brewster's angle imaging apparatus is photographed in Figure 5. Off-axis parabolic mirrors (F/3) were used to focus the incident radiation into a narrow beam incident at the 80 GHz Brewster's angle of 63°. The analogue tile samples were controlled in both horizontal directions using a pair of motorized translation stages, while the emitter and detector positions remained fixed.

Raster images of the analogue tile samples were produced by capturing individual waveforms as the sample was repeatedly repositioned at the focus. Using 1000 averages for each collected waveform, our system produced imaging scans at a rate of approximately 7 seconds per pixel. The coherent detection allowed for temporal alignment of the acquired waveforms in order for a particular depth of reflection to be selected. Temporal alignment was achieved from an estimation of the optical path length of a single-pass reflection of the beam at the first interface at Brewster's angle. Raster waveform data was then used to produce gray scale sample images by first truncating the waveforms to a 25 ps window containing the peak of the reflected pulse, and subsequently extracting the integrated spectral

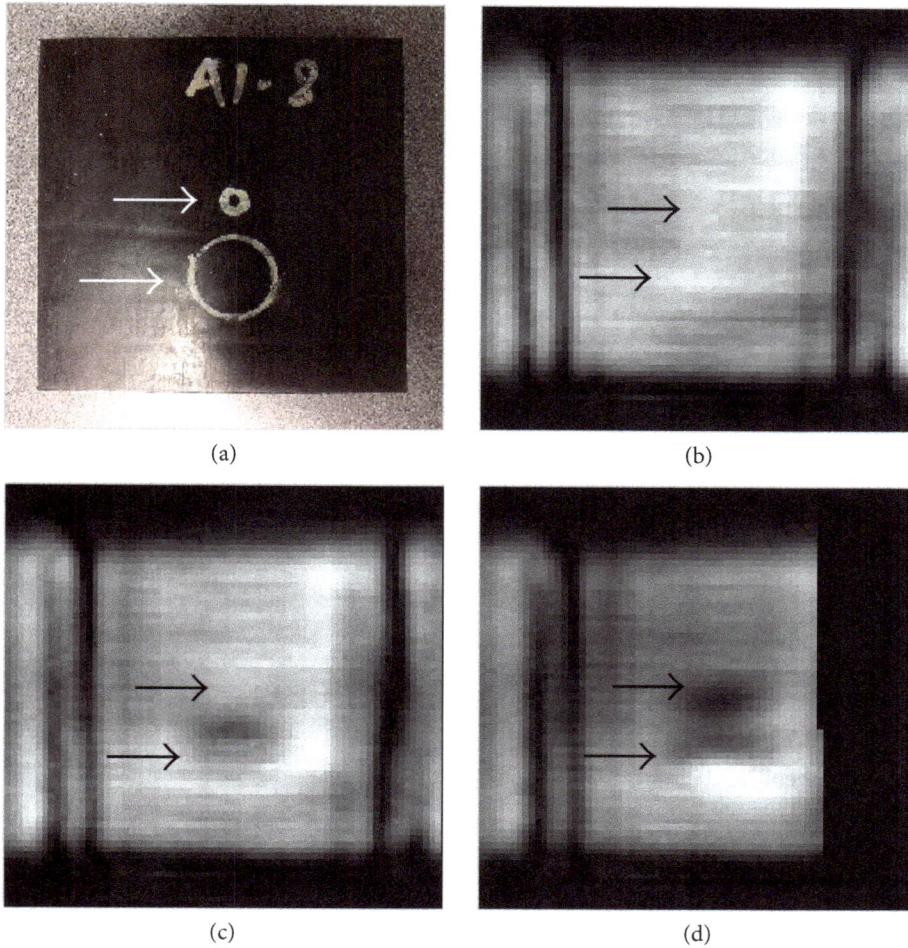

FIGURE 6: Spectral content reflection images (70×70 pixels) of tile sample "A1-8" in the range of 80 GHz. Positions of the voids are highlighted by the arrows. (a) Color photograph. (b) THz reflection image. (c) THz reflection image with larger hole fully injected with water. (d) THz reflection image with both holes fully injected. Unfortunately, a computer fault interrupted scan (d) before completion. The vertical dark bands at the sides of the tiles are edge effects due to the oblique incidence angle of the THz beam, which also distorts the shapes of the hidden defects horizontally. The acquisition time for each image was approximately 10 hours, limited primarily by the electronics of the raster scanning.

content of the signal in the vicinity of 80 GHz through the use of a discrete Fourier transform.

Brewster angle reflection images of the sample tile "A1-8" are shown in Figure 6. Our image progression simulates the imaging effect of water ingress into the internal voids. To achieve this, a syringe was used to inject tap water into the voids until they were completely flooded. The progression in Figure 6 clearly shows the sensitivity of our technique to the detection of flooded internal voids in the sample tiles. The oblong appearance of the flooded circular voids is due to the oblique angle of incidence of the THz beam in the Brewster's angle imaging configuration.

A particular advantage of imaging at the Brewster angle for this application is that those internal voids not contaminated by fluid ingress are invisible. This may be understood by considering the Brewster angle path of the THz beam illustrated in Figure 7 for the cases of rubber-rubber, rubber-air, and rubber-water interfaces. In particular, a beam which penetrates the top rubber layer and enters a void must

exit the cavity ceiling at the angle of incidence (Figure 7(b)), which therefore implies total transmission by the time reversal of the Brewster property. Due to the low dielectric contrast between the tile layers, such near-optimal transmission is also observed at the contiguous rubber-rubber interfaces of the sample (Figure 7(a)). Reflected signals from the first interface therefore do not exhibit contrast between the uncontaminated voids and the contiguous analogue rubber media surrounding them. Upon fluid ingress into the voids, the total transmission property is no longer valid at the cavity interface, and the result is an increased reflected signal from the interface (Figure 7(c)).

We hypothesize that the mostly "dark" appearance of the water-filled voids is due to the destructive Bragg interference between the reflected flooded cavity interface and the front surface reflection, shown by the two dashed rays in Figure 7(c).

Front surface reflections are still encountered even when operating at Brewster's angle in practice and are illustrated by

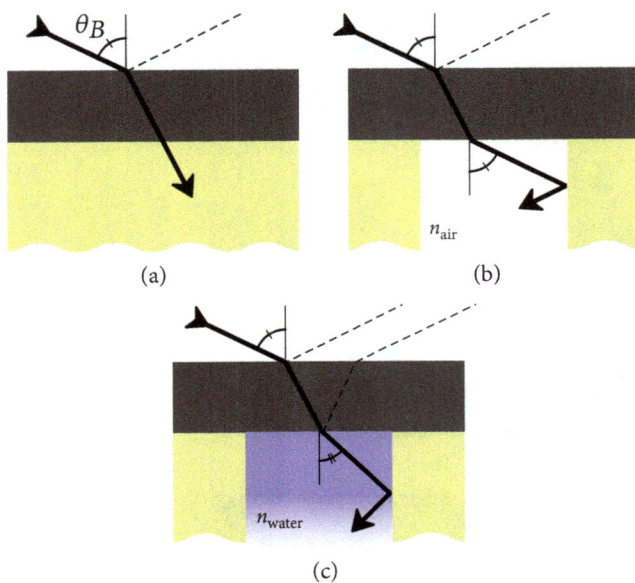

FIGURE 7: Ray diagram illustration of the Brewster's angle reflection geometry for (a) a continuous rubber interface, (b) an air-filled void, and (c) a water-filled void. The dashed lines indicate significant reflected rays.

the topmost dashed rays in Figures 7(a), 7(b), and 7(c). Despite the temporal alignment and truncation of our acquired waveforms, the reflected pulse windows still contained atmospheric "ringing" from the front surface reflected pulse. Interpreting this residual front surface signal as a noise floor for our apparatus, we obtain a signal to noise ratio (SNR) of 1.64 compared to the peak to peak amplitude of the reflected pulse from the void interface. This value is a lower bound for the SNR, as the front surface back reflection does not change significantly over the surface of the samples.

4. Conclusion

We were able to observe hidden water-filled voids through reflection mode imaging without complicated image processing. Successful water-filled void detection was possible at a depth of 0.25 inches beneath a layer of highly absorbing rubber in the laminated analogue tile sample. By operating at the Brewster angle, front surface reflections could be minimized with the added advantage of making uncontaminated voids invisible upon scanning.

A low dielectric contrast between the top and middle layers of the analogue rubber sample allowed for identical transmission between the contiguous rubber and cavity spaces at the depth of the first interface. We note that, for cases in which the dielectric contrast between the top and middle layer is significant, such identical transmission could theoretically be achieved via tuning of the reflection imaging angle through the Fresnel equations.

Acknowledgments

This research was carried out in the course of a larger collaboration between the authors' research group at UNBC, the University of Alberta, McGill University, and Defence Research and Development Canada (DRDC). In addition to the authors, the members of this collaboration included Dr. David Cooke, David Valverde, and Lauren Gingras at McGill University and Rod McGregor at DRDC. The authors wish to acknowledge funding for this project from DRDC, NSERC, and the AITF Strategic Chair Program.

References

[1] K. H. Jin, Y.-G. Kim, S. H. Cho, J. C. Ye, and D.-S. Yee, "High-speed terahertz reflection three-dimensional imaging for nondestructive evaluation," *Optics Express*, vol. 20, no. 23, pp. 25432–25440, 2012.

[2] T. D. Dorney, W. W. Symes, R. G. Baraniuk, and D. M. Mittleman, "Terahertz multistatic reflection imaging," *Journal of the Optical Society of America. A*, vol. 19, no. 7, pp. 1432–1442, 2002.

[3] P. U. Jepsen, D. G. Cooke, and M. Koch, "Terahertz spectroscopy and imaging–Modern techniques and applications," *Laser & Photonics Reviews*, vol. 5, no. 1, pp. 124–166, 2011.

[4] D. Saeedkia, *Handbook of terahertz technology for imaging, sensing and communications*, WP Woodhead Publishing, Oxford, England Philadelphia, Pennsylvania, 2013.

[5] D. Zimdars, J. S. White, G. Stuk, A. Chernovsky, G. Fichter, and S. Williamson, "Large area terahertz imaging and nondestructive evaluation applications," *Insight*, vol. 48, no. 9, pp. 537–537, 2006.

[6] D. Mittleman, R. Jacobsen, and M. Nuss, "T-ray imaging," *IEEE Journal of Selected Topics in Quantum Electronics*, vol. 2, no. 3, pp. 679–692, 1996.

[7] D. Zimdars, J. White, G. Sucha et al., "Terahertz for military and security applications V," in *Conference on Terahertz for Military and Security Applications V*, vol. 6549, p. 54906, 2007.

[8] R. Anastasi and E. Madaras, "Terahertz NDE for under paint corrosion detection and evaluation," in *AIP Conference Proceedings*, vol. 820, pp. 515–522, 2006.

[9] R. F. Anastasi, E. I. Madams, J. P. Seebo et al., "Terahertz NDE application for corrosion detection and evaluation under shuttle tiles," in *Proc. SPIE*, vol. 6531, p. W5310, 2007.

[10] J. B. Jackson, J. Bowen, G. Walker et al., "A Survey of Terahertz Applications in Cultural Heritage Conservation Science," *IEEE Transactions on Terahertz Science and Technology*, vol. 1, no. 1, pp. 220–231, 2011.

[11] Y. L. Hor, J. F. Federici, and R. L. Wample, "Nondestructive evaluation of cork enclosures using terahertz/millimeter wave spectroscopy and imaging," *Applied Optics*, vol. 47, no. 1, pp. 72–78, 2008.

[12] L. Duvillaret, F. Garet, and J.-L. Coutaz, "A reliable method for extraction of material parameters in terahertz time-domain spectroscopy," *IEEE Journal of Selected Topics in Quantum Electronics*, vol. 2, no. 3, pp. 739–746, 1996.

[13] I. Pupeza, R. Wilk, and M. Koch, "Highly accurate optical material parameter determination with THz time-domain spectroscopy," *Optics Express*, vol. 15, no. 7, pp. 4335–4350, 2007.

[14] J. T. Kindt and C. A. Schmuttenmaer, "Far-infrared dielectric properties of polar liquids probed by femtosecond terahertz pulse spectroscopy," *The Journal of Physical Chemistry*, vol. 100, no. 24, pp. 10373–10379, 1996.

[15] D. Hoenig and D. Moebius, "Direct visualization of mono-layers at the air-water interface by Brewster angle microscopy," *The Journal of Physical Chemistry*, vol. 95, no. 12, pp. 4590–4592, 1991.

[16] M. Harke, R. Teppner, O. Schulz, H. Motschmann, and H. Orendi, "Description of a single modular optical setup for ellipsometry, surface plasmons, waveguide modes, and their corresponding imaging techniques including Brewster angle microscopy," *Review of Scientific Instruments*, vol. 68, no. 8, pp. 3130–3134, 1997.

11

Cell Imaging by Spontaneous and Amplified Raman Spectroscopies

Giulia Rusciano,[1,2] **Gianluigi Zito,**[3] **Giuseppe Pesce,**[1] **and Antonio Sasso**[1,2]

[1]*Department of Physics "E. Pancini", University of Naples Federico II, Via Cintia, 80126 Naples, Italy*
[2]*National Institute of Optics (INO), National Research Council (CNR), Via Campi Flegrei 34, 80078 Pozzuoli, Italy*
[3]*Institute of Protein Biochemistry (IBP), National Research Council (CNR), Via Pietro Castellino 111, 80131 Napoli, Italy*

Correspondence should be addressed to Giulia Rusciano; giulia.rusciano@unina.it

Academic Editor: Renata Diniz

Raman spectroscopy (RS) is a powerful, noninvasive optical technique able to detect vibrational modes of chemical bonds. The high chemical specificity due to its fingerprinting character and the minimal requests for sample preparation have rendered it nowadays very popular in the analysis of biosystems for diagnostic purposes. In this paper, we first discuss the main advantages of spontaneous RS by describing the study of a single protozoan (*Acanthamoeba*), which plays an important role in a severe ophthalmological disease (*Acanthamoeba* keratitis). Later on, we point out that the weak signals that originated from Raman scattering do not allow probing optically thin samples, such as cellular membrane. Experimental approaches able to overcome this drawback are based on the use of metallic nanostructures, which lead to a huge amplification of the Raman yields thanks to the excitation of localized surface plasmon resonances. Surface-enhanced Raman scattering (SERS) and tip-enhanced Raman scattering (TERS) are examples of such innovative techniques, in which metallic nanostructures are assembled on a flat surface or on the tip of a scanning probe microscope, respectively. Herein, we provide a couple of examples (red blood cells and bacterial spores) aimed at studying cell membranes with these techniques.

1. Introduction

Optical methods are ideally suited for the analysis of biological systems because of the gentle light-matter interaction that makes them particularly noninvasive. Fluorescence-based techniques have been extensively employed for optical imaging of biological systems due to its high sensitivity. Nevertheless, these techniques suffer significant limitations mainly related to (i) photobleaching and photoblinking of the fluorescent labels, (ii) perturbation of the molecules induced by the fluorophores themselves, and (iii) poor chemical selectivity. Raman spectroscopy (RS) is attracting considerable attention in life sciences [1–3], because it overcomes these limitations and provides additional advantages, related to the recently explored possibility of improving the spatial resolution up to the nanoscale. RS is not affected by fluorescence quenching mechanisms being based on inelastic scattering of light; it does not require any sample preparation (label-free technique), and, further more important, it assesses the chemical composition of the sample by exploiting the unique vibrational structure of molecules (chemical fingerprinting). In principle, Raman spectra provide information similar to infrared (IR) absorption spectra, but with the important advantage of avoiding the strong IR absorption of water, which usually is dominant in biological samples (water, instead, exhibits a low Raman activity). These advantages have led to a remarkable development of the various types of Raman-based spectroscopies and their diffusion in many fields of science (biomedicine [4–6], cultural heritage [7], food control [8], forensics analysis [9], etc.). In addition, the combination of Raman spectroscopy with confocal scanning microscopy has allowed a chemical imaging of tissues or even of single cells (*Raman imaging*) [6]. Moreover, changes in molecular composition of biosystems, detectable using multivariate statistical approaches, enable quantitative diagnosis of diseases with high level of sensitivity and specificity [10, 11].

However, also Raman spectroscopy has several drawbacks. The most severe limit of RS is the relatively low efficiency of the inelastic light-scattering process. As a matter of fact, RS cross section is typically 10^{-30}–10^{-25} cm^2 sr^{-1}, that is, 8–10 orders of magnitude weaker than the fluorescence/absorption cross section. Thus, RS fails in detecting molecules at low concentration (below 0.1 mM). This drawback can be overcome by several Raman-based techniques, including resonant Raman spectroscopy (RRS) [12], coherent anti-Stokes Raman spectroscopy (CARS) [13], or stimulated Raman spectroscopy (SRS) [14], which have been developed to enhance the intensity and improve the spatial resolution with respect to the basic method.

A quite efficient way to enhance the weak signals in spontaneous RS is based on the lucky combination with plasmonics, which takes advantage of the coupling of light to metallic nanostructures. Such combination is at the basis of surface-enhanced Raman scattering (SERS). Historically, SERS birth dates back to 1974 when Fleischmann and coworkers [15] observed an anomalous, huge enhancement of the Raman signal from pyridine molecules adsorbed on a roughened silver electrode. A few years later, two independent papers [16, 17] provided the first physical interpretation based on localized surface plasmon resonances (LSPRs). High local electromagnetic field near the nanostructures provides very high enhancement of Raman scattering. The maximum SERS enhancement factor (EF) occurs at specific positions on the surface called *hot spots*. When molecules are adsorbed to or in close proximity (within a few nm) of a hot spot, a huge Raman enhancement up to 12 orders of magnitude can occur, allowing single-molecule sensitivity [18, 19]. The SERS EF depends on several parameters, such as the real and imaginary parts of the dielectric function of the metal nanostructure, the active substrate's geometry, the metal-molecule affinity, and the possible presence of electronic transition resonances of the analyte molecule at the excitation wavelength. Despite such complexity, SERS has found successful application in cutting-edge biochemical and medical research [20–22]. In particular, SERS technique has proved to be an effective platform for optical probing and imaging of live cells via SERS nanotags or in label-free experiments for surface cell sensing [23–29]. In this context, metal colloids have found wide application as SERS-active substrate, due to the particularly simple technology required for colloid preparation. Nevertheless, colloidal SERS substrates suffer from insufficient spatial reproducibility due to their local geometric heterogeneity that induces an uncontrollable position-dependent enhancement. Although great efforts have been dedicated in the last few decades to the synthesis and fabrication of high performance SERS substrates in terms of sensitivity and reproducibility, this topic is still open.

The unpredictable localization of the hot spots clearly undermines the potential subdiffraction character of SERS which, in fact, remains limited to the size of the focused laser spot. This limit can be overcome by combining the features of SERS with those of scanning probe microscopies, like Atomic Force Microscopy (AFM) and Scanning Tunnelling Microscopy (STM). This technique is called tip-enhanced Raman scattering (TERS) and is implemented by illuminating a metallic (or metaled) tip by a tightly focused laser beam.

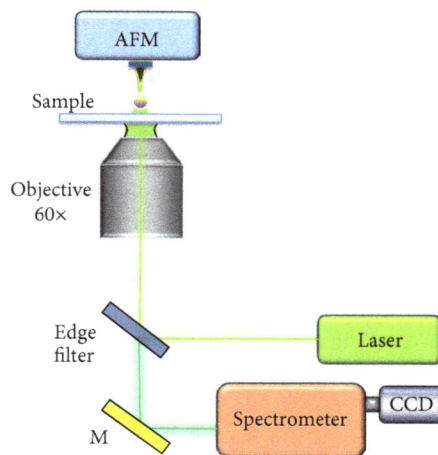

FIGURE 1: Schematic picture of our Raman set-up. Spontaneous RS or SERS measurements are made in backscattering configuration. In the upper part is mounted the AFM head for TERS measurements.

The metal tip acts then as a highly localized nanoantenna able to provide a strong amplification of the Raman signal via LSPR excited at the tip apex. In such a way, TERS analysis provides, simultaneously, the topographical (based on scanning microscopy data) and chemical (based on SERS effect at the tip apex) information of the investigated surface [30, 31]. TERS technique has, in very recent years, allowed a deeper insight into the surface properties of many systems at the nanoscale, including biosystems [32–36].

The scope of this paper is not to provide a description of the physical principles of the mentioned techniques (for that we refer to textbook and papers listed in the references). Rather, this paper reviews some specific cases studied aimed at providing Raman cell imaging at different levels of complexity and spatial resolution. The starting point is the study of a single protozoan based on spontaneous RS; we show how RS, combined with principal component analysis (PCA), is able to distinguish different regions of the cell and their evolution when the cell is exposed to an external stress. Then, after a brief description of a new approach to fabricate high uniform nanostructured metallic substrates, we demonstrate that SERS provides information on the distribution of chemicals on the membrane of red blood cells (RBCs). Finally, we conclude the review by discussing the study of the bacterial spore surface using TERS and demonstrating the high spatial resolution and chemical selectivity of this technique.

2. Material and Methods

2.1. Raman Confocal Microscopy. All the experiments described herein have been performed by using an inverted Raman confocal microscope (WiTec, alpha 300). The system, schematically shown in Figure 1, is equipped with an excitation source at 532 nm, provided by a frequency doubled Nd:YAG laser. Raman scattered light is collected with the same objective lens used to focus the radiation onto the sample and is guided to the spectrograph by a 25 μm core optical fiber, which assures the system confocality. Spectra are acquired using a 600 and 1800 l/mm grating, according to the requested resolution. The system is combined with an Atomic Force Microscope for TERS measurements.

2.2. SERS Substrate Preparation.

The procedure to obtain our SERS substrates is based on the self-assembling of block-copolymer (BCP) micelles of polystyrene-block-poly-4-vinylpyridine (PS-b-P4VP, molecular weight 10400-b-19200) loaded with silver nanoparticles (Ag-NPs) which is described in [37]. It is based on the procedure firstly described in [38], properly modified to increase the nanoislands' packing density (see Figure 2). Briefly, a concentrated Ag-BCP solution was processed with centrifugation, filtration, and slow spin coating over 24 mm × 24 mm commercial coverslips. Polymers were removed from the thin film after UV exposure. In such way, we obtained a self-assembled isotropic nanostructure with characteristics of homogeneity typical of the so-called near hyperuniform disorder [39, 40]. The resulting nanoisland pattern exhibited a hot spot density $\sim 10^4 \mu m^{-2}$. To test the spatial reproducibility and the enhancement factor of such plasmonic structure, we used a common dye (crystal violet) as probe molecule. We found a very high spatial reproducibility (standard deviation < 1% over $2500 \mu m^2$) and a high enhancement factor (EF up to $\sim 10^8$) suitable for single-molecule detection.

3. Cells Preparation and Handling

3.1. Acanthamoeba.

Acanthamoeba cysts (A.) were isolated from the cornea of contact lens user following a procedure described in [41]. Protozoa were suspended in Page's amoeba saline and centrifuged at 800g for 5 min; therefore, cells were isolated in 1.5% nonnutrient agar enriched by Escherichia coli K12. Sample preparation was carried out by dipping PBS solution containing cells into a glass chamber, consisting of two sandwiched coverslips glued by two parafilm strips also acting as spacers. Raman analysis was performed on cells spontaneously adhered onto the bottom glass coverslip.

3.2. Red Blood Cells.

RBCs directly extracted by healthy donors were stored in an isotonic solution for 30 min prior to the experiments with no further treatment. Therefore, cells were infiltrated into a cell chamber, similar to that previously described for A. cysts. For SERS measurements, the chamber bottom coverslip was replaced by a SERS substrate, prepared according to the procedure previously described. For both spontaneous Raman and SERS imaging, RBCs were left to adhere by physisorption to the chamber bottom for ~1 hour before analysis. Importantly, we did not observe hemolytic activity of our immobilized SERS nanostructure on the time scale of the measurements.

3.3. Bacterial Spores.

TERS analysis was performed on Bacillus subtilis (B. subtilis) spores. The sample was prepared by pipetting $50 \mu L$ of a water suspension of mature PY79 wild-type spores at a concentration of 10^5 spores/mL on a clean microscope glass coverslip and left to dry at ambient conditions.

4. Results and Discussion

4.1. Spontaneous Raman Imaging of a Single Protozoan.

Acanthamoeba (A.) keratitis is a severe eye infection, often affecting contact lens wearers, which can lead to permanent visual impairment or even blindness. Diagnosing A. keratitis is a clinical challenge. Nowadays, it is performed by the association of a microscopic examination of the corneal scraping, culture of parasites, and molecular biology assays based on DNA amplification such as Polymerase Chain Reaction (PCR). Clinical prescriptions for A. keratitis often involve the use of Polyhexamethylene Biguanide (PHMB), an antiseptic toxic for the epithelial cells of the cornea. This renders a fundamental timely suspension of the drug therapy when protozoa are eradicated. For this reason, innovative tools to distinguish between living and dead A. are strongly required. We have tested this capability for Raman imaging.

To get a Raman image of a single A., spectra were acquired in a raster scan around the cell. Measurements were performed keeping A. in a simple PBS solution or in presence of PHMB (concentration 1:2 in PBS). The complexity of the obtained spectra, exemplified in Figure 3(a), rendered necessary to analyze signals by PCA [42]. At this purpose, the acquired spectra were background-corrected by removing a fourth-order polynomial and eliminating spurious signals deriving from cosmic ray contributions. Therefore, they were processed by using a homemade MATLAB routine. Figure 3 shows the PC1 score maps reconstructed for the same A., before (Figure 3(b)) and after 6 hours of exposure to PHMB (Figure 3(c)). PC1 loading presents the characteristic DNA bands at 728, 785, and $1570 cm^{-1}$ (data not shown; see [43] for details) indicating therefore that the PC1 score map is related to the nucleic acid distribution. Interestingly, while before treatment with PHMB the high score values are localized in a well-defined region within the cell, after exposure to PHMB, DNA is distributed in a much wider region, therefore suggesting the rupture of nuclear membrane and the successive diffusion of nuclear material throughout the cell. PC2 loading features, instead, can be associated with the cell membrane. In particular, spatial localization of PC2 scores gives evidence of membrane rupture induced by exposition to PHMB.

4.2. SERS Imaging of Red Blood Cell Membrane.

Label-free chemical imaging of live cell membranes can shed light on the molecular basis of cell membrane functionalities and their alterations under pathological conditions. For instance, it is well known that, in the presence of a malignant cell transformation, the membrane biochemical composition is strongly modified. However, spontaneous Raman spectroscopy is unable to reach this goal. As a matter of fact, due to the limit imposed by the diffraction of light, scattered Raman photons originate from a volume whose extension, in the Raman probe propagation direction z, is typically a couple of microns. Hence, the confocal detection region includes a relatively thick cytosolic region, which often renders negligible the contribution of the nanometric membrane layer. Therefore, techniques capable of introducing not only an axial spatial selectivity but also an adequate sensitivity to

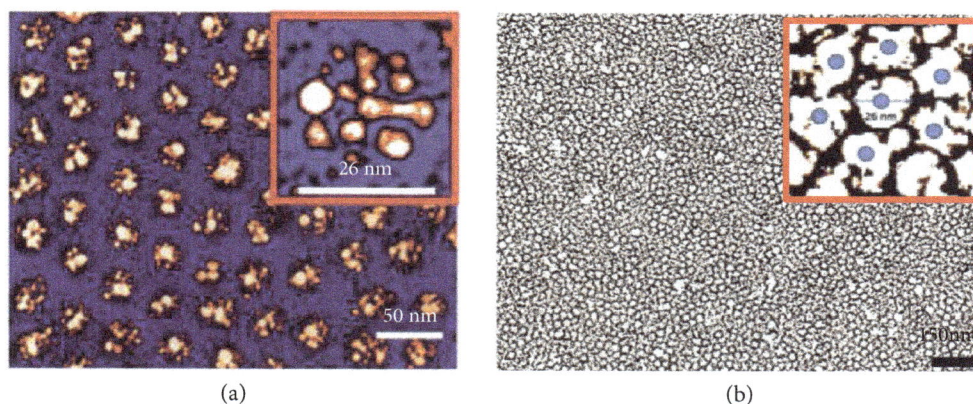

(a)   (b)

FIGURE 2: Characterization of the SERS substrate. (a) TEM micrograph of hexagonal nanostructure obtained following the procedure described in [25]. The inset shows, in detail, a single cluster of Ag-NPs. (b) TEM micrograph of the pattern obtained by applying our implemented protocol. The filling fraction is highly increased reducing nanoisland gaps to 2-3 nm, as shown in the inset. Bar scales are 50 nm in (a) and 150 nm in (b). Bar scale in the inset in (a) is 26 nm. (Adapted from [37] with permission from the Royal Society of Chemistry.)

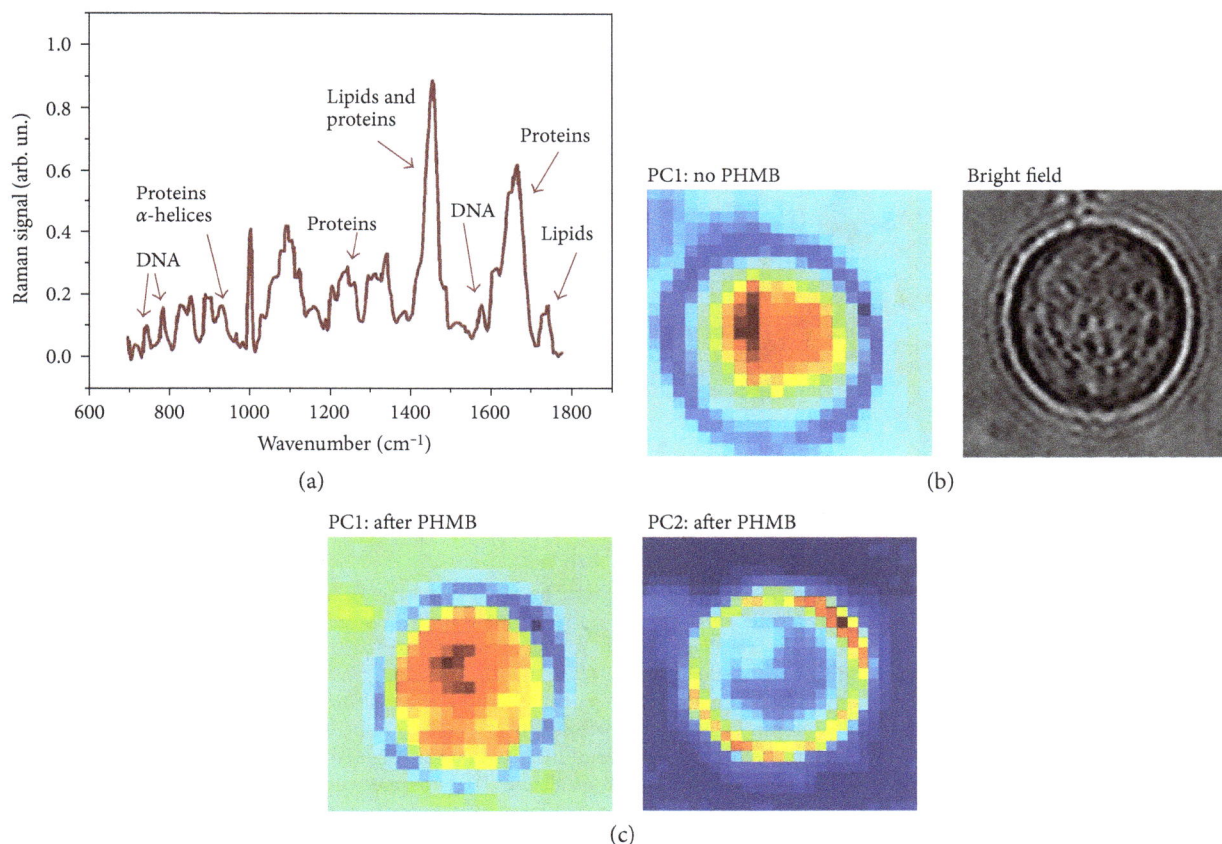

FIGURE 3: (a) A typical Raman spectrum from a live *Acanthamoeba*. (b) Score maps of PC1 component for an *A.*, before exposure to PHMB (left) compared with its bright field image (right). (c) Score map of PC2 (left) and PC3 (right) components after 6 hours of *A.* exposure to PHMB.

reveal a reduced number of molecules are strongly required. SERS fulfills both requirements, thanks to the giant enhancement of the signal corresponding to molecules in close contact with the plasmonic substrate. In order to demonstrate the effectiveness of SERS for membrane analysis, we applied this technique to a challenging *case of study*: the RBC. As a matter of fact, a RBC consists of a closed lipid bilayer membrane (embedding proteins) acting as a bag for hemoglobin

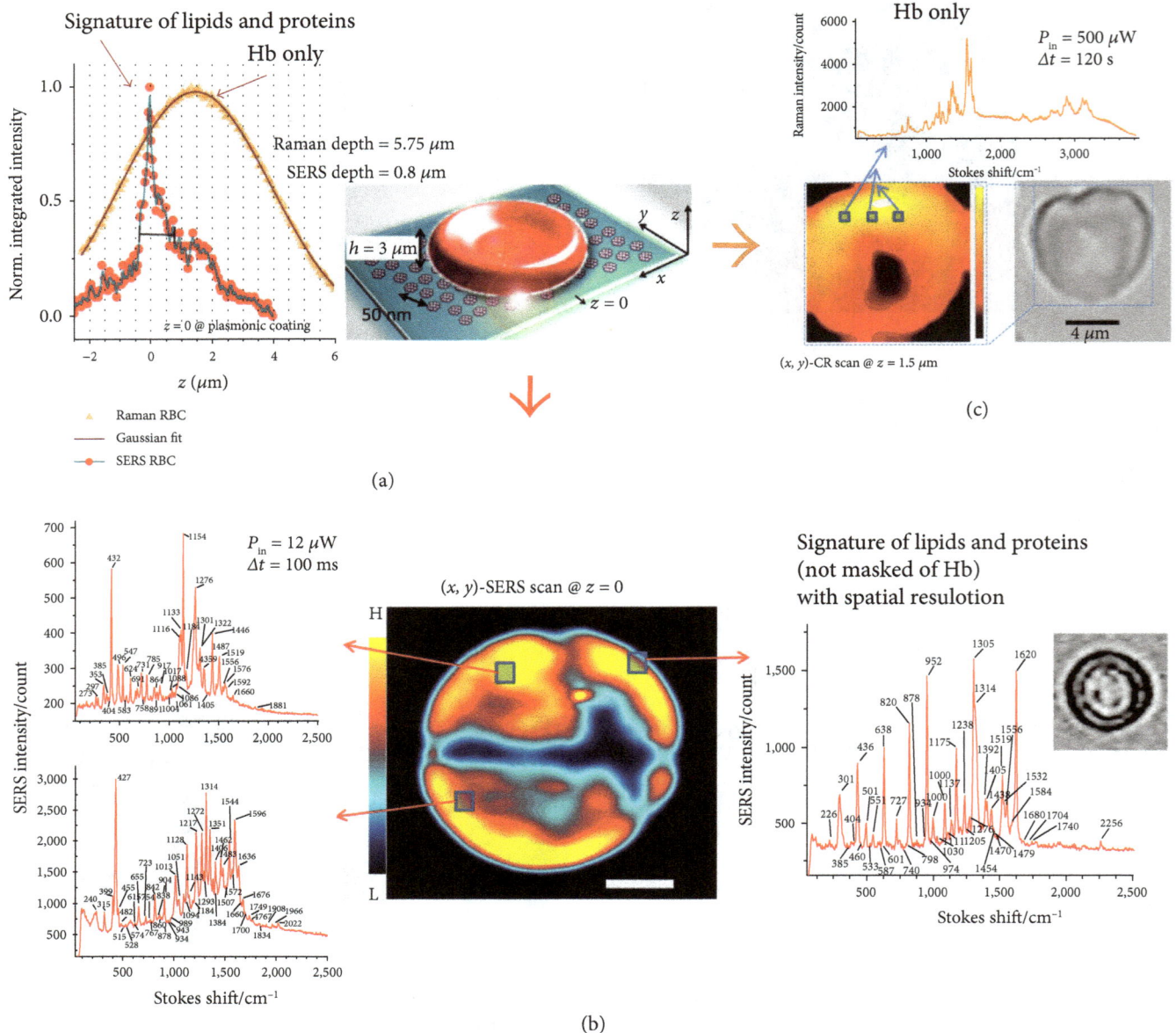

FIGURE 4: Schematic representation of the experiment with red blood cell on a SERS substrate: the SERS signal is peaked at the SERS substrate $z = 0$ in (a) (red points) and is associated with SERS spectra corresponding to lipids and proteins not masked by Hb as shown in (b). In particular, this is achieved with spatially resolved signals revealing sensitivity to the local membrane environment with capability of reconstructing the membrane SERS map of the RBC, in close correlation with the actual morphology of the cell scanned (see optical image in the inset in (b)). In a twin experiment with a normal coverslip as substrate (not shown), the Raman signal is ascribable only to Hb: the spontaneous Raman intensity (integrated between 1400 and 1600 cm^{-1}) versus the axial distance z denotes a broad bell-shaped curve (orange points) to be contrasted to the sharp and localized curve obtained on SERS substrate (red points in (a)). In this case, Raman imaging reveals only the spectral signature of Hb in all the RBC volume with no contribution from membrane, as represented in (c).

(Hb). This latter protein presents spectral resonance for excitation in the visible region, which clearly introduces an additional level of complexity for the spectral detection of scattering from the membrane.

In order to unambiguously assign the observed signals to SERS scattering from the RBC plasma membrane in contact with the Ag nanotextured substrate, we proceeded with two twin experiments. In the first experiment, RBCs

extracted following the protocol described in the previous paragraph were placed on a chamber closed at the bottom with a simple glass coverslip. In the second experiment, the glass coverslip was replaced by a SERS substrate (Figure 4(a)). In both cases, RBCs were allowed to spontaneously adhere to the chamber bottom. Therefore, we repeatedly acquired Raman spectra at different heights by scanning the sample along the z-axial coordinate. Marked differences were found. In particular, for the spontaneous Raman case, spectra were mainly ascribable to the bulk hemoglobin contained inside the cell, presenting typical spectral bands of the heme group [44]. At the same time, the signal intensity (integrated over the 600–1800 cm^{-1} region) versus the z coordinate was characterized by a broad bell shape (~5.8 μm FWHM), consistent with RBC thickness convoluted with the z-axis resolution. On the contrary, in the second case, no Hb contribution was observed and the signal intensity versus z (distance from the substrate) presented a much sharper curve, with a maximum at $z = 0$ and an effective depth (~0.8 μm) consistent with the effective thickness of the confocal scattering volume.

Since membrane thickness is limited to ~5 nm, and whereas the signal decays rapidly with distance from the surface substrate, we believe that we are actually revealing the only contribution from the outermost part of the membrane. Typical SERS spectra are reported in Figure 4(b). The spectra were clearly dominated by spectral features due to lipids, amino acids, and carbohydrates with no typical markers of membrane-bound Hb.

A prerequisite for a *faithful* SERS imaging is the availability of plasmonic architectures with a spatially invariant SERS enhancement factor. This requirement has opened new research area in nanoplasmonic, aimed at producing nanostructured surface showing high efficiency and reproducibility. We faced this problem by taking advantage of the use of the so-called block copolymer (BCP), polymers made up of blocks of different monomeric components. Importantly, when spun on a glass coverslip, BCP tends to self-assemble at nanoscale, giving rise to nanotextured coverslip coating, whose features depend on their concentration and molecular weight. In our case, starting from BCP micelles loaded with silver nanoparticles (Ag-NPs), we obtained a self-assembled isotropic nanostructure with characteristics of homogeneity typical of the so-called near-hyperuniform disorder. The resulting highly dense, homogeneous, and isotropic random pattern consists of clusters of silver nanoparticles with limited size dispersion.

A raster scan SERS imaging was performed by acquiring spectra on a square grid, with a step of 200 nm, close to the spatial resolution in the x-y plane. The result is shown in Figure 4(b) and is indicated as SERS map. We observed variations of the SERS spectrum depending on the membrane position, denoting a sensitivity to the local molecular environment explored in the laser-scattering area. The representative hyperspectral map, integrated in the range 1100–1700 cm^{-1}, reveals a strong correlation with the optical image of the same RBC (see Figure 4(b), optical image on the right). Clearly, such faithful SERS imaging was actually possible because of the excellent uniformity of the enhancement factor of our nanostructure.

4.3. TERS Imaging of Bacterial Spore Membrane. For most of the applications of Raman analysis to biosystems, the intrinsic diffraction-limited spatial resolution of this technique fits the experiment requirements. However, for some specific cases, mainly related to the analysis of single nanometric targets (DNA filaments, fibrils, etc.), it is required to beat the diffraction limit. TERS analysis is the answer for this demand [45].

In this section, we show an application of this technique to a quite interesting biosystem: the *Bacillus subtilis* (*B. subtilis*) spore. These spores have recently been demonstrated to hold potential for many biotechnological applications as the drug delivery [46] or the development of new sources for energy production [47]. We have applied TERS analysis to the spore's outmost layer (the coat), focusing our attention on particular surface-nanosized structures, the so-called spore *ridges* [48], readily distinguishable in topographic AFM images of spores (see Figure 5(a)). The function of these structures is still not fully understood, although it is surely related to the volume changes occurring during spore core dehydration-rehydration cycles.

Figure 5(b) shows a typical TERS spectrum observed in this investigation. It was obtained by a laser power impinging on the sample of 50 μW and an integration time and 2 s. The complexity of this spectrum reflects the intricate arrangements of proteins, lipid, and carbohydrates on the spore surface. Importantly, the acquired spectra exhibited significant point-to-point variations, reflecting the different nanoenvironments explored by the tip. In this condition, obtaining useful information from TERS data is surely a challenge. Even more difficult is to highlight the correlations of TERS data with the observables provided by other nanoscopic techniques. In particular, it is reasonable to expect a strong correlation between AFM phase maps and TERS maps, being both surface representations sensitive to the surface chemical signatures. However, TERS maps reporting the intensity of selected spectral features do not provide useful information and seem to be completely uncorrelated by AFM phase maps. For instance, Figure 5(c) shows a 120 nm × 120 nm phase map, obtained with a scan step of 20 nm. The brighter pixels (corresponding to a positive phase lag) correspond to points on the ridge. Figure 5(d), instead, reports TERS maps obtained by reporting the intensity of assigned spectral TERS bands. Evidently, no apparent correlation can be revealed between these maps and the phase map. However, information-rich TERS maps can be instead obtained by taking advantage of PCA. To test the effectiveness of this statistical tool for unravelling the information provided by TERS spectra of spore surface, we have acquired spectra in a spore-surface zone across a surface ridge clearly visible in the spore AFM phase map. However, a completely different outcome is obtained when TERS data are analyzed by PCA. In particular, as evident in Figure 5(e), both PC2 and PC3 score maps exhibit a clear correlation with the phase map (PC1 components only take into account a residual background, so it was not taken into consideration). Therefore, this study demonstrates that by taking advantage of advanced statistical tools like the principal component analysis, TERS maps can be correlated with AFM phase maps, thus allowing one to pick up some spectroscopic-derived outcomes nonintuitively identifiable. This might be very useful for drawing solid conclusions from the TERS analysis of complex organisms and, therefore, also support the diffusion of this technique in many fields of science, including biology and biomedicine.

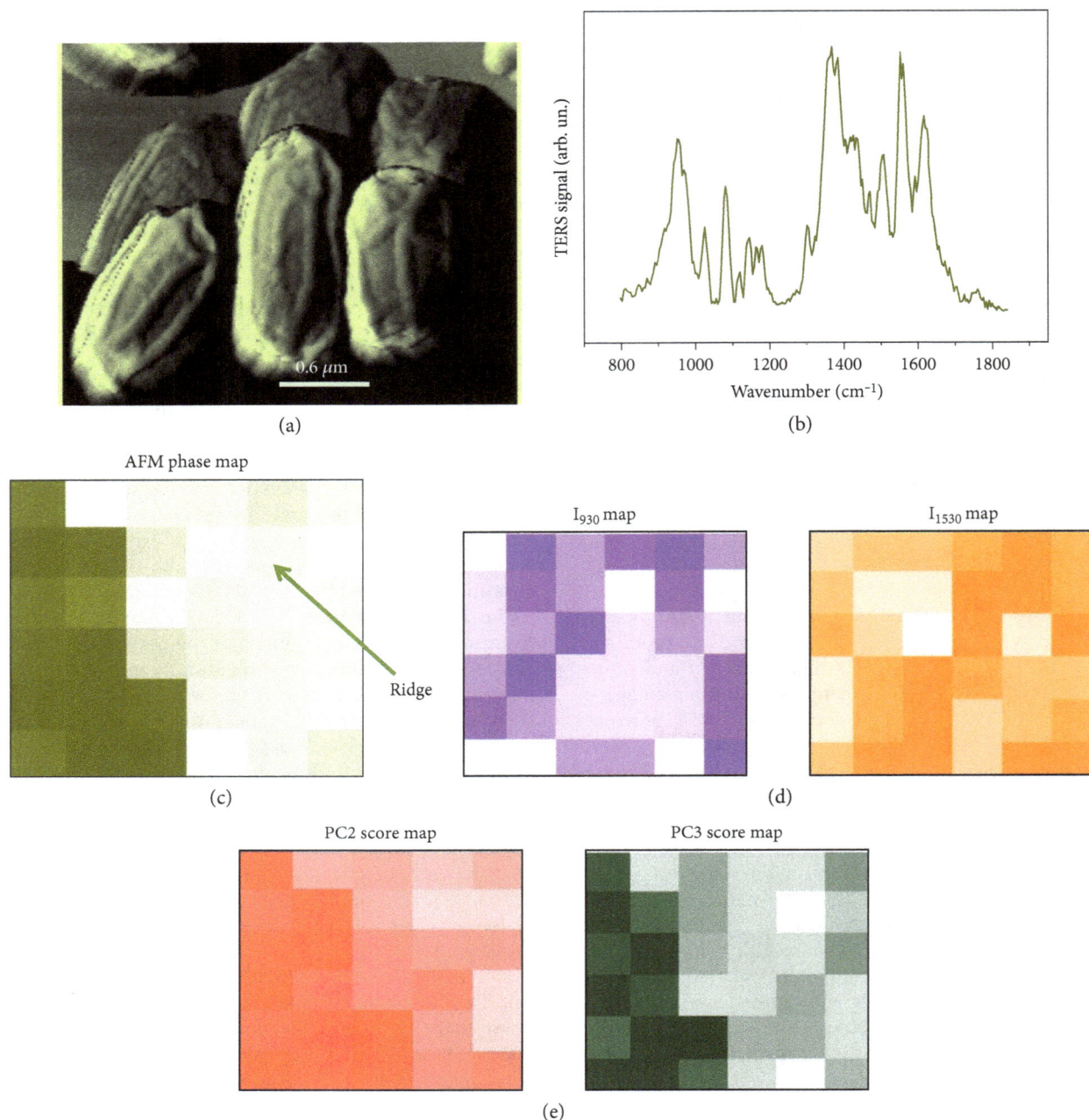

(a)

(b)

AFM phase map

Ridge

(c)

I$_{930}$ map

I$_{1530}$ map

(d)

PC2 score map

PC3 score map

(e)

FIGURE 5: (a) AFM topographical image of fully dehydrated *B. subtilis*. Spore surface exhibits characteristic ridges resulting from volume contraction due to the core dehydration. (b) A typical TERS spectrum from the coat of a *B. subtilis* spore. (c) AFM phase map of a whole spore acquired in tapping mode. Bright (dark) zones correspond to regions of higher (lower) stiffness. (d) TERS maps obtained by reporting the intensity of selected spectral features, indicated in the labels. (e) Second and third PC score maps of TERS data.

5. Conclusions

Single cell Raman spectroscopy, thanks to its noninvasive, label-free, and fingerprint character, provides rich and intrinsic information of the cell (e.g., nucleic acids, protein, carbohydrates, and lipids), reflecting cellular, physiological, and pathological states. This study demonstrates the possibility of applying Raman microspectroscopy for analyzing in vitro single *Acanthamoeba* cells, in order to monitor, in real time, the effect of PHMB, a strong disinfectant and antiseptic also employed to treat *A.* keratitis. A key drawback of Raman spectroscopy is related to the low inelastic scattering cross section, so that spontaneous Raman signals are naturally weak. This review discusses recent research progress in significantly enhancing and improving the signal of spontaneous Raman spectroscopy, including surface-enhanced Raman scattering and tip-enhanced Raman scattering. In particular, the results obtained for two interesting *case studies* are reported. Globally, our results underline the great potential of Raman and plasmon-enhanced Raman techniques for the analysis of biosystems, even for diagnostic purposes.

Competing Interests

The authors declare that they have no competing interests.

Acknowledgments

This research was supported in part by the project FIRB 2012-RBFR12WAPY of the Italian Ministry for Education, University and Research (MIUR).

References

[1] F. Siebert and P. Hildebrandt, *Instrumentation, in Vibrational Spectroscopy in Life Science*, Wiley-VCH Verlag GmbH & Co, KGaA, Weinheim, Germany, 2007.

[2] R. Smith, K. L. Wright, and L. Ashton, "Raman spectroscopy: an evolving technique for live cell studies," *Analyst*, vol. 141, no. 12, pp. 3590–3600, 2016.

[3] G. Rusciano, A. C. De Luca, A. Sasso, and G. Pesce, "Phase-sensitive detection in Raman tweezers," *Applied Physics Letters*, vol. 89, no. 26, Article ID 261116, 2006.

[4] M. Ghomi, Ed., *Applications of Raman Spectroscopy to Biology: from Basic Studies to Disease Diagnosis*, IOS Press, 2012.

[5] S. Managò, C. Valente, P. Mirabelli et al., "A reliable Raman-spectroscopy-based approach for diagnosis, classification and follow-up of B-cell acute lymphoblastic leukemia," *Scientific Reports*, vol. 6, article 24821, 2016.

[6] A. C. De Luca, K. Dholakia, and M. Mazilu, "Modulated Raman spectroscopy for enhanced cancer diagnosis at the cellular level," *Sensors*, vol. 15, no. 6, pp. 13680–13704, 2015.

[7] F. Casadio, C. Daher, and L. Bellot-Gurlet, "Raman spectroscopy of cultural heritage materials: overview of applications and new frontiers in instrumentation, sampling modalities, and data processing," *Topics in Current Chemistry*, vol. 374, no. 5, p. 62, 2016.

[8] Y. S. Li and J. S. Church, "Raman spectroscopy in the analysis of food and pharmaceutical nanomaterials," *Journal of Food and Drug Analysis*, vol. 22, no. 1, pp. 29–48, 2014.

[9] J. M. Chalmers, H. G. M. Edwards, and M. D. Hargreaves, Eds., *Introduction and Scope, in Infrared and Raman Spectroscopy in Forensic Science (eds J. M. Chalmers, H. G. M. Edwards and M. D. Hargreaves)*, John Wiley & Sons, Ltd, Chichester, UK, 2012.

[10] A. Zoubir, Ed., *Raman Imaging: Techniques and Applications*, vol. 168, Springer-Verlag, Berlin Heidelberg, 2012.

[11] I. Barman, N. C. Dingari, G. P. Singh, R. Kumar, S. Lang, and G. Nabi, "Selective sampling using confocal Raman spectroscopy provides enhanced specificity for urinary bladder cancer diagnosis," *Analytical and Bioanalytical Chemistry*, vol. 404, no. 10, pp. 3091–3099, 2012.

[12] A. C. De Luca, G. Rusciano, R. Ciancia et al., "Spectroscopical and mechanical characterization of normal and thalassemic red blood cells by Raman tweezers," *Optics Express*, vol. 16, no. 11, pp. 7943–7957, 2008.

[13] A. Zumbusch, G. R. Holtom, and X. S. Xie, "Three-dimensional vibrational imaging by coherent anti-Stokes Raman scattering," *Physical Review Letters*, vol. 82, no. 20, pp. 4142–4145, 1999.

[14] F. Lu, D. Calligaris, O. Olubiyi et al., "Label-free neurosurgical pathology with stimulated Raman imaging," *Cancer Research*, vol. 76, no. 12, pp. 3451–3462, 2016.

[15] M. Fleischmann, P. J. Hendra, and A. J. McQuillan, "Raman spectra of pyridine adsorbed at a silver electrode," *Chemical Physics Letters*, vol. 26, no. 2, pp. 163–166, 1974.

[16] D. L. Jeanmaire and R. P. Van Duyne, "Surface Raman spectroelectrochemistry, part 1: heterocyclic, aromatic, and aliphatic amines adsorbed on the anodized silver electrode," *Journal of Electroanalytical Chemistry*, vol. 84, no. 1, p. 120, 1977.

[17] M. G. Albrecht and J. A. Creighton, "Anomalously intense Raman spectra of pyridine at a silver electrode," *Journal of American Chemical Society*, vol. 99, no. 15, pp. 5215–5217, 1977.

[18] K. Kneipp, Y. Wang, H. Kneipp et al., "Single molecule detection using surface-enhanced Raman scattering (SERS)," *Physical Review Letters*, vol. 78, no. 9, pp. 1667–1670, 1997.

[19] E. C. Le Ru and P. G. Etchegoin, "Single-molecule surface-enhanced Raman spectroscopy," *Annual Review of Physical Chemistry*, vol. 63, pp. 65–87, 2012.

[20] S. Schlücker, "Surface-enhanced Raman spectroscopy: concepts and chemical applications," *Angewandte Chemie, International Edition*, vol. 53, no. 19, pp. 4756–4795, 2014.

[21] W. Xie and S. Schlücker, "Rationally designed multifunctional plasmonic nanostructures for surface-enhanced Raman spectroscopy: a review," *Reports on Progress in Physics*, vol. 77, no. 11, Article ID 116502, 2014.

[22] G. Rusciano, A. De Luca, G. Pesce et al., "Label-free probing of G-quadruplex formation by surface-enhanced Raman scattering," *Analytical Chemistry*, vol. 83, no. 17, pp. 6849–6855, 2011.

[23] J. H. Granger, N. E. Schlotter, A. C. Crawford, and M. D. Porter, "Prospects for point-of-care pathogen diagnostics using surface-enhanced Raman scattering (SERS)," *Chemical Society Reviews*, vol. 45, no. 14, pp. 3865–3882, 2016.

[24] Z. H. Kim, "Single-molecule surface-enhanced Raman scattering: current status and future perspective," *Frontiers of Physics*, vol. 9, no. 1, pp. 25–30, 2014.

[25] L. Yang, P. Li, and J. Liu, "Progress in multifunctional surface-enhanced Raman scattering substrate for detection," *RSC Advances*, vol. 4, no. 91, pp. 49635–49646, 2014.

[26] S. Ding, J. Yi, J. Li et al., "Nanostructure-based plasmon-enhanced Raman spectroscopy for surface analysis of materials," *Nature Reviews Materials*, vol. 1, 16021 pages, 2016.

[27] A. Bonifacio, S. Cervo, and V. Sergo, "Label-free surface-enhanced Raman spectroscopy of biofluids: fundamental aspects and diagnostic applications," *Analytical and Bioanalytical Chemistry*, vol. 407, no. 27, pp. 8265–8277, 2015.

[28] G. Rusciano, A. C. De Luca, G. Pesce, and A. Sasso, "On the interaction of nano-sized organic carbon particles with model lipid membranes," *Carbon*, vol. 47, no. 13, pp. 2950–2957, 2009.

[29] C. Muehlethaler, M. Leona, and J. R. Lombardi, "Review of surface enhanced Raman scattering applications in forensic science," *Analytical Chemistry*, vol. 88, no. 1, pp. 152–169, 2016.

[30] B. S. Yeo, J. Stadler, T. Schmid, R. Zenobi, and W. H. Zhang, "Tip-enhanced Raman spectroscopy: its status, challenges and future directions," *Chemical Physics Letters*, vol. 472, no. 1, pp. 1–13, 2009.

[31] B. S. Yeo, S. Maedler, T. Schmid, W. H. Zhang, and R. Zenobi, "Tip-enhanced Raman spectroscopy can see more: the case of cytochrome c," *Journal of Physical Chemistry C*, vol. 112, no. 13, pp. 4867–4873, 2008.

[32] E. A. Pozzi, M. D. Sonntag, N. Jiang, J. M. Klingsporn, M. C. Hersam, and R. P. Van Duyne, "Tip-enhanced Raman imaging: an emergent tool for probing biology at the nanoscale," *ACS Nano*, vol. 7, no. 2, pp. 885–888, 2013.

[33] R. Treffer, R. Bohme, T. Deckert-Gaudig et al., "Advances in TERS for biochemical applications," *Biochemical Society Transactions*, vol. 40, no. 4, pp. 609–614, 2012.

[34] T. Schmid, L. Opilik, C. Blum, and R. Zenobi, "Nanoscale chemical imaging using tip-enhanced Raman spectroscopy: a critical review," *Angewandte Chemie, International Edition*, vol. 52, no. 23, pp. 5940–5954, 2013.

[35] G. Sharma, T. Deckert-Gaudig, and V. Deckert, "Tip-enhanced Raman scattering-targeting structure-specific surface characterization for biomedical samples," *Advanced Drug Delivery Reviews*, vol. 89, pp. 42–56, 2015.

[36] Z. D. Schultz, J. Marr, and H. Wang, "Tip enhanced Raman scattering: plasmonic enhancements for nanoscale chemical analysis," *Nanophotonics*, vol. 3, no. 1–2, pp. 91–104, 2014.

[37] G. Zito, G. Rusciano, G. Pesce, A. Dochshanov, and A. Sasso, "Surface-enhanced Raman imaging of cell membrane by a highly homogeneous and isotropic silver nanostructure," *Nanoscale*, vol. 7, no. 18, pp. 8593–8606, 2015.

[38] W. J. Cho, Y. Kim, and J. K. Kim, "Ultrahigh-density array of silver nanoclusters for SERS substrate with high sensitivity and excellent reproducibility," *ACS Nano*, vol. 6, no. 1, pp. 249–255, 2012.

[39] S. Torquato and F. Stillinger, "Local density fluctuations, hyperuniformity, and order metrics," *Physical Review E*, vol. 68, no. 4, Article ID 041113, 2003.

[40] C. De Rosa, F. Auriemma, C. Diletto et al., "Toward hyperuniform disordered plasmonic nanostructures for reproducible surface-enhanced Raman spectroscopy," *Physical Chemistry Chemical Physics*, vol. 17, no. 12, pp. 8061–8069, 2015.

[41] G. Rusciano, P. Capriglione, G. Pesce et al., "Raman microspectroscopy analysis in the treatment of Acanthamoeba keratitis," *PloS One*, vol. 8, no. 8, pp. e72127–e72135, 2013.

[42] G. Rusciano, "Experimental analysis of Hb oxy–deoxy transition in single optically stretched red blood cells," *Physica Medica*, vol. 26, no. 4, pp. 233–239, 2010.

[43] G. Rusciano, P. Capriglione, G. Pesce et al., "Raman-spectroscopy-based biosensing for applications in ophthalmology," *SPIE Optical Sensors Book Series: Proceedings of the SPIE*, vol. 8774, Article ID 87740A, 2013.

[44] B. R. Wood and D. McNaughton, "Raman excitation wavelength investigation of single red blood cells in vivo," *Journal of Raman Specroscopy*, vol. 33, no. 7, pp. 517–523, 2002.

[45] G. Zito, G. Rusciano, A. Vecchione et al., "Nanometal skin of plasmonic heterostructures for highly efficient near-field scattering probes," *Scientific Reports*, vol. 6, p. 31113, 2016.

[46] E. Ricca and S. M. Cutting, "Emerging applications of bacterial spores in nanobiotechnology," *Journal of Nanobiotechnology*, vol. 1, no. 1, p. 6, 2003.

[47] X. Chen, L. Mahadevan, A. Driks, and O. Sahin, "Bacillus spores as building blocks for stimuli responsive materials and nanogenerators," *Nature Nanotechnology*, vol. 9, no. 2, pp. 137–141, 2014.

[48] G. Rusciano, G. Zito, R. Isticato et al., "Nanoscale chemical imaging of Bacillus subtilis spores by combining tip-enhanced Raman scattering and advanced statistical tools," *ACS Nano*, vol. 8, no. 12, pp. 12300–12309, 2014.

12

Study of Molecular and Ionic Vapor Composition over CeI$_3$ by Knudsen Effusion Mass Spectrometry

A. M. Dunaev,[1] V. B. Motalov,[1] L. S. Kudin,[1] M. F. Butman,[1] and K. W. Krämer[2]

[1]*Research Institute of Thermodynamics and Kinetics, Ivanovo State University of Chemistry and Technology, Ivanovo 153000, Russia*
[2]*Department of Chemistry and Biochemistry, University of Bern, 3012 Bern, Switzerland*

Correspondence should be addressed to V. B. Motalov; v.motalov@gmail.com

Academic Editor: Alessandro Longo

The molecular and ionic composition of vapor over cerium triiodide was studied by Knudsen effusion mass spectrometry. In the saturated vapor over CeI$_3$ the monomer, dimer, and trimer molecules and the negative ions I$^-$, CeI$_4^-$, and Ce$_2$I$_7^-$ were identified in the temperature range of 753–994 K. The partial pressures of CeI$_3$, Ce$_2$I$_6$, and Ce$_3$I$_9$ were determined and the enthalpies of sublimation, $\Delta_s H°$(298.15 K) in kJ·mol^{-1}, in the form of monomers (298 ± 9), dimers (415 ± 30), and trimers (423 ± 50) were obtained by the second and third laws of thermodynamics. The enthalpy of formation, $\Delta_f H°$(298.15 K) in kJ·mol^{-1}, of the CeI$_3$ (−371 ± 9), Ce$_2$I$_6$ (−924 ± 30), and Ce$_3$I$_9$ (−1585 ± 50) molecules and the CeI$_4^-$ (−857 ± 19) and Ce$_2$I$_7^-$ (−1451 ± 50) ions were calculated. The electron work function, φ_e = 3.3 ± 0.3 eV, for the CeI$_3$ crystal was evaluated.

1. Introduction

Vaporization thermodynamics of cerium triiodide is the focus of researcher's attention so far. The first measurements of vapor pressure over CeI$_3$ were carried out by Knudsen effusion mass spectrometry (KEMS) [1] and Knudsen effusion Cahn microbalance [2] techniques. Further KEMS studies were performed by Chantry [3, 4], Struck and Feuersanger [5], and Ohnesorge [6]. In addition the vapor pressure of CeI$_3$ was determined by the torsion method [7], by optical absorption spectra [8], and recently by X-ray induced fluorescence [9]. In spite of the numerous experimental results [1–9], data on the vapor composition over CeI$_3$ are very scanty. Moreover, information on the ionic species in saturated vapor over CeI$_3$ is absent so far.

The present work continues our systematic investigations of the molecular and ionic sublimation of lanthanide halides by KEMS; see for example, [10–14]. The composition of the saturated vapor of CeI$_3$ was determined and the thermochemical data of the vapor constituents were refined on the basis of the latest sets of molecular parameters.

2. Experimental

A single-focusing magnetic sector type mass spectrometer MI1201 modified for high-temperature studies was used. The combined ion source allowed carrying out successive measurements in two modes; see Figure 1 (taken from [15]). In addition to a standard mode of electron ionization (EI) for analysis of neutral vapor species, a thermal ion emission (TE) mode was introduced for the analysis of charged vapor constituents formed inside an effusion cell as a result of thermal ionization. In the latter case, the ions are drawn out from the cell by a weak electric field (10^4–10^5 V/m) applied between the cell and the collimator (1). The sample (2) was placed into a molybdenum cell (3) under dry conditions in a glove box and then transferred into the vaporization chamber of the mass spectrometer and evacuated. The lid of the cell had the cylindrical effusion orifice (Ø 0.3 × 0.8 mm). The vaporization-to-effusion area ratio was about 400. A resistance furnace was used for the heating of the cell. Its temperature was measured by a tungsten-rhenium thermocouple calibrated with silver to a ±5 K accuracy in the separate experiment.

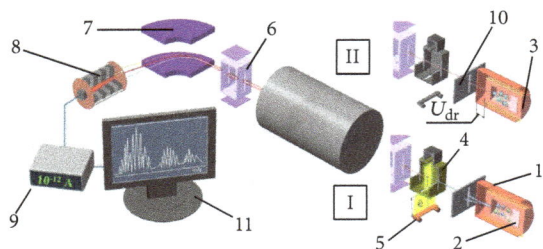

FigurE 1: Scheme of the mass spectrometer (taken from [15]). I—EI mode; II—TE mode. The details are given in text.

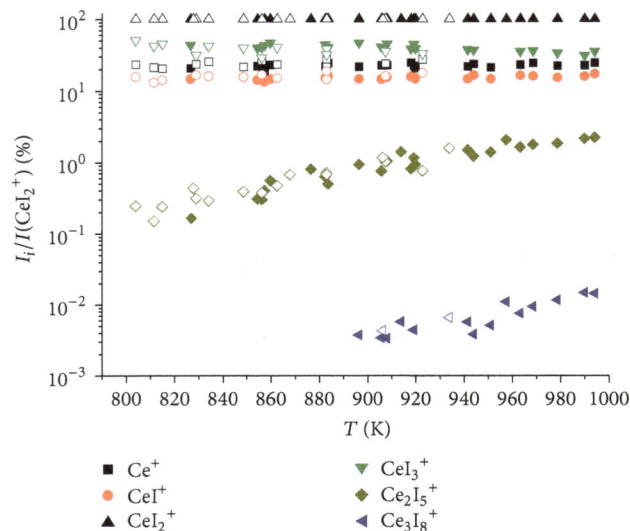

FigurE 2: Temperature dependence of the mass spectra of CeI_3. Heating and cooling runs are marked by solid and open symbols, respectively.

The vapor species effusing from the cell form a molecular beam, which reaches the ionization chamber (4) and intersects with an electron beam of specified energy. The ionization voltage U_e is set by a computer using a programmable power supply AKIP-1125 in the range of 0–150 V with 10 mV resolution. The tungsten ribbon-type cathode (5) is directly heated by an controllable DC source. The current of the cathode was adjusted to provide a constant emission current of 0.25 A.

The ions formed by the collision of molecular species with electrons are extracted from the ionization chamber, focused, and accelerated by a system of electrostatic lenses (6). The electrostatic capacitor mounted after the exit slit of the ion source allows us to study the distribution of ions by the vertical velocity component. The accelerating voltage (3 kV) is applied to the ionization chamber (IE mode) or to the effusion cell (TE mode). The polarity of the high voltage can be reversed with respect to the ground potential. Thus, both positive and negative ions can be analyzed. The ions are separated according to their mass-to-charge ratio in the magnetic field of an electromagnet (7) (90°, 200 mm curvature radius). The magnetic field strength is measured by a Hall probe. The ion current registration system (8, 9) includes a secondary electron multiplier Hamamatsu R595 (8) and a Picoammeter Keithley 6485 with 10 fA resolution and 20 fA typical noise. It allows measuring ion currents down to 10^{-18} A. The movable shutter (10) operated by a computer (11) allows distinguishing signals caused by the effusing species from those of the background. The special software "HTMSLab" was used to control experimental parameters, collect and process the data, and export the results into the database. Further details on the apparatus and experimental procedure can be found elsewhere [16–18].

The cerium triiodide sample was synthesized from cerium metal (99.9%, Metall Rare Earth Ltd.) and iodine (p.a., sublimed, Merck). The elements were sealed in an evacuated silica ampoule and slowly heated to 750°C until the reaction was completed. Afterwards the product was sublimed for purification in a sealed silica ampoule under vacuum at 750°C, that is, slightly below the melting point of CeI_3 at 766°C. The bright yellow CeI_3 is very hygroscopic. Its synthesis was performed under strictly oxygen-free and anhydrous conditions.

3. Results and Discussion

3.1. Neutral Vapor Species. In the IE mass spectra of the saturated vapor over cerium triiodide the $Ce^+(23)$, $CeI^+(16)$, $CeI_2^+(100)$, $CeI_3^+(31)$, $Ce_2I_3^+(0.02)$, $Ce_2I_4^+(0.01)$, $Ce_2I_5^+(2.2)$, $I^+(22)$, and $Ce_3I_8^+(0.02)$ ions, as well as the doubly charged $Ce^{++}(0.5)$, $CeI^{++}(7)$, and $CeI_2^{++}(3)$ ions, were registered in the temperature range of 753–994 K. The relative ion currents are given in parentheses for the temperature of 990 K and the energy of ionizing electrons of 40 eV. The mass spectra were found to be constant over the whole evaporation time; see Figure 2.

To determine the molecular precursors of the ions, the ionization efficiency curves (IEC) (Figure 3) and the temperature dependencies of ion currents (Figure 4) were analyzed. The ionizing electron energy in Figure 3 was corrected by the background signal of HI^+ ($AE = 10.38$ eV [19]). Appearance energies (AE) were determined by vanishing current (VC) and linear extrapolation (LE) methods; average values are given in Table 1. The linear part for the LE method was determined as the segment between two points of inflection on the first-order derivative of the IEC inverse function [20]. The following conclusions were drawn: the ions containing one atom of cerium are formed as a result of direct (CeI_3^+) and dissociative (Ce^+, CeI^+, and CeI_2^+) ionization of the monomer CeI_3 molecules with negligibly small contributions from the fragmentation of more complex molecules; the $Ce_2I_3^+$, $Ce_2I_4^+$, and $Ce_2I_5^+$ ions were produced by the dissociative ionization of the dimer Ce_2I_6 molecules; and the $Ce_3I_8^+$ ion originated from the trimer molecule Ce_3I_9.

Along with the abovementioned ions I^+ was also observed. The determined AE equal to 10.7 ± 0.5 eV (Table 1) points out its origination from atomic iodine (ionization energy $IE(I) = 10.4$ eV [21]), which can be attributed to the partial decomposition of the sample with the formation of

TABLE 1: Ion appearance energies (eV).

Reaction	Ion	AE				
		This work	[1]	[3]	[5]	[30]
$CeI_3 + \tilde{e} = Ce^+ + 3I + 2\tilde{e}$	Ce^+	17.2 ± 0.5	17.7 ± 0.5	16.75 ± 0.15		
$CeI_3 + \tilde{e} = CeI^+ + 2I + 2\tilde{e}$	CeI^+	13.1 ± 0.5	13.6 ± 0.5	13.15 ± 0.15		
$CeI_3 + \tilde{e} = CeI_2^+ + I + 2\tilde{e}$	CeI_2^+	9.8 ± 0.5	9.7 ± 0.5	9.55 ± 0.1	11.2	
$CeI_3 + \tilde{e} = CeI_3^+ + 2\tilde{e}$	CeI_3^+	9.1 ± 0.5	9.6 ± 0.5	9.05 ± 0.1	10.8	9.71
$Ce_2I_6 + \tilde{e} = Ce_2I_5^+ + I + 2\tilde{e}$	$Ce_2I_5^+$	9.3 ± 0.5				
$Ce_3I_9 + \tilde{e} = Ce_3I_8^+ + I + 2\tilde{e}$	$Ce_3I_8^+$	9.1 ± 0.5				
$I + \tilde{e} = I^+ + 2\tilde{e}$	I^+	10.7 ± 0.5	~ 13.5	10.5		

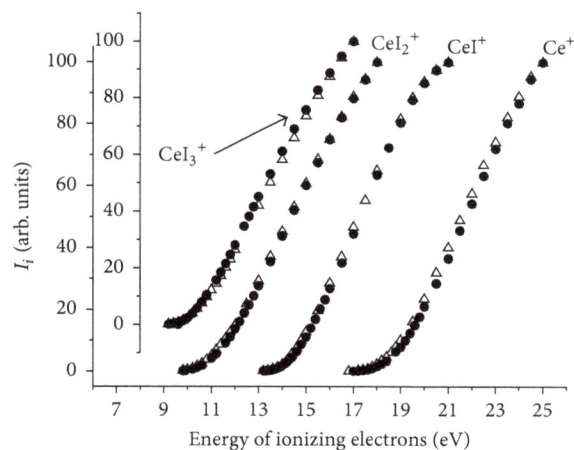

△ 915 K
● 957 K

(a)

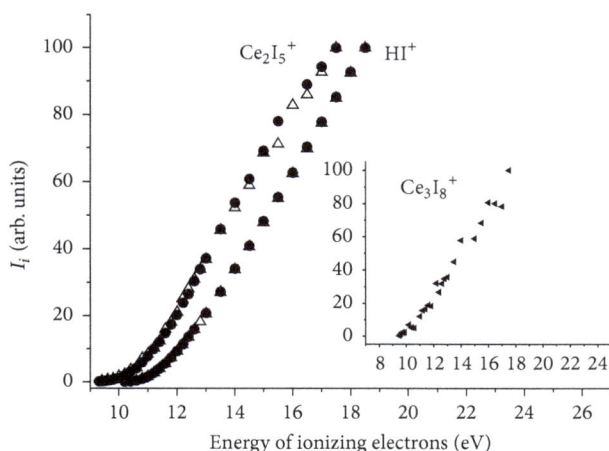

△ 915 K   ◄ 957 K
● 957 K

(b)

FIGURE 3: Ionization efficiency curves.

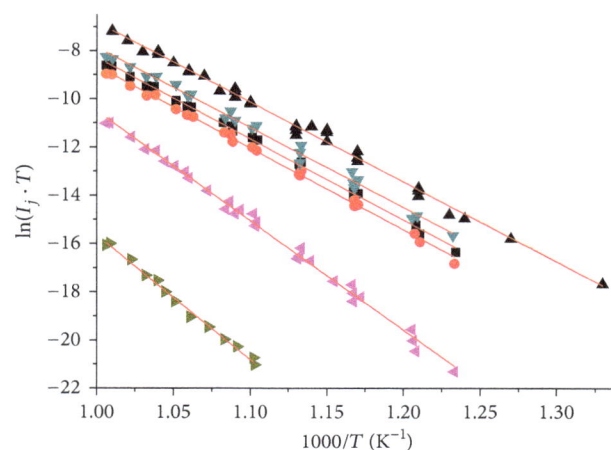

■ Ce^+ ▼ CeI_3^+
● CeI^+ ◄ $Ce_2I_5^+$
▲ CeI_2^+ ► $Ce_3I_8^+$

FIGURE 4: Temperature dependence of ion currents in the EI regime.

CeI_2. Nevertheless the reproducibility of the mass spectra in heating and cooling cycles (Figure 2), the shapes of IECs showing no brakes, and the determined AE values (Table 1) indicate unambiguously the absence of CeI_2 in the vapor. This fact agrees with the expected much lower volatility of CeI_2

compared to that of CeI_3 in the studied temperature range. Therefore it is assumed that the activity of CeI_3 in the solid state was unity.

The partial pressures of molecules (p_j) (see Table 2) were calculated according to the conventional KEMS procedure using the equation

$$p_j = \frac{k \cdot I \cdot T}{\sigma_j}, \qquad (1)$$

where k is the sensitivity constant of mass spectrometer (determined in a separate experiment with Ag; the vapor pressure of silver was taken from [22]), σ_j is the ionization cross section of the jth molecule at the working energy of the ionizing electrons (calculated from the experimentally determined atomic cross sections, σ_{at} [23, 24], by the equation $\sigma_j = 0.75\Sigma\sigma_{at}$ [25]), $I_j = \sum_i (I_{ij}/(a_i \cdot \gamma_i))$ is the total ion current of the ith ion species formed from the jth molecule, a_i is the natural abundance of the measured isotope of the ith ion, γ_i is the ion-electron conversion coefficient of secondary electron multiplier for the ith ion ($\gamma_i \sim M^{-1/2}$ [26], where M is the mass of ion), and T is the temperature of the cell.

TABLE 2: Partial pressures (Pa) of the molecules.

CeI_3		Ce_2I_6		Ce_3I_9	
T, K	$p_j \cdot 10^3$	T, K	$p_j \cdot 10^4$	T, K	$p_j \cdot 10^5$
885	54.5	878	3.59	916	0.85
872	47.8	915	26.8	957	7.45
877	55.4	957	182	932	1.83
877	54.7	933	57.0	907	0.41
914	235	906	14.3	905	0.65
957	1080	866	1.62	921	0.84
934	445	829	0.15	945	4.29
906	150	857	1.27	952	5.24
868	29.8	882	5.80	962	12.5
828	4.32	906	17.0	978	29.9
786	0.52	920	39.8	992	58.1
805	1.23	946	134	907	0.51
753	0.081	853	0.76	923	1.08
872	43.5	952	170	943	2.74
957	1210	963	323	969	15.3
857	20.3	978	566	994	54.2
882	59.5	992	1010		
906	145	907	23.9		
919	274	923	28.8		
942	575	883	4.19		
854	15.8	857	0.90		
951	786	829	0.20		
963	1280	811	0.035		
979	1970	828	0.081		
990	3020	857	0.64		
908	150	884	3.77		
923	244	917	24.0		
883	40.1	943	105		
856	15.5	969	343		
829	4.06	994	999		
811	1.48				
826	3.24				
856	13.7				
883	48.8				
918	189				
944	551				
969	1250				
994	2880				

TABLE 3: Coefficients of (2).

Species	T, K	A	B
CeI_3	753–994	33.35 ± 0.36	34.84 ± 0.41
Ce_2I_6	811–994	44.93 ± 0.62	43.10 ± 0.69
Ce_3I_9	905–994	49.62 ± 1.43	42.43 ± 1.52

The standard deviation is given with a "±" sign.

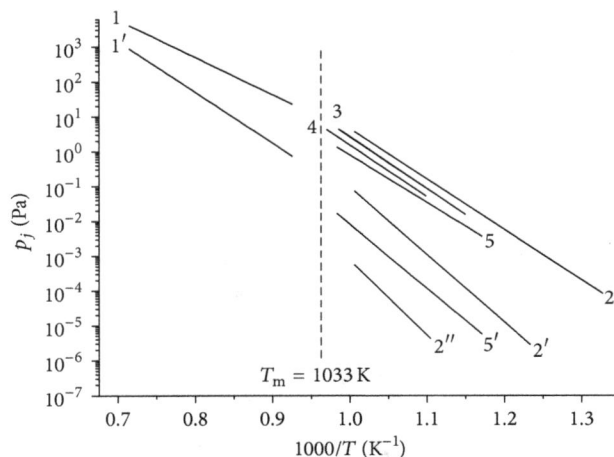

FIGURE 5: Temperature dependencies of the saturated vapor pressure: (1) $p(CeI_3)$ from [9], (2) our data, (3) [2], (4) [7], and (5) [6]; (1′) $p(Ce_2I_6)$ from [9], (2′) our data, and (5′) [6]; (2″) $p(Ce_3I_9)$ our data.

The temperature dependence of the saturated vapor pressures of the monomer and oligomer molecules was approximated by the equation

$$\ln p_i = \frac{-A \times 10^3}{T} + B. \qquad (2)$$

The coefficients of equation (2) are given in Table 3.

The partial pressures of the molecules of cerium triiodide from different references are compared in Figure 5. As one can see, all experimental vapor pressure values are scattered within about one order of magnitude. Temperature

dependencies from the work of Hirayama et al. [2], Villani et al. [7], and Ohnesorge [6] lie below those obtained in this work. The fraction of the dimer molecules measured in this work and in [6] is about the same whereas the absolute pressures differ considerably. The vapor pressure of trimer molecules was determined in this study for the first time.

The enthalpies and entropies of sublimation of cerium triiodide in the form of monomer and oligomer molecules were determined from the temperature dependencies of the partial pressures of the saturated vapor species using the procedure for experimental data processing according to the second and third laws of thermodynamics; see Table 4. The thermodynamic functions required for calculations were taken from [22] for $CeI_{3,cr}$ and evaluated in this work for the monomer and oligomer molecules in the state of an ideal gas (see Appendix).

As it is seen from Table 4, the values of $\Delta_s H°(298.15)$ and $\Delta_s S°(T)$ obtained in this work by the second and third laws are in a fair agreement for both monomer and oligomer molecules. The same can be said about the results for the monomer molecules from [2, 7]. The data of the work [9] are in notably worse agreement, whereas those of [6, 8] do not agree within the given uncertainties. At the same time, the third law values for all the data are in a good consent with each other. The temperature trend of the third law values $\Delta_s H°(298.15)$ is given in Figure 6, from which one can see that the data of this work and [2, 9] do not show a pronounced temperature dependence as compared to those of [6, 7]. Taking into account this analysis, the recommended values

TABLE 4: Enthalpies, $\Delta_s H°$ (kJ·mol^{-1}), and entropies, $\Delta_s S°$ (J·mol^{-1}·K^{-1}), of sublimation of cerium triiodide.

ΔT, K	T, K	N^1	II law^2			III law^3		Ref.
			$\Delta_{s,v}H°(T)$	$\Delta_{s,v}S°(T)$	$\Delta_s H°(298)$	$\Delta_s H°(298)$	$\Delta_{s,v}S°(T)$	
				CeI$_{3,cr,l}$ = CeI$_{3,g}$				
753–994	893	38	277 ± 3	194 ± 3	292 ± 4	294 ± 10	191 ± 10	This work
?	933	?	292 ± 21		307 ± 21			[1]
870–1015	945	32	284 ± 4	197 ± 4	301 ± 4	298 ± 10	189 ± 10	[2]
810–953	877	50	274 ± 2		288 ± 2			[5]
854–1017	936	29	284 ± 1	164 ± 1	278 ± 1	307 ± 10	189 ± 10	[6]
910–1031	970	81	284 ± 3	192 ± 6	301 ± 3	303 ± 10	189 ± 10	[7]
	1000	1				291 ± 10	187 ± 10	[8]
1072–1136	1094	5	201 ± 11	119 ± 10	272 ± 11	296 ± 10	155 ± 10	[8]
1080–1400	1250	?	299 ± 1	119 ± 1	290 ± 1	301 ± 10	123 ± 10	[9]
						297		[22]
						322		[27]
						280		[28]
						300 ± 5		[29]
				$\Delta_s H°$(298.15 K) = **298 ± 9**^{4}				
				2CeI$_{3,cr,l}$ = Ce$_2$I$_{6,g}$				
811–994	893	30	374 ± 6	262 ± 6	395 ± 25	409 ± 30	278 ± 30	This work
825–953	898	50	377 ± 5		400 ± 15			[5]
854–1017	936	17	354 ± 1	219 ± 1	379 ± 25	428 ± 30	276 ± 30	[6]
1080–1400	1250	?	251 ± 1	138 ± 1	413 ± 1	421 ± 30	148 ± 30	[9]
				$\Delta_s H°$(298.15 K) = **415 ± 30**				
				3CeI$_{3,cr}$ = Ce$_3$I$_{9,g}$				
907–994	949	16	413 ± 12	257 ± 13	443 ± 40	423 ± 50	236 ± 50	This work
				$\Delta_s H°$(298.15 K) = **423 ± 50**				

^{1} Number of measurements.
^{2} The original errors are given for the literature data, the standard deviations for this work.
^{3} The uncertainties are mainly determined by those in thermodynamic functions.
^{4} The uncertainty of the recommended values was calculated by Student's method (monomer) and accepted as the third law ones (others).

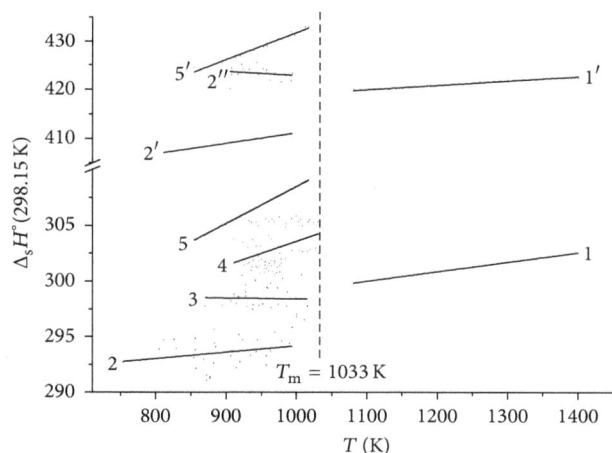

FIGURE 6: Temperature dependencies of the third law $\Delta_s H°$(298.15 K) values: (1) monomer from [9], (2) our data, (3) [2], (4) [7], and (5) [6]; (1′) dimer from [9], (2′) our data, and (5′) [6]; (2″) trimer our data.

were selected and emphasized in bold in Table 4. The early estimates [27, 28] made for monomers differ considerably from the recommended value, while the assessments given in [22, 29] agree with the experimental data.

The *standard formation enthalpies*, $\Delta_f H°$(298.15 K), are equal to $-371 ± 9$ kJ·mol^{-1} (CeI$_3$), $-924 ± 30$ kJ·mol^{-1} (Ce$_2$I$_6$), and $-1585 ± 50$ kJ·mol^{-1} (Ce$_3$I$_9$) and were calculated from the recommended $\Delta_s H°$(298.15 K) values. The formation enthalpy of crystalline cerium triiodide, $\Delta_f H°$(CeI$_{3,cr}$; 298.15 K) = -669 kJ·mol^{-1} [22], was used.

3.2. Charged Vapor Species. In the TE mass spectra in the temperature range of 747–960 K the I$^-$(0.44), CeI$_3^-$(0.18), CeI$_4^-$(100), and Ce$_2$I$_7^-$(0.08) ions were identified with the relative ion currents given in parentheses for $T = 846$ K.

Additional experiments with the CeI$_3$-PrI$_3$ binary system were performed to determine the enthalpy of formation of CeI$_4^-$ ions. The enthalpy of exchange ion-molecular reaction (5) was found to be $1 ± 5$ kJ·mol^{-1}. The enthalpy of formation of PrI$_4^-$ $\Delta_f H°$(298.15 K) = $-860 ± 23$ kJ·mol^{-1} was used as reference value. It was obtained by recalculation of the data ($-865 ± 25$ kJ·mol^{-1}) [13] with the thermodynamic function used in this work. The experimental equilibrium constants of reaction (3) and reaction (4) investigated over pure CeI$_3$ are listed in Table 5:

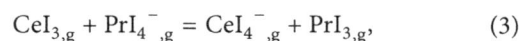

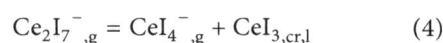

$$CeI_{3,g} + PrI_4^-{}_{,g} = CeI_4^-{}_{,g} + PrI_{3,g}, \tag{3}$$

$$Ce_2I_7^-{}_{,g} = CeI_4^-{}_{,g} + CeI_{3,cr,l} \tag{4}$$

TABLE 5: Experimental data for reactions (3) and (4).

	Reaction (3)		Reaction (4)
T (K)	$\ln K_P^\circ$	T (K)	$\ln K_P^\circ$
889	0.35	868	6.39
865	0.39	862	6.61
835	0.30	898	5.92
846	0.49	932	5.43
907	0.41	960	4.94
934	0.39	924	5.49
966	0.38	892	5.94
956	0.37	846	6.67
929	0.34		
906	0.22		
856	0.34		
806	0.38		
854	0.30		
884	0.38		
902	0.30		
950	0.40		
976	0.30		

The reaction enthalpies were calculated by the second and third laws of thermodynamics; see Table 6. The evaluation of the thermodynamic functions of the CeI_4^- and $Ce_2I_7^-$ ions is described in Appendix.

One can see that the agreement for the enthalpy of reaction (3) obtained by the second and third laws is good. $\Delta_f H^\circ(CeI_4^-, 298.15\,K) = -857 \pm 19\,kJ\cdot mol^{-1}$ was recommended. This value is in agreement with those $-850 \pm 33\,kJ\cdot mol^{-1}$ obtained by Chantry from the enthalpy of the reaction $CeI_4^-{}_{,g} = I^-{}_{,g} + CeI_{3,g}$ and recalculated with our sublimation enthalpy and thermodynamic functions. The selected formation enthalpy of the $Ce_2I_7^-$ ion is $-1451 \pm 50\,kJ\cdot mol^{-1}$ (third law) and was obtained for the first time.

3.3. Thermodynamic Properties Derived from IECs. The enthalpies of ion-molecular reactions were calculated from the differences of the measured AE values given in Table 1; see Table 7. On their basis the standard formation enthalpies of the ions were determined; see Table 8. The accurate values of $\Delta_f H^\circ(Ce^+, 298.15\,K) = 957 \pm 0.1\,kJ\cdot mol^{-1}$ [22] and $\Delta_f H^\circ(I, 298.15\,K) = 106.76 \pm 0.04\,kJ\cdot mol^{-1}$ [22] and the formation enthalpies of the CeI_3, Ce_2I_6, and Ce_3I_9 molecules obtained in this work (see above) were used as references. The comparison of our results with the data obtained by photoelectron spectroscopy [30], computation [31], and assessment [32] shows that all data are in agreement within the given errors; see Table 8.

Analysis of the ion-molecular reaction enthalpies confirmed the weakness of the first cerium iodine bond in comparison with the other two. Probably, it explains the lower enthalpy of the formation of CeI_2^+ ($467 \pm 10\,kJ\cdot mol^{-1}$) compared to CeI_3^+ ($497 \pm 10\,kJ\cdot mol^{-1}$).

The formation enthalpy of gaseous CeI_3 ($-371 \pm 9\,kJ\cdot mol^{-1}$) obtained from the vapor pressure measurement is in a good agreement with those calculated from the appearance energies ($-376 \pm 50\,kJ\cdot mol^{-1}$) and with the value ($-381\,kJ\cdot mol^{-1}$) assessed by Sapegin et al. [33] (Table 9). The coincidence of the formation enthalpy values of $CeI_{3,g}$, as determined from the thermodynamic and threshold approaches, points out a negligible contribution from the excitation and kinetic energy of the fragments. The atomization energy derived from $\Delta_f H^\circ(CeI_3, 298.15\,K) = -371 \pm 9\,kJ\cdot mol^{-1}$ was found to be $1109 \pm 10\,kJ\cdot mol^{-1}$. It yields the average bond strength equal to $370 \pm 10\,kJ\cdot mol^{-1}$.

3.4. Electron Work Function. The mass spectrometric approach for the work function determination is based on the use of thermochemical cycles including desorption enthalpies of ions and sublimation enthalpies of molecules as described elsewhere [34]. The desorption enthalpy of the CeI_4^- ions, $\Delta_{des} H^\circ(851\,K) = 350 \pm 8\,kJ\cdot mol^{-1}$, and the $Ce_2I_7^-$ ions, $\Delta_{des} H^\circ(902\,K) = 430 \pm 18\,kJ\cdot mol^{-1}$, were obtained from the temperature dependence of their ion currents; see Figure 7. The electron work function was calculated in accordance with the following expressions:

$$\varphi_e = -\frac{1}{4}D(CeI_3) + \frac{5}{4}\cdot\Delta_s H^\circ(CeI_3) + EA(I) + \Delta_{diss} H^\circ(CeI_4^-) + \Delta_{des} H^\circ(CeI_4^-), \tag{5}$$

$$\varphi_e = -\frac{1}{4}D(CeI_3) + \frac{9}{4}\cdot\Delta_s H^\circ(CeI_3) + EA(I) + \Delta_{diss} H^\circ(Ce_2I_7^-) + \Delta_{des} H^\circ(Ce_2I_7^-), \tag{6}$$

where EA(I) is the electron affinity of iodine [22], $D(CeI_3)$ is the dissociation enthalpy of the CeI_3 molecule, $\Delta_{diss} H^\circ(CeI_4^-)$ is the enthalpy of the $CeI_4^- = I^- + CeI_3$ reaction (Table 6), and $\Delta_{diss} H^\circ(Ce_2I_7^-)$ is the enthalpy of the $Ce_2I_7^- = I^- + 2CeI_3$ reaction.

The φ_e values equal to $3.3 \pm 0.3\,eV$ and $3.5 \pm 0.5\,eV$ were obtained from (5) and (6), respectively. They turned out to be close to $\varphi_e(LaI_3) = 3.5 \pm 0.3\,eV$ [14].

4. Conclusions

New experimental data on the saturated vapor composition of CeI_3 have been obtained by a Knudsen effusion mass spectrometer. The monomer, CeI_3, dimer, Ce_2I_6, and trimer, Ce_3I_9, molecules, as well as the $[I(CeI_3)_n]^-$ ions ($n = 0-2$), have been observed in the temperature range of 747–994 K. The Ce_3I_9 molecules and the $Ce_2I_7^-$ ions were detected for the first time. The sublimation enthalpies of the monomer and oligomer molecules were calculated by the second and third laws of thermodynamics. Critical analysis of all available data allowed us to recommend the following enthalpies of sublimation: $\Delta_s H^\circ(298.15\,K)$ in $kJ\cdot mol^{-1}$: 298 ± 9 (monomer), 415 ± 30 (dimer), and 423 ± 50 (trimer) and to calculate the formation enthalpies, $\Delta_f H^\circ(298.15\,K)$ in $kJ\cdot mol^{-1}$: -371 ± 9 (CeI_3), -924 ± 25 (Ce_2I_6), -1585 ± 50 (Ce_3I_9), -857 ± 19

TABLE 6: Enthalpies (kJ·mol^{-1}) of reactions (3) and (4).

ΔT, K	T, K	N^1	II law^2		III law^3	Ref.
			$\Delta_r H°(T)$	$\Delta_r H°(298.15\text{ K})$	$\Delta_r H°(298.15\text{ K})$	
			$\text{CeI}_{3,g} + \text{PrI}_4{}^-{}_{,g} = \text{CeI}_4{}^-{}_{,g} + \text{PrI}_{3,g}$			
806–976	897	17	0 ± 3	1 ± 3	1 ± 5	This work,
			$\text{Ce}_2\text{I}_7{}^-{}_{,g} = \text{CeI}_4{}^-{}_{,g} + \text{CeI}_{3,\text{cr,l}}$			
842–960	896	12	-99 ± 5	-113 ± 6	-75 ± 40	This work

^{1}Number of measurements.
^{2}The original errors are given for the literature data, the standard deviations for this work.
^{3}The uncertainties are mainly determined by those in thermodynamic functions.

TABLE 7: Ion-molecular reactions (kJ·mol^{-1}).

Reaction	This work	[1]	[3]	[5]	[30]	[31]
$\text{CeI}^+{}_{,g} = \text{Ce}^+{}_{,g} + \text{I}_g$	391 ± 10	396 ± 50	347 ± 50			393
$\text{CeI}_2{}^+{}_{,g} = \text{CeI}^+{}_{,g} + \text{I}_g$	323 ± 10	376 ± 50	347 ± 50			
$\text{CeI}_3{}^+{}_{,g} = \text{CeI}_2{}^+{}_{,g} + \text{I}_g$	66 ± 10	10 ± 50	48 ± 50	39	62 ± 50	
$\text{CeI}_{3,g} + \text{Ce}_2\text{I}_5{}^+{}_{,g} = \text{Ce}^+{}_{,g} + 2\text{I}_g + \text{Ce}_2\text{I}_{6,g}$	750 ± 10					
$\text{CeI}_{3,g} + \text{Ce}_3\text{I}_8{}^+{}_{,g} = \text{Ce}^+{}_{,g} + 2\text{I}_g + \text{Ce}_3\text{I}_{9,g}$	768 ± 20					

TABLE 8: Ion formation enthalpies (kJ·mol^{-1}).

Ion	This work	[1]	[3]	[32]	[31]
CeI^+	673 ± 10	669 ± 50	717	760	671
$\text{CeI}_2{}^+$	467 ± 10	399 ± 50	476		
$\text{CeI}_3{}^+$	497 ± 10	496 ± 50	535		
$\text{Ce}_2\text{I}_5{}^+$	-131 ± 10				
$\text{Ce}_3\text{I}_8{}^+$	-811 ± 20				

TABLE 9: Formation enthalpies of $\text{CeI}_{3,g}$ (kJ·mol^{-1}).

This work from IEC	This work from $\Delta_s H°$	[1]	[9]	[33]
$> -376 \pm 50$	-371 ± 9	-436 ± 50	-349	-381

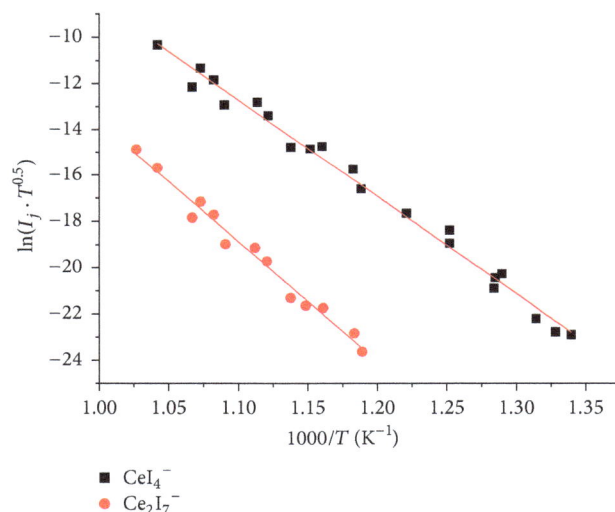

FIGURE 7: Temperature dependencies of $\text{CeI}_4{}^-$ and $\text{Ce}_2\text{I}_7{}^-$ ion currents.

($\text{CeI}_4{}^-$), and -1451 ± 50 ($\text{Ce}_2\text{I}_7{}^-$). The electron work function for cerium triiodide (3.3 ± 0.3 eV) has also been evaluated.

Appendix

A. Description of the Used Thermodynamic Functions of Molecules and Ions

A.1. $CeI_{3,cr,l}$. The thermodynamic functions of CeI$_3$ in the solid and liquid state were taken from [22].

A.2. $CeI_{3,g}$. For the first time the thermodynamic functions of monomer molecules CeI$_3$ in the state of an ideal gas were evaluated by Myers and Graves [35] in the rigid rotator-harmonic oscillator (RRHO) approximation; these were included in the handbooks by Pankratz [36] and Barin [37]. Later the assessment of the functions was made by Osina et al. [38]; they can be found in the IVTANTHERMO database [22]. Recently Solomonik et al. [39] performed quantum-chemical calculations for CeI$_{3,g}$ at a multireference configuration interaction MRCISD+Q level of theory taking into account relativistic effects; a significant spin-orbit coupling effect on the molecular properties was revealed.

The data used in this work are based on those from [22] taking into account the low lying electronic states computed in [39].

A.3. $Ce_2I_{6,g}$. The thermodynamic functions of dimer molecules were calculated in the RRHO approximation using the molecular constants (symmetry D_{2h}) from [40]. The electronic contribution was taken as doubled compared with the monomers in accordance with [41].

TABLE 10: Coefficients of polynomial (A.4).

	T, K	$-a$, 10^{-1}	b, 10^2	$-c$	d, 10^{-3}	$-e$, 10^{-3}	f, 10^{-4}
$CeI_{3,cr,l}$	0–1033	47.08	2.326	3.838	2.391	8.699	1.760
	1033–1500	106.9	26.03	16.90	1.830	3.165	0.289
CeI_3	0–1500	110.0	1.489	4.011	3.480	12.77	2.158
CeI_4^-	0–1500	122.9	1.780	4.748	4.156	15.36	2.604
Ce_2I_6	0–1500	177.9	2.760	7.349	6.477	23.88	4.043
$Ce_2I_7^-$	0–1500	27.52	16.59	20.47	4.390	9.010	0.966
Ce_3I_9	800–1500	992.1	441.8	194.6	2.251	1.601	1.060

The errors in the functions of Gibbs energy estimated in this work are assumed to be equal to ± 5 (CeI_3), ± 25 (Ce_2I_6), ± 40 ($Ce_2I_7^-$), and ± 50 (Ce_3I_9) J mol^{-1} K^{-1}.

A.4. $Ce_3I_{9,g}$. The thermodynamic functions of trimers were assessed by the additive approach with an empiric correction based on the functions of the monomer and oligomer molecules of $LuCl_3$ [42] using the following expressions:

$$TDF_{La_3I_9} = \beta_T \cdot \left(TDF_{LaI_3} + TDF_{La_2I_6} \right),$$

$$\beta_T = \frac{TDF_{Lu_3Cl_9}}{TDF_{LuCl_3} + TDF_{Lu_2Cl_6}}, \qquad (A.1)$$

where TDF means the thermodynamic functions $\Phi°(T)$ or $H°(T) - H°(0)$.

A.5. $CeI_4^-{}_{,g}$. The thermodynamic functions of the CeI_4^- ions were computed in the RRHO approximation assuming a coincidence of the functions for CeI_4^- and CeI_4 based on the molecular constants taken from [43] and the electronic states from [39].

A.6. $Ce_2I_7^-{}_{,g}$. The thermodynamic functions of $Ce_2I_7^-$ ions were evaluated by a comparative method on the basis of those for LnI_3, LnI_4^- (Ln = La, Ce) [22, 44], Ce_2I_6 [40], and $La_2I_7^-$ [14] using the following relation:

$$TDF\left(Ce_2I_7^-\right) = \beta_T' \cdot TDF\left(Ce_2I_6\right), \qquad (A.2)$$

where

$$\beta_T' = \frac{TDF\left(CeI_4^-\right)/TDF\left(CeI_3\right)}{TDF\left(LaI_4^-\right)/TDF\left(LaI_3\right)} \cdot \frac{TDF\left(La_2I_7^-\right)}{TDF\left(La_2I_6\right)}. \qquad (A.3)$$

The thermodynamic functions used in this work were approximated by the polynomial

$$\Phi°(T) = a \ln x + bx^{-2} + cx^{-1} + dx + ex^2 + fx^3,$$

$$\left(x = \frac{T}{1000} \right). \qquad (A.4)$$

The coefficients of (A.4) are listed in Table 10.

Competing Interests

The authors declare that there is no conflict of interests regarding the publication of this paper.

Acknowledgments

This work was supported by the Ministry of Education and Science of the Russian Federation (Project no. 4.1385.2014 K).

References

[1] C. Hirayama and P. M. Castle, "Mass spectra of rare earth triiodides," *Journal of Physical Chemistry*, vol. 77, no. 26, pp. 3110–3114, 1973.

[2] C. Hirayama, J. F. Rome, and F. E. Camp, "Vapor pressures and thermodynamic properties of lanthanide triiodides," *Journal of Chemical & Engineering Data*, vol. 20, no. 1, pp. 1–6, 1975.

[3] P. J. Chantry, "Positive ion appearance potentials measured in CeI_3," *The Journal of Chemical Physics*, vol. 65, no. 11, pp. 4421–4425, 1976.

[4] P. J. Chantry, "Negative ion formation in cerium triiodide," *The Journal of Chemical Physics*, vol. 65, no. 11, pp. 4412–4420, 1976.

[5] C. W. Struck and A. E. Feuersanger, "Knudsen cell measurements of CeI_3(s) sublimation enthalpy," *High Temperature Science*, vol. 31, pp. 127–145, 1991.

[6] M. Ohnesorge, *Untersuchungen zur Hochtemperaturchemie quecksilberfreier Metallhalogenid-Entladungslampen mit keramischem Brenner [Ph.D. thesis]*, Forschungszentrum Jülich, Jülich, Germany, 2005.

[7] A. R. Villani, B. Brunetti, and V. Piacente, "Vapor pressure and enthalpies of vaporization of cerium trichloride, tribromide, and triiodide," *Journal of Chemical and Engineering Data*, vol. 45, no. 5, pp. 823–828, 2000.

[8] C. S. Liu and R. J. Zoilweg, "Complex molecules in cesium-rare earth iodide vapors," *The Journal of Chemical Physics*, vol. 60, pp. 2400–2413, 1970.

[9] J. J. Curry, E. G. Estupiñán, W. P. Lapatovich et al., "Study of CeI_3 evaporation in the presence of group 13 metal-iodides," *Journal of Applied Physics*, vol. 115, no. 3, Article ID 034509, 2014.

[10] D. N. Sergeev, M. F. Butman, V. B. Motalov, L. S. Kudin, and K. W. Krämer, "Knudsen effusion mass spectrometric determination of the complex vapor composition of samarium, europium, and ytterbium bromides," *Rapid Communications in Mass Spectrometry*, vol. 27, no. 15, pp. 1715–1722, 2013.

[11] D. N. Sergeev, A. M. Dunaev, M. F. Butman, D. A. Ivanov, L. S. Kudin, and K. W. Krämer, "Energy characteristics of molecules and ions of ytterbium iodides," *International Journal of Mass Spectrometry*, vol. 374, pp. 1–3, 2014.

[12] L. S. Kudin, D. E. Vorob'ev, and A. E. Grishin, "The thermochemical characteristics of the $LnCl_4^-$ and $Ln_2Cl_7^-$ negative ions," *Russian Journal of Physical Chemistry A*, vol. 81, no. 2, pp. 147–158, 2007.

[13] V. B. Motalov, D. E. Vorobiev, L. S. Kudin, and T. Markus, "Mass spectrometric investigation of neutral and charged constituents in saturated vapor over PrI_3," *Journal of Alloys and Compounds*, vol. 473, no. 1-2, pp. 36–42, 2009.

[14] A. M. Dunaev, L. S. Kudin, V. B. Motalov, D. A. Ivanov, M. F. Butman, and K. W. Krämer, "Mass spectrometric study of molecular and ionic sublimation of lanthanum triiodide," *Thermochimica Acta*, vol. 622, pp. 82–87, 2015.

[15] D. N. Sergeev, *Energy characteristics of molecules and ions of lanthanide bromide (Sm, Eu, Yb) studied by high temperature mass spectrometry [Ph.D. thesis]*, 2011.

[16] A. M. Dunaev, A. S. Kryuchkov, L. S. Kudin, and M. F. Butman, "Automatic complex for high temperature investigation on basis of mass spectrometer MI1201," *Izvestiya Vysshikh Uchebnykh Zavedeniy Seriya "Khimiya I Khimicheskaya Tekhnologiya"*, vol. 54, pp. 73–77, 2011 (Russian).

[17] D. N. Sergeev, A. M. Dunaev, D. A. Ivanov, Y. A. Golovkina, and G. I. Gusev, "Automatization of mass spectrometer for the obtaining of ionization efficiency functions," *Pribory i Tekhnika Eksperimenta*, vol. 1, pp. 139–140, 2014 (Russian).

[18] A. M. Dunaev, V. B. Motalov, and L. S. Kudin, "High temperature mass spectrometric method for the determination of work function of the ionic crystals: triiodide of lanthanum, cerium, and praseodymium," *Russian Chemical Journal*, vol. 59, pp. 85–92, 2015.

[19] K. Kimura, S. Katsumata, Y. Achiba, T. Yamazaki, and S. Iwata, "Ionization energies, Ab initio assignments, and valence electronic structure for 200 molecules," in *Handbook of HeI Photoelectron Spectra of Fundamental Organic Compounds*, p. 268, Japan Scientific Societies Press, Tokyo, Japan, 1981.

[20] D. N. Sergeev, M. F. Butman, V. B. Motalov, L. S. Kudin, and K. W. Krämer, "Extrapolated difference technique for the determination of atomization energies of Sm, Eu, and Yb bromides," *International Journal of Mass Spectrometry*, vol. 348, pp. 23–28, 2013.

[21] D. R. Lide, Ed., *CRC Handbook of Chemistry and Physics*, CRC Press, Boca Raton, Fla, USA, 90th edition, 2009.

[22] L. V. Gurvich, V. S. Iorish, I. V. Veitz et al., Eds., *A Thermodynamic Database of Individual Substances and Software System for the Personal Computer*, IVTANTERMO for Windows, Glushko Thermocenter of RAS, Version 3.0, 2000.

[23] S. Yagi and T. Nagata, "Absolute total and partial cross sections for ionization of free lanthanide atoms by electron impact," *Journal of the Physical Society of Japan*, vol. 70, no. 9, pp. 2559–2567, 2001.

[24] T. R. Hayes, R. C. Wetzel, and R. S. Freund, "Absolute electron-impact-ionization cross-section measurements of the halogen atoms," *Physical Review A*, vol. 35, no. 2, pp. 578–584, 1987.

[25] K. Hilpert, "High temperature mass spectrometry in materials research," *Rapid Communications in Mass Spectrometry*, vol. 5, no. 4, pp. 175–187, 1991.

[26] J. Drowart, C. Chatillon, J. Hastie, and D. Bonnell, "High-temperature mass spectrometry: instrumental techniques, ionization cross-sections, pressure measurements, and thermodynamic data (IUPAC Technical Report)," *Pure and Applied Chemistry*, vol. 77, no. 4, pp. 683–737, 2005.

[27] R. C. Feber, *Heats of Dissociation of Gaseous Halides. TID-4500*, Los Alamos Scientific Laboratory, 40th edition, 1965.

[28] C. W. Struck and J. A. Baglio, "Estimates for the enthalpies of formation of rare-earth solid and gaseous trihalides," *High Temperature Science*, vol. 31, pp. 209–237, 1992.

[29] R. J. M. Konings and A. Kovács, "Thermodynamic properties of the lanthanide (III) halides," in *Handbook on the Physics and Chemistry of Rare Earths*, K. Gschneidner Jr., J.-C. G. Bünzli, and V. Pecharsky, Eds., vol. 33, pp. 147–247, Elsevier Science B.V., 2003.

[30] B. Ruščić, G. L. Goodman, and J. Berkowitz, "Photoelectron spectra of the lanthanide trihalides and their interpretation," *The Journal of Chemical Physics*, vol. 78, no. 9, pp. 5443–5467, 1983.

[31] L. A. Kaledin, M. C. Heaven, and R. W. Field, "Thermochemical properties (D_0° and IP) of the lanthanide monohalides," *Journal of Molecular Spectroscopy*, vol. 193, no. 2, pp. 285–292, 1999.

[32] S. A. Mucklejohn, "Molecular constants and standard enthalpies of formation for the lanthanide monohalide gaseous cations, LnX^+, X = F, Cl, Br, I," *Journal of Light and Visual Environment*, vol. 37, no. 2-3, pp. 78–88, 2013.

[33] A. M. Sapegin, A. V. Baluev, and O. P. Charkin, "Formation enthalpies and atomization energies of gaseous lanthanide halides," *Russian Journal of Inorganic Chemistry*, vol. 32, pp. 318–321, 1987 (Russian).

[34] L. S. Kudin, M. F. Butman, D. N. Sergeev, V. B. Motalov, and K. W. Krämer, "Determination of the work function for europium dibromide by knudsen effusion mass spectrometry," *Journal of Chemical & Engineering Data*, vol. 57, no. 2, pp. 436–438, 2012.

[35] C. E. Myers and D. T. Graves, "Thermodynamic properties of lanthanide trihalide molecules," *Journal of Chemical and Engineering Data*, vol. 22, no. 4, pp. 436–439, 1977.

[36] L. B. Pankratz, "Thermodynamic properties of halides," United States Bureau of Mines, Bulletin 674, 1984.

[37] I. Barin, *Thermochemical Data of Pure Substances*, John Wiley & Sons, 3rd edition, 1995–1999.

[38] E. L. Osina, V. S. Yungman, and L. N. Gorokhov, "Thermodynamic properties of lanthanide triiodide molecules," *Issledovano v Rossii*, vol. 1-4, pp. 124–132, 2000 (Russian).

[39] V. G. Solomonik, A. N. Smirnov, O. A. Vasiliev, E. V. Starostin, and I. S. Navarkin, "Nonemprirical study on the electronic structure of cerium, praseodymium, and ytterbium trihalide molecules," *Izvestiya Vysshikh Uchebnykh Zavedeniy Seriya "Khimiya I Khimicheskaya Tekhnologiya"*, vol. 57, pp. 26–27, 2014 (Russian).

[40] A. Kovács, "Molecular vibrations of rare earth trihalide dimers M2X6 (M=Ce, Dy; X=Br, I)," *Journal of Molecular Structure*, vol. 482-483, pp. 403–407, 1999.

[41] L. N. Gorokhov, G. A. Bergman, E. L. Osina, and V. S. Yungman, "High temperature materials chemistry," in *Proceedings of the 10th International IUPAC Conference*, K. Hilpert, F. W. Froben, and L. Singheiser, Eds., pp. 103–106, Forschungszentrum Juelich, Germany, April 2000.

[42] A. M. Pogrebnoi, L. S. Kudin, A. Yu. Kuznetsov, and M. F. Butman, "Molecular and ionic clusters in saturated vapor over lutetium trichloride," *Rapid Communications in Mass Spectrometry*, vol. 11, no. 14, pp. 1536–1546, 1997.

[43] V. G. Solomonik, A. Y. Yachmenev, and A. N. Smirnov, "Structure, force fields, and vibrational spectra of cerium tetrahalides," *Journal of Structural Chemistry*, vol. 49, no. 4, pp. 613–620, 2008.

[44] V. G. Solomonik, A. N. Smirnov, and M. A. Mileyev, "Structure, vibrational spectra, and energetic stability of LnX_4^- ions (Ln=La, Lu; X=F, Cl, Br, I)," *Russian Journal of Coordination Chemistry*, vol. 31, no. 3, pp. 203–212, 2005.

13

Active Mode Remote Infrared Spectroscopy Detection of TNT and PETN on Aluminum Substrates

John R. Castro-Suarez,[1,2] **Leonardo C. Pacheco-Londoño,**[1,3] **Joaquín Aparicio-Bolaño,**[4] **and Samuel P. Hernández-Rivera**[1]

[1]*ALERT DHS Center of Excellence for Explosives Research, Department of Chemistry, University of Puerto Rico-Mayagüez, Mayagüez, PR 00681, USA*
[2]*Molecular Spectroscopy Research Group, Antonio de Arevalo Technological Foundation, TECNAR, Cartagena, Colombia*
[3]*Environmental Engineering Program, Vice-Rectory for Research, Universidad ECCI, Bogota, Colombia*
[4]*Department of Physics, University of Puerto Rico, Ponce, PR 00732, USA*

Correspondence should be addressed to John R. Castro-Suarez; johncastrosuarez@gmail.com and Samuel P. Hernández-Rivera; samuel.hernandez3@upr.edu

Academic Editor: Christoph Krafft

Two standoff detection systems were assembled using an infrared telescope coupled to a Fourier transform infrared spectrometer, a cryocooled mercury-cadmium telluride detector, and a telescope-coupled midinfrared excitation source. Samples of the highly energetic materials (HEMs) 2,4,6-trinitrotoluene (TNT) and pentaerythritol tetranitrate (PETN) were deposited on aluminum plates and detected at several source-target distances by carrying out remote infrared spectroscopy (RIRS) measurements on the aluminum substrates in active mode. The samples tested were placed at 1–30 m for the RIRS detection experiments. The effect of the angle of incidence/collection of the IR beams on the vibrational band intensities and the signal-to-noise ratios (S/N) were investigated. Experiments were performed at ambient temperature. Surface concentrations from 50 to 400 $\mu g/cm^2$ were studied. Partial least squares regression analysis was applied to the spectra obtained. Overall, RIRS detection in active mode was useful for quantifying the HEMs deposited on the aluminum plates with a high confidence level up to the target-collector distances of 1–25 m.

1. Introduction

The detection and identification of highly energetic materials (HEMs), commonly called explosives, and related devices are an important priority for security and counterterrorism applications [1–4]. Defense and security agencies continuously support research and development strategies for the development of efficient sensing systems that help detect HEM. When used in public places, such as airports, stadiums, maritime, and railway or coach stations, these systems can help prevent or minimize damage that could be caused by terrorist attacks [4].

Investigations on the development of sensors involving analytical methodologies that enable faster, more sensitive, less expensive, and simpler determinations to facilitate the trace identification of explosives in different fields of interest for national defense have increased in recent years [5]. Modern detection systems are routinely used to prevent these events. These are based on ionization techniques accompanied by separation schemes, pyrolysis, gas phase reactions, interaction with radiation, color tests, immunochemical reactions between HEMs and their specific antibodies, and so forth. These techniques have proven to be useful for explosive detection in different phases (solid, liquid, and gas) on various substrates or complex matrixes (such as soil, air, and water) [5–10]. However, in most cases, they require some type of sample preparation for subsequent chemical analysis.

Since each chemical substance has its own distinctive fingerprint spectrum, vibrational techniques such as Raman spectroscopy (RS) and Fourier transform infrared spectroscopy (FT-IRS) exploit this advantage over other analytical techniques and make them optimal for the identification of a large range of high explosives, precursors of homemade explosives, and related compounds. Among the advantages that these vibrational spectroscopy techniques offer are the possibility of analyzing samples with different chemical compositions (organic and inorganic), minimum or no sample preparation, and minute explosive particles which can be readily analyzed. These techniques have been used to characterize, detect, quantify, and discriminate HEMs, biological and chemical agents (or their simulants), toxic industrial compounds, and other threat substances [12–16]. These spectroscopic techniques have the advantage that they can be used remotely in spectral detection mode and hyperspectral imaging mode [4–8]. Remote detection is the operational capability in which the instrumentation and operator remain separated from the sample by some distances (range) while measuring some properties of the target [17]. In remote infrared spectroscopy (RIRS), vibrational signatures can be obtained at distances a few tens of meters to several tens of meters between the target and the observer. This detection modality provides a way of performing real-time analysis, in which no sample preparation or operator contact is required. Rapid cycle times are typical, and enough chemical information on the target HEMs can be obtained to identify, quantify, and discriminate signals from the matrix support or other interfering substances. These capabilities make RIRS a useful technique for sensing for HEMs, and they further prevent or minimize the possibility of harm caused by terrorist action in cases where energetic material is set off [2]. Other techniques available for remote detection of HEM include Surface Enhanced Raman Scattering (SERS), Remote Raman Spectroscopy (RRS), Laser-Induced Breakdown Spectroscopy (LIBS), and Tunable Diode Laser Absorption Spectroscopy (TDLAS). SERS is not feasible for remote detection at long ranges and requires sample preparation. LIBS in comparison with RIRS has a lower selectivity because RIRS can provide vibrational information which is unique for each substance. TDLAS is limited to gases, and moreover, it suffers from the same limitation as LIBS. In comparison with RIRS and RRS, RRS can only analyze particles in remote detection mode. RIRS is amenable to layers, particles, or traces. A comparison between the detection limits is not possible because units of measurement are different: for the molecular layers that can be analyzed by RIRS, this is in mass/area, and for particles, it is the total mass analyzed [11]. If the explosive exists on the surface as a thin layer, the backscattering signal is low. When the explosive was present on the surface as discrete particles (crystals), the backscattered RIRS signal is significantly improved.

RIRS detection is the most versatile of the remote spectroscopy-based technologies because it can measure the presence of many chemicals deposited on substrates at near-trace to trace levels at a distance [18, 19]. Pacheco-Londoño et al. [11] built an active RIRS detection system by coupling a bench FT-IR interferometer to a gold mirror and detector assembly for the detection of trace amounts of TNT and RDX explosives on reflective surfaces in the range

of 1.0–3.7 m. Suter et al. studied the spectral and angular dependence of scattered midinfrared light from surfaces coated with explosive residues (TNT, RDX, and tetryl) detected at a 2 m remote distance [20]. An external cavity quantum cascade laser (QCL) provided tunable excitation between 1250 and 1428 cm^{-1} [19]. Kumar et al. measured the diffuse reflection spectrum of solid samples such as explosives (TNT, RDX, and PETN), fertilizers (ammonium nitrate and urea), and paints (automotive and military grade) at a distance of 5 m using a midinfrared supercontinuum light source with a 3.9 W average output power [21].

In this report, RIRS detection experiments were performed using an open-path FT-IR interferometer. Experiments were carried out in active mode using a telescope-coupled MIR source. The effect of the source-detector-target angle on the IR spectra of PETN was evaluated. Using this sensing modality, partial least squares (PLS) regression calibrations were obtained, and the root mean square errors of cross-validations (RMSECV) and coefficients of determination (R^2) were used as criteria to judge the quality of the detection methodologies.

2. Materials and Methods

2.1. Reagents. The reagents used in this investigation included HEMs and solvents. 2,4,6-Trinitrotoluene (TNT) was acquired from Chem Service, Inc. (West Chester, PA, USA) as a crystalline solid (99%, min. purity; 30% water content). PETN was synthesized and purified in the laboratory according to the methods described by Ledgard [22]. Methanol (99.9%, HPLC grade), dichloromethane (CH$_2$Cl$_2$, HPLC grade), and acetone (99.5%, GC grade) were purchased from Sigma-Aldrich Chemical Co. (Milwaukee, WI, USA) and were used to deposit the HEM samples at various surface concentrations onto aluminum (Al) plates used as substrates.

2.2. Sample Preparation. Sample preparation is an important process in the validation of remote detection experiments of analytes present as trace residues on the substrates. A sample smearing technique was used to deposit HEM samples on metal substrates [15]. Al plates with 1.0 ft. × 1.0 ft. (929 cm^2) dimensions were used as a sample support for the HEM targets. Acetone was used to clean the Al substrates. After cleaning, the plates were allowed to dry before depositing the desired HEM target. A small amount of dichloromethane or methanol was used to dissolve the desired HEM samples to be deposited on the test substrates. Then, a Teflon stub 3 cm × 15 cm was used to smear the samples on the Al plates. The amount of HEM that remained on the Teflon stub after sample smearing was negligible. The nominal surface concentrations obtained by the smearing technique used were 50, 100, 200, 300, and 400 μg/cm^2.

2.3. Experimental Setup. For the remote detection experiments, the HEM-coated Al plates were placed at the target positions, and the ambient temperature, the plate temperature, and the relative humidity were measured during the experiments. Next, active mode RIRS detection experiments were carried out using the two optical systems illustrated in

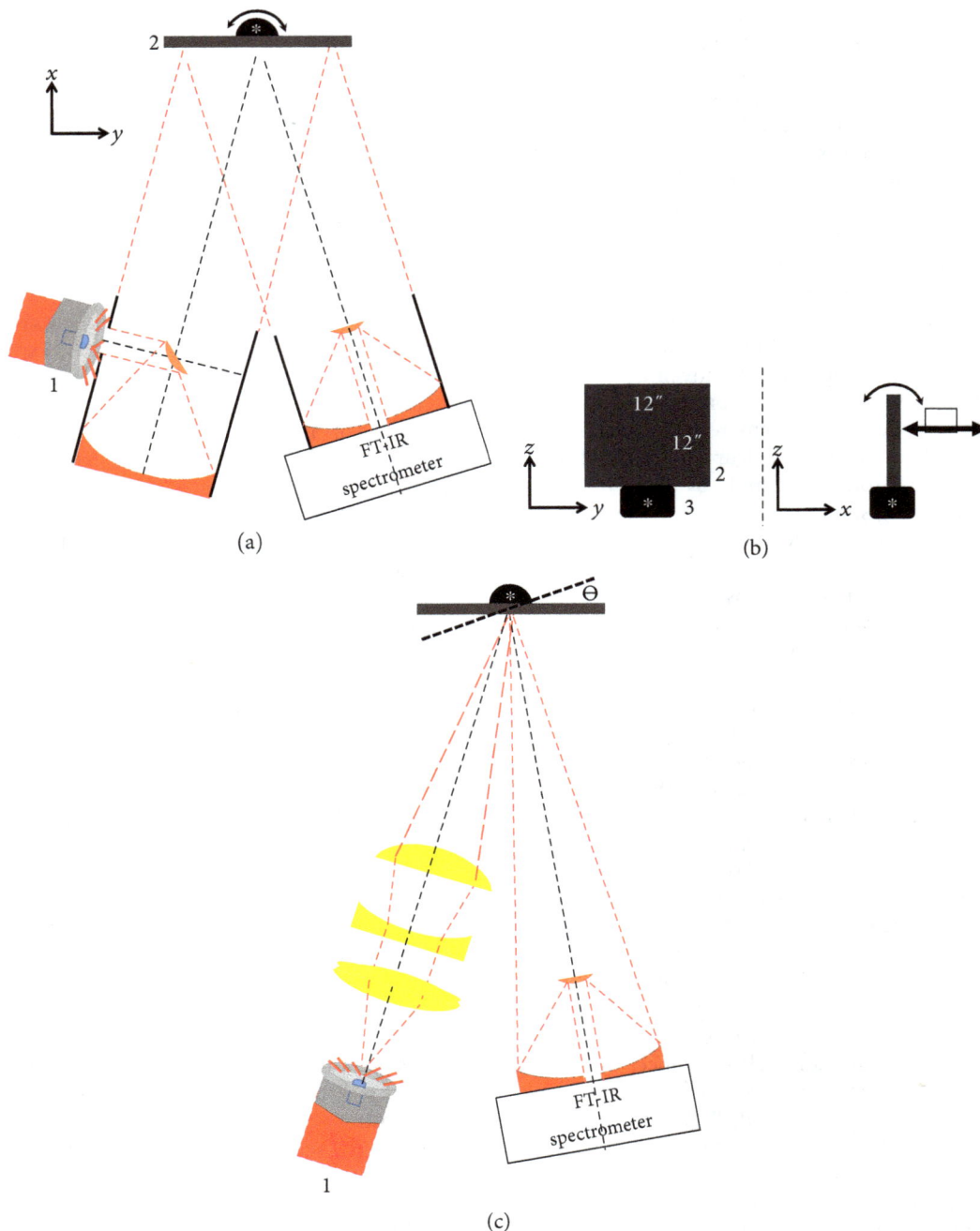

FIGURE 1: FT-IR interferometer configuration: (a) active mode setup for standoff measurements using reflective telescope: 1, IR source, 2, Al plate; (b) plate mount: 3, tilting mount; (c) active mode setup for standoff measurements using refractive telescope.

Figures 1(a) and 1(c). In the first optical system, a midinfrared (MIR) reflective telescope coupled to a heated oxide globar source (Figure 1(a)) was used. In the second optical system, a midinfrared (MIR) refractive telescope coupled to the globar source as shown in Figure 1(c) was used.

In the remote sensing experiments, the IR beam from the globar source was not modulated by the interferometer before interacting with the target and sat side-by-side with the MIR reflective collector telescope that was close coupled to the FT interferometer (Figure 1(a)). The experiments were

conducted in back-reflection mode. The FT-IR spectrometer used was an open-path interferometer, model EM27 (Bruker Optics, Billerica, MA, USA). The optical bench consisted of a compact, enclosed, and desiccated Michelson-type interferometer equipped with ZnSe windows, an internal blackbody calibration source, and a KBr beam splitter. This system had a very fast native focal ratio (f/0.9) and a field of view (FOV) of 30 mrad (1.7°). For the optical system illustrated in Figure 1(a), the transmitter source telescope had a diameter of 6 in., a focal ratio of f/4, and gold-coated mirrors with a

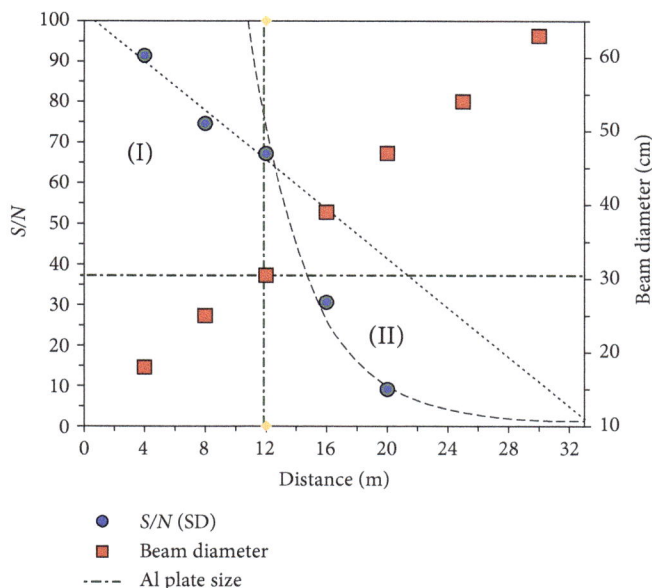

FIGURE 2: Signal-to-noise ratio (principal y-axis) at various distances for active mode RIRS measurements of TNT. IR beam spot size (secondary y-axis) versus range. Noise levels (as standard deviation) were measured at $830–870\,cm^{-1}$ and peak heights were measured for TNT signal at $790\,cm^{-1}$.

FOV ≥ 7.5 mrad (0.43°). The receiver telescope was also 6 in. in diameter and had an $f/3$ focal ratio and gold-coated mirrors with a FOV of 10 mrad (0.57°).

In the second optical system used (Figure 1(c)), the transmitter source telescope consisted of a set of three 4 in. diameter ZnSe lenses of 100 mm focal length (f) for the biconvex lens, $−1000$ mm f for the plano-concave lens, and 1000 mm f for the plano-convex lens. The receiver telescope was the same as in the first optical layout.

In these experimental setups, the targets were carefully aligned with the source and the collector. To accomplish this, the metal plate was placed on a mount that allowed millimeter translations both horizontally and vertically, as illustrated in Figure 1(b). The target-collector distances studied were 1, 4, 8, 12, 16, 20, 25, and 30 m. A total of 10 spectra were taken for each sample at 20 scans/spectrum and a $4\,cm^{-1}$ resolution. The spectra were recorded in the spectral range from 750 to $1400\,cm^{-1}$. The experiments were performed at ambient temperature (~25°C).

IR signals were detected using an MIR closed cycle (Stirling cooled) photoconductive MCT detector with D^{*}_{max} ~$4 \times 10^{10}\,cmHz^{1/2}/W$. Background spectra of the Al plates with no HEM deposited on them were run for every remote distance studied. Data analyses were based on statistical routines using chemometrics. Specifically, partial least squares (PLS) regression analysis was used to perform quantification studies of the surface concentrations of the HEMs at all distances studied.

3. Results and Discussion

Active mode standoff IR measurements of solid HEM dissolved in an appropriate solvent and smeared on Al plates at various surface concentrations were carried out. The spectra were collected in the MIR region ($750–1400\,cm^{-1}$) using the setups described.

3.1. RIRS Detection of HEMs. Signal-to-noise ratios (S/N) were calculated and plotted against the remote detection distance for active mode spectral measurements of TNT. The results for surface concentrations of $400\,\mu g/cm^2$ deposited on the Al plates that were obtained using the optical system illustrated in Figure 1(a) are shown in Figure 2. S/N initially decreased linearly with distance (Figure 2, blue circles; equation: $-3.03 \pm 0.02\,m^{-1}*X + 101.96 \pm 0.01$). However, when the detection distance was larger than 12 m, the signal decreased with an even steeper slope ($Exp(-0.26 \pm 0.04\,m^{-1}*X + 7.4 \pm 0.6) + 1$) reaching a S/N of ~3 (calculated by extrapolation from the exponential fitting of the equation of S/N versus distance, region II) at a distance of 26 m. The decrease in the collected signal did not allow the measurement of S/N for the entire distance range planned (to 60 m), making the measurements accurate up to 25 m. The two linear decreases in the collected signal were calculated from linear fittings and are shown as black and gray dotted lines in Figure 2. The difference in slope in regions I: 4–12 m, and II: 12–25 m, suggested a fundamental reason for the behavior and led to the measurements of the spot size of the MIR beam at the target plane. As shown in the graph, the spot diameter is smaller than the target in region I; it is exactly equal to the target at 12 m and is larger than the target in region II. A linear dependence of the spot diameter with the detection distance is also shown in Figure 2. When an MIR globar source was used to accomplish the spectroscopic measurements in active mode, the peak intensities decreased as the distances increased. At distances longer than 25 m, it was not possible to visually detect some of the TNT vibrational signatures in the spectra obtained. This result was as expected since, as shown in Figure 2, the distance required for the threshold S/N of 3 is 24 m in active mode. Furthermore, the density of infrared radiation that is transferred to the Al plates from the MIR source diminishes as a function of distance, leading to a smaller number of excited molecules at the target. Therefore, the detector could not register the transflection intensities of the low-intensity vibrational modes.

The active mode MIR spectra for TNT deposited on Al plates at several distances and surface concentrations are shown in Figure 3(a). The experiments were performed using the setup illustrated in Figure 1(a). A reference spectrum for TNT is included for comparison purposes and for assignment of the modes of the important nitroaromatic HEM. The latter was obtained by preparing a pellet of 1 mg of microcrystalline TNT in 100 mg KBr and measuring the absorption spectrum in the macro sample chamber of a benchtop interferometer (Bruker Optics IFS-66v). The spectrum of the solid is represented by the black trace labeled "TNT ref." Upon inspection of these spectra, it is evident that the most significant TNT vibrational signatures were detected in the remote sensor measurements. In particular, an intense vibrational band at approximately $908\,cm^{-1}$ was assigned to the –C–N stretching, a vibrational band at

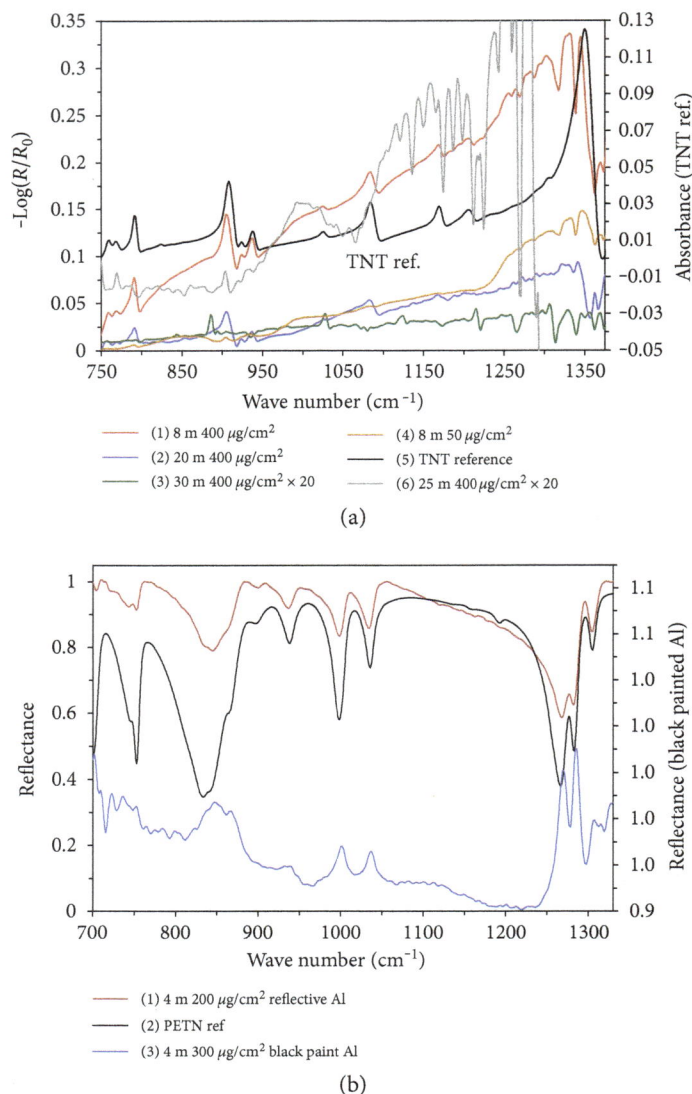

FIGURE 3: (a) Active mode RIRS spectra of TNT deposited on an Al plate measured at 8, 20, 25, and 30 m and surface concentrations of $400\,\mu g/cm^2$ ((1), (2), (3), and (6)) and $50\,\mu g/cm^2$ (4), respectively. (b) Active mode RIRS spectra of PETN deposited on Al plate and black painted Al at 4 m.

$938\,cm^{-1}$ was assigned to the –C–H out-of-plane ring bending, and the symmetric stretch band of the nitro groups appeared at $1350\,cm^{-1}$. All assignments were tentative. These results are all due to the conjugation of the nitro groups with the aromatic ring and agree with results from Pacheco-Londoño et al. [11], Clarkson et al. [23], and Castro-Suarez et al. [24]. These spectra were not submitted to any preprocessing routine, such as offset correction, baseline correction, smoothing, or water vapor rotational line removal. In other words, there was no common baseline for these spectra, and some spectra exhibited positive intensity ramps to higher wave number values. However, an increase in signal intensity as a function of the surface concentrations was clearly displayed without the use of chemometrics routines for distances less than or equal to 20 m. For standoff distance of 25 m, low-intensity TNT

vibrational signals can be appreciated, which can be confirmed with the aid of chemometrics routines.

The spectral band shapes observed in remote detection mode, shown in Figure 3(a), are superimposed on a ramp-shaped background, and the bands themselves exhibit strong transflection band profiles. Since these measurements were performed on a reflective metal substrate, the distortion of the band profiles is expected. Similar effects have been reported in diffuse reflection infrared Fourier transform spectroscopy (DRIFT) [25] and acquired micro-IR spectra of microspheres [26]. In both cases, the distortion of the absorption line shapes is because of the reflective index that undergoes anomalous dispersion within the band. In spectroscopic experiments carried out in reflectance mode, a mixing of the absorptive and dispersive line shapes can occur, resulting in bands that have negative dips at the high wave number side of the peak. This will shift the maximum peak, by up to $15 \, cm^{-1}$, toward lower values [26]. Moreover, the mixing of absorptive and reflective line shapes can be mediated by scattering effects [26], which could also produce significant band distortions. In a paper about the line shape distortion effects in infrared spectroscopy by Miljković et al., the authors address the conditions under which the mixing of reflective and absorptive band shapes will occur. They also discuss the methods that have been developed to correct the spectral distortions [28]. Castro-Suarez et al. demonstrated that the spectra shown in Figure 3(a) have the appearance of "double pass" transmission-reflection (transflectance) spectra [24]. Transflection experiments are usually performed by placing a thin target sample on a non-IR absorbing, reflective substrate such as a polished metal surface, focusing an IR beam onto a region of interest and collecting the radiation that is reflected to the collection optics. The technique is termed transflection because most of the signal intensity collected is a transmission signal as the beam passes through the sample, reflects off the substrate, and passes through the sample again before reaching the detector [27–29].

The active mode MIR spectra for PETN deposited on Al plates at several distances and surface concentrations are shown in Figure 3(b), using the experimental setup illustrated in Figure 1(c). This optical system was designed to overcome the limitations of the setup illustrated in Figure 1(a). Among these was the inability to focus at a certain distance shorter than infinity, which did not allow control over the beam size at the target distance, as is shown in Figure 2. The size of the spot increased as the distance increased. A reference spectrum for PETN, obtained as previously described for TNT, is also included for reference and for facilitating band assignments. The spectra obtained contained the most significant band for identifying the HME. For PETN in Figure 3(b), important vibrational signatures appeared at $703 \, cm^{-1}$ ($-ON$ stretching $+ -NO_2$ rocking), $753 \, cm^{-1}$ ($-ONO_2$ umbrella), $869 \, cm^{-1}$ ($-ON$ stretching), $939 \, cm^{-1}$ ($-CH_2$ torsion), $1003 \, cm^{-1}$ ($-CO$ stretching), $1038 \, cm^{-1}$ ($-NO_2$ rocking), $1272 \, cm^{-1}$ ($-ONO_2$ rocking), $1285 \, cm^{-1}$ ($-NO_2$ stretching), and $1306 \, cm^{-1}$ ($-NO_2$ rocking) [30]. These spectra were not submitted to any preprocessing routine, such as offset correction, smoothing, or water vapor rotational line removal. As for the case of TNT, there was no common baseline for these spectra, and some spectra exhibited positive intensity ramps to higher wave number values.

The remote detection of PETN deposited on pieces of black painted car door is shown in Figure 3(b) (blue trace). This spectrum had upward-looking peaks when compared with the spectrum taken from PETN on a reflective Al substrate (red trace), and the PETN reference spectrum measured in reflectance mode with a bench spectrometer (Bruker Optics IFS-66/v, Bruker Optics) also exhibited downward-looking peaks. Although all spectra were recorded in transflection mode, the spectral patterns observed demonstrate the effect of the nature of the substrate on the spectral profiles of the target HEM. In transflection experiments, such as the one described above, the recorded spectra are a weighted sum of the transmission and reflection characteristics of the samples and substrates. In samples deposited on highly reflective substrates, the

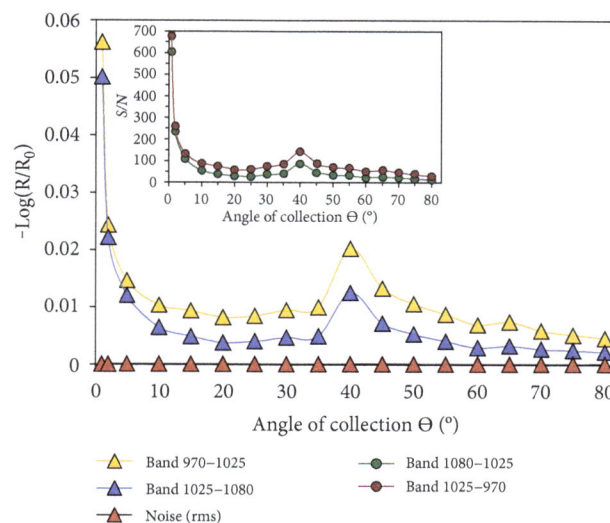

FIGURE 4: PETN ($200 \, \mu g/cm^2$) vibrational band intensities and S/N (inset) at different angles (offset) of collection (Θ).

weighting is such that the transmission signal dominates, and the reflection signal is negligible, producing spectra closely resembling those of a transmission spectrum, in this case, the PETN reference spectrum on Al (black spectrum). However, if the transmission signal is weak or inexistent, the reflection signal will dominate, and a reflectance spectrum is obtained, as shown by the blue trace in Figure 3(b). Castro-Suarez et al. discuss in detail the IRS spectral profile of targets deposited on nonreflective substrates [24].

In active mode RIRS detection of targets residing on metal substrates, the signal strength is highly dependent on the alignment between the IR source, the target, and the detector. Therefore, a study of the effect of the angle of collection (Θ) of the IR beam on the intensity of the vibrational signal was required. In this work, Θ represents the angular offset from the maximum signal intensity, assigned to back reflection alignment ($\Theta = 0$). The optical layout illustrated in Figure 1(c) was used for remote detection at 1 m. The influence on the intensity and the S/N for two vibrational bands of PETN/Al at a surface concentration of $200 \, \mu g/cm^2$ on the angle of collection is shown in Figure 4. Bands analyzed were those at 970–$1025 \, cm^{-1}$ and at 1025–$1080 \, cm^{-1}$. The band intensities were taken as the average height of the five spectra. The reflected intensities (S) were represented as $-\log(R/R_0)$, which is the function proportional to the surface concentration of the analytes. The noise level (N) was taken as the average root mean square (rms) values of the five spectra measured in the range of $1105 \, cm^{-1}$ to $1130 \, cm^{-1}$, that is, the standard deviation of all data points in this region [30]. S/N was calculated as the ratio between the band intensities and the noise level. Intensities of the bands and the S/N were calculated by varying the angle of collection from $1°$ to $80°$, as illustrated in Figure 4. The intensities measured for the two bands decreased by 50% and 25% when the collection angle changed from $1°$ to $2°$ and from $1°$ to $5°$, respectively. For collection angles equal to or larger than $10°$, the intensities decreased considerably compared

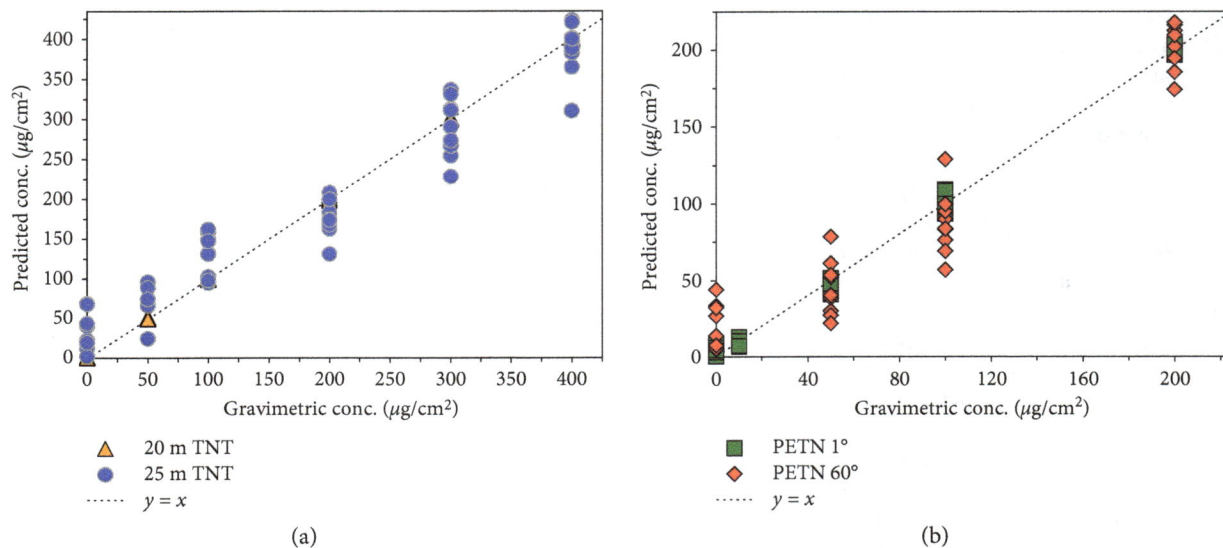

FIGURE 5: (a) Predicted versus true surface concentration for TNT explosives on Al plates at 20 m and 25 m. (b) Predicted versus true surface concentration for PETN explosives on Al plates at 1° and 60°.

to those of 1°. The intensity decrease depended on the spectral band selected. For the band at 1038 cm^{-1}, the intensity of 0.05 ($\Theta = 1°$) was reduced to ~0.004 ($\Theta > 10°$). Thus, the intensity was reduced to 8% when compared to $\Theta = 1°$. However, for the band at 1003 cm^{-1}, the intensity decreased from 0.056 ($\Theta = 1°$) to ~0.008 ($\Theta > 10°$) or 14% of its original value. As shown in Figure 4, at collection angles from 35° to 50°, both spectral bands showed an increase in intensity. This result seems to be due to a geometrical factor. The graph of S/N versus angle of collection, shown as an insert in Figure 4, has a similar behavior to the graph of intensity versus angle of collection. At small angles ($\Theta = 2°$ to 5°), the S/N decreases significantly. However, even at $\Theta = 80°$, S/N is equal to fifteen, so HEM characteristic spectral bands could still be detected at such large angles of incidence.

3.2. HEM Quantification Using PLS Regression.

The multivariate regression algorithm known as partial least squares-1 (PLS-1) was used to find the best correlation function between the HEM spectral information and the surface concentrations [31]. The Quant2™ software of OPUS™ v. 6.0 (Bruker Optics) was used to build the models. Calibrations were performed using PLS-1 in which only one component can be analyzed separately, instead of simultaneously analyzing multiple components, as in the PLS-2 routine of chemometrics. PLS-1 regressions were used to generate chemometrics models for all of the remote distances studied. In addition, cross-validations were made, and the root mean square errors of the cross-validations (RMSECV) and coefficient of determination (R^2) were used as indicators of the quality of the correlations obtained at the various distances studied.

Statistical treatments for the chemometrics models based on PLS-1 regressions were performed using the complete spectral region (from 700 to 1400 cm^{-1}), where the nitro symmetric stretch and aromatic ring –C–H vibrations occur in these compounds. Ten replicate spectra were used for each

tested sample. In the models generated, it was not necessary to eliminate any spectrum. The method of leave-one-out cross-validation (LOOCV) was used, where each of the "n" calibration samples was left out, one at a time, and the resulting calibration models were then used to evaluate the sample, which acts as an independent validation sample and provides an independent prediction of each dependent variable. This process of leaving a sample out was repeated until all of the calibration samples had been left out.

Data preprocessing is an important step in performing chemometrics calibrations. To ensure the reproducibility of the calibration samples, ten spectra of each sample (fixed surface concentration and remote distance) were acquired. The following data preprocessing steps were tested: vector normalization, first derivative, and second derivative. However, no preprocessing of the data worked better than any other preprocessing routine or combination of preprocessing routines. Therefore, no preprocessing steps were applied for achieving the best possible values for RMSECV and R^2 in the spectral region from 700 to 1400 cm^{-1} other than applying mean centering of each spectral variable [30]. Overall, the results indicate that the experimental setup used had good management of the external variables, such as humidity, temperature changes, and the concentrations of the samples deposited on the Al plates.

Figure 5(a) shows the results obtained for the cross-validations carried out for the TNT spectra measured at distances of 20 and 25 m in active mode (represented as orange triangles and blue circles, resp.) using the experimental setup shown in Figure 1(a). Each point of the graph represents ten spectra with a fixed surface concentration (0, 50, 100, 200, 300, or 400 $\mu g/cm^2$). In all of the correlation charts, the predicted surface concentration values (dependent variable) versus the gravimetric surface concentration values (independent variable) for 4, 8, 12, and 16 m were similar to the correlation chart for 20 m distance in Figure 5(a) (orange triangles).

Table 1: PLS calibration parameters for various remote distances analyzed.

Distance (m)	R^2	RMSECV	LVs
4	1	0.457	8
8	1	0.434	7
12	1	0.676	8
16	1	0.736	9
20	0.9999	1.08	9
25	0.9367	34.6	5
30	0.8553	53.5	9

However, as the distance increased above 20 m, some of the spectral information fell below the quantification level causing the spectra for each sample to be slightly different from each other (within experimental error) and making it difficult to predict the surface concentration precisely. The fact that the RMSECV (Figure 5(a); blue circles) is larger at standoff distances greater or equal to 25 m can be due to the limitations of our system, Figure 1(a), in failing to focus the IR beam to a certain size at the target. For this standoff distance (25 m) the size of the IR beam was >200% larger than the size of the target (see Figure 2). Thus, the radiant intensity on the target decreased considerably. This decrease was in addition to the well-established power reduction with the inverse of the standoff distance squared. For the standoff distance of 25 m, very low intensity vibrational signals of TNT can be appreciated, mainly those at 900 cm^{-1} assigned to the –C–N stretching. Low intensity bands measured at 25 m standoff distance are close to $S/N \sim 3$, that is, the detection limit for the experimental setup used (Figure 1(a)) calculated by extrapolation as mentioned above. Table 1 shows the results of RMSECV and R^2 obtained in the PLS models generated.

Figure 6(a) shows the plot of the regression coefficient for the PLS model of TNT detection and quantification on Al plates at 8 m. This regression coefficient spectrum in the spectral range from 750 to 1375 cm^{-1} contains the characteristic bands of TNT that are relevant for predicting the detection/quantification of HEM/Al. In addition, Figure 6(a) also shows the loading (latent variable (LV)) plot used in the PLS model. The LVs are the variables with the largest statistical weights corresponding to the analyte characteristic bands. LV1 for the PLS model at 8 m is enough to describe the principal vibrational bands of TNT. However, at a 25 m remote distance, the important bands of TNT are also contained in LV2 for the PLS model because for detection at longer distances, it was difficult to predict the surface concentration based on only LV1.

The values obtained for RMSECV and R^2 for 25 and 30 m, respectively, are as expected, as already discussed, falling below the threshold of S/N equal to 3 (for distances ≥ 24 m). Taking into account the low values of RMSECV and the high values of R^2 obtained, it is concluded that the models are useful tools for determining the surface concentration of TNT in unknown samples at a maximum distance of 20–24 m with the setup used for the measurements.

Table 2: PLS calibration parameters for various angles of collection analyzed.

Angle of collection Θ (°)	R^2	RMSECV	LVs
1	0.9963	4.26	7
20	0.9628	14.10	7
40	0.9802	10.30	7
60	0.9237	20.40	7

Figure 5(b) shows the results obtained for cross-validations carried out for the PETN spectra measured at 1 m in active mode configuration for two angles of collection (Θ): 1° and 60° (represented as green squares and red diamonds, resp.). The data were acquired using the experimental setup shown in Figure 1(c). Table 2 shows the results of RMSECV and R^2 obtained in the PLS models generated for $\Theta = 1°$, 20°, 40°, and 60°. In these correlation charts, each point represents ten spectra with a fixed surface concentration (0, 10, 50, 100, or 200 μg/cm^2). However, as the angle of collection increased, the intensities of the vibrational bands decreased considerably and made the correlation with the surface concentration difficult.

Figure 6(b) displays the regression coefficient and loading plot (LV1) for the PLS model of PETN explosive detection and the quantification of Al plates at 1 m and an angle of collection of 1°. This regression coefficient spectrum contains the characteristic vibrational bands of PETN that are significant for predicting the detection/quantification of the HEM deposited at near-trace levels on the Al substrates. Figure 6(b) also shows the first loadings used in the PLS model; the variables with the largest statistical weights correspond to the PETN characteristic bands illustrated in Figure 3(b). The values obtained for RMSECV and R^2 for the angles of collection of 1°, 20°, 40°, and 60° demonstrate that it is possible to carry out statistically meaningful RIRS detection experiments with the setups described even at high incident angles. As is shown in Table 2, when Θ is equal to 40°, values of RMSECV and R^2 improve, which is expected from the results shown in Figure 4. Taking into account the low values of RMSECV, the high values of R^2, and the high values of S/N, the models were useful for determining the surface concentration of PETN in unknown samples.

4. Conclusions

Two remote detection IR techniques based on an open-path FT-IR interferometer have been demonstrated and used to obtain spectral information of HEM samples deposited on Al plates. High spectral quality measurements were achieved using MIR reflective and refractive telescopes coupled to remote detection systems. Detection in active mode proved to be successful for detecting TNT vibrational signatures in the range of 4–20 m, and at 25 m with very slight vibrational signals, which could be confirmed with the aid of chemometrics routines. The RIRS detection system worked particularly well for distances smaller than 24 m due to the transmitter telescope characteristics in sensing TNT. At 24 m, the MIR beam size was as large as the target (30 cm),

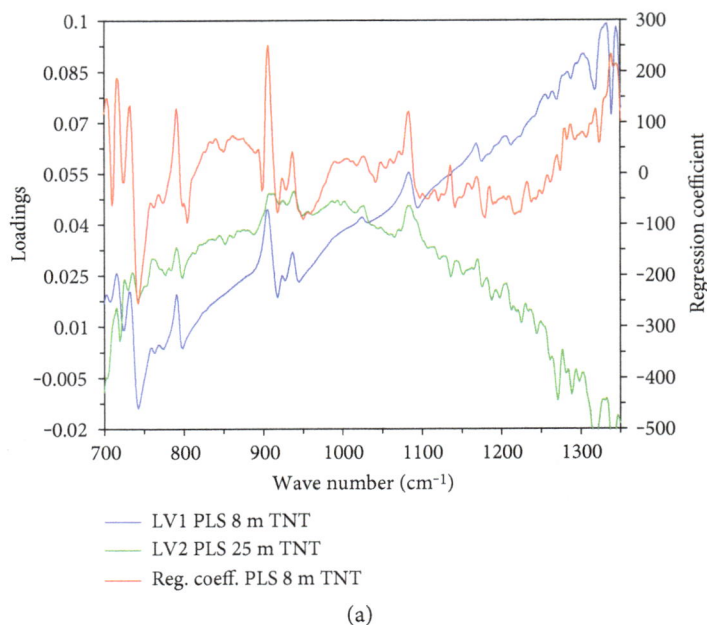

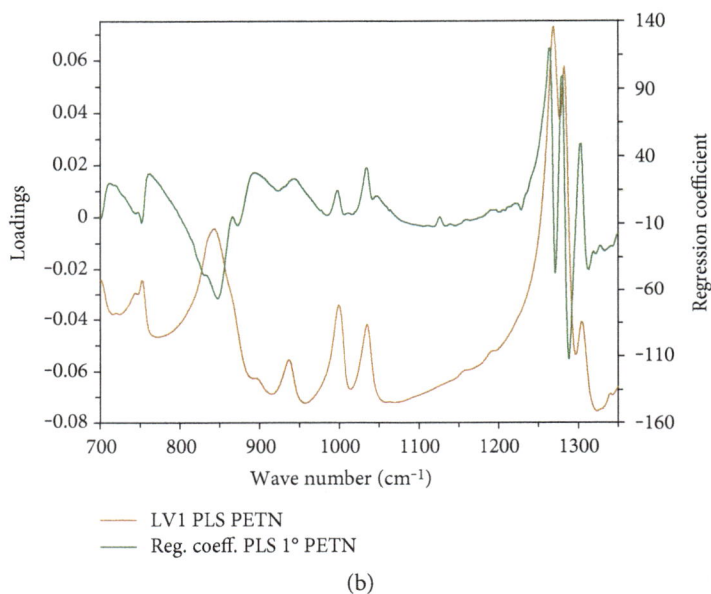

FIGURE 6: (a) Regression coefficient and loading plot for PLS model of detection of TNT explosives on Al plates at 8 m standoff distance. (b) Regression coefficient and loading plot for PLS model of detection of PETN explosives on Al plates at 1 m standoff distance and angle of collection 1°.

and the *S/N* was 3, falling rapidly above 25 m using MIR reflective telescopes. However, it was necessary to align the target carefully with the detector to be able to measure with high accuracy at distances between 25 and 30 m.

The optical system designed for detecting PETN vibrational bands in the range of 1–4 m by varying the angle of collection from 1° to 80° was equally successful. Furthermore, excellent results for RMSECV and R^2 were obtained for models generated based on PLS cross-validations for the active mode experiments.

Disclosure

The views and conclusions contained in this document are those of the authors and should not be interpreted as necessarily representing the official policies, either expressed or implied, of the U.S. Department of Homeland Security.

Acknowledgments

This material is based upon work supported by the U.S. Department of Homeland Security, Science and Technology Directorate, Office of University Programs, under Grant Award 2013-ST-061-ED0001. This contribution was also supported by the U.S. Department of Defense, Agreement no. W911NF-11-1-0152. The authors also acknowledge contributions from Dr. Richard T. Hammond from the Army Research Office, DoD.

References

[1] H. Schubert and A. Rimski-Korsakov, *Stand-Off Detection of Suicide-Bombers and Mobile Subjects*, Springer, Dordrecht-The Netherlands, 2006.

[2] K. Banas, A. Banas, H. O. Moser et al., "Post-blast detection of traces of explosives by means of Fourier transform infrared spectroscopy," *Analytical Chemistry*, vol. 82, no. 7, pp. 3038–3044, 2010.

[3] M. Marshall and J. C. Oxley, *Aspects of Explosives Detection*, Elsevier, Oxford, 2009.

[4] M. López-López and C. García-Ruiz, "Infrared and Raman spectroscopy techniques applied to identification of explosives," *TrAC Trends in Analytical Chemistry*, vol. 54, no. 2, pp. 36–44, 2014.

[5] J. S. Caygill, F. Davis, and S. P. J. Higson, "Current trends in explosive detection techniques," *Talanta*, vol. 88, no. 1, pp. 14–29, 2012.

[6] J. Yinon, "Field detection and monitoring of explosives," *TrAC Trends in Analytical Chemistry*, vol. 21, no. 4, pp. 292–301, 2002.

[7] D. S. Moore, "Instrumentation for trace detection of high explosives," *Review of Scientific Instruments*, vol. 75, no. 8, pp. 2499–2512, 2004.

[8] D. J. Klapec and G. Czarnopys, "Analysis and detection of explosives and explosives residues review: 2010 to 2013," *17th Interpol International Forensic Science Managers Symposium*, vol. 526, pp. 280–435, Lyon, France, October 8–10, 2013.

[9] A. J. Bandodkar, A. M. O'Mahony, J. Ramírez et al., "Solid-state forensic finger sensor for integrated sampling and detection of gunshot residue and explosives: towards 'lab-on-a-finger'," *Analyst*, vol. 138, no. 18, pp. 5288–5295, 2013.

[10] T. Caron, M. Guillemot, P. Montméat et al., "Ultra trace detection of explosives in air: development of a portable fluorescent detector," *Talanta*, vol. 81, no. 1, pp. 543–548, 2010.

[11] L. C. Pacheco-Londoño, W. Ortiz, O. M. Primera, and S. P. Hernández-Rivera, "Vibrational spectroscopy standoff detection of explosives," *Analytical and Bioanalytical Chemistry*, vol. 395, no. 2, pp. 323–335, 2009.

[12] J. L. Gottfried, "Discrimination of biological and chemical threat simulants in residue mixtures on multiple substrates," *Analytical and Bioanalytical Chemistry*, vol. 400, no. 10, pp. 3289–3301, 2011.

[13] W. Ortiz-Rivera, L. C. Pacheco-Londoño, J. R. Castro-Suarez, H. Felix-Rivera, and S. P. Hernandez-Rivera, "Vibrational spectroscopy standoff detection of threat chemicals," in *Proceedings of SPIE 8031, Micro-and Nanotechnology Sensors, Systems, and Applications III*, vol. 8031, Article ID 803129, 2011.

[14] J. Mass, A. Polo, O. Martínez et al., "Identification of explosive substances through improved signals obtained by a portable Raman spectrometer," *Spectroscopy Letters*, vol. 45, no. 6, pp. 413–419, 2012.

[15] J. R. Castro-Suarez, L. C. Pacheco-Londoño, M. Vélez-Reyes, M. Diem, T. J. Tague, and S. P. Hernández-Rivera, "FT-IR standoff detection of thermally excited emissions of trinitrotoluene (TNT) deposited on aluminum substrates," *Applied Spectroscopy*, vol. 67, no. 2, pp. 181–186, 2013.

[16] A. C. Padilla-Jiménez, W. Ortiz-Rivera, J. R. Castro-Suarez, C. Ríos-Velázquez, I. Vázquez-Ayala, and S. P. Hernández-Rivera, "Microorganisms Detection on Substrates Using QCL Spectroscopy, in *Proceedings of SPIE: Chemical, Biological, Radiological, Nuclear, and Explosives (CBRNE) Sensing XIV*, vol. 8710, Article ID 871019, 2013.

[17] J. E. Parmenter, "The challenge of standoff explosives detection," *Proceedings of the 38th Annual 2004 International Carnahan Conference on Security Technology*, pp. 355–358, IEEE, 2004.

[18] P. R. Griffiths, L. Shao, and A. B. Leytem, "Completely automated open-path FT-IR spectrometry," *Analytical and Bioanalytical Chemistry*, vol. 393, no. 1, pp. 45–50, 2009.

[19] J. R. Castro-Suarez, Y. S. Pollock, and S. P. Hernandez-Rivera, "Explosives Detection Using Quantum Cascade Laser Spectroscopy," in *Proceedings of SPIE - The International Society for Optical Engineering, Chemical, Biological, Radiological, Nuclear, and Explosives (CBRNE) Sensing XIV*, vol. 8710 Article ID 871010, 2013.

[20] J. D. Suter, B. Bernacki, and M. C. Phillips, "Spectral and angular dependence of mid-infrared diffuse scattering from explosives residues for standoff detection using external cavity quantum cascade lasers," *Applied Physics B*, vol. 108, no. 4, pp. 965–974, 2012.

[21] M. Kumar, M. N. Islam, F. L. Terry et al., "Stand-off detection of solid targets with diffuse reflection spectroscopy using a high-power mid-infrared supercontinuum source," *Applied Optics*, vol. 51, no. 15, pp. 2794–2807, 2012.

[22] J. Ledgard, *The Preparatory Manual of Explosives*, J. Ledgard, Ed., ISBN-13: 978-0615142906.

[23] J. Clarkson, W. E. Smith, D. N. Batchelder, D. A. Smith, and A. M. Coats, "A theoretical study of the structure and vibrations of 2,4,6-trinitrotoluene," *Journal of Molecular Structure*, vol. 648, no. 3, pp. 203–214, 2003.

[24] J. R. Castro-Suarez, M. Hidalgo-Santiago, and S. P. Hernández-Rivera, "Detection of highly energetic materials on non-reflective substrates using quantum cascade laser spectroscopy," *Applied Spectroscopy*, vol. 69, no. 9, pp. 1023–1035, 2015.

[25] J. M. Chalmers and M. W. Mackenzie, "Some industrial applications of FT-IR diffuse reflectance spectroscopy," *Applied Spectroscopy*, vol. 39, no. 4, pp. 634–641, 1985.

[26] P. Bassan, H. J. Byrne, F. Bonnier, J. Lee, P. Dumas, and P. Gardner, "Resonant Mie scattering in infrared spectroscopy of biological materials—understanding the dispersion artifact," *Analyst*, vol. 134, no. 8, pp. 1586–1593, 2009.

[27] P. Bassan, H. J. Byrne, J. Lee et al., "Reflection contributions to the dispersion artefact in FTIR spectra of single biological cells," *Analyst*, vol. 134, no. 6, pp. 1171–1175, 2009.

[28] M. Miljković, B. Bird, and M. Diem, "Line shape distortion effects in infrared spectroscopy," *Analyst*, vol. 137, no. 17, pp. 3954–3964, 2012.

[29] J. M. Chalmers, "Mid-infrared spectroscopy: anomalies, artifacts and common errors," in *Handbook of Vibrational Spectroscopy*, J. M. Chalmers and P. R. Griffiths, Eds., vol. 3, John Wiley and Sons, Chichester, 2006.

[30] P. R. Griffiths and J. A. De Haseth, *Fourier Transform Infrared Spectrometry, vol. 171*, John Wiley & Sons, Hoboken, NJ, 2nd edition, 2007.

[31] W. F. Perger, J. Zhao, J. M. Winey, and Y. M. Gupta, "First-principles study of pentaerythritol tetranitrate single crystals under high pressure: vibrational properties," *Chemical Physics Letters*, vol. 428, no. 4, pp. 394–399, 2006.

14

Comprehensive Study of a Handheld Raman Spectrometer for the Analysis of Counterfeits of Solid-Dosage Form Medicines

Klara Dégardin, Aurélie Guillemain, and Yves Roggo

F. Hoffmann-La Roche Ltd., Bldg 250 Room 3.504.01, Wurmisweg, 4303 Kaiseraugst, Switzerland

Correspondence should be addressed to Klara Dégardin; klara.degardin@roche.com

Academic Editor: Christoph Krafft

The fight against medicine counterfeiting is a current focus of the pharmaceutical world. Reliable analytical tools are needed to pursue the counterfeiters. Handheld devices present the advantage of providing quick results, with analyses possibly performed on the field. A large number of solid-dosage form medicines have been analyzed with a handheld Raman spectrometer. 33 out of 39 product families could be successfully analysed. The methods were validated with 100% of correct identification. Each product was additionally tested by the methods of the other products and successfully rejected. A second validation was performed using counterfeits, placebos, and generics. All the counterfeits were rejected, with p values close to zero. Some generics presented a similar formulation to the brand products and were then identified as such. One placebo was positively identified, showing that low dosage products are difficult to analyze with Raman. Robustness tests were carried out, showing, for instance, that the operator has no influence on the results and that the analyses might be performed through transparent packaging. The discovery mode was also investigated, which proposes the chemical composition of the samples. The results demonstrated that the Raman handheld device is a reliable tool for the field analysis of counterfeits.

1. Introduction

The pharmaceutical world is increasingly concerned with the counterfeiting of medicines. All types of medicines are targeted, as well as all regions of the globe. The quality of the counterfeits reaching the patients is very variable. In all cases, these products are dangerous for the patients, since they are not respecting the required hygiene for their good manufacturing [1–4]. Quick tools are consequently required for the analysis of suspect products, in order to speed up the investigations and increase the chances to pursue the criminals. Several analytical tools have proved efficient for the reliable and fast analysis of counterfeit medicines. Among them, spectroscopy is one of the most widely used, since it is mostly nondestructive and environment-friendly and requires no sampling [5]. Raman [6–12], near-infrared (NIR) [13–20], and midinfrared (MIR) [21–23] spectroscopy have been especially on the focus for counterfeit analysis.

Technological progress has enabled the development of handheld and portable spectrometers. The aim of these miniaturized devices is to be able to perform measurements on the field rather than having to send all the samples to the lab. Therefore, time can be spared for the analysis by performing a first screening of the samples. Handheld spectrometers are for instance more and more used for the field analysis of pharmaceuticals, agriculture, drugs, explosives, and mineral samples [24–33]. Handheld Raman, NIR, and MIR spectrometers have been especially studied and compared in the context of the analysis of counterfeit medicines [34–38]. Several handheld Raman devices have been evaluated in the literature. A dual-laser handheld Raman was evaluated by Assi et al. for the identification of three medicines with chemometrics [39], and another handheld Raman spectrometer for the authentication of 9 products and 22 test products using principal component analysis (PCA) [34]. Counterfeit medicines of three products were analysed by Luczak and Kalyanaraman [40]. Bate and Hess compared the results obtained with a handheld Raman instrument and the Minilab for the analysis of several antimalarial medicines [41] and for the authentication of antimalarial, antibiotic, and antimycobacterial medicines [42]. A handheld Raman spectrometer was tested on 15 pharmaceutical products by Bugay and Brush [43], and its specificity was evaluated by Hajjou et al. regarding the active pharmaceutical ingredient (API) content of six medicines [44]. Kalyanaraman et al. also presented the general use of such a device for counterfeit detection [45]. Ricci et al. compared the performance of a handheld Raman instrument with a benchtop device and desorption electrospray ionization (DESI) MS for the detection of counterfeits of antimalarials [46]. The use of a handheld Raman device on the field in Nigeria was finally presented by Spink et al. [47].

In this paper, a handheld Raman spectrometer was evaluated for the detection of counterfeits of a high number of pharmaceutical medicines, consisting in tablets, capsules, and powders. Robust methods were developed using a large database of products. In total, there are several batches of 33 product families, and among them, 62 formulations were measured, including the different manufacturing sites of the products. A detailed calibration and validation strategy was proposed, as well as some robustness tests, an evaluation of the discovery mode results for the composition of the confirmed counterfeits and a comparison with a lab instrument.

2. Material and Methods

2.1. Instrument Characteristics and Settings. A handheld Raman spectrometer was used to collect the spectra and perform the calibration. The device is a direct dispersive Raman using a 785 nm laser excitation wavelength. Its spectral resolution varies between 8 and 10.5 cm^{-1} in average, and the Raman shift range is 250 to 2875 cm^{-1}. The integration time, in automatic mode, is of 12 ms minimum.

The tablet sample holder was used for the measurement of the tablets and capsules. Some of the samples had to be measured directly against the cone of the Raman since they were too large for the holder. While the tablets and the capsules were not measured in their packaging, the powders were analysed through their original glass vial.

2.2. Statistical Method. The statistical method used for the identification of the products was the one provided by the manufacturer. The calibration consisted in taking spectra of the genuine products. The device then computed automatically a "signature" of the product with each spectrum recorded. Several signatures of the same product were then gathered into a method. The signatures are based on a multivariate test of equivalence. This test is used to compare a newly acquired Raman spectrum and the reference spectra registered in the database. If the probabilistic value (p value) of the new spectrum is above or equal to 0.05, meaning the spectrum presents a match within the preset 95 percent confidence limit, the measurement is considered consistent with the reference spectra and the device reports a "pass" result. If the p value is inferior to 0.05, the device reports a "fail" result. In the present method, while the validation with the genuine products should provide "pass" results, the counterfeits, generics, and placebos should provide a "fail" result.

The multivariate test of equivalence is computed using all the channels across the detector array region and can be considered a multivariate version of a statistical t-test. The uncertainty is directly modeled by the instrument software, and thus, no modeling has to be performed by the user.

2.3. Calibration and Validation Strategy. The calibration of the methods consisted in taking several spectra of a product, which were automatically converted in signatures, and gathering them into a single method. One method per product was ideally created. For products presenting different formulations between the dosages (different excipients), two or more methods were created per product.

The validation was performed with an independent set of spectra. The samples were tested on the methods in order to check if they were correctly recognized. All tested samples had to deliver a "pass" result with the appropriate method.

Additionally, the products were tested on all the other methods to check that no mismatch could occur for these products.

A second validation was then performed using counterfeits, placebos, and generics in order to test the ability of the handheld Raman spectrometer to detect counterfeits. The generics and the placebos were tested in order to mimic the "best case counterfeits." All these samples had to be rejected (present a "fail" result) by the instrument.

2.4. Samples

2.4.1. Genuine Samples. At least 5 batches per product were measured with the split of 3 batches for the calibration set and 2 batches for the validation set for each of the 62 studied formulations of the 33 product families. For the products manufactured in several sites, more than 5 batches were measured in order to include more variability. One sample per batch and three spectra per sample were measured. The samples were slightly moved (for the capsules and powders) or flipped (for the tablets) between each measurement. 543 signatures were acquired for the calibration, while the validation was performed with 392 spectra.

2.4.2. Placebos, Generics, and Counterfeits. 44 counterfeit samples (from eight different seizures) were tested, which had already been confirmed as counterfeits by previous analyses. In order to enlarge the number of samples, 3 placebos and 14 generics, from 5 different batches, were also tested against the methods. On the whole, 61 spectra were acquired for the second validation samples.

2.5. Robustness Tests. Once the methods were calibrated, validated, and tested against the counterfeits, placebos, and generics, 4 additional tests were performed in order to check the robustness of the methods.

The first robustness test (R1) consisted in measuring the same sample 20 times at the same position in order to see if the laser degraded the sample. Two samples (one tablet and one capsule) were tested.

In the second robustness test (R2), two samples (one tablet and one capsule) were measured 10 times at different positions in the sample holder. The objective was to evaluate

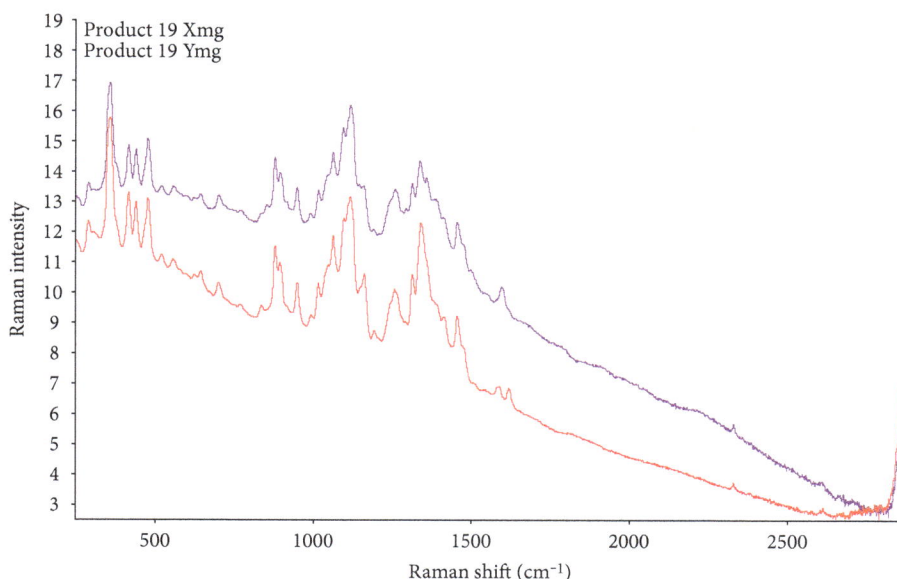

FIGURE 1: Raman spectra of two dosages of the same product family (product 19). Due to differences in the formulation, the product presents heterogeneous spectra.

if the position of the measurement had an influence on the result.

The influence of the sampling was evaluated in the third robustness test (R3). Three samples (one tablet and two capsules) were measured either through a glass vial, through a blister (transparent or white), or directly against the cone of the instrument.

Finally, the fourth robustness test (R4) consisted in measuring two samples (one tablet and one capsule) three times by two operators with the aim of testing if the operators had an influence on the measurement.

3. Results and Interpretation

3.1. Spectrum Interpretation. The spectra obtained by the device were first evaluated. Several challenges to the creation of the calibration could be pointed out.

First of all, it can be observed that some product families present heterogeneous spectra. Two or more dosages of the same product can indeed generate different spectra, when a different formulation—it means a different excipient profile—was used (see Figure 1). Additionally, in rare cases, the formulation can differ for the same product and the same dosage depending on the manufacturing site of the product (see Figure 2). It is thus important to include all the dosages and if possible all the manufacturing sites of a product into the methods. Different methods might then have to be created for one product if the spectra of the product family differ. Products 19 and 23 are indeed concerned by both the heterogeneity within the product family and the different formulations depending on the manufacturing site. Depending on the dosage or the manufacturing site, these products underwent changes in the composition of the sugar excipients, mainly going from lactose to starch. Since these excipients have a big

influence on the spectral pattern of these low-dosage products—0.3%, 0.6%, and 1% API for product 19 and 1.3% and 6% API for product 23—it might be difficult to perform the identification, unless separate methods are created per formulation.

Another challenge to the calibration is that similar excipient profiles can be found between different product families. The resulting spectra will then be similar, like the ones presented in Figure 3, and thus, the products will be hard to differentiate. This is the case for products 18 and 23, which contain the same excipients and have low dosage of API—2% for product 18 and 50% and 83% for product 23. However, small spectral differences can be observed between products 18 and 23 in the range of 1500 and 1700 cm^{-1} and are due to API bands of product 18. The two other products concerned by the similarity of excipient patterns are products 31 and 32, which with their high content of API—83% and 50%—should however be distinguishable.

Some generics present formulations that are almost identical to the ones of the brand products. This will result in similar spectra that are not possible to distinguish from the brand product, like in the example in Figure 4. In these cases, a "pass" result can be expected.

Low-dosage products are difficult to analyse with Raman spectroscopy since the API will be hardly observed in the spectrum of the medicine. The limit of detection of API in Raman spectra is quite difficult to define, especially because it depends on the excipient pattern of the medicine. In the studied products, excipient patterns made of sugar, like starch, cellulose, or glucose, are predominant in the spectra and might mask the API bands. In a previous study, a limit of detection of 0.6% was encountered, while the authors indicated that the limit of detection was usually located around 1 or 2% of API [48]. For this reason, the placebos of

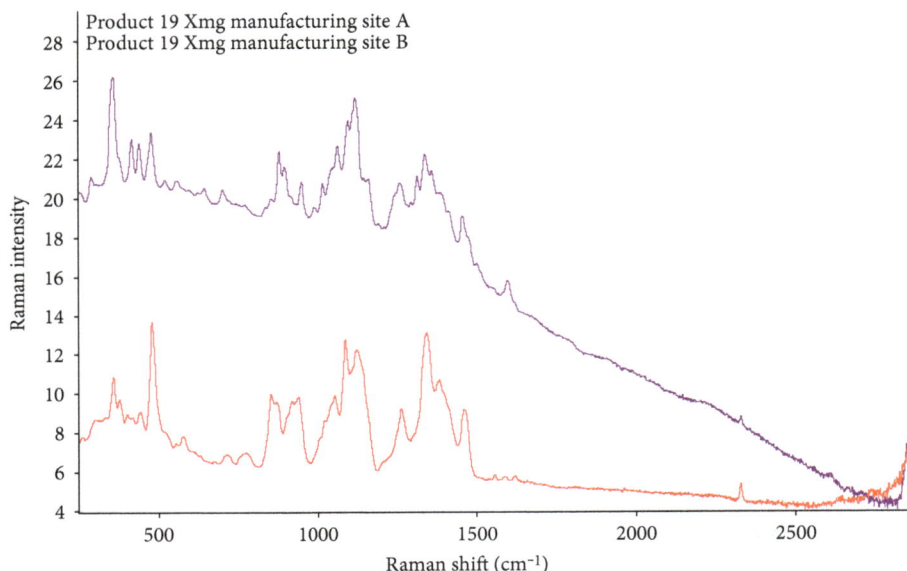

FIGURE 2: Raman spectra of two samples produced at different manufacturing sites and possessing the same dosage (product 19 Xmg). Due to differences in the chosen formulation, the product presents heterogeneous spectra.

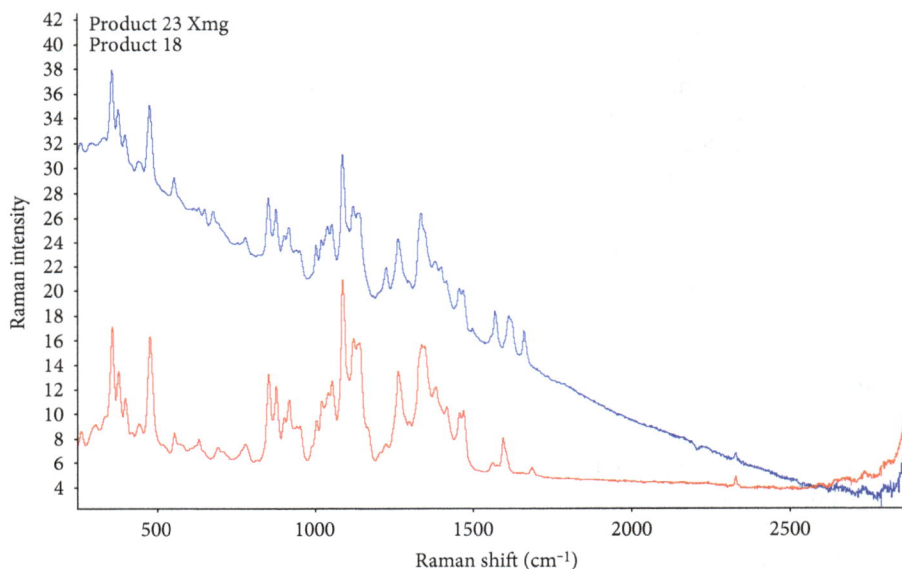

FIGURE 3: Raman spectra of two different products (product 18 and product 23 Xmg). The products present similar excipients profiles and thus close Raman spectra.

low-dosage products might provide spectra that are very similar to the genuine products. This was confirmed for product 26 which, with an API content of 0.6% and an excipient pattern dominated by starch and lactose, has identical spectra between the genuine product and its placebo.

The quality of the spectra might also have an influence on the calibration. The measured spectra present a wide range, from 250 to 2875 cm^{-1}. The end of the range, approximately from 1800 to 2875 cm^{-1}, is often noisy and not necessarily useful since it does not bring additional information about the product. Last but not least, a few samples, like product

32, underwent fluorescence (Figure 5). This phenomenon was observed when special excipients were used in the formulation of either the tablet or the shell of the capsules. In this study, in which a laser wavelength of 765 nm was used, indigo carmine (for product 32), microcrystalline cellulose, hydroxypropyl methylcellulose, and croscarmellose were found responsible for the fluorescence phenomenon. While the fluorescence made the identification harder, most bands of the API and the excipients could still be observed, like in Figure 5. When measuring other products, not presented in this paper, the fluorescence was however so strong that no signature could even be taken and no method created.

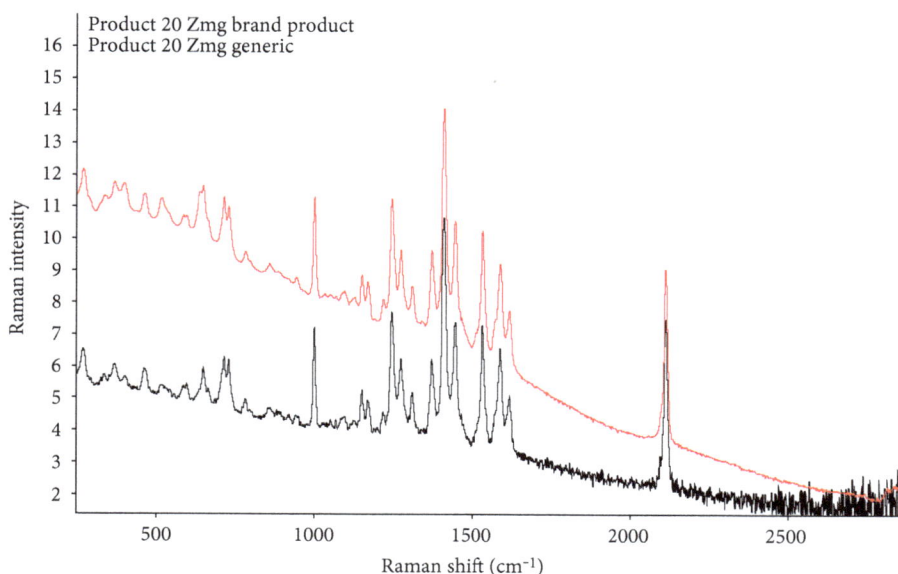

FIGURE 4: Raman spectra of a brand product and one of its generic (product 20 Zmg). Both samples present a similar formulation and the same dosage, and thus similar Raman spectra.

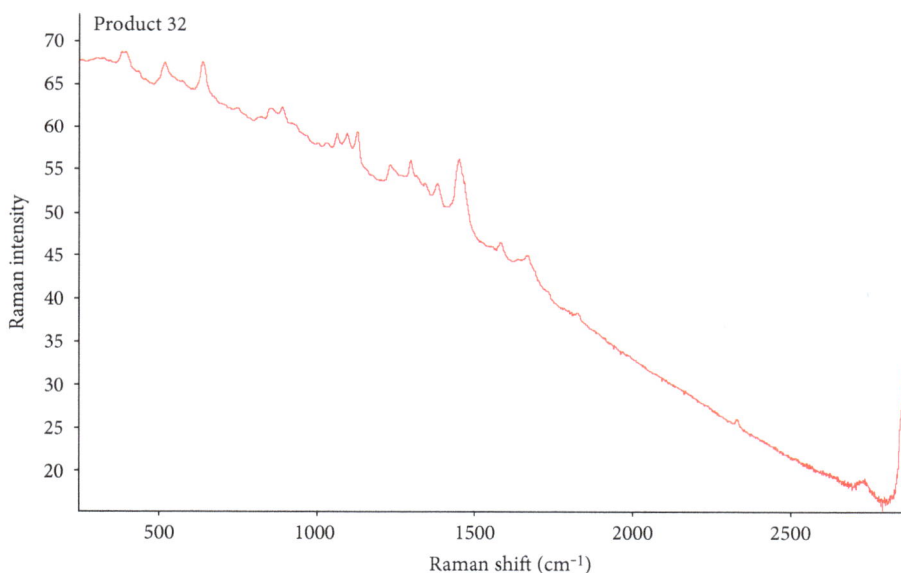

FIGURE 5: Raman spectrum of product 32. While a strong fluorescence phenomenon can be observed, the characteristic bands of the product are still visible, enabling a correct identification.

3.2. Identification Results

3.2.1. Calibration and Validation Results. There are 33 product families, and among them, 62 formulations could be successfully measured and integrated in the database of the instrument. In total, 31 methods have been created for tablets, 6 for capsules, and one for powders. Most methods gather spectra of several dosages per product. For the product families that presented heterogeneous formulations, several methods have been created.

100% of the spectra from the validation set has been correctly identified according to the pVal limit for positive identification (superior to 0.05). According to the manufacturer, a pVal lower than 0.1 is considered too low for a robust method. For all the developed methods, the minimum pVal was higher than 0.1 (see Table 1).

Additionally, no confusion could be observed when testing the products on all the other methods than their dedicated method. 100% of all other products were rejected by the tested method. Table 1 presents the minimum p values obtained for the validation samples of the corresponding method together with the maximum p values of the closest other products. These values show that the methods were successfully validated and are very robust.

TABLE 1: Validation results obtained with the genuine products. The number of signatures measured for each product has been added. The minimum p values (pVal min) obtained for the validation samples of the corresponding method are presented in the table together with the corresponding maximum p values (pVal max) of the closest other products. The device provides a positive identification for p values superior or equal to 0.05.

Sample type	Product (%API)	Number of signatures	Validation	
			pVal min of validation samples	pVal max of closest other products
Tablets	Product 1 (71%)	9	0.1404	0.0000
	Product 2 (36%)	9	0.5973	0.0105
	Product 2 (72%)	9	0.3576	0.0013
	Product 3 (80%, 95%)	18	0.5455	0.0000
	Product 4 (67%, 75%)	12	0.4137	0.0000
	Product 5 (25%, 33%)	18	0.3439	0.0171
	Product 6 (61%)	9	0.4847	0.0000
	Product 7 (57%)	9	0.4490	0.0007
	Product 7 (56%)	9	0.4106	0.0042
	Product 8 (17%)	9	0.4031	0.0083
	Product 9 (3%, 6%, 12%, 30%)	36	0.1292	0.0000
	Product 10 (8%)	9	0.5329	0.0044
	Product 10 (4%)	9	0.4481	0.0010
	Product 11 (3%)	9	0.6458	0.0129
	Product 12 (75%)	9	0.5785	0.0000
	Product 13 (0.4%, 0.8%, 1%, 2%, 3%)	54	0.4089	0.0145
	Product 14 (61%)	9	0.6009	0.0000
	Product 15 (12%)	9	0.4619	0.0167
	Product 16 (51%)	9	0.4299	0.0000
	Product 17 (25%, 45%)	27	0.4416	0.0000
	Product 18 (2%)	9	0.6275	0.0015
	Product 19 (0.3%, 0.6%, 1%)	24	0.1763	0.0009
	Product 20 (24%, 32%)	27	0.3453	0.0000
	Product 21 (75%)	6	0.5781	0.0000
	Product 22 (70%)	9	0.1562	0.0000
	Product 23 (1%, 3%, 6%)	36	0.4444	0.0348
	Product 24 (41%)	9	0.5872	0.0000
	Product 25 (77%, 78%)	18	0.5822	0.0000
	Product 26 (0.8%)	24	0.1807	0.0352
Capsules	Product 27 (83%)	9	0.6223	0.0000
	Product 28 (49%)	9	0.5439	0.0000
	Product 29 (76%)	6	0.5920	0.0000
	Product 30 (49%)	9	0.1669	0.0000
	Product 31 (83%)	18	0.1218	0.0000
	Product 32 (50%)	9	0.1165	0.0000
Powders	Product 33 (81%)	30	0.6139	0.0000

A selectivity test, proposed by the device, was evaluated after the validation. This test consists in evaluating for each method the risks of getting mismatches. The option was tested for the products providing the lowest p values during the validation. For product 26 for instance, the selectivity test suggested that the closest other products would deliver a p value of 0.0026. Tested physically during the validation, the product actually delivered a p value of 0.0352 using the method of product 26. While the selectivity test might be an option to see quickly if possible mismatches might occur, the results delivered are not real. The validation performed in this study is thus considered more completed than the selectivity test. Raman can thus not be considered a universal method for counterfeit detection.

For six other products not included in the 33 presented product families, no method could be created since the

TABLE 2: Second validation results obtained with the challenging samples. The number of samples measured for each product has been added. The maximum p values (pVal) obtained for each counterfeit are presented in the table together with the corresponding pass/fail result.

Sample set	Counterfeited product	Number of samples	pVal max	Identification against the corresponding method
Counterfeits	Product 27	2	0.0000	Fail
	Product 6	6	0.0000	Fail
	Product 28	3	0.0000	Fail
	Product 26	1	0.0000	Fail
	Product 19	5	0.0001	Fail
	Product 23	10	0.0000	Fail
	Product 25	7	0.0000	Fail
	Product 32	10	0.0003	Fail
Placebos	Product 4	1	0.0000	Fail
	Product 32	1	0.0040	Fail
	Product 26	1	0.5257	Pass
Generics	Product 20	1	0.4937	Pass
	Product 25 batch 1	2	0.0224	Fail
	Product 25 batch 2	3	0.6326	Pass
	Product 32 batch 1	1	0.0042	Fail
	Product 32 batch 2	4	0.1548	Pass
	Product 5	2	0.0110	Fail
	Product 33	1	0.1295	Pass

fluorescence prevented the measurement of the signatures. Among these six products, two were tablets, two were capsules, and two were lyophilized products. Since no signatures could be taken with these samples, the results are not displayed in this paper.

3.2.2. Second Validation with Counterfeits, Placebos, and Generics. All the tested counterfeits, consisting in 44 samples, failed for identification. The obtained p values were very low—inferior to 0.0003—which comforted the robustness of the methods for the detection of counterfeits. The details of the results of the second validation are presented in Table 2.

Nine of the tested generics have the same formulation as the genuine brand products, and their spectra are identical, as previously presented in Figure 4. These samples were consequently identified by the device as genuine. Since the objective of the methods is to reject, counterfeits and generics were only used to simulate them; these results are not considered critical. The other five generics were rejected by the methods. The p values obtained are highly dependent on the tested generic.

Two of the three tested placebos were rejected by the methods. The third one, the placebo of product 26, was detected by the instrument as genuine with a maximum p value of 0.5257, which is very high. The method for product 26, a low-dosage product, was then considered not specific enough and was then deleted from the database.

The results of the validation and the tests performed for the second validation are summarized in Table 3.

3.3. Robustness Results. The results of the four robustness tests are presented in Table 4.

According to the first test, a capsule of product 31 and a tablet of product 20 were measured 20 times at the same position to check if the samples were damaged by the laser. The capsule and the tablet were correctly identified even after 20 measurements. Additionally, no important variation of the p value range was observed, which confirms that the laser did not damage the samples. At this point, it should be noted that no soft-gel capsules should be measured with the Raman instrument since their shell is very sensitive to the heat of the laser.

A capsule of product 31 and a tablet of product 20 were each measured 10 times for the robustness test R2. They were moved between each measurement in order to check if the position of the sample had an influence on the results. The samples were correctly identified during each measurement, with little difference in the p values. This means that the position of the samples has low influence on their identification. However, it is recommended to measure the capsules on the side where there is only one layer of gelatin. Also, care should be taken in order not to break the samples with the sample holder. For instance, if a tablet is broken and its coating removed due to the breaking, then a slightly different spectrum can be obtained, which might alter the results.

The robustness test R3 deals with the sampling that is available for the studied medicines. Three products were measured either directly with the sample holder, through transparent glass vials, or through blisters, depending on the available packaging. Two types of blister were evaluated: transparent (plastic) and white (plastic). The samples were all correctly authenticated when measured directly, through transparent glass vials and through transparent blisters. The products could not be correctly identified through white blister. It is consequently recommended to use the tablet

TABLE 3: Overview of the results obtained for the validation and the tests of the second validation. (*) In these tests, "fail" is actually the expected answer, since these products should not be identified by the method.

Step of the method development	Sample set	Results of the identification
Validation	Validation with positive samples (e.g., product 1 tested on the method of product 1)	100% pass
	Validation against all other products (e.g., product 1 tested on the method of product 2)	100% fail*
Second validation	Counterfeits	100% fail*
	Placebos	67% fail*
	Generics	35% fail*

TABLE 4: Results of the four robustness tests. The conditions of the test, like the sampling, have been added together with the number of spectra measured. The results consist in the p value range and the pass/fail answers.

Robustness test	Product	Conditions	Number of spectra	pVal results	Identification against the corresponding method
R1: degradation with the laser	Product 20	No change in sample position	20	[0.5745–0.6669]	Pass
	Product 31	No change in sample position	20	[0.5616–0.7335]	Pass
R2: influence of the sample position	Product 20	Sample position changed	10	[0.5523–0.6746]	Pass
	Product 31	Sample position changed	10	[0.3316–0.6481]	Pass
R3: influence of the sampling	Product 20	Glass vial	3	[0.5386–0.5950]	Pass
		Transparent blister	3	[0.5891–0.6534]	Pass
		Direct	3	[0.5344–0.5746]	Pass
	Product 32	Glass vial	3	[0.1131–0.2471]	Pass
		Transparent blister	3	[0.0814–0.3385]	Pass
		Direct	3	[0.1628–0.4482]	Pass
	Product 27	White blister	3	[0.0175–0.0259]	Fail
		White blister	3	[0.0272–0.1164]	Fail
		White blister	3	[0.0017–0.0731]	Fail
R4: influence of the operator	Product 20	Operator A	3	[0.6815–0.6871]	Pass
		Operator B	3	[0.6143–0.6526]	Pass
	Product 31	Operator A	3	[0.6159–0.6779]	Pass
		Operator B	3	[0.6599–0.7381]	Pass

holder to measure the samples. However, it is interesting to note that according to this test, a product measured on the field through transparent blister or in transparent glass vials might be correctly authenticated.

In order to evaluate the human influence (test R4), a tablet of product 20 and a capsule of product 31 have been measured three times by two different operators. The samples were all correctly identified, and the obtained p values are very close between the operators. It seems that the operators have no influence on the identification, which is an important factor for the use on the field. It should be noted that both operators were properly trained before the robustness test.

3.4. Composition of the Counterfeits. As soon as a counterfeit has been confirmed, it is necessary to determine its chemical composition in order to estimate the impact on the patients' safety. In this frame, the discovery mode of the instrument has been evaluated with the studied counterfeits. This mode proposes, in case of a "fail" result, the chemical composition of the tested sample. This composition is suggested according

to the database furnished by the supplier and integrated in the device. The database contains more than 8000 references of excipients and active substances and was completed by the studied products during the calibration.

Several samples per counterfeit seizures have been analysed with the device and interpreted with the discovery mode. Their chemical composition was previously determined by the lab instruments, a Raman, a near-infrared, and a nuclear magnetic resonance spectrometers.

For 14 samples, no match could be found with the instrument database. This can be explained by the fact that for these samples, the chemical composition was a mixture of at least three components, part of the database, but difficult to detect all with an automatic Raman database search.

The second point is that for 10 counterfeits, a different composition was proposed by the device between the different samples of the same seizure. Once again, it can be explained by the complex mixture of the counterfeits. Some components could be identified, while others not. For instance, for the counterfeits of product 19, made of melatonin, starch, and saccharose, melatonin is missing in the search of the device. While it is observable in the Raman spectra, the bands are too small to be detected by an automatic search.

In three cases (product 29 seizures number 2 and number 3 and product 33), the chemical composition proposed by the discovery mode was correct. The counterfeits are only made of one component, an API, easy to identify by Raman.

It can be concluded that some of the compounds of the counterfeits are common compared to the "right" composition determined by the lab instruments. While the discovery mode can help the analyst determine the chemical composition of the confirmed counterfeits, it cannot be reported as the right composition. It should be noted that in the discovery mode, the operator also has access to the overlay of the tested sample and the encountered matches of the database. The spectra can then be visually compared to evaluate if the proposed matches are correct or not. In this case, the operator should however be a Raman specialist.

3.5. Comparison with a Lab Instrument. In order to evaluate the quality of the data generated by the handheld Raman spectrometer, these were compared with the Raman spectra obtained with a benchtop instrument. The chosen instrument is a Raman spectrometer equipped with a probe generating a 6 mm diameter laser spot. The device possesses a laser of 785 nm, providing a power of 270 mW on the samples, and a spectra range going from 150 to 1890 cm^{-1}.

Figures 6(a) and 6(b) present examples of overlaid spectra generated by the handheld and the benchtop Raman spectrometers. Two types of spectra are represented for the handheld device: the ones generated while taking a signature and the others generated during simple measurements; it means routine analysis or method validation.

The spectra generated by both devices for the signatures are both of good quality and are almost identical. The ones generated by the handheld Raman spectrometer during simple measurements are much noisier. However, the main bands are easily observed and the spectra used for validation all provided 100% of correct identification.

It can also be noted that the range used by the handheld Raman spectrometer, longer than the one of the benchtop device, does not bring much more information and is usually the noisiest part of the spectrum. However, based on the validation results, this additional range (1890–2900 cm^{-1}) does not seem to have a negative impact on the identification.

The results of the calibration obtained by both devices can be partly compared for the tablets while referring to a previous study performed with this lab instrument [48]. In both cases, 100% of the samples used for the calibration and for the validation have been correctly identified. Consequently, the results obtained by the handheld Raman spectrometer are in the present case as good as the ones obtained with the lab instrument.

4. Discussion

In this paper, methods have been presented for the authentication of almost all the solid products produced by the firm.

The database consists indeed in 33 product families, and among them are 62 product formulations. The calibration and validation could be successfully performed for all of these products. Moreover, the methods were validated against all the other products of the database, which were successfully rejected by the studied methods. Several challenges, raised during the spectra analysis, have been encountered during the development of the presented methods. The heterogeneity of the product families and the closeness of formulation between different families could be solved by the high number of signatures taken and their correct gathering in the methods. The importance of taking several spectra, especially of samples from different manufacturing sites, should be underlined. All the tested counterfeits were rejected by the methods, which confirms the dedicated use of the handheld Raman spectrometer for counterfeit detection. Some generics were recognized as genuine brand products, however, only when their formulation was really close. These results were not considered critical since the objective of the methods was to detect counterfeits, and not generic products. The tested placebos were rejected by the methods, apart from one sample, the placebo of product 26. Consequently, product 26 was deleted from the database after the second validation phase. This test suggests that for low-dosage products, an additional analytical technique should be used that would confirm the presence of the API. In these cases, a fingerprint of the whole tablet is nevertheless available by Raman, meaning that at least the right excipients are present in the tested product [48].

Six products, not included in the 33 presented products, could not be measured at all by the instrument since the fluorescence prevented the acquisition of the signatures. Fluorescence constitutes the main drawback of the Raman technology, which can be on other instruments diminished for instance by the use of a fluorescence correction option or the use of a different wavelength like 1084 nm. In the present case, no solution can be found for the products presenting fluorescence. However, a large majority of the tested products could be successfully measured with the handheld Raman spectrometer. Furthermore, the results obtained with the handheld Raman spectrometer were compared to the ones generated by a lab instrument. It resulted that the handheld instrument delivered good results as the lab Raman spectrometer.

The robustness tests performed during the validation demonstrated that the laser did not damage the samples. Moreover, no influence could be observed of the sample position or the operator. However, it was pointed out that the operators had to be trained so that they would position the samples carefully. Also, the samples should not be damaged when placed on the device, for instance by removing involuntarily the coating, or measured through a thick layer of capsule shell. A trained operator could consequently perform the measurement on the field and be able to authenticate the samples without needing further knowledge in Raman spectroscopy. The fourth robustness test showed that a measurement through a white blister did not allow the identification of the capsule. Since the measured product (product 27) contains 83% of API, this confirms that

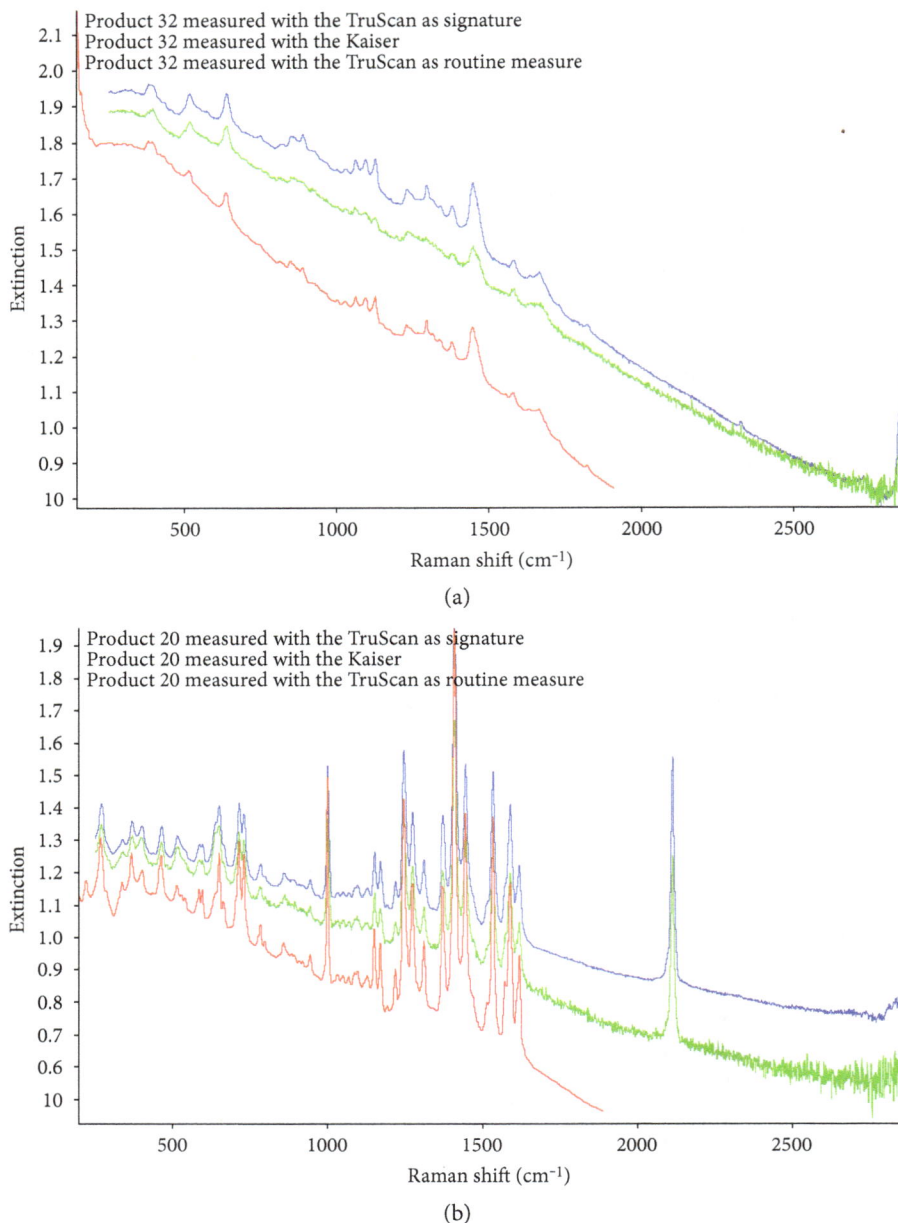

FIGURE 6: Overlay of Raman spectra obtained with the TruScan Thermo® (signature and routine measure) and the RXN1 Kaiser Optical Systems® devices for product 32 capsule (a) and product 20 tablet (b).

this can be attributed to the type of packaging. Furthermore, transparent packaging like glass vial or blister delivered the same results as in direct contact with the instrument in the sample holder. This fact could be of advantage during a measurement on the field, since a nondestructive measurement should always be privileged. While Raman spectroscopy is not destructive in itself, it can be considered destructive if tablets or capsules have to be removed from their packaging for the analysis.

Handheld Raman spectrometers are also interesting for counterfeit detection for the reason that a method can be developed on a master instrument and then easily transferred to several other ones for a local measurement on the field, for instance at airports, at the customs, or in warehouses. In this

frame, the transfer of the presented methods to another instrument is quite easy. The transfer basically only consisted in the copying of files from the master device on the new device. Then, each method can be tested with an independent set of samples in order to check that the spectrometer has no influence on the measurement. At this point, the introduction of all possible manufacturing sites in the database is particularly important if the new instrument is going to be used for the testing in another region of the world. That way, much fewer false negatives should be expected. The use of the handheld devices on the field would help speed up the investigation, by reducing the time spent by sending the samples. It is however recommended to send the samples that failed the identification with the handheld Raman spectrometer to labs in order to confirm the counterfeits and perform the analysis of the whole chemical composition. Indeed, the test conducted in this study concerning the discovery mode of the instrument revealed that in most cases, and based on the previous analyses performed with diverse technologies, the proposed chemical composition was not correct. This mode is able to determine the composition when the counterfeits are only made of one component. For the analysis of mixtures, a manual spectrum interpretation and further analyses are required.

The main advantage of the studied instrument resides in how easy the method calibration is. Indeed, no further knowledge in Raman spectroscopy or chemometrics is required, although knowledge in method validation should be privileged. The proposed algorithm automatically calculates the p values while taking signatures of the products. Nevertheless, this system might be considered a "black box" with no possible way of modifying the calibration of the method. The opportunity of choosing further chemometric tools would be welcome, which seems to be possible in the newest versions of the device. Indeed, the more products are introduced in the database, the more flexibility is needed to develop a robust calibration. Also, it would be of advantage to be able to change the sample exposure during the routine measurements. In fact, the quality of the spectra is worse during this phase than during the signature mode, which might generate less reliable results. The end of the range (1890–2900 cm^{-1}) is additionally not of great interest for the studied products and represents the noisiest part of the spectrum. A compromise might be found between the time of measurement, partly dependent on the sample exposure, and the quality of the spectra. Finally, another possible improvement of the device would be the possibility of importing spectra from another device. Even if the calibration step is as fast as possible concerning the computation, taking signatures of a large database of products and validating all of them require many hours of measurements. The spectra should in this case be transformed with calibration transfer techniques and the methods carefully validated since they are spectrometer dependent, but a lot of time would be spared in the measurement phase.

5. Conclusion

Handheld spectrometers are more and more used for the on-site analysis of pharmaceuticals. In this paper, a handheld Raman spectrometer has been evaluated for the analysis of counterfeits of 33 product families of medicines under the form of tablets, capsules, and powders. Several challenges had to be dealt with during the calibration of the methods, like the heterogeneity of several product families or the closeness of formulation between different products. Nevertheless, the methods could be successfully created and fully validated. The tested counterfeits were rejected by the methods, thus confirming the specificity of the calibration for counterfeit detection. The methods were also proved to be robust in terms of degradation by the laser, sample positioning, and influence of the operators. The samples

could be partly measured through the packaging, depending on their opacity.

Some limits could nevertheless be found concerning the instrumentation, like the positive identification of some generics and one placebo. The device is not adapted to the measurement of low-dosage medicines, for which a complementary method like infrared spectroscopy should be used. Moreover, the analysis of the chemical composition, suggested by the discovery mode of the device, does not provide correct results as soon as more than one component is included in the counterfeit. While the calibration is performed quite quickly, the chemometrics cannot be chosen and the method developer has to rely on the proposed algorithm. Finally, the spectra obtained during the routine measurements are quite noisy and might generate false negative results. Apart from these encountered limits, the handheld Raman spectrometer performed very well for the identification of the 33 product families of medicines and is uncontestably a reliable tool for the analysis of counterfeit medicines on the field. More and more handheld devices are being used in this context, which should be encouraged in order to speed up the investigations and provide better chances to pursue the criminals. While the handheld Raman spectrometer might be used as a first screening tool on the field, the counterfeits detected can be sent to a lab for confirmation and determination of the chemical composition for the evaluation of the patient's health risk.

References

[1] C. del Castillo Rodriguez and M. J. Lozano Estevan, "Counterfeit medicine. A threat to health. Legal situation," *Anales de la Real Academia Nacional de Farmacia*, vol. 81, no. 4, pp. 329–333, 2015.

[2] K. Dégardin, Y. Roggo, and P. Margot, "Understanding and fighting the medicine counterfeit market," *Journal of Pharmaceutical and Biomedical Analysis*, vol. 87, pp. 167–175, 2014.

[3] J. Harris, P. Stevens, and J. Morris, *Keeping It Real. Combating the Spread of Fake Drugs in Poor Countries*, International Policy Network, 2009, April 2014, http://www.africanliberty.org/pdf/Keepingitreal.pdf.

[4] E. Medina, E. Bel, and J. M. Sune, "Counterfeit medicines in Peru: a retrospective review (1997-2014)," *BMJ Open*, vol. 6, no. 4, pp. 1–11, 2016.

[5] S. Garrigues and M. de la Guardia, "Non-invasive analysis of solid samples," *Trends in Analytical Chemistry*, vol. 43, pp. 161–173, 2013.

[6] F. Adar, E. Lee, A. Whitley, and M. Witkowski, *Single-Point Analysis and Raman Mapping of Tablet Dosage Formulation as a Means for Detecting and Sourcing Counterfeit Pharmaceuticals*, Pharmaceutical Online, 2007, March 2011, http://www.pharmaceuticalonline.com/download.mvc/Single-Point-Analysis-And-Raman-Mapping-Of-Do-0001?user=20.

[7] M. Boiret, D. Rutledge, N. Gorretta, Y. M. Ginot, and J. M. Roger, "Utilisation de la Microscopie Raman et des Méthodes

Chimiométriques pour la Détection de Comprimés Contrefaits," *Spectra Analyse*, vol. 298, pp. 74–80, 2014.

[8] K. Dégardin, Y. Roggo, F. Been, and P. Margot, "Detection and chemical profiling of medicine counterfeits by Raman spectroscopy and chemometrics," *Analytica Chimica Acta*, vol. 705, no. 1-2, pp. 334–341, 2011.

[9] K. Kwok and L. S. Taylor, "Analysis of the packaging enclosing a counterfeit pharmaceutical tablet using Raman microscopy and two-dimensional correlation spectroscopy," *Vibrational Spectroscopy*, vol. 61, pp. 176–182, 2012.

[10] J. K. Mbinze, P.-Y. Sacré, A. Yemoa et al., "Development, validation and comparison of NIR and Raman methods for the identification and assay of poor-quality oral quinine drops," *Journal of Pharmaceutical and Biomedical Analysis*, vol. 111, pp. 21–27, 2015.

[11] S. Neuberger and C. Neusüss, "Determination of counterfeit medicines by Raman spectroscopy: systematic study based on a large set of model tablets," *Journal of Pharmaceutical and Biomedical Analysis*, vol. 112, pp. 70–78, 2015.

[12] J. Peters, A. Luczak, V. Ganesh, E. Park, and R. Kalyanaraman, "Raman spectral fingerprinting for biologics counterfeit drug detection," *American Pharmaceutical Review*, vol. 19, no. 2, 2016.

[13] P. De Peinder, M. J. Vredenbregt, T. Visser, and D. De Kaste, "Detection of Lipitor® counterfeits: a comparison of NIR and Raman spectroscopy in combination with chemometrics," *Journal of Pharmaceutical and Biomedical Analysis*, vol. 47, no. 4-5, pp. 688–694, 2008.

[14] F. E. Dowell, E. B. Maghirang, F. M. Fernandez, P. N. Newton, and M. D. Green, "Detecting counterfeit antimalarial tablets by near-infrared spectroscopy," *Journal of Pharmaceutical and Biomedical Analysis*, vol. 48, no. 3, pp. 1011–1014, 2008.

[15] J. Dubois, J. C. Wolff, J. K. Warrack, J. Schoppelrei, and E. N. Lewis, "NIR chemical imaging for counterfeit pharmaceutical products analysis," *Spectroscopy*, vol. 22, pp. 36–41, 2007, http://www.iesmat.com/iesmat/upload/file/Malvern/Product os-MAL/NIR-NIR%20Chemical%20Imaging%20for%20Cou nterfeit%20Pharmaceutical%20Products%20Analysis.pdf.

[16] R. da Silva Fernandes, F. S. L. da Costa, P. Valderrama, P. H. Março, and K. M. G. de Lima, "Non-destructive detection of adulterated tablets of glibenclamide using NIR and solid-phase fluorescence spectroscopy and chemometric methods," *Journal of Pharmaceutical and Biomedical Analysis*, vol. 66, pp. 85–90, 2012.

[17] M. B. Lopes and J. C. Wolff, "Investigation into classification/ sourcing of suspect counterfeit Heptodin™ tablets by near infrared chemical imaging," *Analytica Chimica Acta*, vol. 633, no. 1, pp. 149–155, 2009.

[18] S. H. F. Scafi and C. Pasquini, "Identification of counterfeit drugs using near infrared spectroscopy," *The Analyst*, vol. 126, no. 12, pp. 2218–2224, 2001.

[19] K. Dégardin, A. Guillemain, N. Viegas Guerreiro, and Y. Roggo, "Near infrared spectroscopy for counterfeit detection using a large database of pharmaceutical tablets," *Journal of Pharmaceutical and Biomedical Analysis*, vol. 128, pp. 89–97, 2016.

[20] J. R. Lucio-Gutierrez, J. Coello, and S. Maspoch, "Expeditious identification and semi-quantification of Panax ginseng using near infrared spectral fingerprints and multivariate analysis," *Analytical Methods*, vol. 5, no. 4, pp. 857–865, 2013.

[21] E. Deconinck, P. Y. Sacré, D. Coomans, and J. De Beer, "Classification trees based on infrared spectroscopic data to discriminate between genuine and counterfeit medicines," *Journal of Pharmaceutical and Biomedical Analysis*, vol. 57, pp. 68–75, 2011.

[22] C. Ricci, C. Eliasson, N. Macleod, P. Newton, P. Matousek, and S. Kazarian, "Characterization of genuine and fake artesunate anti-malarial tablets using Fourier transform infrared imaging and spatially offset Raman spectroscopy through blister packs," *Analytical and Bioanalytical Chemistry*, vol. 389, no. 5, pp. 1525–1532, 2007.

[23] K. Dégardin and Y. Roggo, "Counterfeit analysis strategy illustrated by a case study," *Drug Testing and Analysis*, vol. 8, no. 3-4, pp. 388–397, 2015.

[24] H. Ayvaz and L. E. Rodriguez-Saona, "Application of hand-held and portable spectrometers for screening acrylamide content in commercial potato chips," *Food Chemistry*, vol. 174, pp. 154–162, 2014.

[25] A. Keil, N. Talaty, C. Janflet et al., "Ambient mass spectrometry with a handheld mass spectrometer at high pressure," *Analytical Chemistry*, vol. 79, no. 20, pp. 7734–7739, 2007.

[26] P. Leary, G. S. Dobson, and J. A. Reffner, "Development and applications of portable gas chromatography-mass spectrometry for emergency responders, the military, and law-enforcement organizations," *Applied Spectroscopy*, vol. 70, no. 5, pp. 888–896, 2016.

[27] J. D. Dunn, C. M. Gryniewicz-Ruzicka, J. F. Kauffman, B. J. Westenberger, and L. F. Buhse, "Using a portable ion mobility spectrometer to screen dietary supplements for sibutramine," *Journal of Pharmaceutical and Biomedical Analysis*, vol. 54, no. 3, pp. 469–474, 2010.

[28] C. R. Appoloni and F. L. Melquiades, "Portable XRF and principal component analysis for bill characterisation in forensic science," *Applied Radiation and Isotopes*, vol. 85, pp. 92–95, 2014.

[29] D. Lauwers, A. Candeias, A. Coccato, J. Mirao, L. Moens, and P. Vandenabeele, "Evaluation of portable Raman spectroscopy and handheld X-ray fluorescence analysis (hXRF) for the direct analysis of glyptics," *Spectrochimica Acta Part A: Molecular and Biomolecular Spectroscopy*, vol. 157, pp. 146–152, 2015.

[30] M. D. Hargreaves, K. Page, T. Munshi, R. Tomsett, G. Lynch, and H. G. M. Edwards, "Analysis of seized drugs using portable Raman spectroscopy in an airport environment - a proof of principle study," *Journal of Raman Spectroscopy*, vol. 39, no. 7, pp. 873–880, 2008.

[31] E. L. Izake, "Forensic and homeland security applications of modern portable Raman spectroscopy," *Forensic Science International*, vol. 202, no. 1–3, pp. 1–8, 2010.

[32] J. Zheng, S. Pang, T. P. Labuza, and L. He, "Evaluation of surface-enhanced Raman scattering detection using a handheld and a bench-top Raman spectrometer: a comparative study," *Talanta*, vol. 129, pp. 79–85, 2014.

[33] J. Jehlicka, P. Vítek, H. G. M. Edwards, M. Heagraves, and T. Capoun, "Application of portable Raman instruments for fast and non-destructive detection of minerals on outcrops," *Spectrochimica Acta Part A: Molecular and Biomolecular Spectroscopy*, vol. 73, no. 3, pp. 410–419, 2009.

[34] S. Assi, "Investigating the quality of medicines using handheld Raman spectroscopy," *European Pharmaceutical Review*, vol. 19, no. 5, pp. 56–60, 2014.

[35] A. J. O'Neil, R. D. Jee, G. Lee, A. Charvill, and A. C. Moffat, "Use of a portable near infrared spectrometer for the

authentication of tablets and the detection of counterfeit versions," *Journal of Near Infrared Spectroscopy*, vol. 16, no. 3, pp. 327–333, 2008.

[36] J. E. Polli, S. W. Hoag, and S. Flank, "Comparison of authentic and suspect pharmaceuticals," *Pharmaceutical Technology*, vol. 33, no. 8, pp. 46–52, 2009.

[37] M. Alcalà, M. Blanco, D. Moyano et al., "Qualitative and quantitative pharmaceutical analysis with a novel hand-held miniature near infrared spectrometer," *Journal of Near Infrared Spectroscopy*, vol. 21, no. 6, pp. 445–457, 2013.

[38] K. Dégardin, Y. Roggo, and P. Margot, "Evaluation de Spectromètres Portables Raman, Infrarouge et Proche Infrarouge pour la Détection de Contrefaçons de Médicaments," *Spectra Analyse*, vol. 276, pp. 46–52, 2010.

[39] S. Assi, R. A. Watt, and A. C. Moffat, "Identification of counterfeit medicines from the internet and the world market using near-infrared spectroscopy," *Analytical Methods*, vol. 3, no. 10, pp. 2231–2236, 2011.

[40] A. Luczak and R. Kalyanaraman, "Portable and benchtop Raman technologies for product authentication and counterfeit detection," *American Pharmaceutical Review*, pp. 1–9, 2014.

[41] R. Bate and K. Hess, "Anti-malarial drug quality in Lagos and Accra - a comparison of various quality," *Malarial Journal*, vol. 9, no. 1, p. 157, 2010.

[42] R. Bate, R. Tren, K. Hess, L. Mooney, and K. Porter, "Pilot study comparing technologies to test for substandard drugs in field settings," *African Journal of Pharmacy and Pharmacology*, vol. 3, no. 4, pp. 165–170, 2009.

[43] D. E. Bugay and R. C. Brush, "Chemical identity testing by remote-based dispersive raman spectroscopy," *Applied Spectroscopy*, vol. 64, no. 5, pp. 467–475, 2010.

[44] M. Hajjou, Y. Qin, S. Bradby, D. Bempong, and P. Lukulay, "Assessment of the performance of a handheld Raman device for potential use as a screening tool in evaluating medicines quality," *Journal of Pharmaceutical and Biomedical Analysis*, vol. 74, pp. 47–55, 2012.

[45] R. Kalyanaraman, M. Ribick, and G. Dobler, "Portable Raman spectroscopy for pharmaceutical counterfeit detection," *European Pharmaceutical Review*, vol. 17, no. 5, pp. 35–39, 2012.

[46] C. Ricci, L. Nyadong, F. Yang et al., "Assessment of hand-held Raman instrumentation for in situ screening for potentially counterfeit artesunate antimalarial tablets by FT-Raman spectroscopy and direct ionization mass spectrometry," *Analytica Chimica Acta*, vol. 623, no. 2, pp. 178–186, 2008.

[47] J. Spink, D. C. Moyer, and M. R. Rip, "Addressing the risk of product fraud: a case study of the Nigerian combating counterfeiting and sub-standard medicines initiatives," *Journal of Forensic Science and Criminology*, vol. 4, no. 2, pp. 1–13, 2016.

[48] Y. Roggo, K. Dégardin, and P. Margot, "Identification of pharmaceutical tablets by Raman spectroscopy and chemometrics," *Talanta*, vol. 81, no. 3, pp. 988–995, 2010.

15

Bromide-Assisted Anisotropic Growth of Gold Nanoparticles as Substrates for Surface-Enhanced Raman Scattering

Melissa A. Kerr and Fei Yan

Department of Chemistry, North Carolina Central University, Durham, NC 27707, USA

Correspondence should be addressed to Fei Yan; fyan@nccu.edu

Academic Editor: Nikša Krstulović

We report herein a one-step synthesis of gold nanoparticles (Au NPs) of various shapes such as triangles, hexagons, and semispheres, using 5-hydroxyindoleacetic acid (5-HIAA) as the reducing agent in the presence of potassium bromide (KBr). Anisotropic Au NPs have received ever-increasing attention in various areas of research due to their unique physical and chemical properties. Numerous synthetic methods involving either top-down or bottom-up approaches have been developed to synthesize Au NPs with deliberately varied shapes, sizes, and configurations; however, the production of templateless, seedless, and surfactant-free singular-shaped anisotropic Au NPs remains a significant challenge. The concentrations of hydrogen tetrachloroaurate ($HAuCl_4$), 5-HIAA, and KBr, as well as the reaction temperature, were found to influence the resulting product morphology. A detailed characterization of the resulting Au NPs was performed using ultraviolet-visible (UV-Vis) spectroscopy, scanning electron microscopy (SEM), and Raman spectroscopy. The as-prepared Au NPs exhibited excellent surface-enhanced Raman scattering (SERS) properties, which make them very attractive for the development of SERS-based chemical and biological sensors.

1. Introduction

Raman spectroscopy, based on molecular vibrational transitions, has long been regarded as a valuable tool for the identification and quantification of chemical and biological species [1, 2]. While it is well known that signals in normal Raman spectroscopy are extremely weak, great progress has been made since the discovery of surface-enhanced Raman scattering (SERS) [3–5]. A great deal of recent research has been delving into how effective Au NPs are as SERS substrates. The surface chemistry of gold is quite versatile, as it can easily be manipulated into a variety of sizes and shapes [6–10]. The electron cloud plasmons and the wavelength where that metal nanoparticle absorbs photons are directly affected by the type, shape, size, and structure of the metal nanoparticle. If it is possible to match the plasmon absorption wavelength of Au NPs with the excitation wavelength of the laser, resonance Raman, yet another boost to the Raman signal, is possible. Therefore, being able to manipulate the size and shape of a gold nanoparticle would allow for simple and easy modifications to analytical devices that utilize SERS technology that would greatly improve sensitivity and selectivity.

It has been shown that Au NPs can be synthesized in any number of sizes or shapes quite easily. The Au NPs that have been synthesized are nanoflowers [1, 11], planar triangles (or prisms) [12–14], planar hexagons [12], nanorods [13, 15, 16], spheres [17, 18], octahedra, and cubes [19], to name just a few. Even though the toxicity of Au NPs has been widely studied, they have not been shown to be toxic to humans [9]. With this in mind, Au NPs can be utilized for a broad range of biomedical sensing, diagnostics, or other biomedical-based applications [12].

Herein, we report a facile approach for the synthesis of anisotropic Au NPs of various shapes such as triangles, hexagons, and semispheres, using 5-hydroxyindoleacetic acid (5-HIAA) as the reducing agent in the presence of potassium bromide (KBr). The concentrations of hydrogen tetrachloroaurate ($HAuCl_4$), 5-HIAA, and KBr, as well as the reaction temperature, were found to influence the resulting product morphology. A detailed characterization of the resulting Au NPs was performed using ultraviolet-visible (UV-Vis) spectroscopy, scanning electron microscopy (SEM), and Raman spectroscopy. The as-prepared Au NPs exhibited excellent SERS properties, which make them very attractive for the development of SERS-based chemical and biological sensors.

2. Materials and Methods

2.1. Chemicals. Hydrogen tetrachloroaurate (1% w/v) was obtained from Ricca Chemical Company, 5-hydroxyindole-3-acetic acid (5-HIAA, 99%) and crystal violet were obtained from Acros Organics, and potassium bromide (KBr, 99% FT-IR grade) was obtained from Sigma-Aldrich. All solutions were prepared with high purity deionized water (resistivity $\geq 18 \, M\Omega \cdot cm$) from a Picopure®2 ultrapure water purification system (Hydro, Inc.).

2.2. Synthesis of Anisotropic Gold Nanoparticles in the Absence of KBr. A 0.75 mM solution of 5-HIAA was created by adding 7.17 mg of 5-HIAA to 50 mL of diH$_2$O (DI). The solution was sonicated for about 1 minute until there was no longer any visual evidence of the solid 5-HIAA. On a stir plate, 44.42 mL of DI, 5 mL of 0.75 mM 5-HIAA, and 579.2 μL of 1% HAuCl$_4$ were mixed at approximately 300 rpm for 30 minutes under ambient conditions. The color turned gray about a minute after addition of the 5-HIAA. At the end of the 30-minute period, the sample started to turn to a salmon-pink color.

2.3. Synthesis of Anisotropic Gold Nanoparticles in the Presence of KBr. A 0.75 mM solution of 5-HIAA was created by adding 7.17 mg of 5-HIAA to 50 mL of diH$_2$O (DI). The solution was sonicated for about 1 minute until there was no longer any visual evidence of the solid 5-HIAA. A 0.60 mM solution of KBr was made by adding 3.57 mg of KBr to 50 mL of DI. Synthesis was performed using increasing volumes of KBr. In a beaker, 4 mL, 5 mL, and 6 mL of 0.60 mM KBr, 5 mL of 0.75 mM 5-HIAA, and 579.2 μL of HAuCl$_4$ (1% w/v) were mixed with DI water in a solution of 50 mL. The solutions were mixed on a stir plate set at approximately 300 rpm for 30 minutes under ambient conditions. The color turned yellow about a minute after addition of the 5-HIAA and then to gray and purple after about 5 minutes.

2.4. UV-Vis Spectrophotometry. The UV-Vis spectra were obtained using an Evolution 220 UV-Vis spectrophotometer. The spectra measured absorbance from 400 to 750 nm, with a 100% T baseline correction. Samples were analyzed in a standard quartz cuvette.

2.5. Scanning Electron Microscopy (SEM) Analysis. SEM images were obtained using a FEI XL-30 Field Emission SEM. The accelerating voltage used was 30 kV with a spot size of 2. A working distance of around 7 mm was used to optimize the magnification using the through-the-lens detector (TLD). Samples were prepared on a silicon wafer.

2.6. Raman Spectroscopy. Raman spectra were acquired using a Renishaw RM1000 Raman microspectrometer (with a 20x objective). The Raman system was coupled to an Olympus BH-2 microscope and equipped with a 785 nm diode laser, an edge filter with 200 cm^{-1} cutoff, 1200-line/mm grating, and a thermoelectrically cooled CCD detector. The system is operated using Renishaw WIRE™ software (version 3.3).

3. Results and Discussion

3.1. Effect of Varied Order and Manner of Addition. In order to determine the best order and manner of addition, a series of 4 samples all began by adding 44.42 mL of DI on a stir plate. Sample 1 was performed by adding 5 mL of 0.75 mM 5-HIAA, followed by simple addition of 579.2 μL of 1% HAuCl$_4$. Sample 2 was the same as sample 1, but with dropwise addition of HAuCl$_4$. Sample 3 was performed by adding 579.2 μL of 1% HAuCl$_4$, followed by dropwise addition of 5 mL of 5-HIAA. Sample 4 was the same as sample 3, but with simple addition of 5-HIAA. From the resulting SEM images seen in Figure 1, it was observed that all of the variations were successful in creating gold nanoparticles. The simple addition of HAuCl$_4$ to the solution of 5-HIAA was the method that garnered the most gold nanoparticles.

3.2. Effect of Varied Mixing Parameters. In order to determine the optimal mixing parameters, the two conditions tested were to gently agitate synthesis solution on a platform agitator and to vigorously agitate on a stir plate set at around 300 rpm. From the resulting SEM images seen in Figure 2, it was observed that both variations were successful in creating gold nanoparticles of the same size and shape. In general, the results showed that there were more nanoparticles seen using the vigorous mixing parameter.

3.3. Effect of Reaction Temperatures. The gold nanoparticle synthesis was performed with variances in the temperature at which the synthesis was performed, either in a 15°C bath, at room temperature, or in a 45°C bath. From the resulting SEM images seen in Figure 3, it was observed that all three variations were successful in creating gold nanoparticles, but the shapes and sizes were different in each temperature set. The lower temperature set did not create as many gold nanoparticles, and there seemed to be some residual unreacted material. The heated synthesis showed more nanoparticles, but the edges and forms were less defined. The results showed that the room temperature synthesis was the most successful in creating gold nanoparticles.

3.4. Effect of Varied Concentrations of Reactants. The gold nanoparticle synthesis was performed with variances in the concentrations of both HAuCl$_4$ and 5-HIAA. Specifically, the first concentration variation was to the 0.75 mM of 5-HIAA. The initial volumes of DI were varied to compensate for the variations of 5-HIAA added. In samples 1 through 5, 10 mL, 5 mL, 2.5 mL, 1.67 mL, and 1.25 mL of 0.75 mM 5-HIAA were added. The volume of 1% HAuCl$_4$ added remained constant at 579.2 μL. A second set of syntheses were performed as stated

(a)

(b)

(c)

(d)

FIGURE 1: SEM images showing results from gold nanoparticle synthesis with variances in manner and order of addition. (a) Simple addition of $HAuCl_4$ to solution of 5-HIAA. (b) Dropwise addition of $HAuCl_4$. (c) Dropwise addition of 5-HIAA to solution of $HAuCl_4$. (d) Simple addition of 5-HIAA to solution of $HAuCl_4$.

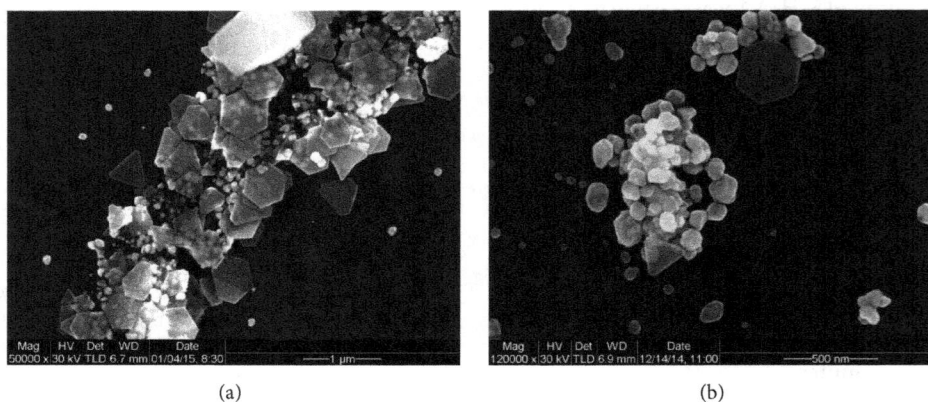

(a)

(b)

FIGURE 2: SEM images showing results from gold nanoparticle synthesis with variances in mixing parameters. (a) Vigorous mixing at around 250 rpm and (b) gentle mixing on a platform agitator.

in Section 2.2, but with a variation in the concentrations of $HAuCl_4$. The initial volumes of DI were varied to compensate for the variations of $HAuCl_4$ added, while the volume of 0.75 mM 5-HIAA remained constant at 5 mL. In samples 1 through 4, 678.0 μL, 579.2 μL, 482.6 μL, and 386.1 μL of 1% $HAuCl_4$ were added. The resulting SEM images seen in Figure 4 were from the variances in 5-HIAA. It was

observed that all variations were successful in creating gold nanoparticles, but the shapes and sizes were different for each concentration of 5-HIAA.

A higher concentration seemed to hinder the reaction, where the concentration in the literature resulted in gold nanoparticles more consistent with previous results. The lower concentrations created gold nanoparticles with less

(a)

(b)

(c)

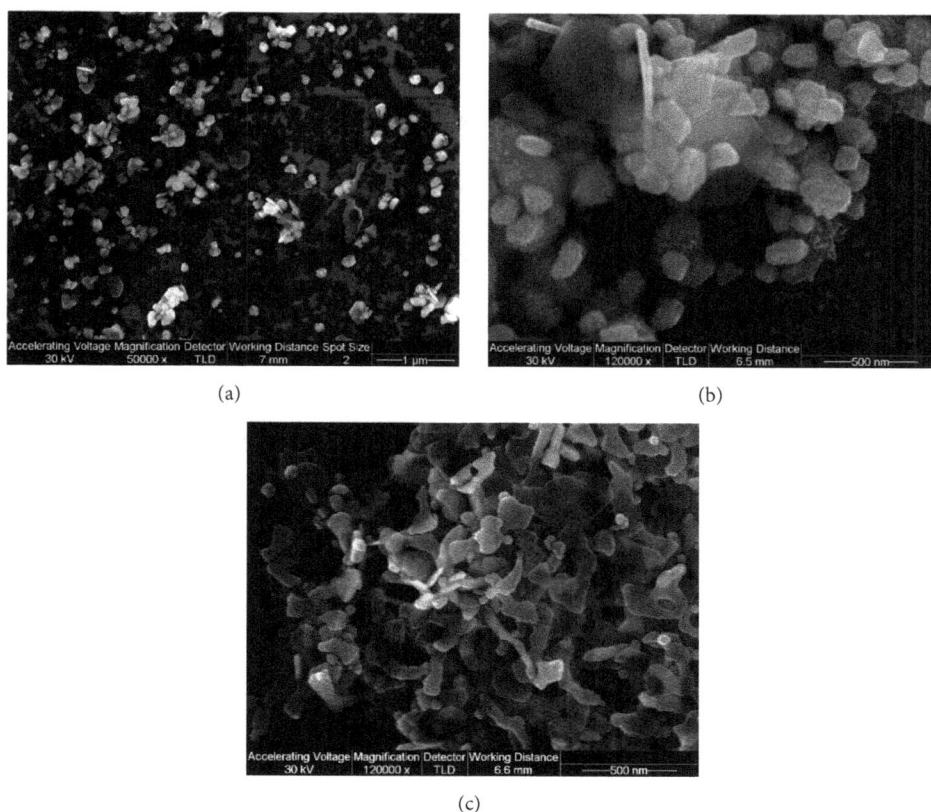

FIGURE 3: SEM images showing results from gold nanoparticle synthesis with variances in the temperature at which the synthesis was performed. (a) 15°C bath, (b) room temperature, and (c) 45°C bath.

defined edges and inconsistent shapes. The resulting SEM images seen in Figure 5 were from the variances in HAuCl$_4$. It was observed that all variations were successful in creating gold nanoparticles of similar shapes and sizes. The concentration of the HAuCl$_4$ did not seem to be as strong of a determining factor in the synthesis. There does seem to be a lower threshold, as the lowest concentrations still created gold nanoparticles, but with less of a population.

3.5. *Effect of Potassium Bromide.* Figure 6 shows the SEM images of the resulting gold nanoparticles when increasing concentrations of KBr were added to the synthesis. All three variations show that gold nanoparticles were successfully created. The lower concentration of KBr seemed to hinder the synthesis. The higher concentration of KBr created many planar shapes, but the edges of the nanoparticles were worn away, as seen in the inset picture of Figure 6(c). The synthesis involving the introduction of 0.075 mM KBr showed a great increase of planar shapes as compared to the synthesis in the absence of KBr. The synthesis with the addition of 0.075 mM KBr seemed to lead to the intended direction of more planar shapes as compared to the original synthesis. KBr is, in and of itself, a reducing agent [19]. The dependence of nanoparticle morphology on bromide concentration may be explained based on the preferential adsorption of bromide on the (111)

crystal facet of Au, as it was previously shown that halide ions adsorb on gold surfaces with binding energies that scale with crystal facet ((111) > (110) > (100)) [18, 20].

3.6. *Characterization and Analytical Applications of Anisotropic Au NPs.* Figure 7 shows two aspects of the synthesis: (1) visual images of the resulting gold nanoparticle solutions in the absence of KBr (left) and in the presence of 0.075 mM KBr (right) and (2) their corresponding UV-Vis spectra. Typical UV-visible absorption values for this kind of gold nanoparticle synthesis are around 540 nm [21]. The synthesis without KBr was close to this value, with the KBr-mediated synthesis nanoparticles showing absorption at wavelengths a bit longer. UV-Vis absorption spectrum could not be obtained for the as-prepared Au NP solution with the addition of 0.090 mM KBr, as all the Au NPs sank to the bottom of the cuvette.

Figure 8 shows the Raman spectra of crystal violet (1 mM) obtained from these two syntheses with 785 nm excitation. Top spectrum was measured from Batch 1 Au NPs solution which contained 0.075 mM KBr and gives a much higher signal for this excitation wavelength. An estimated enhancement factor of 5.4×10^6 was obtained on the basis of the intensity of the band centered at 724 cm^{-1} in the crystal

FIGURE 4: SEM images showing results from gold nanoparticle synthesis with variances in concentration of 5-HIAA. (a) 0.150 mM, (b) 0.075 mM, (c) 0.038 mM, (d) 0.025 mM, and (e) 0.019 mM.

violet Raman spectra. Postsynthesis isolation of individual shapes of gold nanoprisms via techniques such as sucrose gradient separation [21] is currently underway in our group.

4. Conclusions

We have demonstrated the facile synthesis of anisotropic Au NPs, which did not involve the use of any templates, seeds, surfactants, or polymers. Temperature, order and manner of addition, and ratio of chloroauric acid to reducing agent are extremely important. The most successful procedure which

led to the production of the largest number of Au nanoprisms within a batch of nanoparticles is as follows. 5 mL of 5-HIAA (0.75 mM), 579.2 μL of HAuCl$_4$ (1% w/v), and 5 mL of KBr (0.75 mM) were mixed with DI water to form a solution of a total volume of 50 mL. The solutions were mixed on a stir plate set at approximately 300 rpm for 30 minutes under ambient conditions. The as-prepared Au NPs exhibited excellent Raman enhancement when tested with a Raman-active compound (i.e., crystal violet) and showed great potential as a novel substrate for SERS-based chemical and biological sensors.

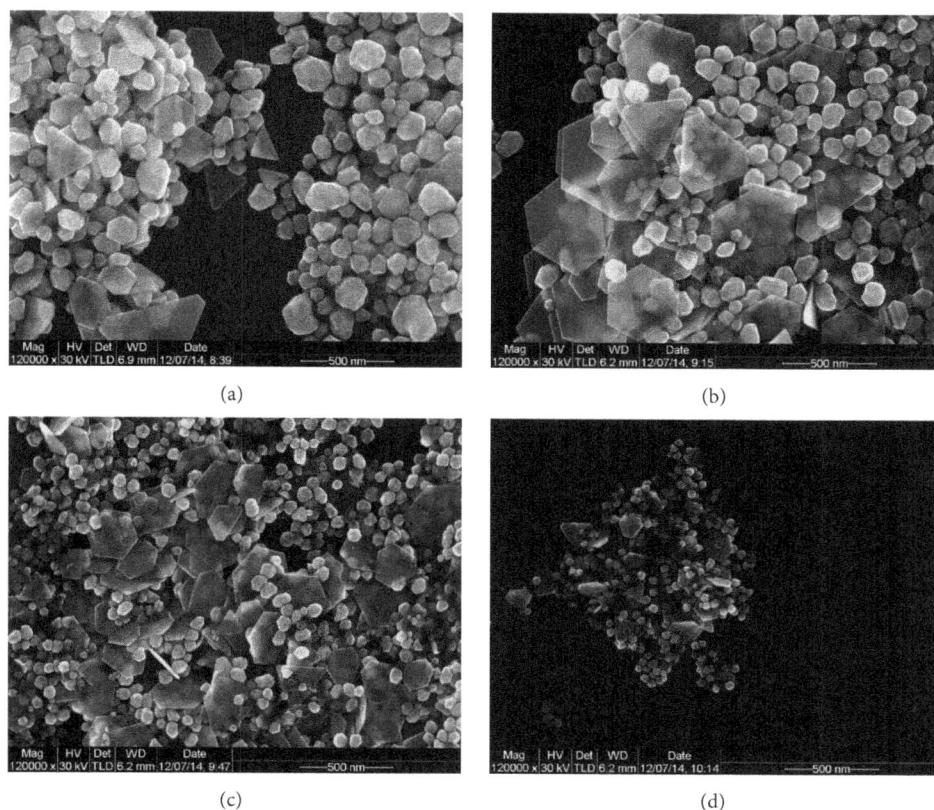

FIGURE 5: SEM images showing results from gold nanoparticle synthesis with variances in concentration of $HAuCl_4$. (a) 0.40 mM, (b) 0.30 mM, (c) 0.25 mM, and (d) 0.20 mM.

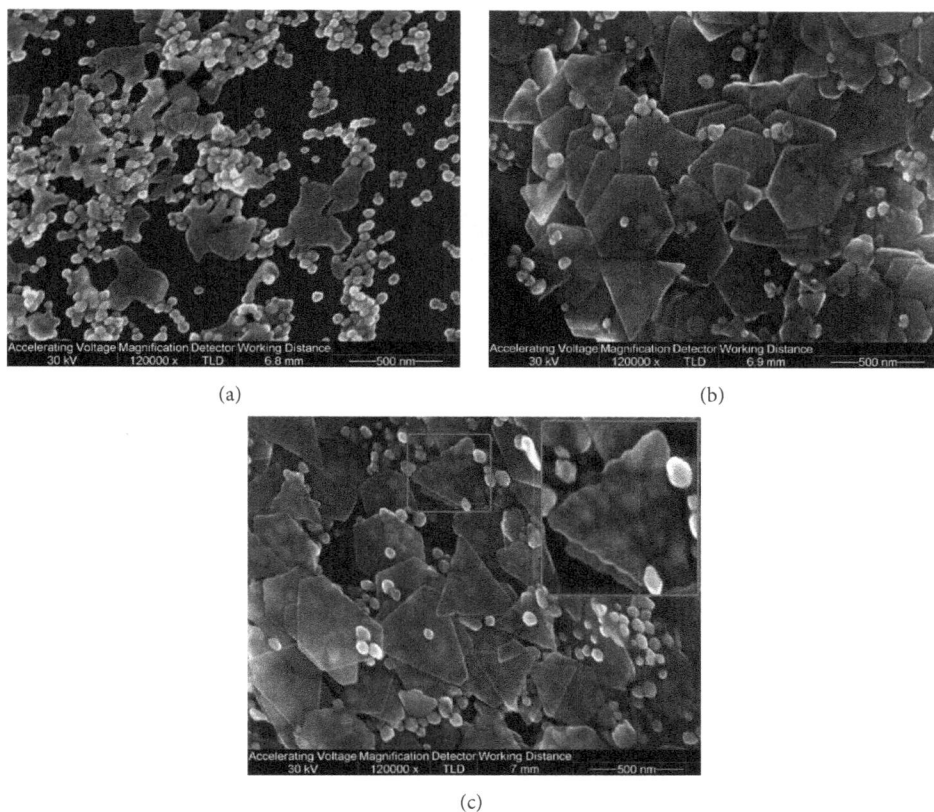

FIGURE 6: SEM images showing results from gold nanoparticle synthesis with variances in KBr concentrations. (a) 0.060 mM KBr, (b) 0.075 mM KBr, and (c) 0.090 mM KBr with inset of zoomed-in section.

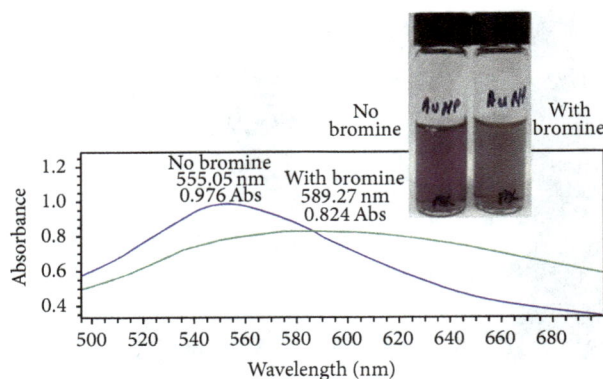

FIGURE 7: Image of the resulting gold nanoparticle solutions prepared in the absence of KBr (left) and in the presence of 0.075 mM KBr and their corresponding UV-Vis spectra.

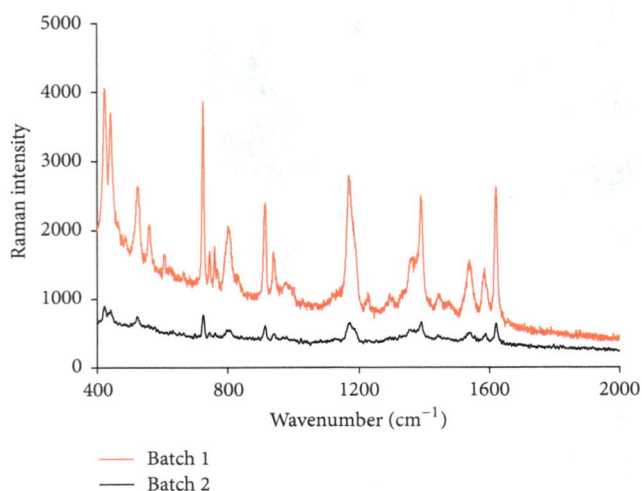

FIGURE 8: SERS spectra of crystal violet (1 mM) from different batches of Au NPs. Batch 1: Au NPs prepared in the presence of 0.075 mM KBr. Batch 2: Au NPs prepared in the absence of KBr.

Competing Interests

The authors declare that there are no competing interests regarding the publication of this paper.

Acknowledgments

This work was supported by the National Science Foundation (Awards nos. HRD-1238441 and DMR-1523617) and North Carolina Space Grant Consortium. The authors thank the Shared Materials Instrumentation Facility (SMIF) at Duke University and Dr. Marvin Wu at the Department of Mathematics and Physics at NCCU for granting them access to the scanning electron microscopes.

References

[1] J. R. Ferraro and K. Nakamoto, *Introductory Raman Spectroscopy*, Academic Press, Boston, Mass, USA, 2nd edition, 1994.

[2] C. V. Raman and K. S. Krishnan, "A new type of secondary radiation," *Nature*, vol. 121, no. 3048, pp. 501–502, 1928.

[3] T. von Foerster, "Surface-enhanced Raman effect," *Physics Today*, vol. 33, no. 4, pp. 18–20, 1980.

[4] A. Campion and P. Kambhampati, "Surface-enhanced Raman scattering," *Chemical Society Reviews*, vol. 27, no. 4, pp. 241–250, 1998.

[5] M. Fleischmann, P. J. Hendra, and A. J. McQuillan, "Raman spectra of pyridine adsorbed at a silver electrode," *Chemical Physics Letters*, vol. 26, no. 2, pp. 163–166, 1974.

[6] B. K. Jena, S. Ghosh, R. Bera, R. S. Dey, A. K. Das, and C. R. Raj, "Bioanalytical applications of Au nanoparticles," *Recent Patents on Nanotechnology*, vol. 4, no. 1, pp. 41–52, 2010.

[7] M.-C. Daniel and D. Astruc, "Gold nanoparticles: assembly, supramolecular chemistry, quantum-size-related properties, and applications toward biology, catalysis, and nanotechnology," *Chemical Reviews*, vol. 104, no. 1, pp. 293–346, 2004.

[8] K. Saha, S. S. Agasti, C. Kim, C. X. Li, and X. V. M. Rotello, "Gold nanoparticles in chemical and biological sensing," *Chemical Reviews*, vol. 112, no. 5, pp. 2739–2779, 2012.

[9] D. A. Giljohann, D. S. Seferos, W. L. Daniel, M. D. Massich, P. C. Patel, and C. A. Mirkin, "Gold nanoparticles for biology and medicine," *Angewandte Chemie-International Edition*, vol. 49, no. 19, pp. 3280–3294, 2010.

[10] Y. K. Shrestha and F. Yan, "Determination of critical micelle concentration of cationic surfactants by surface-enhanced Raman scattering," *RSC Advances*, vol. 4, no. 70, pp. 37274–37277, 2014.

[11] B. K. Jena and C. R. Raj, "Seedless, surfactantless room temperature synthesis of single crystalline fluorescent gold nanoflowers with pronounced SERS and electrocatalytic activity," *Chemistry of Materials*, vol. 20, no. 11, pp. 3546–3548, 2008.

[12] P. K. Jain, K. S. Lee, I. H. El-Sayed, and M. A. El-Sayed, "Calculated absorption and scattering properties of gold nanoparticles of different size, shape, and composition: applications in biological imaging and biomedicine," *Journal of Physical Chemistry B*, vol. 110, no. 14, pp. 7238–7248, 2006.

[13] S. Jain, D. G. Hirst, and J. M. O'Sullivan, "Gold nanoparticles as novel agents for cancer therapy," *British Journal of Radiology*, vol. 85, no. 1010, pp. 101–113, 2012.

[14] J. Z. Zhu, Y. Shen, A. Xie, L. Qiu, Q. Zhang, and S. Zhang, "Photoinduced synthesis of anisotropic gold nanoparticles in room-temperature ionic liquid," *The Journal of Physical Chemistry C*, vol. 111, no. 21, pp. 7629–7633, 2007.

[15] H. E. Cramer, L. Giri, M. H. Griep, and S. P. Karna, "Shape-controlled gold nanoparticle synthesis," Tech. Rep. ARL-TR-6662, U.S. Army Research Laboratory, Aberdeen, Md, USA, 2013.

[16] B. K. Jena and C. R. Raj, "Shape-controlled synthesis of gold nanoprism and nanoperiwinkles with pronounced electrocatalytic activity," *Journal of Physical Chemistry C*, vol. 111, no. 42, pp. 15146–15153, 2007.

[17] B. Xiong, J. Cheng, Y. Qiao, R. Zhou, Y. He, and E. S. Yeung, "Separation of nanorods by density gradient centrifugation," *Journal of Chromatography A*, vol. 1218, no. 25, pp. 3823–3829, 2011.

[18] P. J. Straney, C. M. Andolina, and J. E. Millstone, "Seedless initiation as an efficient, sustainable route to anisotropic gold nanoparticles," *Langmuir*, vol. 29, no. 13, pp. 4396–4403, 2013.

[19] J. Clark, "Redox reactions involving halide ions and sulphuric acid," 2002, http://www.chemguide.co.uk/inorganic/group7/halideions.html.

[20] O. M. Magnussen, "Ordered anion adlayers on metal electrode surfaces," *Chemical Reviews*, vol. 102, no. 3, pp. 679–725, 2002.

[21] S. H. Lee, B. K. Salunke, and B. S. Kim, "Sucrose density gradient centrifugation separation of gold and silver nanoparticles synthesized using *Magnolia kobus* plant leaf extracts," *Biotechnology and Bioprocess Engineering*, vol. 19, no. 1, pp. 169–174, 2014.

16

Raman Spectroscopic Study of As-Deposited and Exfoliated Defected Graphene Grown on (001) Si Substrates by CVD

T. I. Milenov,[1] E. Valcheva,[2] and V. N. Popov[2]

[1]*"E. Djakov" Institute of Electronics, Bulgarian Academy of Sciences, 72 Tzarigradsko Chaussee Blvd., 1784 Sofia, Bulgaria*
[2]*Faculty of Physics, University of Sofia, 5 James Bourchier Blvd., 1164 Sofia, Bulgaria*

Correspondence should be addressed to T. I. Milenov; teddymilenov@abv.bg

Academic Editor: Tino Hofmann

We present here results on a Raman spectroscopic study of the deposited defected graphene on Si substrates by chemical vapor deposition (thermal decomposition of acetone). The graphene films are not deposited on the (001) Si substrate directly but on two types of interlayers of mixed phases unintentionally deposited on the substrates: a diamond-like carbon (designated here as DLC) and amorphous carbon (designated here as αC) are dominated ones. The performed thorough Raman spectroscopic study of as-deposited as well as exfoliated specimens by two different techniques using different excitation wavelengths (488, 514, and 613 nm) as well as polarized Raman spectroscopy establishes that the composition of the designated DLC layers varies with depth: the initial layers on the Si substrate consist of DLC, nanodiamond species, and C_{70} fullerenes while the upper ones are dominated by DLC with an occasional presence of C_{70} fullerenes. The αC interlayer is dominated by turbostratic graphite and contains a larger quantity of C_{70} than the DLC-designated interlayers. The results of polarized and unpolarized Raman spectroscopic studies of as-grown and exfoliated graphene films tend to assume that single- to three-layered defected graphene is deposited on the interlayers. It can be concluded that the observed slight upshift of the 2D band as well as the broadening of 2D band should be related to the strain and doping.

1. Introduction

Graphene is a one-atom-thick layered material that consists of completely sp^2-bonded carbon atoms tightly packed into a honeycomb lattice. It has a lot of unique properties promising a huge number of possible applications (see, e.g., [1]). A lot of different ways of synthesizing graphene were experimentally tested during the last decade; however, only thermally and plasma-assisted chemical vapor deposition (CVD/PECVD) on metal substrates (copper, nickel, etc.) [2, 3] as well as epitaxial growth on SiC substrates and so on [4, 5] were developed for industrial application. The latter method is based on C (or Si) termination of the $(0001)_C$ (or $(0001)_{Si}$) SiC surface and requires high temperature and expensive SiC substrates. The CVD method is based on the plasma-enhanced thermal decomposition of a carbon-containing precursor on a catalytic metal surface. This method provides high reliability and relatively high quality of graphene films, and now, there are a lot of suppliers of reactors for PECVD of graphene. The most preferred precursor is methane (CH_4) as the chemical bond in CH_4 is relatively strong and prevents fast decomposition of the reagent at temperatures below 1000°C (see, e.g., [6]). However, production for microelectronic applications requires transfer of the graphene layers on an insulating surface and, consequently, a large number of additional defects affecting the properties of graphene can be introduced. Therefore, the problem with the deposition of graphene on silicon (or surfaces compatible with silicon technology such as SiO_2) still remains unsolved. We demonstrated the possible application of acetone as a precursor in a thermally assisted CVD and showed that few-layered defected graphene/folded graphene can be deposited on commercially available metal foils—Ni, $(Cu_{0.5}Ni_{0.5})$, μ-metal, and stainless steel SS 304 in a recently published work [7]. Further, we established (see [8]) by Raman spectroscopy, scanning electron microscopy (SEM), X-ray diffraction (XRD), and grazing incidence X-ray diffraction (GIXRD) as well as by X-ray photoelectron spectroscopy

(XPS) the presence of single- to few-layered defected graphene on two different types of interlayers deposited on (100) Si surface: (i) a diamond-like carbon (DLC) layer with some SiC contents (in the range below 5w%) and some residual quantities of SiO_2, and (ii) a complex amorphous carbon layer consisting of a mixture of sp^2- and sp^3-hybridized carbon as well as very small amount of fullerenes, SiO, and so on.

Here, we focus our experimental study on the Raman spectroscopic characterization of defected as-deposited graphene layers (including polarized spectroscopy) as well as graphene flakes exfoliated from similar specimens by two different ways using 488, 514, and 633 nm excitation laser wavelengths aiming at unambiguous confirmation of the graphene deposition of as well as the identification of the exact composition of the interlayers between the Si substrate and graphene layer/s.

2. Experimental

2.1. CVD Process.
We use 2 inches in diameter (001) Si substrates and a horizontal tube quartz CVD reactor with an internal diameter of approximately 70 mm. The experimental setup also consists of a gas-supply system (inlet and outlet parts), a thermostat with acetone evaporation alert/indication system, a quartz substrate holder, and a resistive heating furnace. The CVD process is based on thermal decomposition of acetone in Ar main gas flow. The deposition temperature was in the range 1150–1160°C. The temperature of the thermostat was kept at 12°C. In order to prevent the supersaturation in the high-temperature zone of the reactor, we used a "pulsed" regime in experiments by alternating the flow of the gas mixture of Ar + C_3H_6O) for 3 min on top of the main flow of pure Ar of about 150–180 cm^3/min for 1.5 min for each pulse. The optimal results (predominantly single-layered graphene) were obtained after two deposition pulses.

2.2. Exfoliation.
We exfoliated the carbon films deposited on (001) Si substrates by the following two different techniques:

(i) The Scotch tape method (see, e.g., [9]): we put tightly the adhesive Scotch tape on the multilayered graphene side of the specimens. After peeling the tape off the specimen, a single- to few-layered graphene remains on the tape's surface and the interlayer between the upper few layers of graphene and the substrate becomes accessible for spectroscopic examination. Then, we put tightly the Scotch tape with graphene flakes either on 320 nm SiO_2/Si or on glass substrate. About 30–50% of the graphene flakes remain adhered to the SiO_2 or glass substrate after removing the tape due to the Van der Waals force.

(ii) We also adhered the multilayered graphene side of the specimens to epoxy resin. After careful cleavage, the most part of the graphene layer/s remains on the surface of the resin. Then, the adhered to the resin graphene film becomes accessible for spectroscopic examination. The Raman spectrum of the epoxy resin does not contain strong peaks around the 2D band of graphene (the area around 2630–2670 cm^{-1}). We

established that the 2D band of graphene is clearly distinguishable for graphene regions lying on gas bubbles close to the surface of the resin; otherwise, the 2D band of graphene is weak.

2.3. Characterization.
The Raman measurements were carried out in backscattering geometry at a micro-Raman HORIBA Jobin Yvon Labram HR 800 visible spectrometer equipped with a Peltier-cooled CCD detector with a He-Ne (633 nm wavelength and 0.5 mW) laser excitation. The 514 nm (about 23 mW) as well as 488 nm (about 24 mW) lines of an external Ar laser were also used. The laser beam was focused on a spot of about 1 μm in diameter, the spectral resolution being 0.5, 0.7, and 1 cm^{-1}, respectively, or better.

The Raman spectrum of graphene is a clearly established fingerprint of this 2D material (see [10]). The main first-order features in the Raman spectra of graphene and defect-infested graphene excited at 633 nm wavelength are the following:

(i) G band (~1582 cm^{-1}) is the only band in graphene allowed by selection rules for first-order Raman effect; it is ascribed to optical (iTO and LO) doubly degenerate phonons of E_{2g} symmetry at the Γ point (initially described by Tuinstra and Koenig [11]).

(ii) D band (~1330 cm^{-1}) is due to breathing-like bands of C hexagonal rings (corresponding to transverse optical phonons near the K point) and requires a defect for its activation via an intervalley double-resonance Raman process (see [12]).

(iii) D' band (at about 1615 cm^{-1}; defect induced similarly to the D band) occurs via an intravalley double-resonance process (see, e.g., [13]).

(iv) D" band (at about 1145 cm^{-1}) is resulting from double-resonance intervalley scattering of LA phonons on defects (see [14]). The intensity of this band should be about 100 times lower than that of the D band.

Overtones and combination bands:

(i) 2D band (historically known from graphite and carbon nanotube-related literature as G'- peak) appears at about 2648–2665 cm^{-1}. It is clearly shown [15–20] that the shape and width of the 2D band can be used for the identification of the mono-, bi-, and three-layered graphene.

(ii) The overtone of the D'- peak (2D') and combination G* (G* = (iTA + LA) phonons), as well as (D+D') bands, occur around 3230, 2450, and 2930 cm^{-1}, respectively (see [21]).

3. Results and Discussion

Two areas with different surface morphologies are observed by optical microscopy (Figure 1(a)): a clear relief (ridge-like

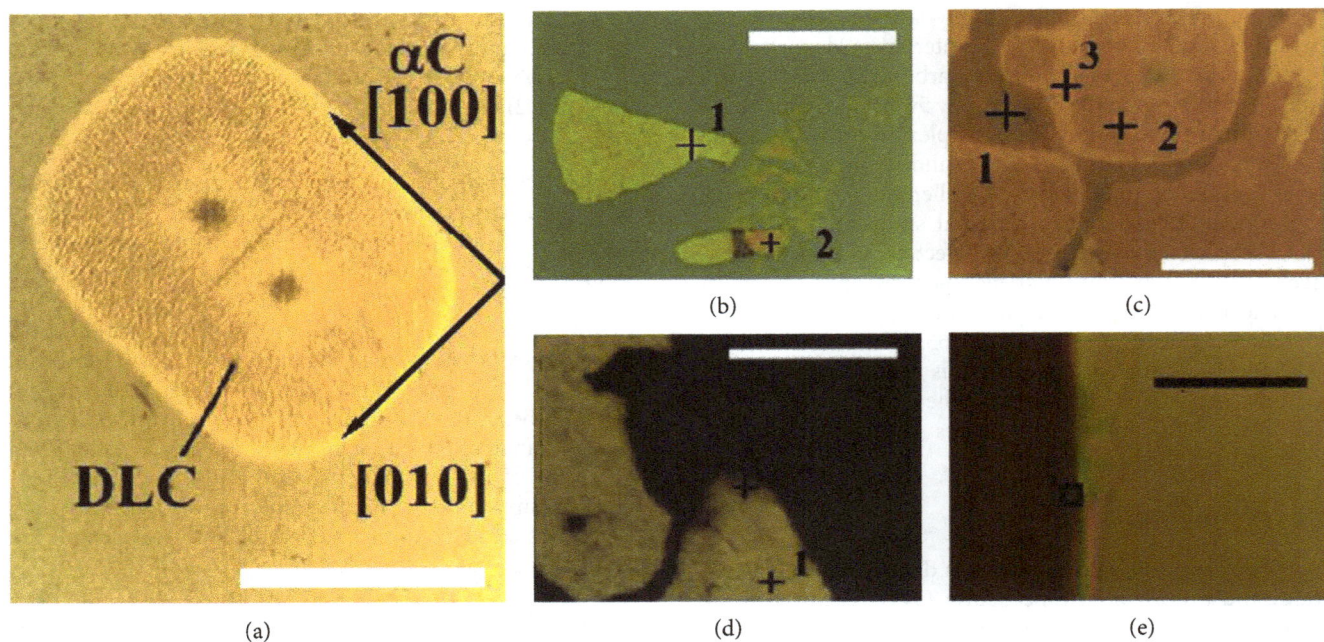

FIGURE 1: Optical microscopy image of the surface morphology of (a) as-deposited graphene and graphene-related phases on diamond-like carbon (DLC) and amorphous carbon (αC) interlayers. The arrows remarked [100] and [010] directions of the Si substrate. The marker represents 20 μm. (b) The exfoliated and transferred graphene flakes on 320 nm SiO$_2$. The Raman spectra are taken from "+"-marked positions. The marker represents 30 μm. (c) The layers remaining on the surface of the substrate after exfoliation by Scotch tape. The Raman spectra are taken from the "+"-marked positions near points 1, 2, and 3. The marker represents 30 μm. (d) The exfoliated and transferred graphene flakes on glass substrate. The Raman spectra are taken from the "+"-marked positions near points 1 and 2. The marker represents 20 μm. (e) A graphene flake on air bubble near the epoxy resin surface. The Raman spectra are taken from the square-marked position. The marker represents 20 μm.

formations) lying along <001> directions covers the first area denoted as DLC while the second area (denoted as αC) is covered by an inhomogeneous film with a constant depth. It should be also mentioned that optical inhomogeneities are observed on the DLC as well as αC-marked areas.

It should be recalled that the Raman spectrum (excited at 633 nm laser wavelength) taken from αC- and DLC-denoted areas (see [8]) contains all features typical for graphene: symmetric and clearly pronounced 2D band with full width at a half maximum (FWHM) of 40–56 cm^{-1}, I$_{2D}$/I$_{G}$ ratio between 2 and 3.5, and I$_{2D}$/I$_{D}$ ratio between 2 and 4. However, the 2D band appears at about 2660–2668 cm^{-1} (for single- and bilayered graphene, respectively), that is, it is blueshifted by about 20 cm^{-1} relative to the results presented in [15, 22, 23].

Due to the double-resonance origin of most of the monitored spectral features, we perform a Raman spectroscopy examination of as-deposited defected graphene at 488, 514, and 633 nm excitation wavelengths and the results are presented in Figure 2. The 2D bands are blueshifted by about 20 cm^{-1} and can be typically deconvoluted into (a) a single Lorentzian with FWHM of about 40-41 cm^{-1} (see Figure 3(a)); (b) four Lorentzians (FWHM of 22 ($\pm$1) cm^{-1}) for 2D band with total width of 45–56 cm^{-1} (see Figure 3(b)); and (c) six Lorentzians (Figure 3(c)) for 2D band with total width larger than 56 cm^{-1}. The results of the deconvolution indicate the presence of single-, bi-, and three-layered defected graphene, respectively (see [15–20]). We did not establish a clear difference between the

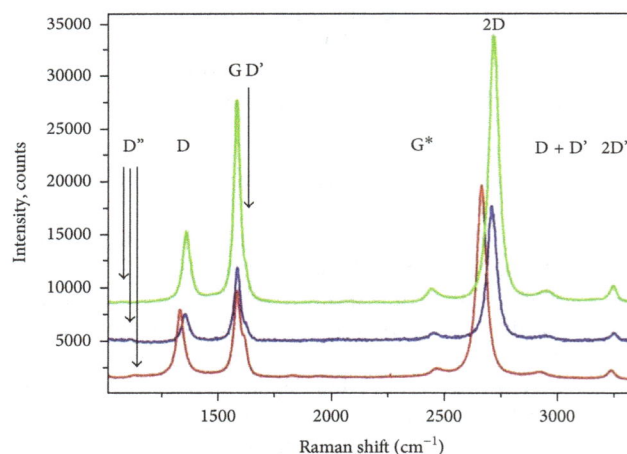

FIGURE 2: Raman spectra taken from as-grown films excited at 633 nm (red trace), 514 nm (blue trace), and 488 nm (green trace) laser wavelengths.

graphene layers deposited on αC and DLC interlayers; however, bi- and three-layered areas were more frequently observed on DLC interlayers. The results for predominantly single-layered (SL) and bilayered (BL) defected graphene (according to the deconvolution of 2D bands) are summarized in Table 1.

In order to access the interlayers as well as graphene flakes for Raman examination, the so-called Scotch tape

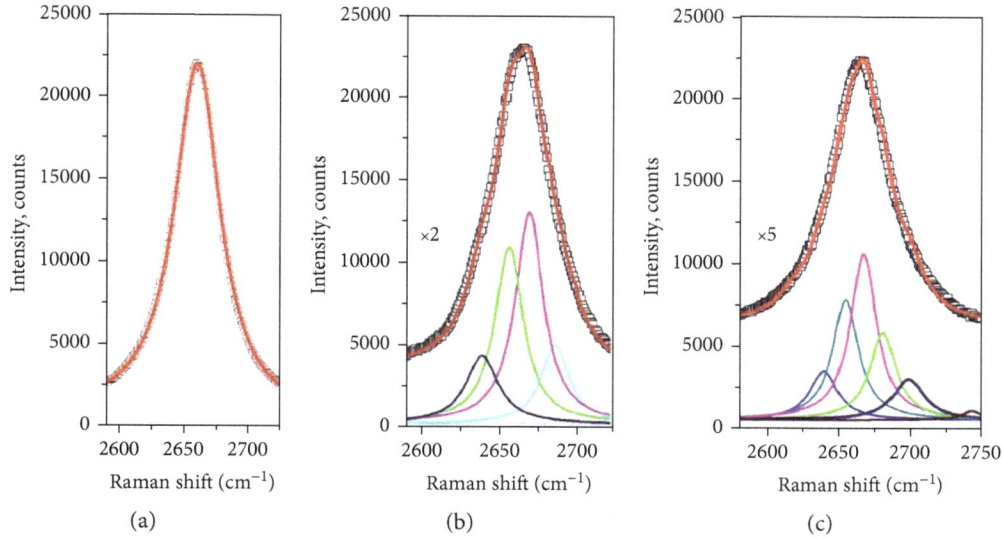

FIGURE 3: Deconvolution of 2D band identified as coming from single-layered (a), bilayered (b), and three-layered defected graphene deposited on αC. The spectrum is excited at 633 nm laser wavelength.

TABLE 1: Summarized results of Raman spectroscopy examination of as-deposited defected graphene films.

Excitation wavelengths	D band, cm^{-1}	G band, cm^{-1}	2D band, cm^{-1}	FWHM 2D Band, cm^{-1}; assignment
488 nm	1358	1581	2717	40–42; SL 48–54; BL
514 nm	1353	1582	2707	40–42; SL 48–54; BL
633 nm	1330	1581	2660	40–42; SL 48–54; BL

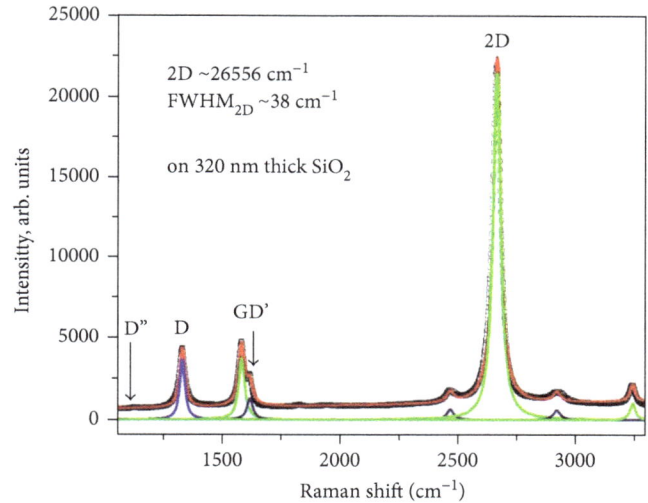

FIGURE 4: The Raman spectrum of defected 1-2-layered graphene transferred on 320 nm SiO$_2$. The 2D band is symmetric and appears at 2658-2659 cm^{-1} with FWHM of 38–40 cm^{-1} (measured in point 2) and 40–42 cm^{-1} (measured in point 1).

method was initially used for exfoliation. The Raman spectra of the graphene flakes exfoliated in this way with some occasional amorphous (αC) interlayers transferred to Si/SiO$_2$ (300 nm) or glass substrate are shown in Figures 4 and 5, respectively.

A lot of flakes (of the order of 10^2) were transferred on Si/SiO$_2$ and examined by Raman spectroscopy. The Raman spectra are enhanced due to interference effects caused by the SiO$_2$ 300 nm layer over the Si substrate, and I$_{2D}$/I$_G$ varies in the range 3.5-6.0. However, it was impossible to isolate single-layered graphene flake (or to obtain clear Raman response of single-layered graphene) in this way. The exfoliated flakes were never transparent (see point 1 in Figure 1(b)). The best spectra were recorded from the points in a darker contrast (point 2 in Figure 1(b)), but the FWHM of 2D Raman band remains >35 cm^{-1}. Moreover, the D" band slightly overlaps with the second order of Si substrate when the spectrum is excited at 514 as well as 488 nm laser wavelengths.

After peeling the tape off the specimen, the interlayer between the upper flake and the substrate is accessed. The remaining interlayers have different optical contrasts (see Figure 1(c)) and Raman spectra: the spectrum of typically retained interlayer (point 1 in Figure 1(c)) in Figure 5 is very similar to that of turbostratic graphite (see [24]), but weak peaks of C$_{70}$ fullerenes (the features observed at about 1450 and 1530 cm^{-1} (see [25, 26])) are also clearly distinguished (Figure 5). The strong modes of fullerenes C$_{70}$ at about 1180 and 1568 cm^{-1} are merged with D" and G bands.

The Raman spectra (Figure 6) taken from points 2 and 3 (Figure 1(c)) are similar as they contain the most prominent modes of C$_{70}$ peaks at 1160, 1220, 1454, 1526, and 1565 cm^{-1} [25, 26], nanodiamond (Nd) peaks at 1330 and 1620 cm^{-1} (see [27]), and turbostratic graphite. The D, G, and D' bands are found at 1335, 1590, and 1612 cm^{-1}, respectively, but in a different proportion: the spectrum from point 3 is dominated

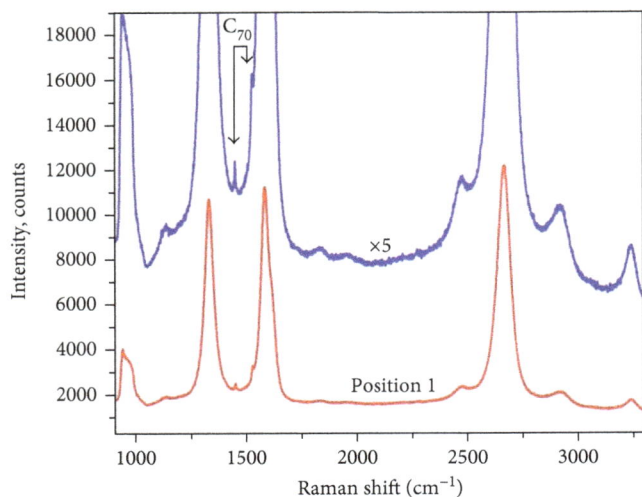

FIGURE 5: The Raman spectrum of the interlayer (point 1, Figure 1(c)) of αC after exfoliation by Scotch tape. The features observed at about 1450 and 1530 cm^{-1} are typical for C_{70} fullerenes.

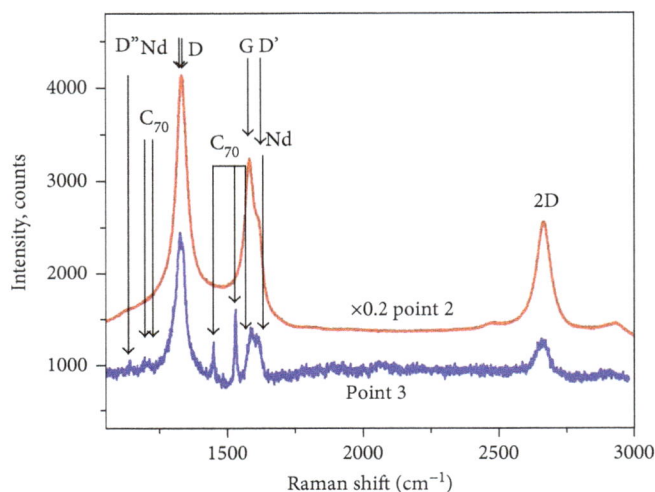

FIGURE 7: The Raman spectra of as-grown graphene on αC excited at 488 nm (green trace) and 633 nm (black trace) wavelengths. The similar spectra of exfoliated graphene transferred on a glass substrate excited at 488 nm (blue trace) and 633 nm (red trace). The inset: magnified part of the region 900–1200 cm^{-1}.

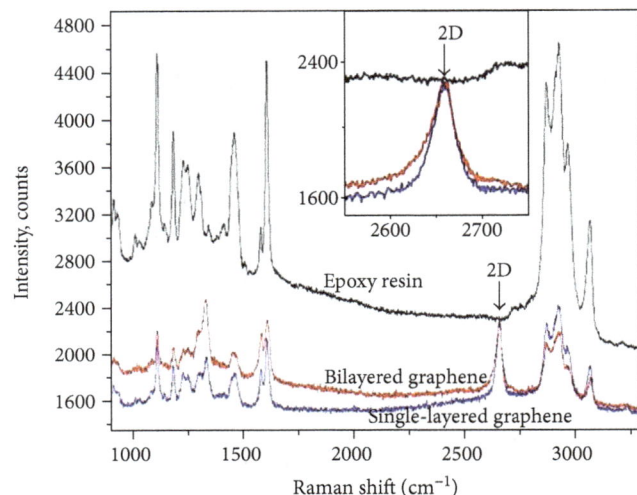

FIGURE 6: Raman spectra of the interlayer that remains on the substrate after exfoliation by the Scotch tape method taken from points 2 and 3 (Figure 1(c)).

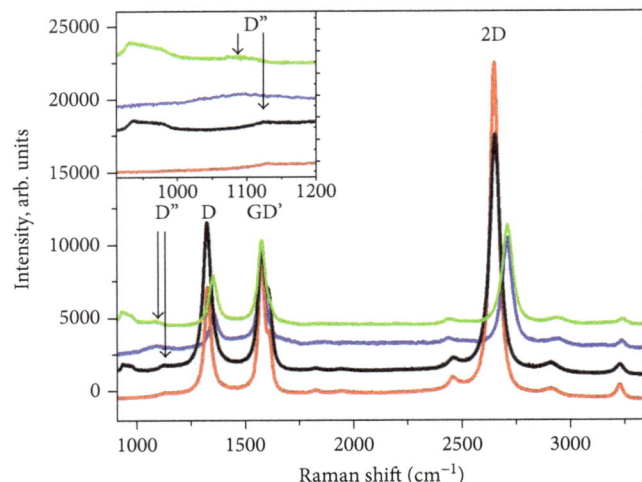

FIGURE 8: Raman spectra of graphene films situated on air bubbles/cavities. The 2D band is situated at 2655 cm^{-1} and has FWHM ~28 cm^{-1} (i.e., it corresponds to single-layered graphene—blue trace).

by the peaks of C_{70} and Nd while the spectrum from point 2 is dominated by turbostratic graphite (see Figure 6). It should be also remarked that features of C_{60} fullerenes (see, e.g., [25, 26]) were not observed.

As it was mentioned above, the D" band overlaps with the second-order band of Si substrate especially when the spectrum is excited at 488 and 514 nm laser wavelengths. In order to distinguish the dispersion of the D" band of several Scotch tape methods, exfoliated flakes were transferred on glass substrates. The flakes have very similar surface morphology to those transferred on SiO_2/Si substrates (Figure 1(d)). The Raman spectrum of such flakes is not significantly different from that of the as-deposited layers (Figure 7); however, the D" band appears at 1096 (for 488 nm excitation) and at

1135 cm^{-1} (for 633 nm excitation), respectively, that is, they coincide with the data of Herziger et al. [14].

According to the above results, we conclude that the exfoliation by the Scotch tape method does not enable splitting up between the defected graphene and the interlayers (especially the αC-designated one). Another way for exfoliation was probed (by exfoliation on epoxy resin), and the optical micrograph image of the area of the edge of a resin bubble and the Raman spectrum taken from this area (excited at 633 nm) are shown in Figures 1(e) and 8, respectively. The Raman spectrum of epoxy resin does not contain any features in the 2D region of graphene (upper

(a) (b)

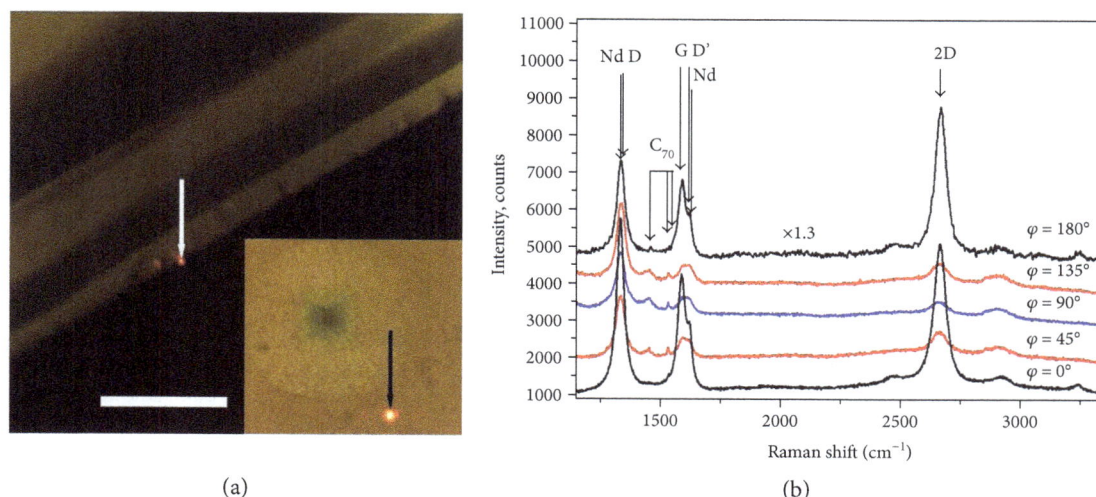

FIGURE 9: (a) Optical photography of the specimen in $X(YY)X$ geometry (in Porto notations). The inset: optical photography of the specimen in $Z(YY)Z$ geometry (in Porto notations). The arrow-remarked laser spots are eye guide showing the real area of the laser spot during measurements. The marker represents $10\,\mu m$. (b) Spatially resolved Raman spectra of as-deposited defected graphene at 633 nm excitation.

trace in Figure 8); hence, 2D bands of a single- and bilayered graphene were identified at the edge of a lot of bubbles on the surface of the resin (Figure 1(e)). It should be clearly remarked that the measured FWHM of the 2D band of such single-layered graphene is about 27–29 cm^{-1}, but it is situated at 2654–2656 cm^{-1}, that is, it remains upshifted with about 10–15 cm^{-1}.

Recently, Li et al. [28] established that the intensity of 2D band varies as a cosine to the fourth power when the laser propagation direction is parallel to the graphene layer and the polarization is rotated around it. They also derived the orientation distribution function of monolayered graphene as well as that of graphene paper and highly oriented pyrolytic graphite. We perform similar measurements in $X(Y_\varphi Y_\varphi)X$ geometry, φ being the angle between the incident laser beam polarization and the graphene layer plane; Z is the axis perpendicular to the graphene plane, and the laser beam propagates transversely to the graphene layer along the X direction (see Figure 9(a)). The excitation laser beam was focused in a manner to comprise no more than 30% of the edge of the Si substrate and graphene film (Figure 9(a)). The parallel scattering geometry was used. The measurements were performed starting from $\varphi = 0°$ (corresponding to $X(YY)X$ in Porto notations) and finished at $\varphi = 180°$. The preliminary results of these rotational angle-dependent Raman measurements of as-deposited specimen are presented in Figure 9. The signal significantly drops upon changing the angle from 0° to 90° and increases again in the interval between 90 and 180° which resembles indeed the cos^4 law. At 90° (corresponding to $X(ZZ)X$ in Porto notations), the Raman signal is very weak but still observable (Figure 9), and the rotational angle-independent features of C$_{70}$ fullerenes and nanodiamond (Nd) dominate the spectrum. The residual features in the Raman spectra taken at $\varphi = 90°$ point out that the measured polarized Raman spectra are taken from graphene deposited on DLC interlayer. The measurements in this scattering geometry ($X(YY)X$ in Porto notations) access measurements of the interlayer/s

without exfoliation. On the other hand, the polarized Raman study confirms the deposition of graphene because the intensities of the most prominent Raman features of graphite (D, G, and 2D bands) show similar behavior in similar conditions as those of graphene. However, the intensity of the Raman features of graphene decreases significantly slower than those of graphene as it is shown in [28].

It is worth noting that the 2D band from the single-layered graphene regions is symmetric and strong, but it is somewhat broadened with FWHM of about 40–42 cm^{-1} and is blueshifted by 15–20 cm^{-1} in as-grown specimens. It is well known that such behavior is usually related to strain (see [29–32]) and doping [33]. Moreover, Lee et al. [34] and Bouhafs et al. [35] experimentally studied the influence of these parameters on the position and FWHM of G and 2D bands in single- and bi-/multilayered graphene, respectively. The deduced simple plot of the 2D versus G band positions enables distinguishing the influence of doping and strain on the positions of G and 2D bands. In our single-layered specimens, the G band is slightly uphifted by 1–2 cm^{-1} while the 2D band is more significantly blueshifted and broadened by 10–20 cm^{-1}. Therefore it can be assumed that the 2D band blueshift and broadening are due to the lattice strain predominantly as well as to the doping. It can be suggested that the lattice strain is due to the bonding between graphene and the interlayers while the doping should be related to charge transfer from the interlayers/interfaces to graphene as well as to different intrinsic (grain boundaries, etc.) and extrinsic (trapped nitrogen, oxygen, and impurities during the deposition) defects, that is, it can be related to the influence of the interlayers/substrates as well as of the deposition process.

4. Conclusions

We extended the analysis of defected graphene deposited by CVD as well as the two types of interlayers between the defected graphene layer/s and Si substrates by both unpolarized and polarized Raman spectroscopy. The performed Raman spectroscopy examination of as-deposited defected graphene at 488, 514, and 633 nm excitation wavelengths enables the most of the monitored spectral features of double-resonance origin (D, D", and 2D bands). The Raman studies of exfoliation by the so-called Scotch tape method revealed that (a) the composition of the designated DLC interlayers varies with depth: the initial layers on the Si substrate consist of a mixed phase of turbostratic graphite, nanodiamond/diamond-like carbon, and C_{70} fullerenes while the upper ones are dominated by diamond-like carbon and some C_{70} fullerenes and (b) the amorphous carbon interlayer is dominated by turbostratic graphite and contains a larger quantity of C_{70} than the DLC-designated interlayers. Single- and bilayered defected graphene flakes were exfoliated on epoxy resin. The preliminary results of polarized Raman experiments show that the intensity of the 2D band varies as a cosine to the fourth power when the laser propagation direction is parallel to the graphene layer and the polarization is rotated around it which is an additional indication of the deposition of single-layered graphene. The results of Raman spectroscopic studies of as-grown and exfoliated graphene films tend to assume that the observed slight upshift of the 2D band as well as the broadening of 2D band is due to the strain and can be related to the bonding between the graphene and the interlayers, that is, it could be regarded as an influence of the interlayers between the defected graphene and the Si substrates.

References

[1] A. Ferrari, F. Bonaccorso, V. Fal'ko et al., "Science and technology roadmap for graphene, related two-dimensional crystals, and hybrid systems," *Nanoscale*, vol. 7, pp. 4598–4810, 2015.

[2] K. S. Kim, Y. Zhao, H. Jang et al., "Large-scale pattern growth of graphene films for stretchable transparent electrodes," *Nature*, vol. 457, pp. 706–710, 2009.

[3] A. Reina, X. Jia, J. Ho et al., "Large area, few-layer graphene films on arbitrary substrates by chemical vapor deposition," *Nano Letters*, vol. 9, pp. 30–35, 2009.

[4] C. Berger, Z. Song, T. Li et al., "Ultrathin epitaxial graphite: 2D electron gas properties and a route toward graphene-based nanoelectronics," *The Journal of Physical Chemistry*, vol. 108, pp. 19912–19916, 2004.

[5] C. Berger, Z. Song, T. Li et al., "Electronic confinement and coherence in patterned epitaxial graphene," *Science*, vol. 312, pp. 1191–1196, 2006.

[6] R. Muñoz and C. Gómez-Aleixandre, "Review of CVD synthesis of graphene," *Chemical Vapor Deposition*, vol. 19, pp. 297–322, 2013.

[7] T. I. Milenov, I. Avramova, E. Valcheva, and S. S. Tinchev, "Deposition of graphene/graphene-related phases on different substrates by thermal decomposition of acetone," *Optical & Quantum Electronics*, vol. 48, p. 135-1-12, 2016.

[8] T. I. Milenov, I. Avramova, E. Valcheva et al., "Deposition of defected graphene on (001) Si substrates by thermal decomposition of acetone," *Superlattices and Microstructures*, In press.

[9] K. S. Novoselov, D. Jiang, F. Schedin et al., "Two-dimensional atomic crystals," *PNAS*, vol. 102, pp. 10451–10453, 2005.

[10] A. C. Ferrari and D. M. Basko, "Raman spectroscopy as a versatile tool for studying the properties of graphene," *Nature Nanotechnology*, vol. 8, pp. 235–246, 2013.

[11] F. Tuinstra and J. L. Koenig, "Raman spectrum of graphite," *The Journal of Chemical Physics*, vol. 53, pp. 1126–1130, 1970.

[12] C. Thomsen and S. Reich, "Double resonant Raman scattering in graphite," *Physical Review Letters*, vol. 85, pp. 5214–5217, 2000.

[13] A. C. Ferrari, "Raman spectroscopy of graphene and graphite: disorder, electron–phonon coupling, doping and nonadiabatic effects," *Solid State Communications*, vol. 143, pp. 47–57, 2007.

[14] F. Herziger, C. Tyborski, O. Ochedowski, M. Schleberger, and J. Maultzsch, "Double-resonant LA phonon scattering in defective graphene and carbon nanotubes," *Physical Review B*, vol. 90, p. 245431-1-6, 2014.

[15] A. C. Ferrari, J. C. Meyer, V. Scardaci et al., "Raman spectrum of graphene and graphene layers," *Physical Review Letters*, vol. 97, pp. 187401–187404, 2007.

[16] L. M. Malard, M. A. Pimenta, G. F. Dresselhaus, and M. S. Dresselhaus, "Raman spectroscopy in graphene," *Physics Reports*, vol. 473, pp. 51–87, 2009.

[17] A. K. Gupta, T. J. Russin, H. R. Gutiérrez, and P. C. Eklund, "Probing graphene edges via Raman scattering," *ACS Nano*, vol. 3, pp. 45–52, 2009.

[18] Y. Hao, Y. Wang, L. Wang et al., "Probing layer number and stacking order of few-layer graphene by Raman spectroscopy," *Small*, vol. 6, pp. 195–200, 2010.

[19] S. Chen, W. Cai, R. D. Piner et al., "Synthesis and characterization of large-area graphene and graphite films on commercial Cu–Ni alloy foils," *Nano Letters*, vol. 11, pp. 3519–3525, 2011.

[20] J. U. Lee, N. M. Seck, D. Yoon, S. M. Choi, Y. W. Son, and H. Cheong, "Polarization dependence of double resonant Raman scattering band in bilayer graphene," *Carbon*, vol. 72, pp. 257–263, 2014.

[21] V. N. Popov and P. Lambin, "Theoretical polarization dependence of the two-phonon double-resonant Raman spectra of graphene," *European Physical Journal B*, vol. 85, p. 418, 2012.

[22] P. Klar, E. Lidorikis, A. Eckmann, I. A. Verzhbitskiy, A. C. Ferrari, and C. Casiraghi, "Raman scattering efficiency of graphene," *Physical Review B*, vol. 87, p. 205435-1-12, 2013.

[23] P. Poncharal, A. Ayari, T. Michel, and J.-L. Sauvajol, "Raman spectra of misoriented bilayer graphene," *Physical Review B*, vol. 78, p. 113407-1-4, 2008.

[24] P. H. Tan, C. Y. Hu, J. Dong, W. C. Shen, and B. F. Zhang, "Polarization properties, high-order Raman spectra, and frequency asymmetry between Stokes and anti-Stokes scattering of Raman modes in a graphite whisker," *Physical Review B*, vol. 64, p. 214301-1-12, 2001.

[25] K. A. Wang, P. Zhou, A. M. Rao, P. C. Eklund, R. A. Jishi, and M. S. Dresselhaus, "Intramolecular-vibrational-mode softening in alkali-metal-saturated C70 films," *Physical Review B: Condensed Matter*, vol. 48, pp. 3501–3506, 1993.

[26] P. M. Rafailov, V. G. Hadjiev, H. Jantoljak, and C. Thomsen, "Raman depolarization ratio of vibrational modes in solid C 60," *Solid State Communications*, vol. 112, pp. 517–520, 1999.

[27] S. Prawer, K. W. Nugent, D. N. Jamieson, J. O. Orwa, L. A. Bursill, and J. L. Peng, "The Raman spectrum of nanocrystalline diamond," *Chemical Physics Letters*, vol. 332, pp. 93–97, 2000.

[28] Z. Li, R. J. Young, I. A. Kinloch et al., "Quantitative determination of the spatial orientation of graphene by polarized Raman spectroscopy," *Carbon*, vol. 88, pp. 215–224, 2015.

[29] M. Mohr, J. Maultzsch, and C. Thomsen, "Splitting of the Raman 2 D band of graphene subjected to strain," *Physical Review B*, vol. 82, p. 201409-1-4 R, 2010.

[30] O. Frank, M. Mohr, J. Maultzsch et al., "Raman 2D-band splitting in graphene: theory and experiment," *ACS Nano*, vol. 5, pp. 2231–2239, 2011.

[31] V. N. Popov and P. Lambin, "Theoretical 2 D Raman band of strained graphene," *Physical Review B*, vol. 87, p. 155425-1-7, 2013.

[32] V. N. Popov and P. Lambin, "Theoretical Raman intensity of the G and 2D bands of strained graphene," *Carbon*, vol. 54, pp. 86–93, 2013.

[33] A. Das, S. Pisana, B. Chakraborty et al., "Monitoring dopants by Raman scattering in an electrochemically top-gated graphene transistor," *Nature Nanotechnology*, vol. 3, pp. 210–215, 2008.

[34] J. E. Lee, G. Ahn, J. Shim, Y. S. Lee, and S. Ryu, "Optical separation of mechanical strain from charge doping in graphene," *Nature Communications*, vol. 3, p. 1024-1-8, 2012.

[35] C. Bouhafs, A. A. Zakharov, I. G. Ivanov et al., "Multi-scale investigation of interface properties, stacking order and decoupling of few layer graphene on C-face 4H-SiC," *Carbon*, vol. 116, pp. 722–732, 2017.

17

Longitudinal Raman Spectroscopic Observation of Skin Biochemical Changes due to Chemotherapeutic Treatment for Breast Cancer in Small Animal Model

Myeongsu Seong,[1] NoSoung Myoung,[2] Songhyun Lee,[1] Hyeryun Jeong,[1] Sang-Youp Yim,[2] and Jae Gwan Kim[1,3]

[1]Department of Biomedical Science and Engineering, Institute of Integrated Technology, Gwangju Institute of Science and Technology, 123 Cheomdan Gwagiro, Bukgu, Gwangju 61005, Republic of Korea
[2]Advanced Photonics Research Institute, Gwangju Institute of Science and Technology, 123 Cheomdan Gwagiro, Bukgu, Gwangju 61005, Republic of Korea
[3]School of Electrical Engineering and Computer Science, Gwangju Institute of Science and Technology, 123 Cheomdan Gwagiro, Bukgu, Gwangju 61005, Republic of Korea

Correspondence should be addressed to Sang-Youp Yim; syim@gist.ac.kr and Jae Gwan Kim; jaekim@gist.ac.kr

Academic Editor: Yusuke Oshima

The cancer field effect (CFE) has been highlighted as one of indirect indications for tissue variations that are insensitive to conventional diagnostic techniques. In this research, we had a hypothesis that chemotherapy for breast cancer would affect skin biochemical compositions that would be reflected by Raman spectral changes. We used a fiber-optic probe-based Raman spectroscopy to perform preliminary animal experiments to validate the hypothesis. Firstly, we verified the probing depth of the fiber-optic probe (~800 μm) using a simple intravenous fat emulsion-filled phantom having a silicon wafer at the bottom inside a cuvette. Then, we obtained Raman spectra during breast cancer treatment by chemotherapy from a small animal model in longitudinal manner. Our results showed that the treatment causes variations of biochemical compositions in the skin. For further validation, the Raman spectra will have to be collected from more populations and spectra will need to be compared with immunohistochemistry of the breast tissue.

1. Introduction

Cancer field effect (CFE) has been highlighted since it could unveil minute chemical changes in the region of interest that cannot be captured by clinical or histological diagnostic techniques. Since first finding of CFE from oral cancers in 1953 [1], studies related to CFE have been gradually increased [2, 3]. While other techniques including microscopy, polymerase chain reaction (PCR), and immunohistochemistry were utilized to study CFE [3–6], Raman spectroscopy has been spotlighted as a useful tool since it has been shown that the spectroscopy can detect changes that are insensitive to histological evaluation implying that chemical changes are not local and proceed earlier than anatomical or physical changes [7–9]. Such studies proved the ability of Raman spectroscopy as a dominant candidate for a sensitive and minimally invasive optical biopsy tool. Based on the results from the previous studies, we hypothesized that breast cancer treatment may induce biochemical changes in the skin because they are adjacent to each other. The objective of this study was to find the possibility utilizing skin biochemical alternations to monitor breast cancer treatment. To our best of knowledge, there has been no report about skin biochemical alternation due to breast cancer treatment to date. In this paper, we show that the probe has limited maximum probing depth of about 800 μm (shallower than the thickness of the skin of a rat, 1.5 mm~ [10]) and longitudinal biochemical changes by breast cancer treatment in a small animal model ($n = 4$).

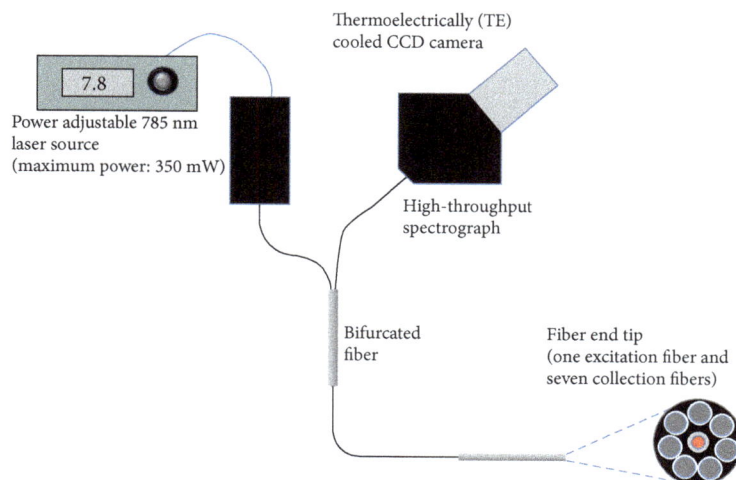

FIGURE 1: Schematic of the Raman system. One centered excitation fiber and seven surrounding collection fibers were configured as a bifurcated fiber. A power adjustable 785 nm laser source (maximum power: 350 mW) coupled with the excitation fiber was used as the excitation source, and a thermoelectrically (TE) cooled CCD camera coupled to a high-throughput spectrograph collects spontaneous Raman signals transmitted through the seven collection fibers.

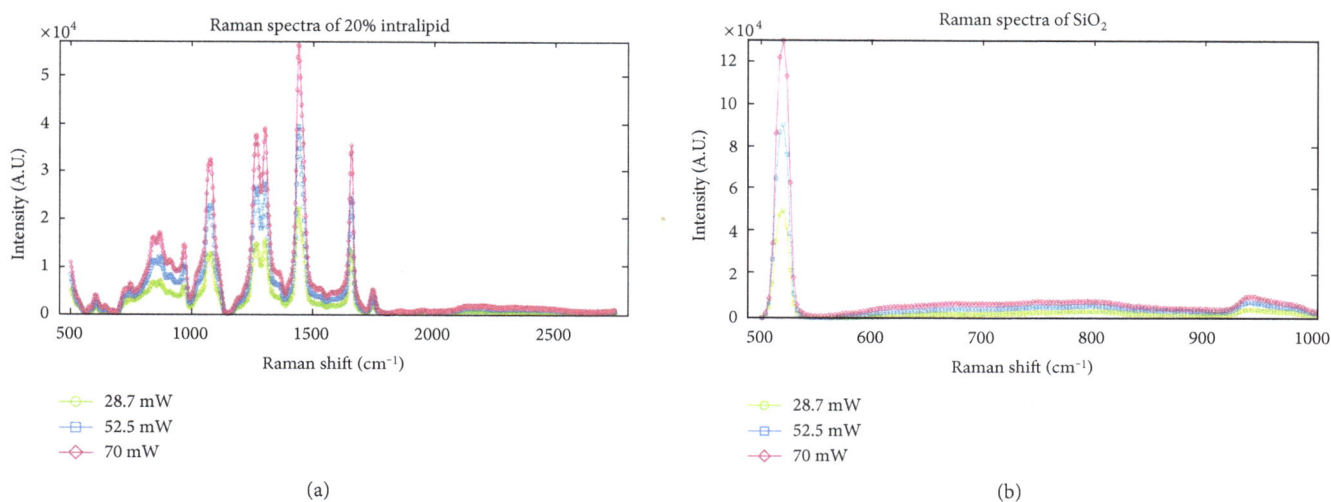

(a)

(b)

FIGURE 2: (a) Spectra of 20% intralipid solution after autofluorescence suppression and (b) spectra of a silicon wafer (SiO_2) measured with laser optical powers of 28.7 mW (empty diamond, magenta), 52.5 mW (empty square, blue), and 70 mW (empty circle, green). No peak shift or spectral change was observed from power adjustment.

2. Materials and Methods

2.1. Configuration and Calibration of a Fiber-Optic Raman System.
Figure 1 shows the schematic of a fiber-optic Raman system used in the study. The Raman system has a power adjustable 785 nm laser diode (FC-785-350-MM2-PC-1-0-RM, 350 mW of maximum power, RGBLase) as a laser source, a high-throughput spectrograph (XPE85-NIR, F/1.4, Nanobase) coupled with a thermoelectrically (TE) cooled charge coupled device (CCD) camera (iDus 401 BR-DD, Andor) for light collection and data transfer. A fiber-optic probe (Emvision *LLC*) with one centered excitation fiber (200 μm core) and seven surrounding collection fibers (300 μm core) and a plano-convex lens that covers in front

of the fibers limiting the working distance of the probe to be 0–400 μm specified in the specification sheet. The excitation fiber is coupled to the laser source, and seven surrounding collection fibers are coupled with a spectrograph for back-scattered spontaneous Raman signal delivery. In order to assure transmission of the wavelength of 785 nm through excitation path and the wavelengths longer than 785 nm through collection path, a band pass filter and a long pass filter were placed in the paths, respectively. Raman shifts measured range approximately from 500 cm^{-1} to 2800 cm^{-1}. The spectral resolution of the device was 11 cm^{-1}.

2.2. Probing Depth Validation Using a Simple Liquid Phantom.
In order to make sure its probing depth is limited

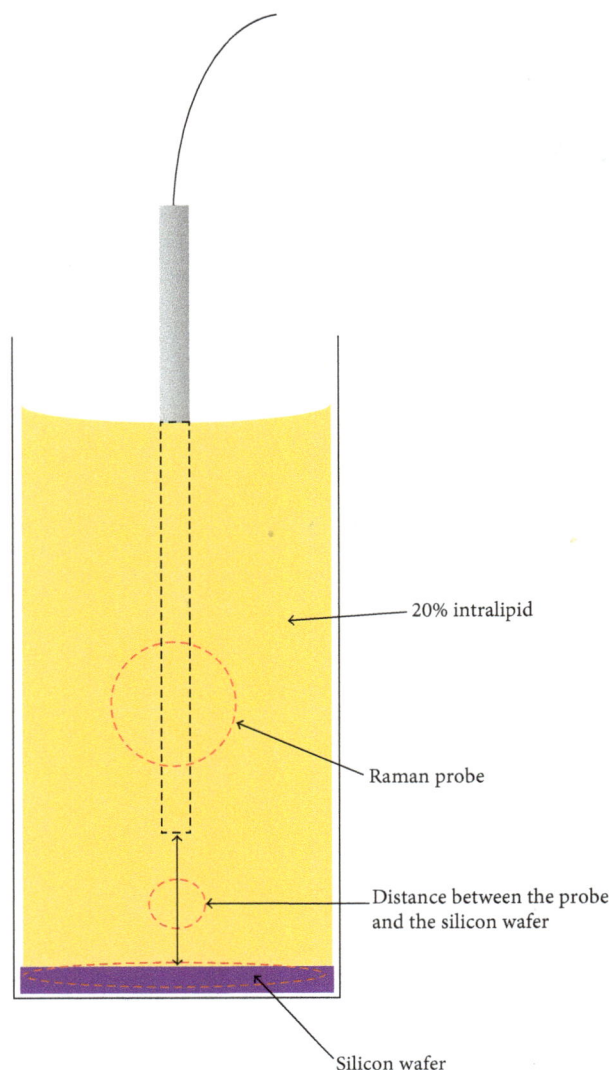

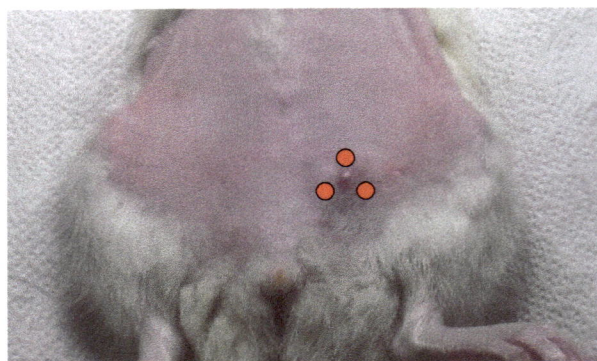

FIGURE 4: An example of shaved and depilated animal. Among eight nipples, one abdominal nipple with breast tumor was monitored in the study. Raman signals were measured in triangular shape around a tumor (red colored dots). Below the right nipple, breast cancer cells (~1 million) were inoculated into mammary fat pad subcutaneously.

FIGURE 3: Schematic of a setup to check the sensitive depth of the Raman probe. A silicon wafer was placed at the bottom in a cuvette. The cuvette was filled with 20% intralipid solution. Raman probe was immersed in the intralipid solution. The distance between the silicon wafer and the probe end tip was varied from $0\,\mu m$ to $1000\,\mu m$ incrementing the distance by $100\,\mu m$ using a one-axis linear translation stage.

probe, Raman spectra were collected varying distances between the silicon wafer and the probe from $0\,\mu m$ (fully contact) to $1000\,\mu m$ with a step of $100\,\mu m$ distance increment. The laser power was set to 28.7 mW for the validation. The distance was carefully varied by using a one-axis linear translation stage (MT01, Thorlabs).

2.3. Animal Model. Fischer 344 female rats (160 g–220 g, Japan SLC) were used in the study. After baseline measurement, about one million 13762 MAT B-III breast cancer cells (CRL-1666, ATCC) were inoculated to lateral caudal abdominal breast (right nipple on Figure 4) of each rat to induce breast tumor growth. When the tumor size became 8 mm, single high dose (100 mg/kg) of cyclophosphamide (C0768-5G, Sigma Aldrich) solution, an alkylating agent that interferes cancer cell growth by preventing deoxyribonucleic acid (DNA) replication and ribonucleic acid (RNA) creation, was intraperitoneally injected as chemotherapy treatment. During the experimental period, the tumor size was measured by a caliper. The breast cancer animal model was chosen since the model is a well-established animal model in breast cancer study (13762 MAT B-III breast cancer cell line is highly tumorigenic in Fischer 344 since the cell line is derived from the strain of Fischer 344 itself) [12–14]. Cyclophosphamide was used because the chemical has been used as a chemotherapeutic agent for treating human [15–17] and rat [14, 18] breast tumors in previous researches. All the animal experiments were approved by the Gwangju Institute of Science and Technology Institutional Review Board.

2.4. Longitudinal Animal Measurements. For all the animals, fur on the belly was shaved and depilated to reduce scattering due to fur. Figure 4 shows an example of a shaved and then depilated belly of an animal. During measurement, each animal was under general anesthesia by 1% to 2% of isoflurane mixed with breathing gas (21% O_2 and 79% N_2). In case of animals with breast tumor, Raman spectra from three sites around the tumor were acquired in each measurement. In case of animals without the tumor, the Raman spectra were

to the specific depth, we made a simple intravenous fat emulsion-filled liquid (intralipid 20%, Fresenius Kabi) phantom having a silicon oxide wafer as the bottom. The intralipid was utilized since that has been used in previous Raman researches as a human tissue mimetic phantom [11]. Also, the silicon wafer was used because intralipid has no characteristic peak nearby 520 cm^{-1} while the silicon wafer has strong characteristic peak nearby 520 cm^{-1}. Spectra of 20% intralipid and a silicon wafer were acquired with different optical power of the source (28.7 mW, 52.5 mW, and 70 mW) as the references. Figure 2 shows the measured spectra of 20% intralipid (Figure 2(a)) and silicon wafer (Figure 2(b)). Figure 3 shows a schematic of measurement setup for probing depth validation. For determining the working distance of the Raman

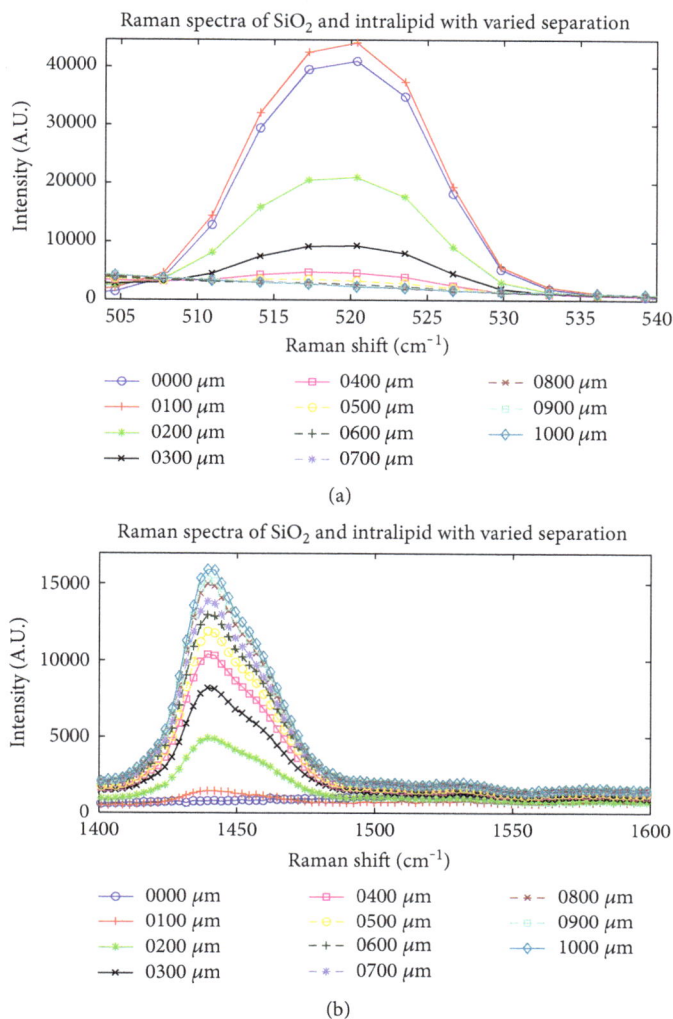

FIGURE 5: Decay of silicon wafer's characteristic peak and increase of intralipid's one characteristic peak while the distance between the silicon wafer and the probe end tip increases in the phantom measurement as shown in Figure 3. The laser power was fixed to 28.7 mW. Notice that from full contact to 100 μm separation, the Raman characteristic peak of the silicon increases and it might happen in the combination of the mismatch of refractive indices between intralipid, silicon wafer, and the probe and the effect of the plano-convex lens in front of the excitation and detection fibers. Mixture of characteristic peaks of the silicon and intralipid still exists till the distance of 800 μm, and it implies that when measuring highly scattering medium in near-infrared (NIR) range such as the biological tissues, the working distance of the probe would be longer than the one described in the specification.

acquired around the nipple three times with triangular shape. Multiple measurements were done to minimize the effect of heterogeneity of the tissue. Figure 4 also shows an example of Raman signal acquisition points. The total integration time for one Raman signal measurement was 10 seconds (100 times summation of Raman signal acquired in 0.1 second). Animals were grouped into three groups: rat group ($n = 4$), no chemo group ($n = 3$), and chemo only group ($n = 2$). Breast tumor growth was induced for both rat group and no chemo group by inoculating about one million breast cancer cells to the mammary fat pad. The tumor was treated by cyclophosphamide injection when tumor size reached to about 8 mm for rats in rat group and the Raman measurement was started. For no chemo group, the Raman spectra were acquired when tumor size became

TABLE 1: The tumor volume variation from the four rats in the rat group.

	Rat 1 (tumor volume, mm^3)	Rat 2 (tumor volume, mm^3)	Rat 3 (tumor volume, mm^3)	Rat 4 (tumor volume, mm^3)
Baseline	432.63	172.76	656.57	467.24
Day 1	603.85	615.19	1017.24	631.27
Day 2	1065.68	569.64	1350.13	789.19
Day 3	781.20	434.11	865.37	862.11
Day 4	539.50	390.18	864.33	606.29

about 8 mm and no chemotherapy was performed. Chemo only group did not have breast cancer cell inoculation while cyclophosphamide was administered to observe effects of

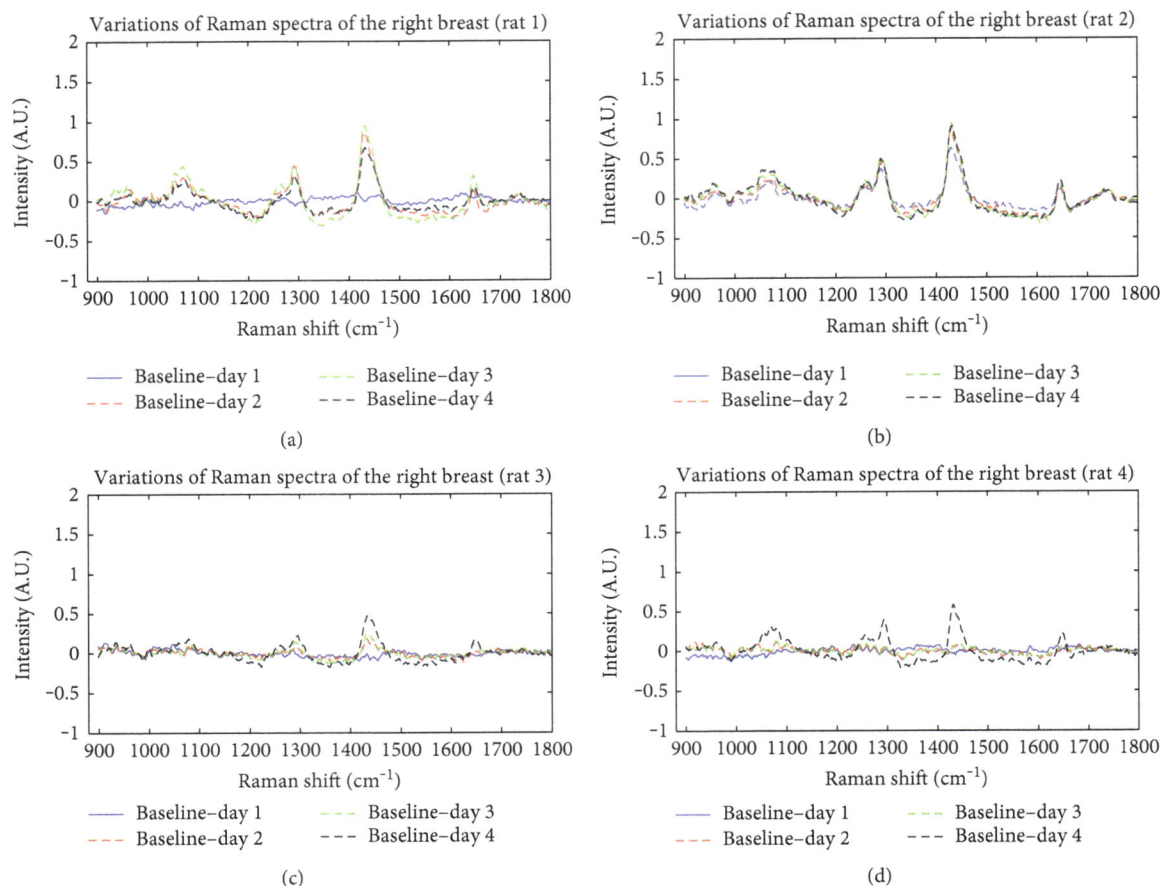

FIGURE 6: Variations of Raman spectra from rat group with regard to the baseline (baseline means the day high-dose chemotherapy was started). The spectrum for each day was subtracted from the baseline signal. Positive values mean that the intensity was decreased, and the negative values mean that the intensity was increased with respect to the baseline signal. Day n means n day postchemotherapy.

only cyclophosphamide on skin chemical compositions. The laser power was adjusted lower than 28 mW not to damage the tissue. In case of tumor-bearing rats, the tumor volume was calculated by an equation of ellipsoid volume.

2.5. Preprocessing and Data Analysis.
Raw bio-Raman signals are mostly mixed signals of autofluorescence and pure Raman signals when 785 nm of excitation beam is used. Since the autofluorescence obscures Raman spectrum, all the spectra were fitted using modified polyfit method, and the fitted autofluorescence was subtracted following fluorescence subtraction procedure by Lieber and Mahadevan-Jansen [19] except Raman spectra of silicon wafer. Also, all the animal Raman spectra were smoothed by a Savitzky-Golay filter (third order, window size: 5). The smoothed Raman spectra were mean centered by subtracting one's mean intensity from each spectrum's intensity and normalized with respect to the mean intensity. All the preprocessing scripts were written and performed using Mathematica 9.0 (Wolfram) and MATLAB R2013b (Mathworks). The spectral range from $900\,\mathrm{cm}^{-1}$ to $1800\,\mathrm{cm}^{-1}$ was used since the range has rich chemical information of lipid, protein, and more [20]. Note that no chemometric technique was used in the study due

TABLE 2: The tumor volume variation from the three rats in the no chemo group.

	No chemo 1 (tumor volume, mm^3)	No chemo 2 (tumor volume, mm^3)	No chemo 3 (tumor volume, mm^3)
Baseline	58.99	22.86	51.62
Day 1	125.34	48.49	107.23
Day 2	216.03	105.32	169.59
Day 6	2161.48	1096.82	2509.94

to limited number of samples and objective of the study (longitudinal observation of biochemical composition changes).

3. Results and Discussion

3.1. Probing Depth Validation Using a Simple Liquid Phantom.
The characteristic peaks of intralipid and silicon wafer are shown in Figure 2(a) and Figure 2(b), respectively. Source power changes did not cause any effect on the characteristic peaks except intensity increments. Figure 5 shows the characteristic peaks of silicon wafer (Figure 5(a)) and intralipid (Figure 5(b)) gradually decrease and increase as the distance between the silicon wafer and the probe end becomes longer, respectively. When the distance reached $800\,\mu m$, the

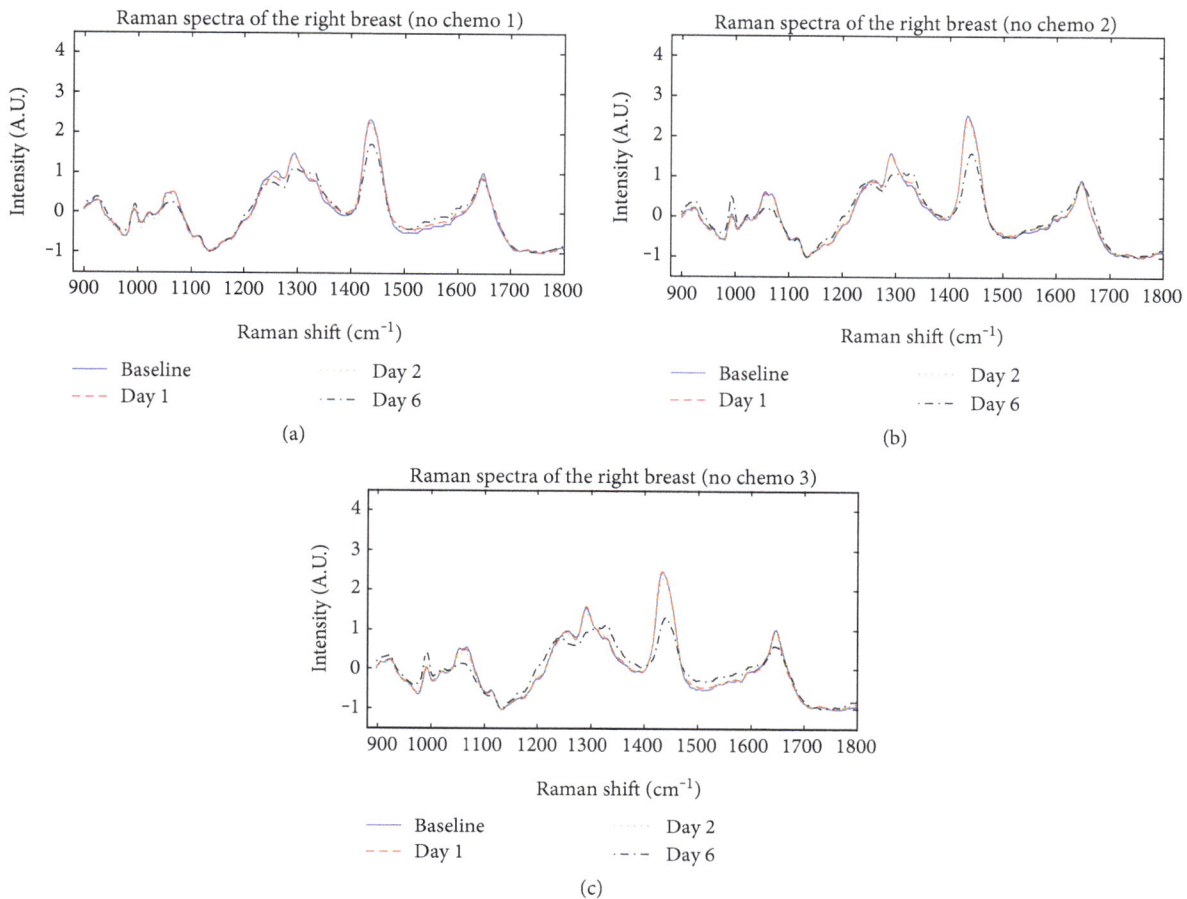

FIGURE 7: Variation of Raman spectra with breast tumor growth and without breast cancer treatment (no chemo group, baseline means the day high-dose chemotherapy was started). Day n means n day postchemotherapy.

characteristic peaks of the silicon wafer and intralipid reached their extrema showing no changes with longer separations. The result shows that the probing depth of the Raman probe used in this study is mostly limited to 0–800 μm. The range of the working distance was longer than the values in the specification (0–400 μm). This may come from scattering dominant nature of phantom [21–23]. We also notice that while the separation varies from 0 μm to 100 μm, the Raman characteristic peak of the silicon increases and it might happen in the combination of the mismatch of refractive indices between intralipid, silicon wafer, and the probe and the effect of the plano-convex lens in front of the excitation and detection fibers.

3.2. Longitudinal Animal Measurements

3.2.1. Rat Group. Table 1 and Figure 6 show the tumor volume changes and relative variations of Raman spectra from the baseline, respectively. The Raman spectra before subtraction are not shown here because the variation was minute. Baseline means day 0 from the chemotherapy. In other words, the chemotherapeutic agent was injected on the day 0. Tumor size for rat 2 regressed earliest and the Raman spectrum was immediately changed on day 1 (Figure 6(b)). Rat 3 had the biggest tumor size when

chemotherapy was started, and the Raman spectra showed minute change as the days passed (Figure 6(c)). Also, the Raman spectrum on day 1 did not show distinct change with respect to the baseline signal. Rat 4 showed delayed response regressing the tumor size from the day 4. The Raman spectra variation was matched as the tumor regression day showing the clear change of Raman spectrum on day 4 while there was no significant Raman spectral variation from day 1 to day 3. Among rat 1 to rat 4, rat 1 showed moderate response that the tumor regressed from day 3. The significant variation was happening around 1064 cm^{-1} (lipid), 1265–1267 cm^{-1} (collagen, protein, and lipid), 1299–1303 cm^{-1} (lipid, fatty acids, collagen, and protein), 1440 cm^{-1} (lipid), 1445 cm^{-1} (protein), and 1654–1656 cm^{-1} (protein and lipid) [20, 24]. The peaks or Raman shifts are known as the positions related to lipid and protein components [20]. It reflects that tumor regression due to chemotherapeutic agent causes variations of lipid and protein components in the skin because the probing depth for the probe is limited to the depth of skin as shown in the Section 3.1.

3.2.2. No Chemo Group. Table 2 and Figure 7 show the tumor volume changes and the Raman spectra on the measurement days, respectively. While day 1 and day 2 Raman spectra did not show any significant variation with respect to the baseline

Variations of Raman spectra of the right breast (no chemo 1)

Variations of Raman spectra of the right breast (no chemo 2)

Variations of Raman spectra of the right breast (no chemo 3)

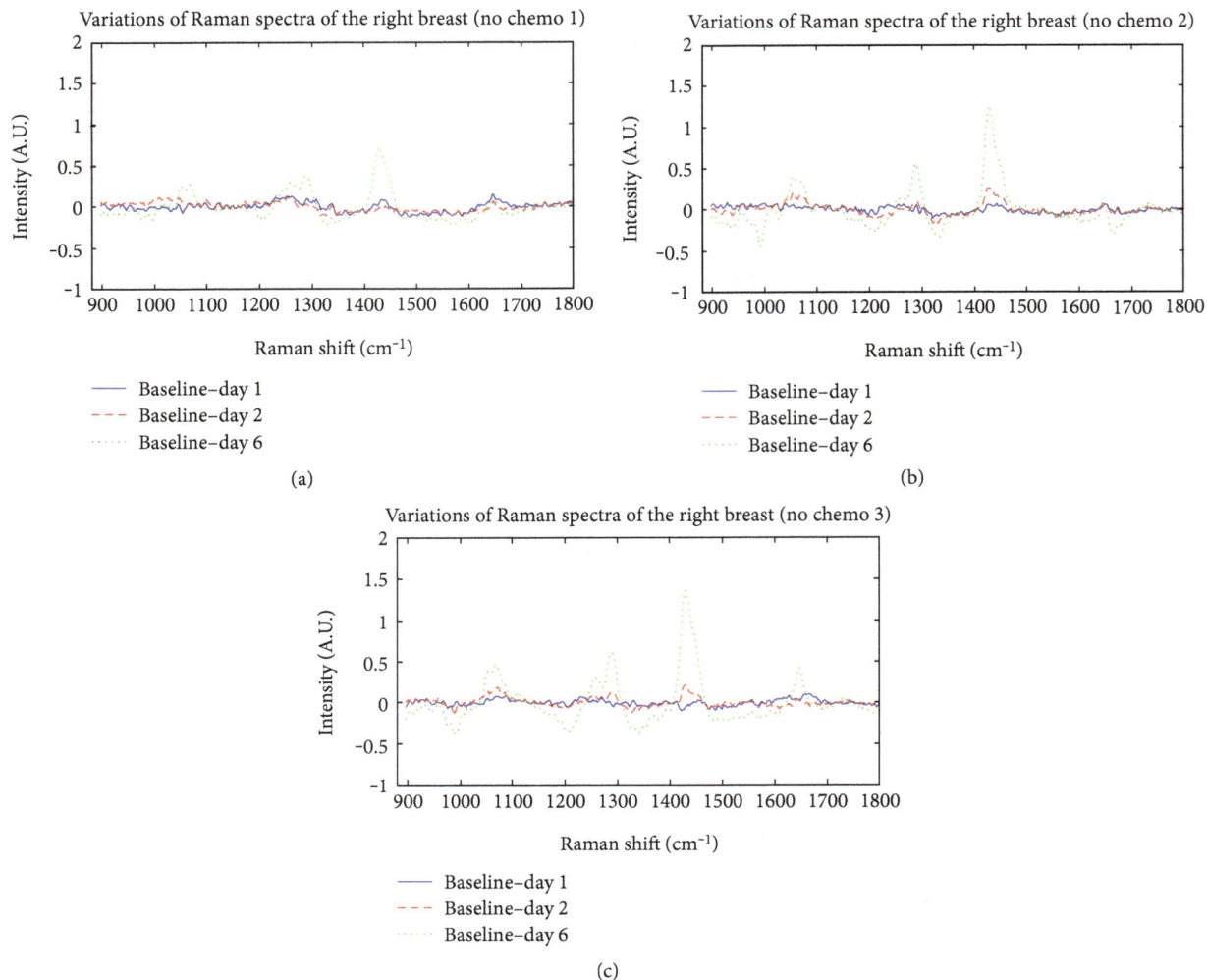

FIGURE 8: Variations of Raman spectra with regard to the baseline (no chemo group). The spectrum for each day was subtracted from the baseline signal. Positive values mean that the intensity was decreased, and the negative values mean that the intensity was increased with respect to the baseline signal. Day n means n day postchemotherapy.

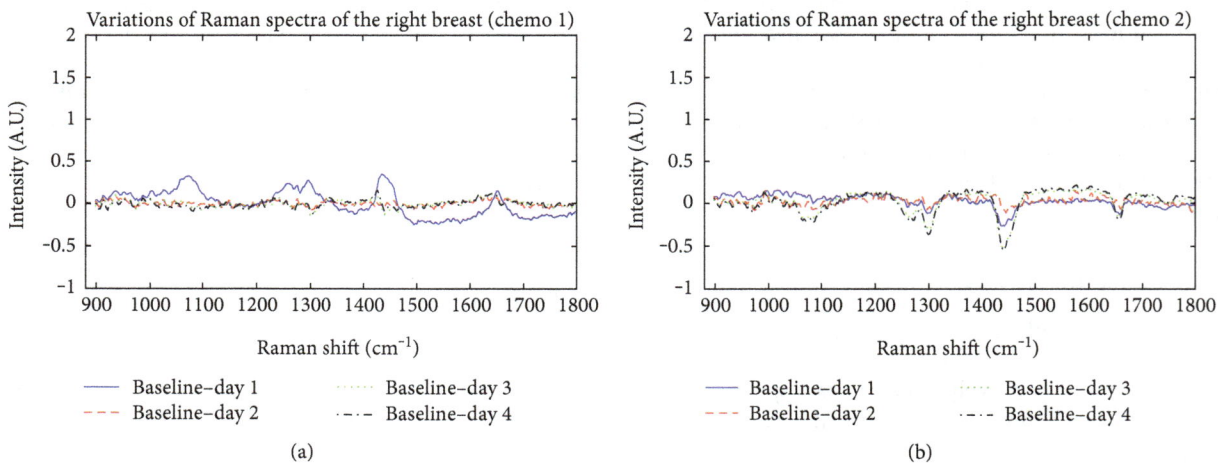

Variations of Raman spectra of the right breast (chemo 1)

Variations of Raman spectra of the right breast (chemo 2)

FIGURE 9: Variations of Raman spectra with regard to the baseline (chemo only group, baseline means the day high-dose chemotherapy was started). No breast tumor growth was induced. Day n means n day postchemotherapy.

signal, Raman spectra on day 6 from the baseline show degradation of peaks around 1299–$1303\,\mathrm{cm}^{-1}$ and the peak height around $1440\,\mathrm{cm}^{-1}$ and $1445\,\mathrm{cm}^{-1}$. When tumor size reached to similar size as the one in the rat group, the Raman spectra vary in a different way from the spectra in the rat group showing the possibility of monitoring breast cancer treatment from the external tissue. Because there were studies showing tissue composition changes either $1\,\mathrm{cm}$ or $7\,\mathrm{cm}$ away from the tumor region [3, 5], Raman spectral variations on the skin due to breast cancer treatment and breast cancer growth without treatment are also possible. Figure 8 shows the variations of Raman spectra from the baseline. Comparing Figure 6 and Figure 8, significant difference in Raman spectral region of 1000–$1003\,\mathrm{cm}^{-1}$ (protein) [24] between the rat group and no chemo group can be found.

3.2.3. Chemo Only Group. Figure 9 shows variations of Raman spectra from day 0 (chemotherapeutic agent injection) for two rats. The Raman spectral variations were distinct from the changes in the rat group and no chemo group. Our data show that administration of cyclophosphamide causes the peak variations in 1090–$1100\,\mathrm{cm}^{-1}$ (lipid/phospholipid DNA backbone), 1280–$1282\,\mathrm{cm}^{-1}$ (collagen), and 1447–$1450\,\mathrm{cm}^{-1}$ (DNA/RNA protein lipid) [24]. There were previous researches showing that high dose of cyclophosphamide affects the skin especially for protein/collagen components [25, 26], and our data correlate well with the previous studies.

Meanwhile, chemo only group did not show any variation in $1200\,\mathrm{cm}^{-1}$ while rat group shows the peak variation nearby $1200\,\mathrm{cm}^{-1}$ (protein) [24]. In addition, the two rats showed different response from the same chemotherapeutic agent injection. This discrepancy might come from the physiological differences between the rats. The different response is another subject that needs further investigation to seek any possibility as the early prediction factor of chemotherapeutic efficacy for a specific chemotherapeutic agent.

4. Conclusion

Here, we had hypothesized that breast tumor treatment would influence neighboring tissue (here skin) varying chemical compositions of the tissue and showed the skin Raman spectral changes while breast cancer was treated. The study suggests the possibility of monitoring tumor treatment located relatively deep such as breast cancer. From our results, we could distinguish the three groups from the peak variations in the Raman shifts of 1000–$1003\,\mathrm{cm}^{-1}$, 1125–$1132\,\mathrm{cm}^{-1}$, and $1200\,\mathrm{cm}^{-1}$. To the authors' best of knowledge, there was no study about variations of chemical compositions in neighboring tissue due to breast cancer treatment; this preliminary result will require additional validations such as comparison of the Raman spectra with immunohistochemistry.

Authors' Contributions

Myeongsu Seong and NoSoung Myoung contributed equally to this work.

Acknowledgments

This work is supported by the Research on Advanced Optical Science and Technology grant (Sang-Youp Yim) funded by the Gwangju Institute of Science and Technology (GIST) and "Biomedical Integrated Technology Research" Project through a grant (Jae Gwan Kim) provided by GIST in 2017, the National Research Foundation of Korea Grant 2012K1A2B1A03000757 (Jae Gwan Kim), and Basic Science Research Program through the NRF funded by the Ministry of Education, Science and Technology 2013R1A1A2013625 (Jae Gwan Kim) and 2013R1A1A2058746 (NoSoung Myoung).

References

[1] D. P. Slaughter, H. W. Southwick, and W. Smejkal, "Field cancerization in oral stratified squamous epithelium. Clinical implications of multicentric origin," *Cancer*, vol. 6, no. 5, pp. 963–968, 1953.

[2] Z. J. Smith, T. R. Huser, and S. Wachsmann-Hogiu, "Raman scattering in pathology," *Analytical Cellular Pathology*, vol. 35, no. 3, pp. 145–163, 2012.

[3] H. Chai and R. E. Brown, "Review: field effect in cancer-an update," *Annals of Clinical and Laboratory Science*, vol. 39, no. 4, pp. 331–338, 2009.

[4] R. K. Bista, P. Wang, R. Bhargava et al., "Nuclear nanomorphology markers of histologically normal cells detect the 'field effect' of breast cancer," *Breast Cancer Research and Treatment*, vol. 135, no. 1, pp. 115–124, 2012.

[5] C. M. Heaphy, M. Bisoffi, C. A. Fordyce et al., "Telomere DNA content and allelic imbalance demonstrate field cancerization in histologically normal tissue adjacent to breast tumors," *International Journal of Cancer*, vol. 119, no. 1, pp. 108–116, 2006.

[6] M. C. Risk, B. S. Knudsen, I. Coleman et al., "Differential gene expression in benign prostate epithelium of men with and without prostate cancer: evidence for a prostate cancer field effect," *Clinical Cancer Research*, vol. 16, no. 22, pp. 5414–5423, 2010.

[7] E. Vargis, E. M. Kanter, S. K. Majumder et al., "Effect of normal variations on disease classification of Raman spectra from cervical tissue," *Analyst*, vol. 136, pp. 2981–2987, 2011.

[8] M. D. Keller, E. M. Kanter, C. A. Lieber et al., "Detecting temporal and spatial effects of epithelial cancers with Raman spectroscopy," *Disease Markers*, vol. 25, no. 6, pp. 323–337, 2008.

[9] S. P. Singh, A. Sahu, A. Deshmukh, P. Chaturvedi, and C. M. Krishna, "In vivo Raman spectroscopy of oral buccal mucosa: a study on malignancy associated changes (MAC)/cancer field effects (CFE)," *Analyst*, vol. 138, pp. 4175–4182, 2013.

[10] B. Wagner, C. Tan, J. L. Barnes et al., "Nephrogenic systemic fibrosis: evidence for oxidative stress and bone marrow-derived fibrocytes in skin, liver, and heart lesions using a 5/6 nephrectomy rodent model," *The American Journal of Pathology*, vol. 181, no. 6, pp. 1941–1952, 2012.

[11] M. Agenant, M. Grimbergen, R. Draga, E. Marple, R. Bosch, and C. van Swol, "Clinical superficial Raman probe aimed for epithelial tumor detection: phantom model results," *Biomedical Optics Express*, vol. 5, no. 4, pp. 1203–1216, 2014.

[12] A. K. Laust, B. W. Sur, K. Wang, B. Hubby, J. F. Smith, and E. L. Nelson, "VRP immunotherapy targeting neu: treatment efficacy and evidence for immunoediting in a stringent rat mammary tumor model," *Breast Cancer Research and Treatment*, vol. 106, no. 3, pp. 371–382, 2007.

[13] T. P. Archer, P. Bretscher, and B. Ziola, "Immunotherapy of the rat 13762SC mammary adenocarcinoma by vaccinia virus augmentation of tumor immunity," *Clinical & Experimental Metastasis*, vol. 8, no. 6, pp. 519–532, 1990.

[14] V. K. Todorova, Y. Kaufmann, and V. S. Klimberg, "Increased efficacy and reduced cardiotoxicity of metronomic treatment with cyclophosphamide in rat breast cancer," *Anticancer Research*, vol. 31, no. 1, pp. 215–220, 2011.

[15] A. Howell, W. D. George, D. Crowther et al., "Controlled trial of adjuvant chemotherapy with cyclophosphamide, methotrexate, and fluorouracil for breast cancer," *Lancet*, vol. 324, no. 8398, pp. 307–311, 1984.

[16] C. Zhou, R. Choe, N. Shah et al., "Diffuse optical monitoring of blood flow and oxygenation in human breast cancer during early stages of neoadjuvant chemotherapy," *Journal of Biomedical Optics*, vol. 12, no. 5, article 51903, 2013.

[17] M. Ayers, W. F. Symmans, J. Stec et al., "Gene expression profiles predict complete pathologic response to neoadjuvant paclitaxel and fluorouracil, doxorubicin, and cyclophosphamide chemotherapy in breast cancer," *Journal of Clinical Oncology*, vol. 22, no. 12, pp. 2284–2293, 2004.

[18] J. P. Sleeman, U. Kim, J. LePendu et al., "Inhibition of MT-450 rat mammary tumour growth by antibodies recognising subtypes of blood group antigen B," *Oncogene*, vol. 18, no. 31, pp. 4485–4494, 1999.

[19] C. A. Lieber and A. Mahadevan-Jansen, "Automated method for subtraction of fluorescence from biological Raman spectra," *Applied Spectroscopy*, vol. 57, no. 11, pp. 1363–1367, 2003.

[20] L. Raniero, R. A. Canevari, L. N. Z. Ramalho et al., "In and ex vivo breast disease study by Raman spectroscopy," *Theoretical Chemistry Accounts*, vol. 130, no. 4–6, pp. 1239–1247, 2011.

[21] S. Jacques, "Optical properties of intralipid," January 2017, http://omlc.org/spectra/intralipid/.

[22] S. T. Flock, S. L. Jacques, B. C. Wilson, W. M. Star, and M. J. C. van Gemert, "Optical properties of intralipid: a phantom medium for light propagation studies," *Lasers in Surgery and Medicine*, vol. 12, no. 5, pp. 510–519, 1992.

[23] H. J. van Staveren, C. J. M. Moes, J. van Marie, S. A. Prahl, and M. J. C. van Gemert, "Light scattering in intralipid-10% in the wavelength range of 400–1100 nm," *Applied Optics*, vol. 30, no. 31, p. 4507, 1991.

[24] R. E. Kast, S. C. Tucker, K. Killian, M. Trexler, K. V. Honn, and G. W. Auner, "Emerging technology: applications of Raman spectroscopy for prostate cancer," *Cancer Metastasis Reviews*, vol. 33, no. 2-3, pp. 673–693, 2014.

[25] H. Wie, L. B. Engesaeter, and E. I. Beck, "Effects of cyclophosphamide on mechanical properties of bone and skin in rats," *Acta Orthopaedica Scandinavica*, vol. 50, pp. 629–634, 1979.

[26] H. Wie and E. I. Beck, "Synthesis and solubility of collagen in rats during recovery after high-dose cyclophosphamide administration," *Acta pharmacologica et toxicologica*, vol. 48, no. 4, pp. 294–299, 2009.

18

Combining Near-Infrared Spectroscopy and Chemometrics for Rapid Recognition of an Hg-Contaminated Plant

Bang-Cheng Tang,[1] Hai-Yan Fu,[2] Qiao-Bo Yin,[2] Zeng-Yan Zhou,[1] Wei Shi,[1]
Lu Xu,[1,2] and Yuan-Bin She[3]

[1]Research Institute of Applied Chemistry, College of Material and Chemical Engineering, Tongren University, Tongren,
 Guizhou 554300, China
[2]The Modernization Engineering Technology Research Center of Ethnic Minority Medicine of Hubei Province, College of Pharmacy,
 South-Central University for Nationalities, Wuhan 430074, China
[3]College of Chemical Engineering, Zhejiang University of Technology, Hangzhou 310014, China

Correspondence should be addressed to Wei Shi; wzlswdbd@163.com, Lu Xu; lxchemo@163.com,
and Yuan-Bin She; sheyb@zjut.edu.cn

Academic Editor: Paulo R. G. Hein

The feasibility of rapid recognition of an Hg-contaminated plant as a soil pollution indicator was investigated using near-infrared spectroscopy (NIRS) and chemometrics. The stem and leave of a native plant, *Miscanthus floridulus* (Labill.) Warb. (MFLW), were collected from Hg-contaminated areas ($n1 = 125$) as well as from regular areas ($n2 = 116$). The samples were dried and crushed and the powders were sieved through an 80-mesh sieve. Reference analysis of Hg levels was performed using inductively coupled plasma-atomic emission spectrometry (ICP-AES). The actual Hg contents of contaminated and normal samples were 16.2–30.5 and 0.0–0.1 mg/Kg, respectively. The NIRS measurements of impacted sample powders were collected in the mode of reflectance. The DUPLEX algorithm was utilized to split the NIRS data into representative training and test sets. Different spectral preprocessing methods were performed to remove the unwanted and noncomposition-correlated spectral variations. Classification models were developed using partial least squares discrimination analysis (PLSDA) based on the raw, smoothed, second-order derivative (D2), and standard normal variate (SNV) data, respectively. The prediction accuracy obtained by PLSDA with each data preprocessing option was 100%, indicating pattern recognition of Hg-contaminated MFLW samples using NIRS data was in perfect consistence with the ICP-AES results. NIRS combined with chemometrics will provide a tool to screen the Hg-contaminated MFLW, which can be potentially used as an indicator of soil pollution.

1. Introduction

The growing development of agricultural, industrial, and urban activities has largely increased the release of toxic substances such as heavy metals and organic compounds to environmental systems [1–3]. In particular, toxic heavy metals in air, water, soil, and plants have caused severe public environmental concern because of their severe adverse influences on human health [4–6]. It is well known that soil is a major sink for heavy metal pollutants, which can be accumulated and transferred to water, air, plants, and animals. It was estimated that 20% of the total farmland in China had been contaminated, which directly threatens the safety of food production [7].

Numerous research efforts have been devoted to evaluation of the level of soil contamination with heavy metals caused by human activities, including electroplating industry, mining, smelting, coal-fired power stations, steel and iron manufacturing, waste incineration, leather industry, and cement production [8–12]. Most of these researches focused on direct determination of heavy metal levels in the soil. Various analytical methods have been developed and used to quantify the levels of heavy metals in soil, plants, and animals [13, 14], including inductively coupled plasma-atomic emission spectroscopy (ICP-AES), inductively coupled mass spectroscopy (ICP-MS), atomic fluorescence spectrometer (AFS), X-ray fluorescence spectrometer (XRF), neutron activation analysis (NAA), DC argon plasma multielement atomic emission spectrometer (DCP-MAES), atomic absorption spectrometer (AAS), and scanning electron microscopy with energy dispersive X-ray (SEM–EDX). Although accurate evaluation of heavy metals can be obtained, most of these techniques generally require laborious preconcentration of analytes and sample pretreatment, which have made the analysis time-consuming.

It is well known that excessive adsorption and accumulation of certain pollutants can influence the growth and metabolism of native plants [15, 16]. A traditional method to recognize and evaluate soil pollution is by examining the morphological variations of plant indicators caused by soil pollution, which are sensitive to the presence of certain pollutants. Although the level of soil pollution could only be qualitatively evaluated using plant indicators, it is more convenient and economic compared with direct methods by analysis of pollutants in soil. However, the use of plant indicators for soil pollution can be limited for some reasons. Firstly, because the plant species in an area can be influenced by many factors, such as geographical and climatic conditions, usually it is not ready to have a well-studied and suitable plant indicator in certain areas. Secondly, in some seriously polluted areas, the plants sensitive to soil pollution would perish and be gradually replaced with the dominant species, which have adapted to the pollution and whose morphological changes are not significant enough to be exactly recognized by the naked eye. Therefore, rather than examining the plants by the naked eye, it is more reasonable and reliable to characterize the changes in chemical compositions of polluted plants using instrumental techniques.

Near-infrared spectroscopy (NIRS) has been widely applied to analysis of various food and agricultural products [17–22]. The feasibility of using NIRS for quantitative analysis of heavy metals in environmental samples has been extensively evaluated [23–26]. Although in some cases NIRS demonstrates potential for quantitative analysis of heavy metals, in many cases, the sensitivity is lower than by using other methods and time-consuming sample preparation is required to obtain reliable results. Some studies also indicate NIRS is very useful for qualitative analysis of heavy metals [27]. NIRS can provide a powerful tool to simultaneously characterize the multicompounds in a complex system, which could be combined with pattern recognition methods [28, 29] to perform rapid classifications of different types of samples.

Therefore, the objective of this paper was to investigate the feasibility of rapid recognition of a native Hg-contaminated plant *Miscanthus floridulus* (Labill.) Warb. (MFLW) from normal MFLW using NIRS and chemometrics. Special attention was made on the experimental design and data collection to avoid obtaining artifacts caused by factors other than Hg-contamination.

2. Materials and Methods

2.1. Collection and Preparation of MFLW Samples. MFLW samples were collected with leaves cut off from the upper end with a length of 25 cm. The Hg-contaminated MFLW samples ($n1 = 125$) were collected around a mercury mining factory

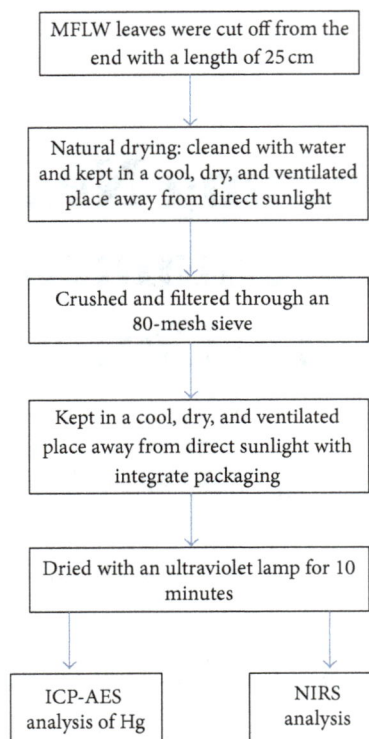

FIGURE 1: The preparation of MFLW samples.

in Huashi, Tongren, China, within a range of 3 Kilometers; normal samples ($n2 = 116$) were collected from an area about 10 Kilometers away (Chuandong, Tongren, China). All the MFLW samples were cleaned with water and kept in a cool, dry, and ventilated place away from direct sunlight to remove the moisture. Each sample (leaves and stalks) was crushed by a disintegrator and then the powders were sieved through an 80-mesh sieve. The dried, crushed, and filtered samples were kept with integrate packaging. An ultraviolet lamp was used to dry each sample for 10 minutes before NIR analysis and Hg reference analysis. The flowchart of sample preparation is shown in Figure 1.

2.2. NIRS Measurements. Impacted MFLW powders were analyzed in a quartz sample cup using an Antaris II Fourier transform-NIR spectrometer (Thermo Electron Co., Waltham, Massachusetts, USA) using the RESTLT 3.0 software in the reflectance mode. The spectra were measured using a PbS detector with an internal gold background as the reference. The working range of spectrometer was $4000–10000\ cm^{-1}$. Each sample was measured triply while being stirred and impacted before each measurement and the average spectra were obtained. The number of scans for each measurement was 32. The instrumental resolution was $8\ cm^{-1}$ with a scanning interval of $3.857\ cm^{-1}$, so each raw spectrum had 1557 wavelengths. The temperature was kept at around $25°C$ and the humidity was kept at a stable level during analysis. In order to avoid artificial spectral variations between different types of samples, the order of analysis for all the samples was permuted randomly.

TABLE 1: The programmed microwave-assisted digestion conditions for ICP-AES analysis of MFLW samples.

Step	Power (W)	Temperature (°C)	Heating time (min)	Holding time (min)
1	800	80	30	5
2	1280	120	30	7
3	1600	160	30	5

2.3. Reference Analysis of Hg Using Inductively Coupled Plasma-Atomic Emission Spectroscopy (ICP-AES). The total Hg contents in MFLW were analyzed according to the national standard (GB5009.17-2014). The MFLW powders were digested using the CEM Mars 5 Microwave Accelerated Reaction System (CEM Corp., Matthews, USA). About 0.4 g of homogenized samples was digested in Teflon vessels with 8.0 mL of nitric acid (HNO_3) (V/V, 10%) overnight and kept at 150°C for 5 h. The programmed digestion conditions are summarized in Table 1. Hg contents were analyzed using an Agilent 725 ICP-AES system (Agilent, Victoria, Australia). The precision of ICP-AES analysis was verified by triplication of the samples. Pearson's r of the standard curve was over 0.9999. The average relative standard deviation (RSD) was less than 5.0% and the recovery rate was 96.1~104.5%. The limit of detection (LOD) was calculated to be 0.0025 mg/Kg according to the IUPAC method, where the signal of 3σ of 11 blank solutions was calibrated using the standard curve.

2.4. Chemometrics Analysis. The data analysis was performed on MATLAB 7.0.1 (Mathworks, Sherborn, MA). In order to remove the unwanted variation in NIRS data, smoothing [30], taking second-order derivative (D2) [30], and standard normal variate (SNV) [31] were performed on the raw data. The DUPLEX algorithm [32] was used to divide the measured samples into a representative training set and test set.

Partial least squares discriminant analysis (PLSDA) [33] was used to develop classification models to distinguish the Hg-contaminated from the regular samples. For PLSDA, a dummy response vector was constructed using +1 and −1 to represent the regular and Hg-contaminated samples, respectively. The number of PLSDA components was estimated using Monte Carlo cross validation (MCCV) [34]. The number of PLS components was determined as to obtain the lowest error rate of MCCV (ERMCCV):

$$\text{ERMCCV} = \frac{\sum_{i=1}^{B} m_i}{\sum_{i=1}^{B} N_i}, \tag{1}$$

where B is the number of MCCV data splitting and m_i and N_i are the numbers of misclassified and leave-out samples, respectively. For prediction, a cutoff value of zero was used to assign a new sample to one of the two classes.

For prediction, the overall accuracy (ACCU) was computed to evaluate the performance of classification models:

$$\text{ACCU} = \frac{\text{TP} + \text{TN}}{\text{TP} + \text{FN} + \text{TN} + \text{FP}}, \tag{2}$$

where TP, TN, FN, and FP represent the numbers of true positives, true negatives, false negatives, and false positives, respectively. In this work, regular and Hg-contaminated MFLW samples were seen as "positives" and "negatives," respectively. Another two usually used indices, sensitivity (SENS) and specificity (SPEC), were also adopted to evaluate the classification performance:

$$\begin{aligned} \text{SENS} &= \frac{\text{TP}}{\text{TP} + \text{FN}} \\ \text{SPEC} &= \frac{\text{TN}}{\text{TN} + \text{FP}}. \end{aligned} \tag{3}$$

SENS and SPEC describe the model ability to correctly accept the "positives" and to correctly reject the "negatives," respectively.

3. Results and Discussion

According to the analytical results of ICP-AES, the Hg contents of regular and contaminated MFLW objects ranged from 0.0 to 0.1 mg/Kg and 16.2 to 30.5 mg/Kg, respectively, indicating an obvious Hg-contamination of soil surrounding the mercury mining areas. The NIR spectra of regular and Hg-contaminated MFLW samples are shown in Figure 2. Seen from Figure 2, the raw spectra of regular and Hg-contaminated MFLW samples have verysimilar absorbance peaks in the range of 4000–10000 cm^{-1}. The peaks can be mainly assigned as follows [35]: 8377 cm^{-1} (the second overtones of C–H stretching), 6823 cm^{-1} (overlapping of the first overtone of O–H stretching and N–H stretching), 5662 cm^{-1} (the first overtones of C–H stretching), 5184 cm^{-1} (the combination of the baseband of O–H stretching and the first overtone of C–O deformation), and 4748 cm^{-1} (combination of N–H stretching and deformation of peptide groups). Some bands (8377 cm^{-1}, 5662 cm^{-1}, and 4748 cm^{-1}) are very weak and the peak resolution is very low. Figure 2 also demonstrates the NIRS data preprocessed by smoothing and taking D2 and SNV transformation. Even with data preprocessing, the spectral difference between regular and Hg-contaminated MFLW samples is still very subtle and is difficult to be distinguished by the naked eye. Therefore, it is necessary to develop chemometric models to extract the relevant information for classification of regular and Hg-contaminated MFLW samples.

In order to obtain representative data sets for developing and validating classification models, the DUPLEX algorithm was adopted to divide the collected samples into training and prediction objects. Considering the regular and Hg-contaminated MFLW samples have different distributions, the DUPLEX algorithm was performed separately on the two classes. The 116 regular samples were split into 80 training and 36 test samples; the 125 Hg-contaminated objects were split into 85 training and 40 test samples. For model building, the training and test samples from the two classes were combined to form the final training and test sets, so 165 (80+85) training samples and 76 (36 + 40) test samples were obtained.

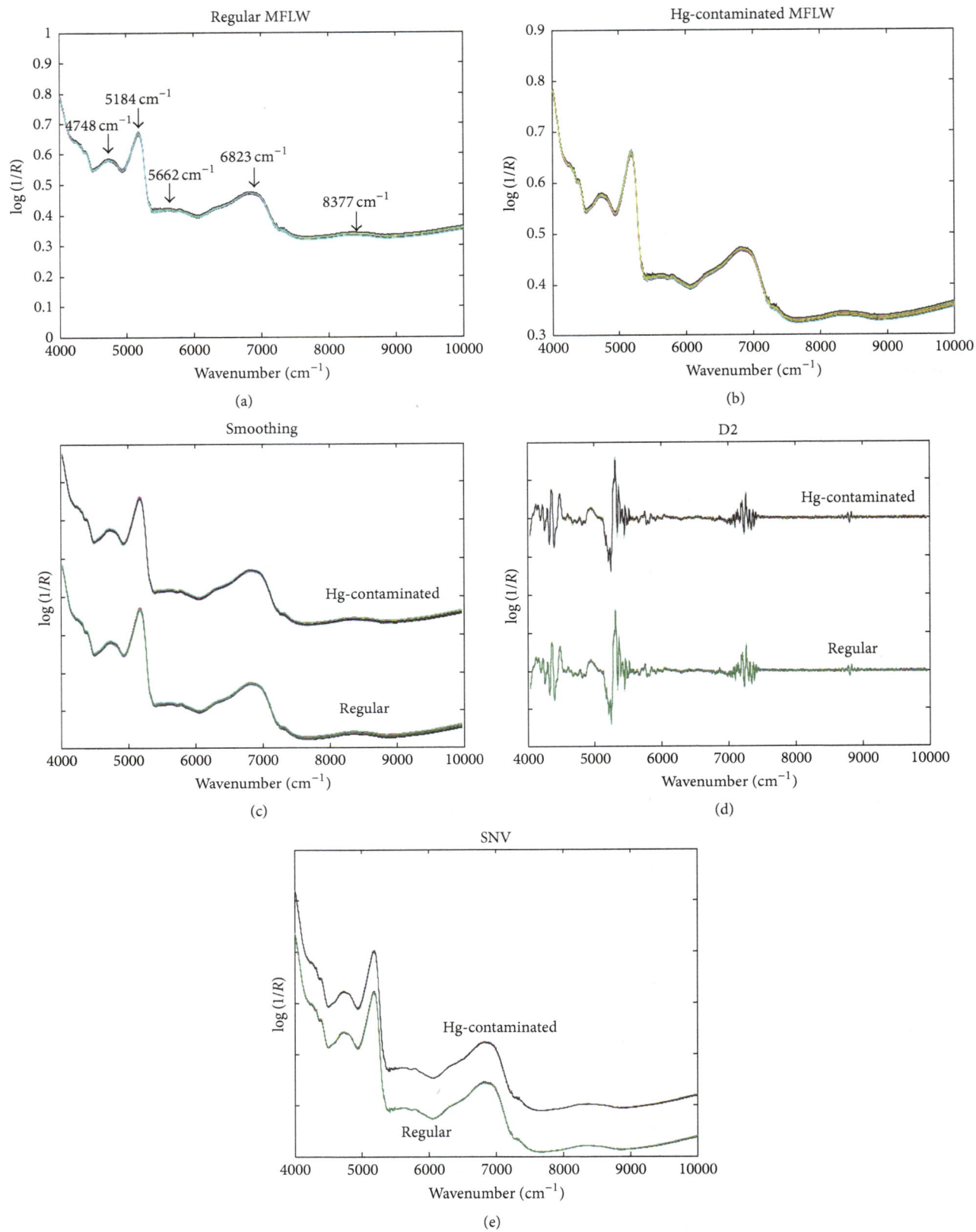

FIGURE 2: Some of the raw, smoothed, D2, and SNV spectra of regular and Hg-contaminated MFLW samples. An artificial shift was included to distinguish the spectra of regular and Hg-contaminated samples.

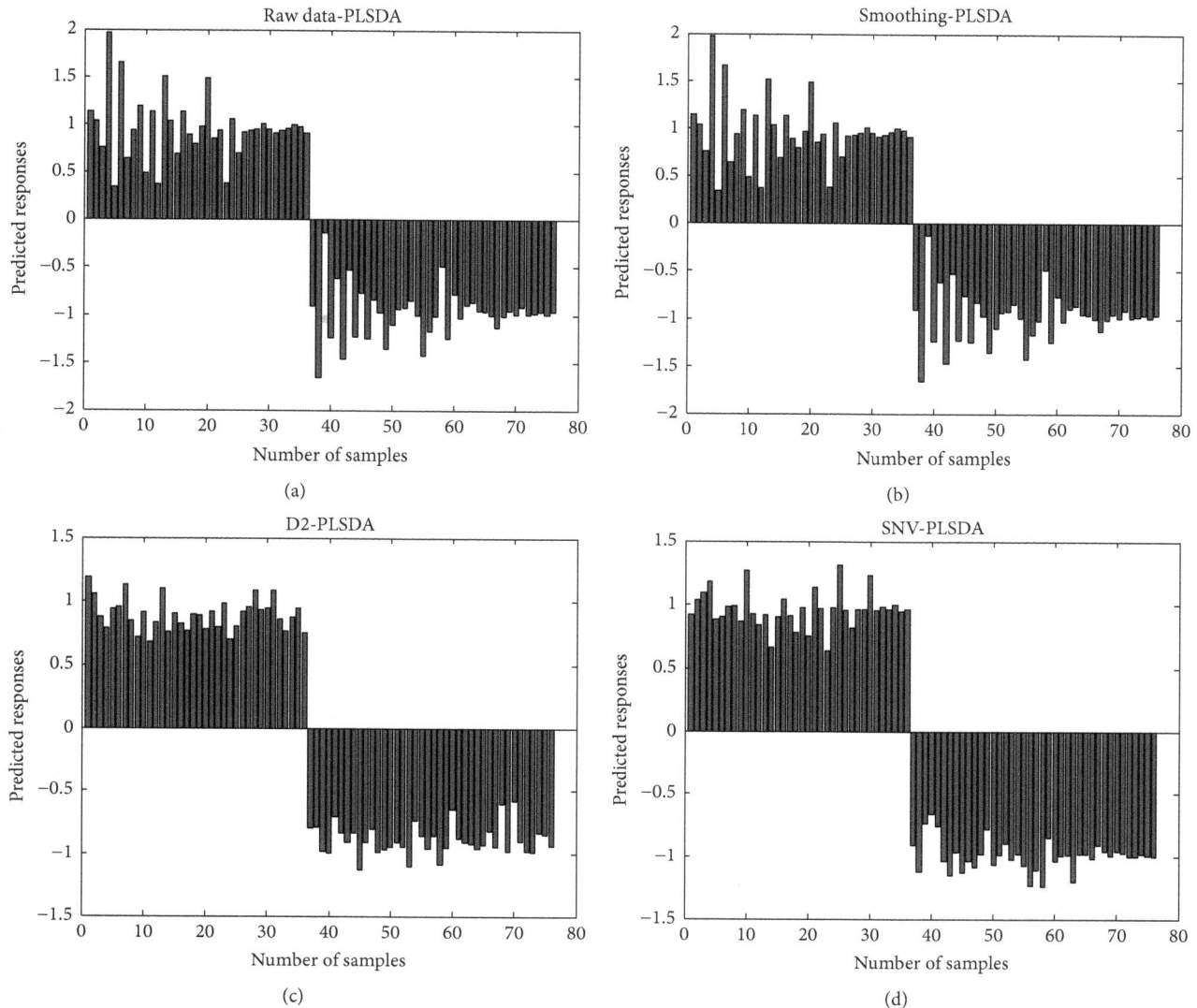

FIGURE 3: The predicted responses by PLSDA with different data preprocessing; test samples 1–36 are regular MFLW samples (positives) and 37–76 are Hg-contaminated (negatives).

PLSDA models were developed with the raw and preprocessed spectra. With different numbers of PLSDA components, ERMCCV was computed and the model complexity was determined as to minimize the ERMCCV value. The number of MCCV data splitting was set to be 100 in this work. Considering the size of training set is moderate, in each MCCV data splitting, 30% of the training set was randomly left out for prediction and the other 70% training samples were used for model development. Based on different data preprocessing options, the model parameters and prediction performance are shown in Table 2. It can be seen that, with each data preprocessing option and even without data preprocessing, PLSDA could obtain perfect classification of regular and Hg-contaminated samples and the accuracy, sensitivity, and specificity were all 1, indicating data preprocessing was not necessary to develop an accurate model. Moreover, all the PLSDA models had 2 latent variables

and the low model complexity means that the models would provide good generalization performance. The prediction results by PLSDA with different data preprocessing are shown in Figure 3, indicating distinct classification of regular and Hg-contaminated MFLW samples by PLSDA despite the kind of data preprocessing. By examining and comparing the predicted responses by PLSDA models with different preprocessing methods, the results by PLSDA with raw data and smoothed spectra were very similar, which were obviously different from those obtained by PLSDA with D2 and SNV spectra. Moreover, the prediction errors (with references to the dummy response vector of +1 and −1) obtained by PLSDA with D2 and SNV spectra were much lower than those obtained by PLSDA with the raw and smoothed spectra. Although all the four PLSDA models could achieve a classification accuracy of 1, D2 and SNV were still necessary to remove some unwanted spectral variations

TABLE 2: Discrimination of regular and Hg-contaminated MFLW by PLSDA of NIRS data.

Data	LV	ERMCCV	Accuracy	SENS	SPEC
Raw data	2	0	1	1	1
Smoothing	2	0	1	1	1
D2	2	0	1	1	1
SNV	2	0	1	1	1

to ensure the generalization performance of PLSDA when predicting new samples.

4. Conclusions

The feasibility of using NIRS for rapid classification of regular and Hg-contaminated MFLW samples was investigated. Classification accuracies of 1 were obtained with low model complexity despite the option of data preprocessing. D2 and SNV were demonstrated to be useful to improve training accuracy by removing unwanted spectral variations. Rapid recognition of the Hg levels in the native plant MFLW would provide a useful alternative indicator of Hg-contaminated soil, which can be used for rapid and economic screening of Hg-contamination. Our future research would be focused on investigating the feasibility of other plants as soil indicators as well as on developing the relationship between the levels of heavy metals in plants and soil.

Competing Interests

Bang-Cheng Tang, Hai-Yan Fu, Qiao-Bo Yin, Zeng-Yan Zhou, Wei Shi, Lu Xu, and Yuan-Bin She declare no competing interests.

Authors' Contributions

Hai-Yan Fu and Bang-Cheng Tang equally contributed to this work.

Acknowledgments

Bang-Cheng Tang is grateful to the financial support from the Research Projects of Guizhou Science and Technology (no. QKHJZLKT[2012]15). Lu Xu is financially supported by the Open Research Program (no. GCTKF2014007) of State Key Laboratory Breeding Base of Green Chemistry Synthesis Technology (Zhejiang University of Technology), the Research Fund for the Doctoral Program of Tongren University (no. trxyDH1501), the Open Research Program (no. 2015ZY006) from the Modernization Engineering Technology Research Center of Ethnic Minority Medicine of Hubei Province (South-Central University for Nationalities), and the research funds from the Education Department of Guizhou Province (no. QJHKYZ[2015]498). Zeng-Yan Zhou is financially supported by Science and Technology Department of Guizhou Province (no. QKHLHZ[2015]7245).

References

[1] B. J. B. Nyarko, S. B. Dampare, Y. Serfor-Armah, S. Osae, D. Adotey, and D. Adomako, "Biomonitoring in the forest zone of Ghana: the primary results obtained using neutron activation analysis and lichens," *International Journal of Environment and Pollution*, vol. 32, no. 4, pp. 467–476, 2008.

[2] J. Aznar-Márquez and J. R. Ruiz-Tamarit, "Environmental pollution, sustained growth, and sufficient conditions for sustainable development," *Economic Modelling*, vol. 54, pp. 439–449, 2016.

[3] R. van Stigt, P. P. J. Driessen, and T. J. M. Spit, "Steering urban environmental quality in a multi-level governance context. How can devolution be the solution to pollution?" *Land Use Policy*, vol. 50, pp. 268–276, 2016.

[4] J. A. Acosta, A. Faz, S. Martínez-Martínez, R. Zornoza, D. M. Carmona, and S. Kabas, "Multivariate statistical and GIS-based approach to evaluate heavy metals behavior in mine sites for future reclamation," *Journal of Geochemical Exploration*, vol. 109, no. 1–3, pp. 8–17, 2011.

[5] Y. Ma, P. Egodawatta, J. McGree, A. Liu, and A. Goonetilleke, "Human health risk assessment of heavy metals in urban stormwater," *Science of the Total Environment*, vol. 557-558, pp. 764–772, 2016.

[6] Y.-G. Gu, Y.-P. Gao, and Q. Lin, "Contamination, bioaccessibility and human health risk of heavy metals in exposed-lawn soils from 28 urban parks in southern China's largest city, Guangzhou," *Applied Geochemistry*, vol. 67, pp. 52–58, 2016.

[7] C. Y. Wei and T. B. Chen, "Hyperaccumulators and phytoremediation of heavy metal contaminated soil: a review of studies in China and abroad," *Acta Ecologica Sinica*, vol. 21, no. 7, pp. 1196–1203, 2001.

[8] G. S. Senesi, M. Dell'Aglio, R. Gaudiuso et al., "Heavy metal concentrations in soils as determined by laser-induced breakdown spectroscopy (LIBS), with special emphasis on chromium," *Environmental Research*, vol. 109, no. 4, pp. 413–420, 2009.

[9] X. Zeng, X. Xu, H. M. Boezen, and X. Huo, "Children with health impairments by heavy metals in an e-waste recycling area," *Chemosphere*, vol. 148, pp. 408–415, 2016.

[10] T. Sarkar, M. M. Alam, N. Parvin et al., "Assessment of heavy metals contamination and human health risk in shrimp collected from different farms and rivers at Khulna-Satkhira region, Bangladesh," *Toxicology Reports*, vol. 3, pp. 346–350, 2016.

[11] C. Gutiérrez, C. Fernández, M. Escuer et al., "Effect of soil properties, heavy metals and emerging contaminants in the soil nematodes diversity," *Environmental Pollution*, vol. 213, pp. 184–194, 2016.

[12] R. A. Wuana and F. E. Okieimen, "Heavy metals in contaminated soils: a review of sources, chemistry, risks and best available strategies for remediation," *ISRN Ecology*, vol. 2011, Article ID 402647, 20 pages, 2011.

[13] R. K. Soodan, Y. B. Pakade, A. Nagpal, and J. K. Katnoria, "Analytical techniques for estimation of heavy metals in soil ecosystem: a tabulated review," *Talanta*, vol. 125, no. 11, pp. 405–410, 2014.

[14] Y. N. Vodyanitskii and I. O. Plekhanova, "Biogeochemistry of heavy metals in contaminated excessively moistened soils (Analytical review)," *Eurasian Soil Science*, vol. 47, no. 3, pp. 153–161, 2014.

[15] A. R. A. Usman, S. S. Lee, Y. M. Awad, K. J. Lim, J. E. Yang, and Y. S. Ok, "Soil pollution assessment and identification of hyperaccumulating plants in chromated copper arsenate (CCA) contaminated sites, Korea," *Chemosphere*, vol. 87, no. 8, pp. 872–878, 2012.

[16] W. Meng, Z. Wang, B. Hu, Z. Wang, H. Li, and R. C. Goodman, "Heavy metals in soil and plants after long-term sewage irrigation at Tianjin China: a case study assessment," *Agricultural Water Management*, vol. 171, pp. 153–161, 2016.

[17] A. Alishahi, H. Farahmand, N. Prieto, and D. Cozzolino, "Identification of transgenic foods using NIR spectroscopy: a review," *Spectrochimica Acta A: Molecular and Biomolecular Spectroscopy*, vol. 75, no. 1, pp. 1–7, 2010.

[18] L. E. Agelet, P. R. Armstrong, J. G. Tallada, and C. R. Hurburgh Jr., "Differences between conventional and glyphosate tolerant soybeans and moisture effect in their discrimination by near infrared spectroscopy," *Food Chemistry*, vol. 141, no. 3, pp. 1895–1901, 2013.

[19] P. M. Santos, E. R. Pereira-Filho, and L. E. Rodriguez-Saona, "Rapid detection and quantification of milk adulteration using infrared microspectroscopy and chemometrics analysis," *Food Chemistry*, vol. 138, no. 1, pp. 19–24, 2013.

[20] J. U. Porep, D. R. Kammerer, and R. Carle, "On-line application of near infrared (NIR) spectroscopy in food production," *Trends in Food Science & Technology*, vol. 46, no. 2, pp. 211–230, 2015.

[21] M. Schmutzler and C. W. Huck, "Simultaneous detection of total antioxidant capacity and total soluble solids content by Fourier transform near-infrared (FT-NIR) spectroscopy: a quick and sensitive method for on-site analyses of apples," *Food Control*, vol. 66, pp. 27–37, 2016.

[22] S. D. Afandi, Y. Herdiyeni, L. B. Prasetyo, W. Hasbi, K. Arai, and H. Okumura, "Nitrogen content estimation of rice crop based on near infrared (NIR) reflectance using artificial neural network (ANN)," *Procedia Environmental Sciences*, vol. 33, pp. 63–69, 2016.

[23] T. Kemper and S. Sommer, "Estimate of heavy metal contamination in soils after a mining accident using reflectance spectroscopy," *Environmental Science & Technology*, vol. 36, no. 12, pp. 2742–2747, 2002.

[24] D. Cozzolino, "Near infrared spectroscopy as a tool to monitor contaminants in soil, sediments and water—state of the art, advantages and pitfalls," *Trends in Environmental Analytical Chemistry*, vol. 9, pp. 1–7, 2016.

[25] L. Galvez-Sola, J. Morales, A. M. Mayoral et al., "Estimation of parameters in sewage sludge by near-infrared reflectance spectroscopy (NIRS) using several regression tools," *Talanta*, vol. 110, pp. 81–88, 2013.

[26] D. F. Malley, "Near-infrared spectroscopy as a potential method for routine sediment analysis to improve rapidity and efficiency," *Water Science and Technology*, vol. 37, no. 6-7, pp. 181–188, 1998.

[27] C. Palmborg and A. Nordgren, "Partitioning the variation of microbial measurements in forest soils into heavy metal and substrate quality dependent parts by use of near infrared spectroscopy and multivariate statistics," *Soil Biology and Biochemistry*, vol. 28, no. 6, pp. 711–720, 1996.

[28] R. G. Brereton, "Pattern recognition in chemometrics," *Chemometrics and Intelligent Laboratory Systems*, vol. 149, pp. 90–96, 2015.

[29] P. K. Hopke, "Chemometrics applied to environmental systems," *Chemometrics and Intelligent Laboratory Systems*, vol. 149, pp. 205–214, 2015.

[30] A. Savitzky and M. J. E. Golay, "Smoothing and differentiation of data by simplified least squares procedures," *Analytical Chemistry*, vol. 36, no. 8, pp. 1627–1639, 1964.

[31] R. J. Barnes, M. S. Dhanoa, and S. J. Lister, "Standard normal variate transformation and de-trending of near-infrared diffuse reflectance spectra," *Applied Spectroscopy*, vol. 43, no. 5, pp. 772–777, 1989.

[32] R. D. Snee, "Validation of regression models: methods and examples," *Technometrics*, vol. 19, no. 4, pp. 415–428, 1977.

[33] M. Barker and W. Rayens, "Partial least squares for discrimination," *Journal of Chemometrics*, vol. 17, no. 3, pp. 166–173, 2003.

[34] L. Xu, J.-H. Jiang, Y.-P. Zhou, H.-L. Wu, G.-L. Shen, and R.-Q. Yu, "MCCV stacked regression for model combination and fast spectral interval selection in multivariate calibration," *Chemometrics and Intelligent Laboratory Systems*, vol. 87, no. 2, pp. 226–230, 2007.

[35] B. M. Nicolaï, K. Beullens, E. Bobelyn et al., "Nondestructive measurement of fruit and vegetable quality by means of NIR spectroscopy: a review," *Postharvest Biology and Technology*, vol. 46, no. 2, pp. 99–118, 2007.

19

A Feasibility Study on the Potential Use of Near Infrared Reflectance Spectroscopy to Analyze Meat in Live Animals: Discrimination of Muscles

J. J. Roberts, J. Ch. Motin, D. Swain, and D. Cozzolino

School of Health, Medical and Applied Sciences, Central Queensland Innovation and Research Precinct (CQIRP), Central Queensland University (CQU), Bruce Highway, North Rockhampton, QLD 4701, Australia

Correspondence should be addressed to D. Cozzolino; d.cozzolino@cqu.edu.au

Academic Editor: Davide Ferri

Near infrared (NIR) spectroscopy has been proposed as a potential method to analyze different properties in live animals and humans, as infrared light has the ability to penetrate living tissues. This study evaluated the potential use of NIR spectroscopy to identify and analyze beef muscles through the skin nondestructively. The results from this study demonstrated that the NIR region has the potential to noninvasively monitor some properties of meat associated with either fat or muscle characteristics and to differentiate either muscle or fat tissue analyzed through the skin. At present, there are no rapid and noninvasive tools to monitor and assess any characteristic or property in live beef animals. Although these results look promising, more experiments and research need to be carried out before recommending the beef industry using this technology in live animals.

1. Introduction

Research in animal feeding and nutrition often requires the evaluation of live animals, postmortem carcass assessment, or meat composition to assess the nutritional value of feeds on animal performance [1, 2]. However, conventional analytical methods currently in use by the food and beef industries are usually destructive (e.g., postmortem evaluation), time consuming, and expensive and require of the use of different chemical and physical methods [3–5].

Near infrared (NIR) spectroscopy has been proposed as a potential method to analyze different properties in live animals and humans, as infrared light penetrates living tissues [4, 6–9]. This technique has been applied in the agricultural industries since the 1960s in routine measurements of agricultural commodities and has expanded rapidly due to its relatively low cost, speed of analysis, and the ability to assess various quality aspects simultaneously [10–12].

It has been reported that NIR spectroscopy can detect tissue oxygenation changes in the brain and muscle of humans where the detection of tissue oxygenation is possible because of the relative transparency of biological tissues to NIR wavelengths [9]. This type of application uses the oxygen-dependent absorption spectra of the main tissue chromophores—haemoglobin, myoglobin, and cytochrome c oxidase at different wavelengths in the NIR region [9]. It has been also reported that pigmented tissues can reduce the NIR signal intensity and since animals have both extensive hair covering and heavily pigmented epidermal tissues, these might interfere with the NIR spectra of the tissue analyzed [13–15].

Short wavelengths (SW) in the NIR (700 to 1100 nm; 13,900 to 9400 cm^{-1}) range can penetrate deeply into the skin, offering a potential spectral window for the analysis of animal and human tissues [2, 4, 16]. In recent years, researchers in Spain have demonstrated the potential of NIR spectroscopy as a very easy and quick tool for the quantitative and qualitative analysis of quality parameters in meat and as a tool for carcass or product grading [2, 4, 16]. In addition, these authors have indicated that in vivo or on-the-skin NIR measurements at the farm or slaughterhouse have shown a detrimental effect on the calibration performance (e.g., standard error of prediction) for the prediction of meat quality parameters compared to those developed using meat cuts or homogenised samples [4].

One of the main reasons for the low performance in the NIR measurements in living animals is related with the amount of incident light (e.g., reflectance). Using this methodology, light is highly scattered by the complex structure of the skin and its main layers (epidermis (blood-free layer), dermis (vascularised layer with dense irregular connective tissue with collagenous fibers), and hypodermis (subcutaneous adipose tissue layer composed of two sublayers separated by thin connective tissue)) [4, 17, 18]. In several in vivo or live applications of NIR spectroscopy, different studies have highlighted that the light has to travel through all these layers of the skin before reaching the layer of interest (e.g., muscle), where the subcutaneous adipose tissue influencing the collected reflectance signal from the sample [4, 17, 18]. Therefore, understanding the interactions between the skin and between the different tissues and the light propagation through different layers of the tissue will be essential for the correct interpretation of results obtained using this approach in live animals [4, 18]. Although reports can be found in the literature on the use of NIR spectroscopy to measure different properties of meat or fat composition in pigs through the skin, no information is available on the use of this technique to measure beef meat composition in live animals or on the evaluation of meat quality through the skin [3–5].

This study evaluated the potential use of NIR spectroscopy to analyze muscle and fat characteristics in beef cattle through the skin. The optical properties of tissues analyzed through the skin and the application of pattern recognition methods were used in order to demonstrate the ability of this method to analyze different tissues through the skin.

2. Materials and Methods

2.1. Samples. In order to study the optical properties or spectra of different tissues, two commercial meat cuts, namely, rump (*Gluteous medius*) and steak (*Semitendinosus* and *Semimembranosus*), were analyzed with and without the presence of the skin. Five combinations of tissue and the skin were analyzed as follows: fat, muscle, and the skin alone and either muscle or fat plus skin. A total of 90 samples were created by combining the skin, meat, and fat tissues and the two commercial meat cuts.

2.2. Near Infrared Reflectance Spectroscopy. The diffuse reflectance spectra of the samples were recorded from the flat surface using the purpose built contact probe attached by a fibre optic (10 mm diameter) cable to a Fourier transform (FT) NIR instrument (Antaris II, Thermo, USA). The instrument records spectra with resolution of 1 nm for the wavelength region 14,000–4000 cm^{-1} (700–2500 nm). Data collection and processing was achieved using the Thermo interface software enabling automation of data collection provided by the Antaris II instrument. The instrument was set to average 10 readings internally for each spectrum saved. A ceramic tile or reference panel provided by the

instrument manufacturer was used as a white reference between each measurement.

2.3. Chemometric Analysis. Spectra were exported from the Thermo software in GRAMS format (*.spc) into The Unscrambler software (version X CAMO ASA, Oslo, Norway) for chemometric analysis. Principal component analysis (PCA) was performed with full cross validation (leave one out). The optimum numbers of terms in the PCA models were indicated by the lowest number of factors that gave the minimum value of the prediction residual error sum of squares (PRESS) in cross validation in order to avoid over fitting in the models. Preprocessing was achieved using second derivative (Savitzky-Golay derivative, 20 data points smoothing and second polynomial order) [19, 20].

Discrimination models were developed using partial least squares discriminant analysis regression (PLS-DA). The PLS-DA regression technique is a variant of PLS regression, where for each class, a model ($C = T.q$), where T is the PLS scores obtained from the original data using the PLS algorithm, q is the vector, and C is the class membership function [21, 22]. In this study, PCA and PLS-DA models were developed using full cross validation (leave one out) and defined by the PRESS (prediction residual error sum of squares) function in order to avoid overfitting of the models (The Unscrambler, CAMO AS, version 10.3, Oslo, Norway).

3. Results and Discussion

Figure 1 shows the second derivative of the NIR mean spectrum of fat, muscle, fat plus skin, muscle plus skin, and skin. In the short NIR wavelength region, differences were observed around 10,750 cm^{-1} and 10,288 cm^{-1}, associated with water (O-H bonds) [23, 24]. Around 11,900 cm^{-1}, 11,655 cm^{-1}, 10,929 cm^{-1}, and 9911 cm^{-1} wavenumbers, values can be related with differences in water (O-H bonds) content, as well associated with C-H$_2$ third overtones [23, 24]. Absorption bands between 10,929 and 10,800 cm^{-1} can be associated with lipid content of the tissue analyzed where additional bands associated with the CH stretch overtones were observed between 11,655 and 10,929 cm^{-1} [23, 24]. Absorption bands between 10,400 and 9911 cm^{-1} are mainly associated with water (O-H bonds) [23, 24].

The most relevant changes were observed in the long NIR wavelength region where differences were observed at 7100 cm^{-1} (O-H and C-H), 5263 cm^{-1} (O-H, mainly related with water/moisture), and between 4762–4350 cm^{-1} (CH combination tones). The same wavenumber range (4762–4350 cm^{-1}) was reported by other authors to be associated with cartilage when pure bovine cartilage samples were analyzed using NIR spectroscopy [25–27]. These wavenumbers were associated with C-H combination and asymmetric first overtones of C-N-C stretch and N-H of proteins [25–27]. This region appears to be unique between the different tissues (e.g., fat, muscle, and skin) analyzed, indicating that differences in the absorbance values can be attributed to different protein and collagen ratios between tissues. In the same region, shifts were also observed around 4329 cm^{-1} (e.g., lipids, collagen, and proteoglycans) and between 4762–

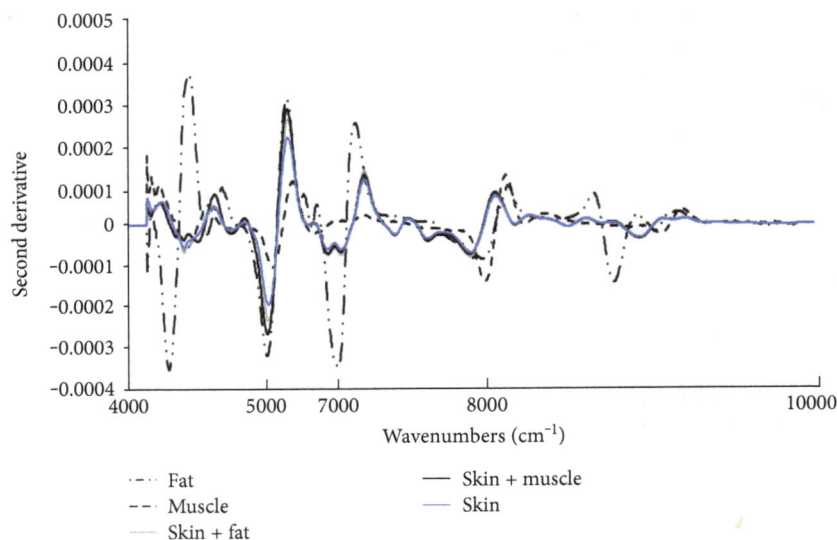

FIGURE 1: Second derivative of the average of the near infrared mean spectrum of skin, muscle, fat, skin plus fat, and skin plus muscle.

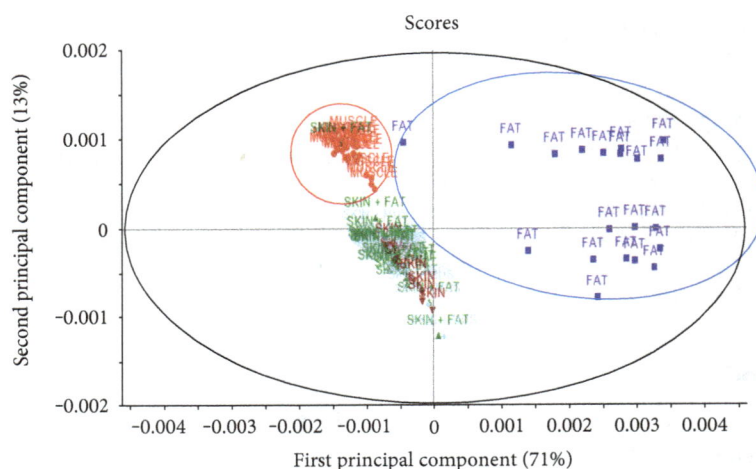

FIGURE 2: Score plot of fat, muscle, skin, skin plus fat, and skin plus muscle samples analyzed using near infrared spectroscopy.

$4350\,cm^{-1}$ (N-H and O-H) [6, 17, 23–27]. Overall, the visual inspection of the wavenumber region between 5263 and $4175\,cm^{-1}$ indicated that long wavelengths in the NIR region can be useful to identify differences between tissues. However, the use of NIR spectroscopy to analyze in vivo samples or live animals might be limited to the NIR spectral range between 5200 and $4370\,cm^{-1}$ due to the large effect of water ($5263\,cm^{-1}$). Other authors reported similar results when different animal tissues were analyzed using NIR spectroscopy [6, 17, 27, 28].

In order to further test and evaluate the interactions between different tissues (skin, muscle, and fat) the NIR spectra were analyzed using PCA (see Figure 2). Separation between "fat samples" and the other samples (skin and skin + muscle) was observed along PC1, while separation between muscles was achieved in PC2. The amount of variation explained for each PC was 71% and 13% for PC1 and

PC2, respectively. Examination of the PCA loadings is very important as they might indicate specific wavelengths or regions in the NIR range related with the variability observed in the PCA score plot. Figure 3 shows the PCA loadings of the first two principal components (PCs) associated with the PCA shown in Figure 2. Wavenumbers associated with lipids and moisture content were the most relevant in explaining the separation between samples [23, 24]. Clear differences between the loadings in PC1 and PC2 were observed, indicating that different spectral information might be associated with the different tissues or combination of samples analyzed (see Figures 4 and 5).

The PLS-DA models were used to further evaluate the ability of NIR spectroscopy to identify different tissue types analyzed through the skin (see Figure 6). Even though the classification results obtained were between 60 and 70% (correct classification), the overall results indicated that the

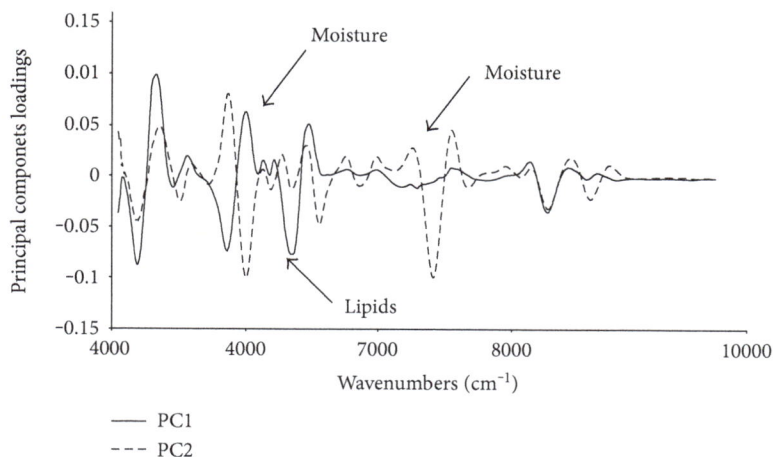

FIGURE 3: Loadings derived from the PCA used to analyze fat, muscle, skin, skin plus fat, and skin plus muscle samples analyzed using near infrared spectroscopy.

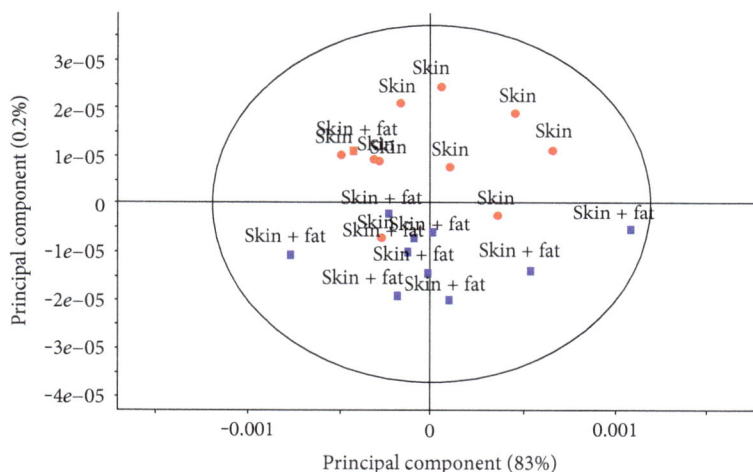

FIGURE 4: Score plot of fat and skin plus fat in the meat samples analyzed using near infrared spectroscopy.

NIR method evaluated in this study is capable to identify different tissues through the skin.

Zamora-Rojas et al. [4] showed that information related with the optical properties of pigskin (epidermis and dermis) might be limited to the short NIR wavelength range (below $6250 \, cm^{-1}$) [29]. However, other studies indicated that the optical properties of dermis tissues from pigs can be observed in the $10{,}929$–$6850 \, cm^{-1}$ NIR range, while studies of epidermis and dermis using wavenumbers between $10{,}000$ and $6200 \, cm^{-1}$ were reported by other authors using the same animal species [30–32]. Other characteristics of the skin such as the presence of the skin can also contribute to the optical properties of the system. Hair structure, type, and pigmentation are important factors that determine the coloration of a species, which influences the spectral reflectance and absorption for any given species [13, 15]. Most terrestrial mammals have two hair types, namely, guard hairs and underfur [13, 15]. Guard hairs are typically longer

and thicker and have a complex physical structure. Underfur is short, fine, and dense, with a simple physical structure and little variability in coloration [13, 15].

It is well known that light penetration depth in a specific sample is a function of the geometry of the optical probe, the scattering and the absorption characteristics of the sample [33, 34]. Typical penetration depth within biological tissues is at least 10 mm or longer depending upon the tissue studied [34]. When light enters the tissues, it is either absorbed or scattered, based on wavelength and differences in chemical composition of the tissue [33, 34]. Since different types of molecules vibrate at different frequencies, different compounds in the tissue can be identified based on the frequencies of light that are detected by the instrument. For example, in the $13{,}900$–$9386 \, cm^{-1}$ wavenumber range, second and third overtones of the OH, CH, and NH vibrational transitions are detected in addition to the combination bands from other types of vibration [23, 33]. The peak at

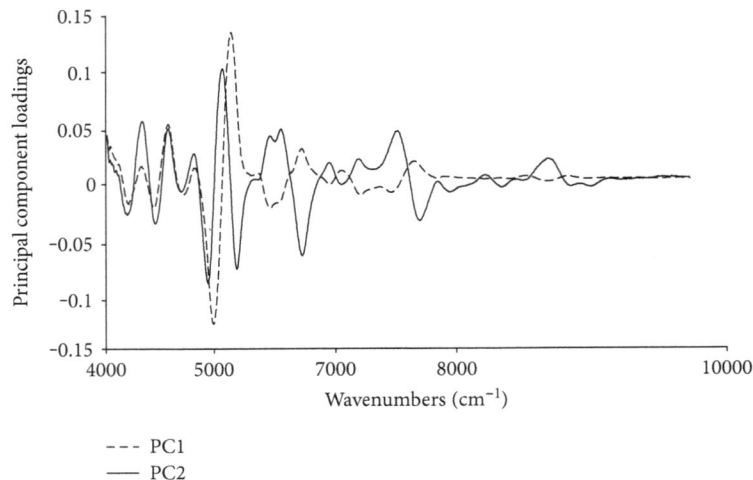

FIGURE 5: Loadings derived from the PCA used to analyze fat and skin plus fat in the meat samples analyzed using near infrared spectroscopy.

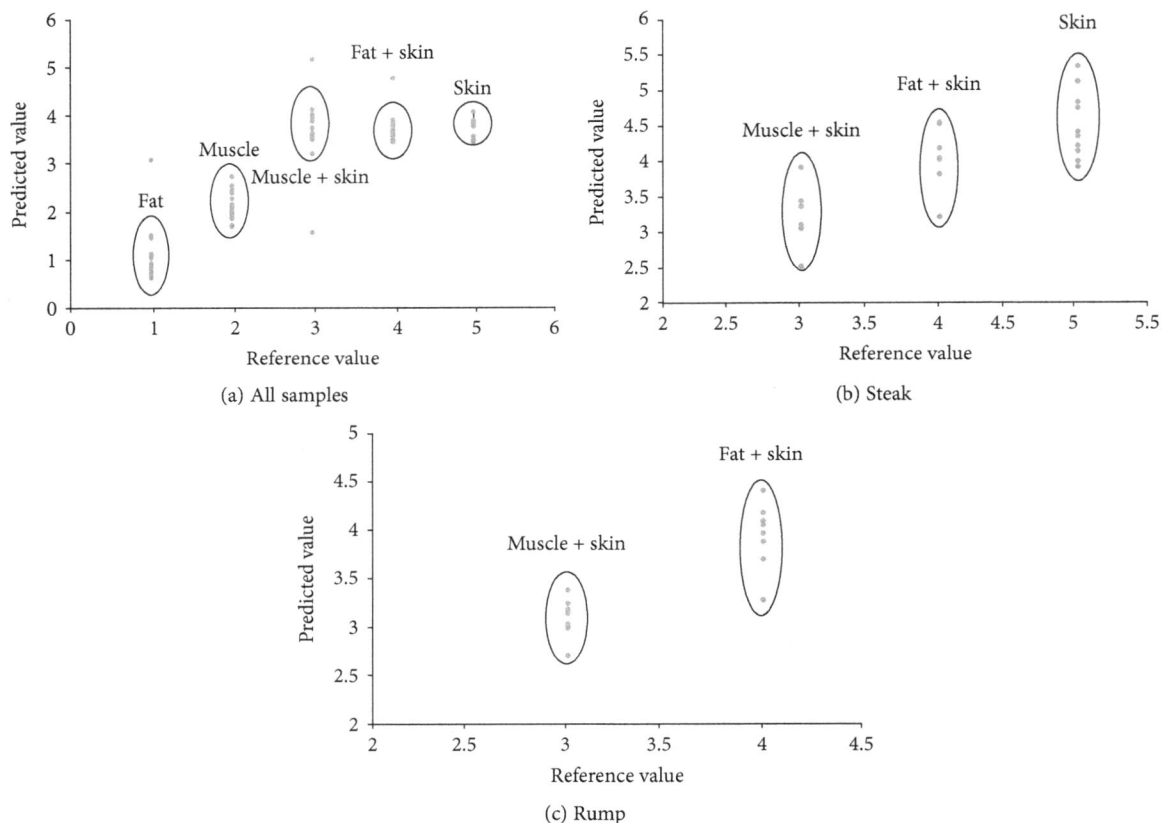

(a) All samples

(b) Steak

(c) Rump

FIGURE 6: Partial least squares discriminant analysis of different types of beef tissues analyzed using near infrared reflectance spectroscopy.

$10,750\,cm^{-1}$, which predominates on the fat spectrum, corresponds to that reported for fat, while peaks in the 10,400 to $9911\,cm^{-1}$, region for the other tissues corresponds to that reported for water [33, 35]. These authors indicated that by dissecting and reconstructing the tissue layers of varying thickness, a change in composition could be detected by NIR spectroscopy to a depth of approximately 1 cm. For the estimation of body composition in pigs, the spectrum for

the tissue composite appears to be influenced predominately by the composition of the skin layer and to a lesser extent by the presence of the fat and muscle layers [33, 36].

4. Conclusions

The results from this study demonstrated that the NIR region has the potential to monitor some beef meat properties associated with fat or muscle noninvasively through the skin. The proposed method would provide beef producers or industry with a powerful tool to monitor more closely the effects of health, nutrition, or product characteristics with the goal of producing a consistently uniform product and maximize profits assuring consumer with a product of high quality. Although these results look promising, more experiments and research need to be carried out before recommending the beef industry to use this technology in live animals.

References

[1] D. Alomar, C. Gallo, M. Castaneda, and R. Fuchslocher, "Chemical and discriminant analysis of bovine meat by near infrared reflectance spectroscopy (NIRS)," *Meat Science*, vol. 63, pp. 441–450, 2003.

[2] N. Prieto, R. Roehe, P. Lavín, G. Batten, and S. Andrés, "Application of near infrared reflectance spectroscopy to predict meat and meat products quality: a review," *Meat Science*, vol. 83, pp. 175–186, 2009.

[3] D. Pérez-Marín, E. De Pedro, J. E. Guerrero-Ginel, and A. Garrido-Varo, "A feasibility study on the use of near-infrared spectroscopy for prediction of the fatty acid profile in live Iberian pigs and carcasses," *Meat Science*, vol. 83, pp. 627–633, 2009.

[4] E. Zamora-Rojas, B. Aernouts, A. Garrido-Varo, D. Pérez-Marín, J. E. Guerrero-Ginel, and W. Saeys, "Double integrating spheres measurements for estimating optical properties of pig subcutaneous adipose tissue," *Innovative Food Science and Emerging Technologies*, vol. 19, pp. 218–226, 2013.

[5] E. Zamora-Rojas, A. Garrido-Varo, E. De Pedro-Sanz, J. E. Guerrero-Ginel, and D. Pérez-Marín, "Prediction of fatty acid content in pig adipose tissue by near infrared spectroscopy: at-line versus in-situ analysis," *Meat Science*, vol. 95, no. 3, pp. 503–511, 2013.

[6] H. Azizian, J. K. G. Kramer, S. B. Heymsfield, and S. Winsborough, "Fourier transform near infrared spectroscopy: a newly developed, non-invasive method to measure body fat non-invasive body fat content measurement using FT-NIR," *Lipids*, vol. 43, pp. 97–103, 2008.

[7] A. N. Bashkatov, E. A. Genina, V. I. Kochubey, and V. V. Tuchin, "Optical properties of human skin, subcutaneous and mucous tissues in the wavelength range from 400 to 2000 nm," *Journal of Physics D: Applied Physics*, vol. 38, pp. 2543–2555, 2005.

[8] A. N. Bashkatov, E. A. Genina, V. I. Kochubey, and V. V. Tuchin, "Optical properties of the subcutaneous adipose tissue in the spectral range 400–2500 nm," *Optics and Spectroscopy*, vol. 99, no. 5, pp. 836–842, 2005.

[9] S. L. Jacques, "Optical properties of biological tissues: a review," *Physics in Medicine and Biology*, vol. 58, pp. R37–R61, 2013.

[10] E. Dufour, *Principles of Infrared Spectroscopy*, Academic Press, Elsevier Science & Technology Books, San Diego, CA, USA, 2008.

[11] B. G. Osborne, T. Fearn, and T. Davies, "Practical NIR spectroscopy with applications in food and beverage analysis," in *Food and Beverage Analysis*, B. G. Osborne, T. Fearn and P. T. Hindle, Eds., pp. 11–35, Longmand, Essex, 1993.

[12] J. Workman, "A review of process near infrared spectroscopy: 1980-1994," *Journal of Near Infrared Spectroscopy*, vol. 1, no. 4, p. 16, 1993.

[13] C. D. Hutchinson, T. E. Allen, and F. B. Spence, "Measurement of the reflectance for solar radiation of the coats of live animals," *Comparative Biochemistry and Physiology*, vol. 52A, pp. 343–349, 1975.

[14] J. Pringle, C. Roberts, M. Kohl, and P. Lekeux, "Near infrared spectroscopy in large animals: optical path length and influence of hair covering and epidermal pigmentation," *The Veterinary Journal*, vol. 158, pp. 48–52, 1999.

[15] P. Terletzky, R. D. Ramsey, and C. M. U. Neale, "Spectral characteristics of domestic and wild mammals," *GIScience and Remote Sensing*, vol. 49, pp. 597–608, 2012.

[16] M. Prevolnik, M. Candek-Potokar, and D. Skorjank, "Ability of NIR spectroscopy to predict chemical composition and quality – a review," *Czech Journal of Animal Science*, vol. 49, pp. 500–510, 2004.

[17] H. Ding, J. Q. Lu, K. M. Jacobs, and X. Hu, "Determination of refractive indices of porcine skin tissues and intra lipid at eight wavelengths between 325 and 1557 nm," *Journal of the Optical Society of America*, vol. 22, no. 6, pp. 1151–1157, 2005.

[18] D. Renaudeau, M. Leclercq-Smekens, and M. Herin, "Differences in skin characteristics in European (large white) and Carribbean (Creole) growing pigs with reference to thermoregulation," *Animal Research*, vol. 55, pp. 207–219, 2006.

[19] W. Hruschka, *Data Analysis: Wavelength Selection Methods*, American Association of Analytical Chemists, 2nd edition, 2001.

[20] A. Savitzky and M. J. E. Golay, "Smoothing and differentiation of data by simplified least squares procedures," *Analytical Chemistry*, vol. 36, pp. 1627–1632, 1964.

[21] R. G. Brereton, *Applied Chemometrics for Scientist*, Wiley, 2008.

[22] R. G. Brereton, "Introduction to multivariate calibration in analytical chemistry," *The Analyst*, vol. 125, pp. 2125–2154, 2000.

[23] L. G. Weyer, "Near-infrared spectroscopy of organic substances," *Applied Spectroscopy Reviews*, vol. 21, pp. 1–43, 1985.

[24] J. Workman and L. Weyer, *Practical Guide to Interpretive near-Infrared Spectroscopy*, CRC Press Taylor and Francis Group, Boca Raton, 2008.

[25] C. P. Brown, C. Jayadev, S. Glyn-Jones et al., "Characterization of early stage cartilage degradation using diffuse reflectance near infrared spectroscopy," *Physics in Medicine and Biology*, vol. 56, pp. 2299–2307, 2011.

[26] C. P. Brown, A. Oloyede, R. W. Crawford, G. E. Thomas, A. J. Price, and H. S. Gill, "Acoustic, mechanical and near-infrared profiling of osteoarthritic progression in bovine joints," *Physics in Medicine and Biology*, vol. 57, pp. 547–559, 2012.

[27] U. P. Palukuru, C. M. McGoverin, and N. Pleshko, "Assessment of hyaline cartilage matrix composition using near infrared spectroscopy," *Matrix Biology*, vol. 38, pp. 3–11, 2014.

[28] M. V. Padalkar, R. G. Spencer, and N. Pleshko, "Near infrared spectroscopic evaluation of water in hyaline cartilage," *Annals of Biomedical Engineering*, vol. 41, no. 11, pp. 2426–2436, 2013.

[29] A. N. Bashkatov, E. A. Genina, and V. V. Tuchin, "Optical properties of skin, subcutaneous and muscle tissues: a review," *Journal of Innovative Optical Health Sciences*, vol. 4, no. 1, pp. 9–38, 2011.

[30] A. Cain, T. Milner, S. Telenkov, K. Schuster, K. Stockton, and D. Stolarski, "Porcine skin thermal response to near-IR lasers using a fast infrared camera," in *Laser Interaction with Tissue and Cells*, S. L. Jacques and P. W. Roach, Eds., pp. 313–324, SPIE, Bellingham, WA, 2004.

[31] Y. Du, X. H. Hu, M. Cariveau, X. Ma, G. W. Kalmus, and J. Q. Lu, "Optical properties of porcine skin dermis between 900 nm and 1500 nm," *Physics in Medicine and Biology*, vol. 46, pp. 167–181, 2001.

[32] X. Ma, J. Q. Lu, H. Ding, and X. H. Hu, "Bulk optical parameters of porcine skin dermis at eight wavelengths from 325 to 1557 nm," *Optics Letters*, vol. 30, no. 4, pp. 412–414, 2005.

[33] F. Martelli, "An ABC of near infrared photon migration in tissues: the diffuse regime of propagation," *Journal of Near Infrared Spectroscopy*, vol. 20, pp. 29–42, 2012.

[34] I. Murray and I. Cowe, "Sample preparation in near infrared spectroscopy in agriculture," American Society of Agronomy, Crop Science Society of America, Soil Science Society of America, Madison, WI, USA, 2004.

[35] J. M. Conway, K. H. Norris, and C. E. Bodwell, "A new approach for the estimation of body composition: infrared interactance," *The American Journal of Clinical Nutrition*, vol. 40, pp. 1123–1130, 1984.

[36] A. D. Mitchell, A. M. Scholz, and M. B. Solomon, "Estimation of body composition of pigs by a near infrared interactance probe technique," *Archiv Tierzucht Dummerstorf*, vol. 48, pp. 580–591, 2005.

20

Using a Spectrofluorometer for Resonance Raman Spectra of Organic Molecules

Vadivel Masilamani,[1,2] Hamid M. Ghaithan,[1] Mamduh J. Aljaafreh,[1] Abdullah Ahmed,[1] Reem al Thagafi,[1] Saradh Prasad,[1,2] and Mohamad S. Alsalhi[1,2]

[1]Department of Physics and Astronomy, College of Science, King Saud University, P.O. Box 2455, Riyadh 11451, Saudi Arabia
[2]Research Chair on Laser Diagnosis of Cancers, Department of Physics and Astronomy, College of Science, King Saud University, Riyadh 11451, Saudi Arabia

Correspondence should be addressed to Vadivel Masilamani; masila123@gmail.com

Academic Editor: Jau-Wern Chiou

Scattering (Rayleigh and Raman) and fluorescence are two common light signals that frequently occur together, confusing the researchers and graduate students experimenting in molecular spectroscopy laboratories. This report is a brief study presenting a clear discrimination between the two signals mentioned, employing a common spectrofluorometer such as the PerkinElmer LS 55. Even better, the resonance Raman signal of a molecule (e.g., acetone) can be obtained elegantly using the same instrument.

1. Introduction

When photons fall on organic molecules (benzene, acetone, coumarin, etc.), some get scattered, while others are absorbed. Among the scattered signals, some have the same wavelength of the incident photon (termed Rayleigh scattering signals), some have wavelength longer than the incident light (called Raman Stokes signal), and only very few photons have a shorter wavelength (called Raman anti-Stokes signal) [1].

Some molecules which absorb the photons produce fluorescence; this is a reemission process, a phenomenon from one quantum state to another one. A molecule (i.e., acetone) is at the ground state electronic level S_0, which consists of manifold vibro-roto energy sublevels (see Figure 1(a)). When a photon of suitable wavelength interacts with it, and gets absorbed, the molecule goes from the S_0, $v = 0$, level to the electronically excited state S_1, $v = 1, 2, 3$, level depending upon the Franck-Condon factor [2]. The molecule then undergoes thermalization by colliding with other molecules or the container walls and comes down to the S_1, $v = 0$, level, which is the lowest vibrational state of the first excited state S_1. It remains here for 5 to 10 ns and then goes down to the lower ground electronic state S_0 and any one of the vibrational states $v = 1, 2, 3$, and so on, once again determined by the Franck-Condon factor. During this process, it may emit a photon (fluorescence) or may not do so (internal conversion). As the molecule has lost a substantial fraction of its excitation energy in the S_1 due to thermalization, the fluorescence occurs with longer wavelengths (or lower energy) than the incident light. As the S_1 and S_0 energy levels are quasi continuum, absorption and fluorescence can occur across a range of wavelengths (about 50–100 nm); also from the time of absorption, all the thermalizations occur in a picosecond time scale, while the fluorescence occurs in a nanosecond time scale, always starting from S_1, $v = 0$, independent of excitation wavelength; in other words, the fluorescence spectral profile is independent of the excitation wavelength, although the intensity is dependent. This is the Kasha-Vavilov rule [2].

Such absorption and reemission (fluorescence) are reasonably strong only in a few specific set of molecules, like dyes, which have a chain of alternate single and double levels, called conjugation. Most spectrofluorometers, commonly employed in research labs, are used only for the studies mentioned above.

In contrast, scattering is generally nonspecific, implying that light of any wavelength can interact with any molecule

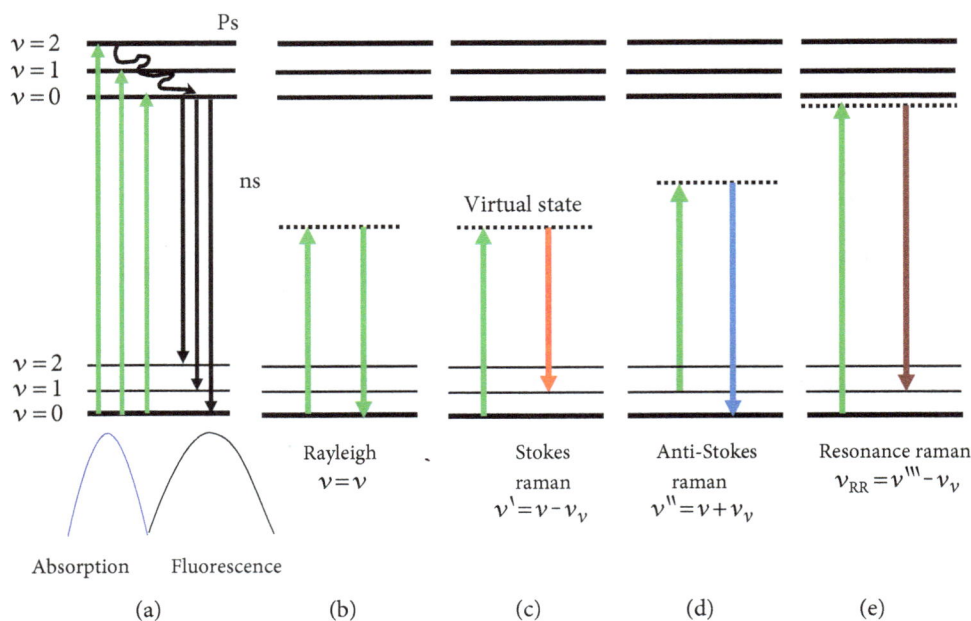

FIGURE 1: Schematic of energy level diagram of a molecule. S_0 and S_1 are electronic energy levels of a molecule, with vibration levels $v = 0, 1, 2$. Rotational levels are within each vibrational level, but not shown here. (a) For Rayleigh scattering, (b) for Raman scattering Stokes, (c) for Raman scattering anti-Stokes, and (d) for resonance Raman scattering.

and undergo scattering. In this instance, if a photon of frequency v interacts with a molecule in the S_0, $v = 0$, state, the molecule goes to a virtual state, a nonexistent state represented by a dotted line (Figure 1(b)) and comes back very quickly (in a picosecond time scale) to the S_0, $v = 0$, state. In this case, the photon does not lose any energy. The scattered photon has the same frequency v as the incident one. This is called Rayleigh scattering. However, if the molecule returns to the S_1, $v = 1$, state, (a vibrationally excited level) the scattered photon has less energy ($v' = v - v_v$, where v_v is the frequency of the vibration of the molecule). This happens only occasionally; therefore, the scattered radiation of frequency $v' = v - v_v$ is rather difficult to detect (see Figure 1(c)). There is one more possibility, the molecule may already be in a vibrationally excited state (S_0, $v = 1$), and the incident photon of v can move it to another virtual level (Figure 1(d)); when this comes down to the S_0, $v = 0$, level, the scattered photon has a higher energy and frequency of $v'' = v + v_v$. The extra energy has been conferred by the molecule. As this is a very rare occurrence, the radiation with $v'' = v + v_v$ is even more difficult to detect. Roughly, as a rule of thumb, if 100 photons produce Rayleigh scattering, only 10 will produce $v' = v - v_v$ (called Stokes Raman) and only two will produce $v' = v + v_v$ (called anti-Stokes Raman). An observed Indian scientist, Sir CV Raman, the first to detect such molecular scattering of radiation with quantized energy v, $v \pm v_v$, was awarded the Nobel Prize in 1930.

It is very significant that for a given molecule, both types of scattering can be observed by employing UV, VIS, or IR radiation, because the virtual level can be found anywhere. On the other hand, the shift in the scattered radiation $v - v'' = \pm v_v$ is the characteristic fingerprint of the molecule, as its unique vibrational frequency is dependent on the

configuration of the molecule. The whole of Raman spectroscopy, as an analytical tool, rests on this concept.

Scattering is always from a real quantum state to a virtual state whereas fluorescence is a phenomenon that occurs between two quantum states; hence, the latter is 10–100 times stronger than the former.

In order to obtain the Raman signal reasonably comparable to the fluorescence signal, one must resort to resonance Raman scattering [3, 4]. In this instance, one must launch a radiation of wavelength v that would raise the molecule from the real quantum state S_0, $v = 0$, to a virtual state that almost overlaps the real quantum state of S_1, $v = 0$, as shown in (Figure 1(e)). This could be done by groping and exploring the region between the absorption and fluorescence bands as evident from the following experiment.

Many new reviews shows ever increasing applications of Raman scattering in fields such as material science from bulk to nano. In nanotechnology, new techniques such as nano-Raman techniques based on aperture scanning near-field optical microscopy (SNOM) are maturing and diversifying [5], which enables researchers to do Raman studies at nano scales. In chemistry, it gives a thumbprint to identify the molecules [6–8]. The Raman spectroscopy has immense application in other fields such as clinical spectroscopy [9], biological samples analysis [10], and clinical instrumentation [11]. Resonance Raman scattering (RRS) is an important technique and is found in many applications [12].

2. Experiment

The PerkinElmer model LS 55 was employed throughout this experiment; however, any other commonly available

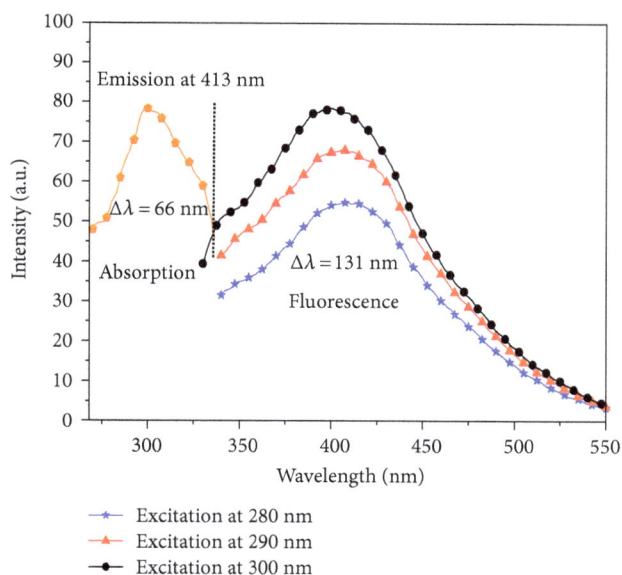

FIGURE 2: The fluorescence spectra of acetone with a peak at 413 nm for different wavelengths of excitation; on its left side is the excitation spectra with a peak at 327 nm.

spectrofluorometer (Horiba, Hitachi, or Shimatsu), with dual gratings, one for excitation wavelength selection and another for emission band scanning, will be equally suitable.

2.1. Fluorescence Spectra. Acetone (analytical grade) quality was taken in a transparent quartz cuvette of 1 cm path length and loaded into the cell compartment of the instrument. Excitation wavelength was arbitrarily selected as 330 nm; width was 10 nm in excitation as well as emission slits. The emission (synonymous with fluorescence in this case) spectrum was scanned from 350 nm to 600 nm and is presented in Figure 2, which shows a smooth band of emission with a peak at 413 nm of intensity 120 in the arbitrary units. This has a full width at half maximum (FWHM) of 102.4 nm. Two more emission spectra were taken with excitation at 280 and 290 nm, respectively. In both cases, the emission bands were similar, the only difference being the emission intensity was low. This confirms Kasha's rule (discussed earlier).

2.2. Excitation Spectra. In this case, the emission grating was fixed at 413 nm, which was the emission peak of acetone and the excitation grating was rotated from 250 nm to 390 nm to obtain the excitation band of acetone, which had a peak at 327 nm with FWHM of 70 nm. This corresponded to the S_0-S_1 absorption spectrum of acetone, obtainable from any UV-VIS spectrophotometer.

The excitation band indicates that if the excitation had been induced at 327 nm, the emission peak would occur at 413 nm, with the spectral profile of emission band becoming a mirror image of the excitation band, with the dotted line acting as the mirror as shown in Figure 2. For the other wavelengths of excitation, the emission intensity would be markedly low.

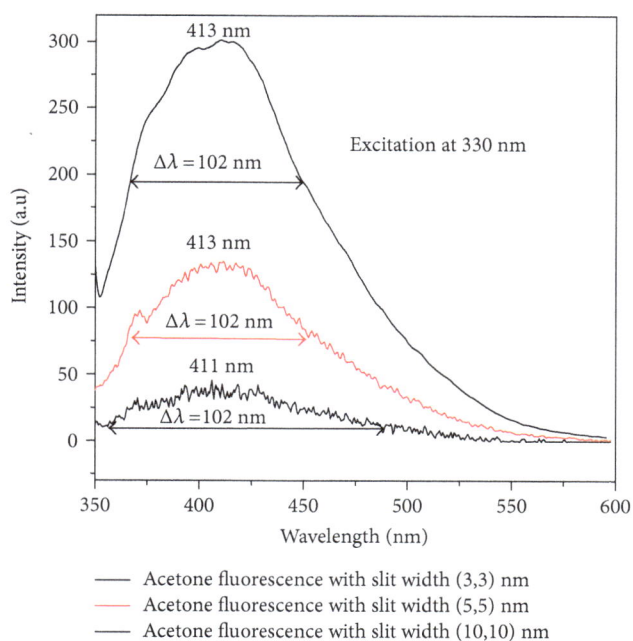

FIGURE 3: Acetone fluorescence with different slit widths.

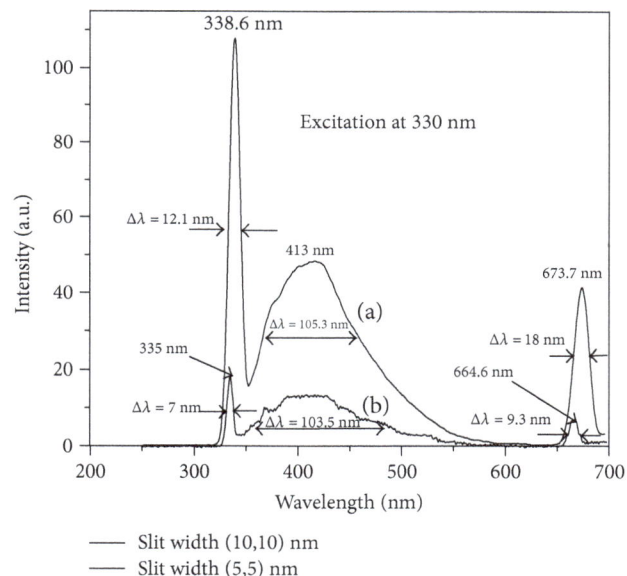

FIGURE 4: Rayleigh scattering and fluorescence signal with different slit widths (for acetone). (a) Black line and (b) blue line.

2.3. Fluorescence Spectra for Different Slit Widths. Utilizing the same acetone sample and excitation at 330 nm, emission spectra were obtained for 10, 5, and 3 nm spectral widths, in excitation as well as emission slits. The spectra in Figure 3 shows the same FWHM of 102 nm in each case; only the intensity has decreased twice or thrice. All the features mentioned above would be markedly different for scattering spectra.

2.4. Rayleigh Scattering. To obtain the Rayleigh scattering spectra of acetone, the excitation grating was fixed at 330 nm;

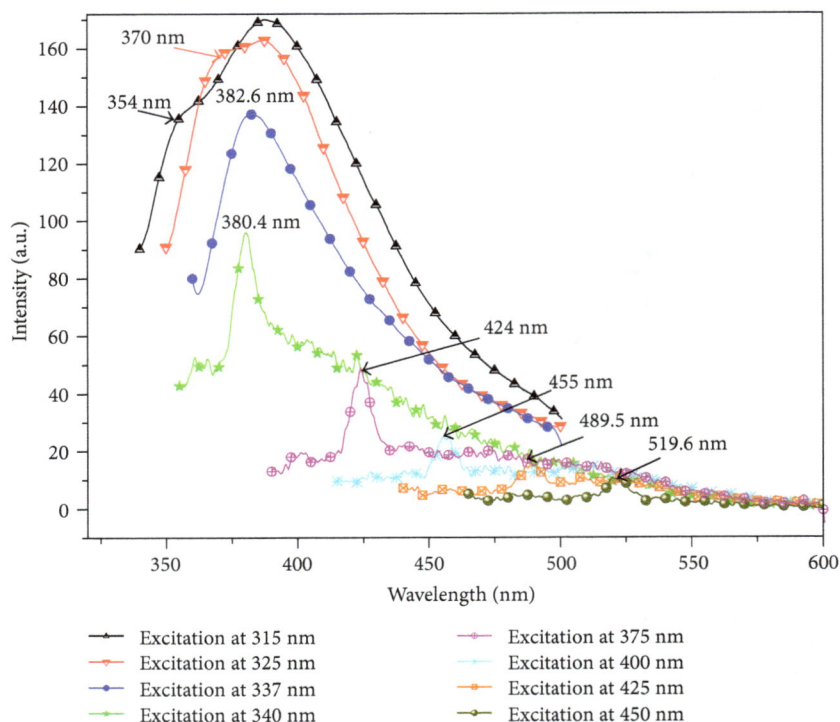

FIGURE 5: Resonance Raman spectra for acetone.

TABLE 1: Resonance Raman spectra of acetone.

Number	Excitation wavelength λ_e (nm)	Excitation wave number ν_e (cm^{-1})	Raman signal In wavelength λ_R (nm)	Raman signal In wave number ν_R (cm^{-1})	Raman shift ν_ν (cm^{-1})	Intensity (a.u.)
1	450	22222	519.6	19246	2976	10.7
2	425	23529	489.5	20429	3100	17
3	400	25000	455	21978	3022	25.5
4	375	26667	424	23585	3082	49
5	340	29412	380.4	26288	3124	95
6	337	29673	382.6	26136	3537	138
7	325	30769	370	27027	3742	158
8	315	31746	354	28248	3498	135.3
Average					3260 ± 10	

the slit width was 10 nm in each slit and the emission grating was scanned from 200 nm to 700 nm. The spectrum, as shown in Figure 4(a), indicates a sharp peak at 338 nm with FWHM = 12 nm, another broad band at 410 nm, and the next sharp peak at 673 nm (FWHM = 18 nm).

The first band, without experiencing any change in wavelength, is the Rayleigh scattering (please note, it should have occurred at 330 nm but instead was seen at 338 nm due to instrumental artifacts); the 410 nm band is the fluorescence band of acetone; the band at 673 nm is due to the Rayleigh scattering occurring in the second order of grating (this is an experimental unavailable artifact; if only a prism had been used instead of grating for dispersion, it would have disappeared).

For the same sample and scattering setup arrangement, the experiment was repeated with slit widths = 5 nm in each.

As shown in Figure 4(b), the spectra of Figure 4(a and b) are very similar barring two major differences. The intensity in all the bands had decreased (as expected); but the FWHM of Rayleigh scattering also dropped from 12 nm to 7 nm. The fluorescence spectra of acetone, however, continued to remain at 102 nm. This is the first major difference between scattering and fluorescence.

Scattering is, in fact, an irregular reflection at the molecular surface, very similar to a tennis ball bouncing off a racquet. What comes, goes off in a different direction, with minimal or no interaction. In contrast, during fluorescence, an incident photon is absorbed, and the energy gets thermalized in a few picoseconds and is then given off by the molecule, with its fingerprint, within the time span of a few nanoseconds.

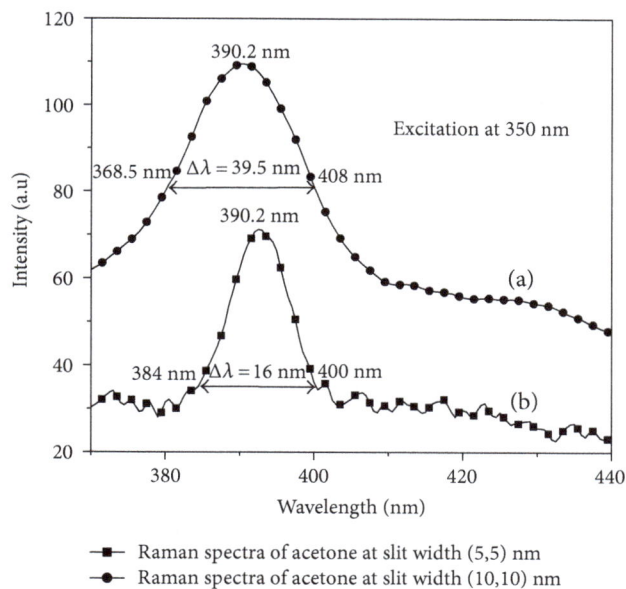

FIGURE 6: Raman spectra of acetone at different slit widths. (a) Blue line and (b) black line.

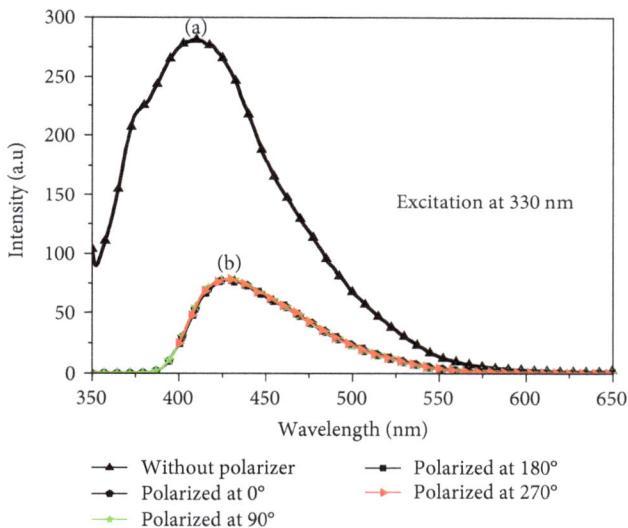

FIGURE 7: Acetone fluorescence spectra for different positions of polarizer.

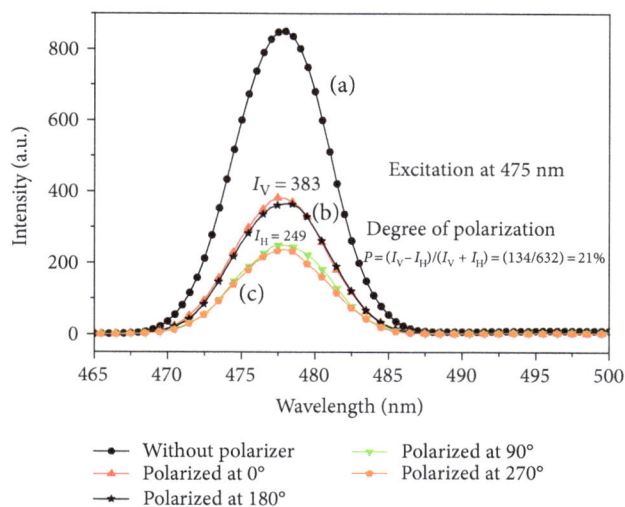

FIGURE 8: Polarization of Rayleigh scattering of acetone. (a) Black line; (b) blue and red line; (c) green and orange line.

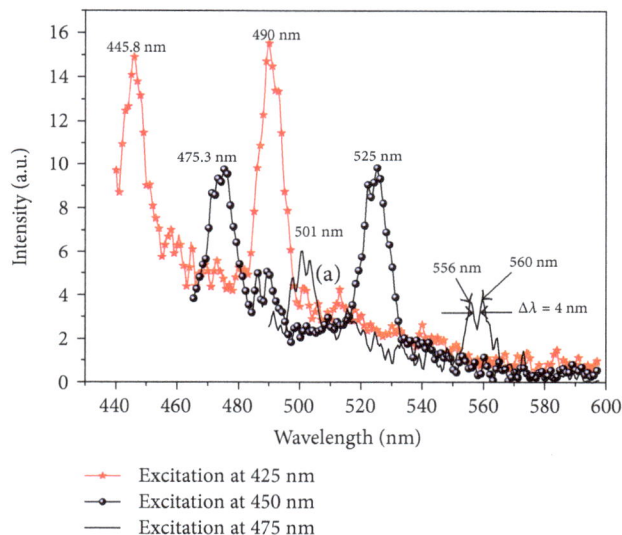

FIGURE 9: Raman spectra of benzene at different excitation wavelengths. (a) is black line.

2.5. Resonance Raman Scattering.

For measuring the Raman spectral features of acetone, the excitation wavelength was fixed at 450 nm and the emission grating was scanned from 485 nm to 600 nm. As evident from Figure 5, a small sharp signal was observed at 519.6 nm, corresponding to the Raman shift of $(1/450 \text{ nm} - 1/519.6 \text{ nm}) = 2976 \text{ cm}^{-1}$. This 2976 cm^{-1} represents the C-H bond vibrational frequency of the acetone molecule which causes the Raman shift.

To confirm this, the experiment was repeated with the excitation at 425 nm and scanning was done from 440 nm to 600 nm. The Raman signal now appeared at 489.5 nm and the Raman shift was $(1/425 \text{ nm} - 1/489.5 \text{ nm}) = 3100 \text{ cm}^{-1}$.

This implies that the Raman shift is a constant and characteristic of the molecule, but the Raman signal moves with the incident light. The experiment was repeated by shifting the excitation to 400, 375, and 340 nm. In each case, the Raman scattered signal got shifted to 455, 424, and 380 nm; but the shift in their wave number approximately corresponded to 3060 cm^{-1}, confirming the Raman effect.

The most significant feature of this part of the experiment is as follows: as the excitation was moved from 450 nm to 315 nm, over a range of 135 nm, the Raman signal intensity changed from 9 to 160 (in arbitrary unit), an increase by 17 times. Apparently, 340 nm represents the near resonance Raman spectra of acetone. When the excitation goes to 325 nm, the Raman signal occurring at 350 nm was

TABLE 2: Raman spectra for benzene.

Number	Excitation wavelength λ_e (nm)	Excitation wave number ν_e (cm^{-1})	Raman signal In wavelength λ_R (nm)	Raman signal In wave number ν_R (cm^{-1})	Raman shift ν_ν (cm^{-1})	Intensity (a.u.)
1	475	21053	501	19960	1093	6
			556	17986	3067	3.7
2	450	22222	475.3	21039	1183	9.9
			525	19048	3174	9.8
3	425	23529	445.8	22472	1057	14.9
			490	20408	3121	15.5

comparable in intensity to the fluorescence at 413 nm, representing the resonance Raman excitation; for excitation at 315, the Raman signal occurred at 352 nm, with the intensity of 120. For excitations with a wavelength lower than this, the acetone produced only a fluorescent broad band with a peak of 413 nm, as discussed earlier. Or, in other words, the best region to obtain resonance Raman is the valley between the excitation and emission bands.

Table 1 summarizes the results mentioned above. (Note also, the Raman shift gradually changed from 2970 to 3500 cm^{-1} as the excitations moved from longer to shorter wavelengths due to certain molecular properties.)

2.6. Raman Spectra for Different Slit Widths. The next experiment was performed with reduced slit widths. Figure 6 shows the Raman spectra with excitation at 350 nm, which produced a signal at 390 nm. In Figure 6(a), slit width 10 nm (in excitation and emission slits) is shown, and Figure 6(b) shows the spectra with slit width 5 nm each. As the slit widths were reduced, the intensity of signal was observed to decrease; more important was the drop in the FWHM from 39.5 nm to 16 nm. This is again very characteristic of any scattering phenomena, regardless of whether Rayleigh or Raman.

In fact, when the Nd:YAG laser (at 532 nm) was used to produce Raman signal from acetone, a sharp signal at 630 nm of FWHM of 1 nm or less was obtained [13, 14]; but when Rhodamine dye solution was excited, a broad fluorescence was observed with the peak at 630 nm and FWHM of 20 nm [15].

2.7. Polarization Properties. Another interesting experiment was performed to highlight the differences in the polarization properties.

Figure 7 shows the fluorescence of acetone when excited at 330 nm. Figure 7(a) reveals the fluorescence without any polarizer; whereas Figure 7(b) shows the fluorescence with the polarizer maintained at 0, 90, 180, and 270 degrees of orientation. (The polarizer was inserted between the emission slit and detector.) It is very obvious that the polarizer by itself has cut off the signal by four times due to the absorption by the material around 350 to 600 nm but such a reduction was independent of the orientation of the polarizer proving that the fluorescence from the molecules in the solution get depolarized. The fluorescence is a process involving three steps, the absorption of a photon as a

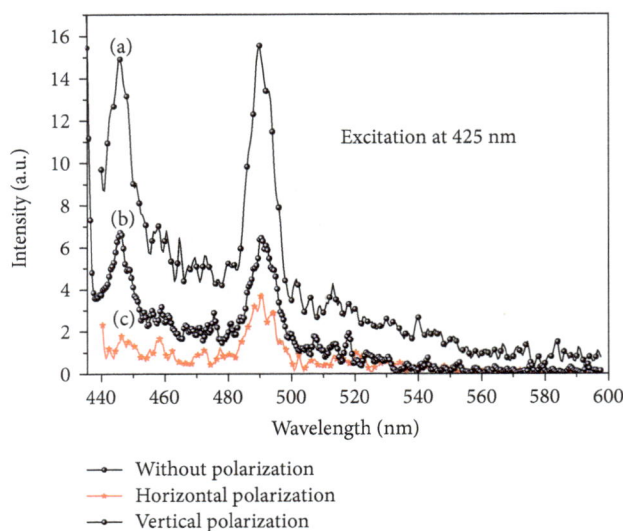

FIGURE 10: Polarized Raman spectra for benzene.

whole, thermalization, and reemission; the polarization feature is generally lost, particularly in a liquid medium where the molecules are in a state of constant collision among themselves.

Figure 8 shows the Rayleigh scattering signal for the excitation of acetone at 475 nm. The resultant signal occurred at 478 nm (due to experimental artifacts). Figure 8(a) shows a signal of 800 units without any polarizer. In Figure 8(b), the signal is obvious with the polarizer at 0 and 180 degrees and an intensity of 383; Figure 8(c) shows the signal of intensity 249 for the polarizer at 90 and 270 degrees. Thus, it is clearly evident that the degree of polarization for the Rayleigh scattering is 0.21.

Figure 9(a) shows the Raman spectra drawn for benzene, with excitation at 475 nm. The Raman signals occurred at 501 nm and 560 nm, which corresponded to $\nu_\nu = 1100$ cm^{-1} and 3010 cm^{-1}, due to skeleton breathing and the stretching of the benzene C-H bond. This is again confirmed by exciting at 450 nm and observing the two Raman signals mentioned above, at 475.3 and 525 nm. Another excitation was with the 425 nm produced Raman signals at 445.8 and 490 nm. (The instrument is so sensitive that even the rotational features of the Raman spectra were observable;

e.g., the Raman signal for excitation at 475 nm produced two lines at 556 and 560 nm, a separation of $18 \, cm^{-1}$; however, it got blurred for the other excitation wavelengths.) Table 2 summarizes the results given above.

Figure 10(a) shows the Raman signal of benzene as excited at 425 nm, without any polarizer. Figure 10(b) shows the same with the polarizer placed horizontally, and Figure 10(c) shows the same with the polarizer rotated to 90°. Both bands are clearly polarized differently. The $1100 \, cm^{-1}$ band appears more strongly polarized than the other, with the degrees of polarization $p = 0.6$ and 0.2, respectively.

It is noteworthy that all scattering exhibits a high degree of polarization, because it changes both direction and polarization.

3. Conclusion

Utilizing a spectrofluorometer commonly available in any graduate laboratory, a distinction between most of the spectral features of fluorescence and scattering has been clearly demonstrated (Rayleigh and Raman). It has elegantly been shown that the valley between the two mountains of absorption and emission is the best region to grope for the resonance Raman spectra. It has also been shown that the scattered signals are strongly polarized and that their spectral widths are determined most often by the width of the incident signal. In contrast, fluorescence is neither polarized nor dependent on the incident signal. Further, scattering is many times more common than fluorescence in general; however, for a few select, conjugated molecules like dyes, the fluorescence is tremendously high that it is almost impossible to observe the scattering signals. On the other hand, in the weakly fluorescent molecules like acetone or benzene, the scattering signals are comparable to, and sometimes even higher than, the fluorescence.

Acknowledgments

The authors would like to extend their sincere appreciation to the Deanship of Scientific Research at King Saud University for its funding of this research through the Research Group Project no. "RGP-223."

References

[1] C. V. Raman and K. S. Krishnan, "A new type of secondary radiation," *Nature*, vol. 121, pp. 501–502, 1928.

[2] M. Kasha, "Characterization of electronic transitions in complex molecules," *Discussions of the Faraday Society*, vol. 9, pp. 14–19, 1950.

[3] P. M. Kroneck and M. Sosa Torres, *Sustaining Life on Planet Earth: Metalloenzymes Mastering Dioxygen and Other Chewy Gases*, Springer International Publishing, Switzerland, 2015.

[4] S. Hu, K. M. Smith, and T. G. Spiro, "Assignment of protoheme resonance Raman spectrum by heme labeling in myoglobin," *Journal of the American Chemical Society*, vol. 118, no. 50, pp. 12638–12646, 1996.

[5] R. C. Prince, R. R. Frontiera, and E. O. Potma, "Stimulated Raman scattering: from bulk to nano," *Chemical Reviews*, 2016.

[6] N. Jiang, N. Chiang, L. R. Madison et al., "Nanoscale chemical imaging of a dynamic molecular phase boundary with ultra-high vacuum tip-enhanced Raman spectroscopy," *Nano Letters*, vol. 16, no. 6, pp. 3898–3904, 2016.

[7] S. Nie and S. R. Emory, "Probing single molecules and single nanoparticles by surface-enhanced Raman scattering," *Science*, vol. 275, no. 5303, pp. 1102–1106, 1997.

[8] C. L. Evans, E. O. Potma, M. Puoris' Haag, D. Côté, C. P. Lin, and X. S. Xie, "Chemical imaging of tissue in vivo with video-rate coherent anti-stokes Raman scattering microscopy," *Proceedings of the National Academy of Sciences of the United States of America*, vol. 102, no. 46, pp. 16807–16812, 2005.

[9] O. Stevens, I. E. I. Petterson, J. C. Day, and N. Stone, "Developing fibre optic Raman probes for applications in clinical spectroscopy," *Chemical Society Reviews*, vol. 45, no. 7, pp. 1919–1934, 2016.

[10] S. Laing, K. Gracie, and K. Faulds, "Multiplex in vitro detection using SERS," *Chemical Society Reviews*, vol. 45, no. 7, pp. 1901–1918, 2016.

[11] I. Pence and A. Mahadevan-Jansen, "Clinical instrumentation and applications of Raman spectroscopy," *Chemical Society Reviews*, vol. 45, no. 7, pp. 1958–1979, 2016.

[12] B. Robert, "Resonance Raman spectroscopy," *Photosynthesis Research*, vol. 101, no. 2-3, pp. 147–155, 2009.

[13] A. M. Azzeer and V. Masilamani, "A new, efficient laser source at 630 nm for photodynamic therapy and pulsed hologram," *Japanese Journal of Applied Physics*, vol. 42, no. 2R, p. 471, 2003.

[14] V. Masilamani and A. Aldwayyan, "Evidence of superexciplex in dye molecules," *Japanese Journal of Applied Physics*, vol. 41, no. 9R, p. 5801, 2002.

[15] R. Pratesi, *Optronic Techniques in Diagnostic and Therapeutic Medicine*, Springer Science & Business Media, 2012.

21

The Clinical Application of Raman Spectroscopy for Breast Cancer Detection

Pin Gao,[1] Bing Han,[1] Ye Du,[1] Gang Zhao,[1] Zhigang Yu,[2] Weiqing Xu,[3] Chao Zheng,[2,4] and Zhimin Fan[1]

[1]*Department of Breast Surgery, The First Hospital of Jilin University, Changchun 130021, China*
[2]*Department of Breast Surgery, The Second Hospital of Shandong University, Jinan 250033, China*
[3]*State Key Laboratory for Supramolecular Structure and Materials, Jilin University, Changchun 130012, China*
[4]*Department of Mechanical Engineering, Johns Hopkins University, Baltimore, MD 21218, USA*

Correspondence should be addressed to Chao Zheng; czheng5@jhu.edu and Zhimin Fan; fanzhimn@163.com

Academic Editor: Christoph Krafft

Raman spectroscopy has been widely used as an important clinical tool for real-time in vivo cancer diagnosis. Raman information can be obtained from whole organisms and tissues, at the cellular level and at the biomolecular level. The aim of this paper is to review the newest developments of Raman spectroscopy in the field of breast cancer diagnosis and treatment. Raman spectroscopy can distinguish malignant tissues from noncancerous/normal tissues and can assess tumor margins or sentinel lymph nodes during an operation. At the cellular level, Raman spectra can be used to monitor the intracellular processes occurring in blood circulation. At the biomolecular level, surface-enhanced Raman spectroscopy techniques may help detect the biomarker on the tumor surface as well as evaluate the efficacy of anticancer drugs. Furthermore, Raman images reveal an inhomogeneous distribution of different compounds, especially proteins, lipids, microcalcifications, and their metabolic products, in cancerous breast tissues. Information about these compounds may further our understanding of the mechanisms of breast cancer.

1. Introduction

Breast cancer is the most common cancer in women worldwide. In 2012, nearly 1.7 million people were diagnosed with breast cancer [1]. Reducing the incidence and death of breast cancer is a major public health priority. Once a mass is found in the breast, ultrasonography, diagnostic mammography, and/or magnetic resonance imaging (MRI) will be recommended by a physician. If the mass is possibly malignant, a core needle biopsy and histopathological techniques are necessary [2]. Unfortunately, this invasive examination cannot ensure a 100% correct diagnosis. If the lesion is malignant or the possibility of malignancy cannot be excluded, the patient will undergo surgery. Surgery for breast cancer includes lesion resection and axillary lymph node resection. In breast surgery, breast-conserving surgery and sentinel lymph node biopsy (SLNB) are two breakthroughs. SLNB results in less lymphedema of the upper arm [3]. Both techniques require an

intraoperative pathological diagnosis, which usually takes at least 30 min. After mastectomy for breast cancer, some patients also need chemotherapy, radiotherapy, and endocrine therapy. The necessity of molecular-targeted therapy is based on the result of fluorescence in situ hybridization (FISH). After operation, [(18)F]-2-fluoro-2-deoxy-D-glucose positron emission tomography/computed tomography (FDG-PET/CT) can be used for detecting the recurrence and metastasis [4]. Despite advances in the diagnosis and treatment of breast cancer, some limitations remain:

(1) Neither ultrasound nor mammography can make a qualitative diagnosis, so invasive biopsy or even surgery remains the gold standard in breast cancer.

(2) Sentinel lymph nodes in intraoperative frozen sections cannot identify the metastasis of the lymph node with 100% accuracy [5].

(3) After the lesion is removed, there is no way to evaluate the effect of chemotherapy or molecular-targeted therapy.

(4) Examinations such as pathology in current clinical application cannot provide information at the molecular or cellular level.

This review introduces some of the latest progress of Raman spectroscopy, a promising technique in breast cancer diagnosis. This optical technique has the potential to solve the problems listed above. First, Raman spectra can provide information of molecules (e.g., lipids, DNA, and proteins) to distinguish cancerous lesions from noncancerous lesions. Molecular changes may also indicate the mechanism of cancer development. Due to its real-time characteristics, Raman spectroscopy is a powerful intraoperative diagnostic technique. Furthermore, when combined with a nanotag, Raman spectroscopy can reveal whether cancer is sensitive or resistant to an anticancer drug.

2. Brief Overview of Raman Spectroscopy

When a photon interacts with a molecule, elastic or inelastic collision happens. During inelastic collision, there is an exchange of energy between the photon and the molecule. The photon changes its motion direction after the collision (Figure 1). Meanwhile, the photon passes part of the energy to the molecule or obtains energy from the molecule to change the photon's frequency. This process is called Raman scattering. In this process, if the incident photons lose energy, the scattered photons have a lower frequency. In contrast, if the incident photons get more energy from the environment, the scattered photons have a higher frequency. If the difference of the energy between the incident and scattered photons is monitored, information on the frequency and vibration can be obtained and used to analyze the sample compounds [6]. After Raman spectral imaging is obtained, mathematical analyses are necessary to extract the useful information from the test. Many spectral unmixing algorithms are applied currently, including principal component analysis (PCA), vertex component analysis (VCA), hierarchical cluster analysis (HCA), support vector machine (SVM), and SVM-recursive feature elimination (SVM-RFE) [7]. PCA is more commonly used for quantitative analysis in detecting cancers such as melanoma, breast cancer, and cervical cancer [8–10]. As for HCA, it is used for cancer diagnosis such as esophageal cancer [11] and some cancer cells [12–14]. Besides, SVM-RFE can be used to analyze and identify the most distinctive spectral features and, thus, has higher sensitivity and specificity [7].

The advantages of Raman spectroscopy include almost all those of optical detections. Raman spectroscopy has high resolution; it is safe, noninvasive, and nonradiative; and it can continuously monitor soft tissues. Compared with fluorescence spectrum, the Raman spectrum can reflect the molecular structures and biochemical compositions without the need to process samples in advance [15]. Although Raman spectroscopy offers highly specific information on

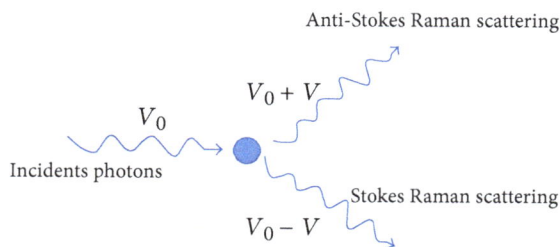

FIGURE 1: Schematic of inelastic collision showing anti-Stokes Raman scattering and Stokes Raman scattering.

the molecular features of tissues, its use has remained limited because of its relatively low ability to produce distinguishable signals. However, conventional Raman signals are less sensitive than fluorescence. To solve this problem, surface-enhanced Raman spectroscopy (SERS) was developed. When some nanoparticles absorbed to the surface of some rough metal, such as gold and silver, the Raman signal is amplified. This phenomenon is called SERS. One of the most important mechanisms of SERS is the enhancement of the electromagnetic field through the localization of optical fields in metallic nanostructures [16]. Although SERS can amplify Raman signals, it has some limitations [6]. First, the colloidal aggregation added to samples has not been proven able to amplify Raman signals. Second, there may be side reactions in the process. The spectra of the sample and nanoparticle may overlap. Third, when the nanoparticle is added to the analyte, there may be more toxic reactions. Despite these limitations, SERS remains a very important research direction of Raman spectroscopy.

In clinical application, Raman spectroscopy has been a rapidly developed technology that monitors signals at different levels. At the molecular level, it can be used to analyze DNA sequences and gene products. At the cellular level, Raman spectra can monitor the process of malignant cells or relative intercellular processes in the peripheral blood of cancer patients [6]. The information can then be used to evaluate cancer cell metabolism, proliferation, and even metastasis. Furthermore, Raman spectroscopy provides a promising way to detect biomarkers that are expressed on the surfaces of tumor cells [17, 18]. All of this information can be used to help doctors to diagnose diseases, determine the appropriate treatment, and even evaluate prognosis. To date, Raman technology has been used to diagnose many cancers, including cervical cancer, malignant skin tumors, gastric cancer, colorectal cancer, and breast cancer [19–22]. This article aims to review the application of Raman spectra in breast cancer and describe the newest progress in this field.

3. Application of Raman Spectroscopy in Breast Cancer

Raman spectroscopy is noninvasive and has high specificity, making its clinical application very promising. Nowadays, studies focus on using Raman spectroscopy to differentiate between benign and malignant tissues in place of traditional biopsy [23]. Raman spectroscopy can provide real-time information, making its intraoperative use possible.

Monitoring sentinel lymph node and surgical margins for breast-conserving surgery using Raman spectra can shorten the operative time. If focusing on a single cell, it no longer requires several days to detect biomarkers on the cell surface as with FISH. To determine the efficacy of anticancer drugs, Raman spectra are used to monitor cancer cell metabolism and composition. This is a new research direction that aims to elucidate how DNA, protein, lipid, and calcification contribute to cell proliferation in vivo that will further our understanding of the pathogenesis of cancer.

3.1. Raman Spectroscopy in Tissues.
Generally, Raman bands associate with the vibration of a special chemical bond or a distinct functional group in the molecule. Therefore, based on distinctive spectral features and intensity differences, we could discriminate normal, benign, and malignant. In recent years, many research papers have also shown the availability of Raman spectroscopy to classify diseased tissue in the breast [7, 24] and especially for intraoperative margin assessment and SLNB for breast cancer.

3.1.1. Qualitative Diagnosis of Breast Disease with Raman Spectroscopy.
Many studies have shown that the spectral profiles of normal breast (NB) tissue peaks appear at 1004, 1080, 1158, 1266, 1304, 1444, 1518, 1660, 1750, 2674, 2727, 2854, 2888, 2926, and 3009 cm^{-1}, while those of malignant tissue peaks appear at 1004, 1259, 1444, 1660, 2854, 2888, 2926, 3009, and 3311 cm^{-1}. In normal tissues, the peaks at 1004, 1158, and 1518 cm^{-1} belong to carotenoids (C-C, C=C), whereas in malignant tissues, the peaks at 1080, 1259, 1444, 1660, 2854, and 2940 cm^{-1} are more obvious. Peaks at 2854 cm^{-1} belong to lipids, while those at 2940 cm^{-1} are believed to represent mixtures of lipids and proteins [26]. Shell-isolated nanoparticle-enhanced Raman spectroscopy (SHINERS) is a powerful SERS substrate. After SHINERS is added to a sample, Raman spectral intensity is amplified. Knowing the magnification of different peaks is helpful in diagnosing breast diseases. Researchers have also utilized the Au@SiO2 SHINERS technique to detect NB tissues, fibroadenoma (FD), atypical ductal hyperplasia (ADH), ductal carcinoma in situ (DCIS), and invasive ductal carcinoma (IDC). Compared with traditional Raman spectroscopy, SHINERS can provide more spectral feature information (Figure 2) [24, 25].

Studies of cells in response to extracellular signals using two-dimensional (2D) culture conditions, including cultures of isolated, cloned, and immortal cells that grow on rigid surfaces, date back to the 1900s. The limitation of this type of model is that it does not consider the extracellular matrix (ECM), a very important factor in cell phenotypes [27]. In 1977, Emerman and Pitelka found that a deformable three-dimensional ECM (3D-ECM) environment is necessary for maintaining the tissue structure of NB tissue [28]. Because 3D-ECM cultures can better mimic in vivo environments, Damayanti et al. first combined 3D-ECM cultures with Raman spectroscopy to study breast cancer [29]. In the 3D-ECM culture model, peaks of precancerous lesions were observed at 1451 and 1650 cm^{-1}, while peaks of noncancerous tissues were observed at 1650 cm^{-1}. Different from

traditional Raman spectroscopy, the signal intensity in precancerous lesions is stronger than that in noncancerous tissues. In addition, the sensitivity and specificity are higher in 3D-ECM cultures than in 2D cultures. Microenvironment is a major factor in modulating cell morphology and proliferation [30]. Changes in the surrounding stroma may be an initial factor of precancerosis.

3.1.2. Intraoperative Raman Spectroscopy.
Considering its real-time nature without the need for preparation, Raman spectroscopy is a promising new technique in intraoperative margin assessment and SLNB.

(1) Raman Spectroscopy of Surgical Margins. Haka and colleagues first used Raman spectra to examine surgical margins in nine patients during partial mastectomy surgeries [31]. Compared with pathological examinations after surgeries, the algorithm results in an overall accuracy of 93.3%. Conventional Raman spectroscopy has a depth limitation. Spatially offset Raman spectroscopy (SORS) is a new technique that was first applied by Matthew for cancer diagnosis. SORS can detect deeper tissues at least 0.5–2 mm below the surface [32]. The use of SORS in intraoperative margin assessment has 95% sensitivity and 100% specificity [33]. In recent years, Mohs et al. developed a handheld spectroscopic pen device (termed SpectroPen) using both a fluorescent contrast agent (indocyanine green) and a SERS contrast agent (pegylated colloidal gold). The team gained real-time information on tumor margins. In that study, the tissue penetration depth was 5–10 mm depending on the tissue's optical properties and the ability to resolve contrast agent signals [34]. They also constructed a fiber optic spectroscopic system with stable alignment. The SpectroPen has also been used to detect bioluminescent 4T1 breast tumors of in vivo mouse models. In the intraoperative measurement, the SpectroPen is helpful for distinguishing tumor margins and detecting whether any tumor remains.

(2) Raman Spectroscopy in SLNB. Another important application of Raman spectra is in SLNB. In traditional operations, surgeons use dyes or radionuclides Te-99 to localize sentinel lymph nodes. However, the rates of localizing sentinel lymph nodes through these methods are 73.8% and 94.1%, respectively [35]. After lymph node removal, surgeons must wait for the immunohistochemical analysis (IHA) results to decide if axillary lymph node dissection is necessary, which usually takes 30–60 min. In some cases, a second operation cannot be avoided because the specificity of frozen section rapid pathologic examination cannot reach 100% and the false negative rate is at least 4% [36]. Isabelle et al. first used Raman spectra and infrared images to detect the lymph node excision during head and neck tumor surgeries [37] and showed that lymph nodes with metastatic cancer have greater nucleic acid and less lipid and carbohydrate content than normal lymph nodes. These results suggest that lymph nodes with metastatic cancer grow faster and are much more poorly differentiated than normal lymph nodes.

Smith et al. used mapping Raman spectra to accurately judge lymph node metastases [38]. Although metastatic

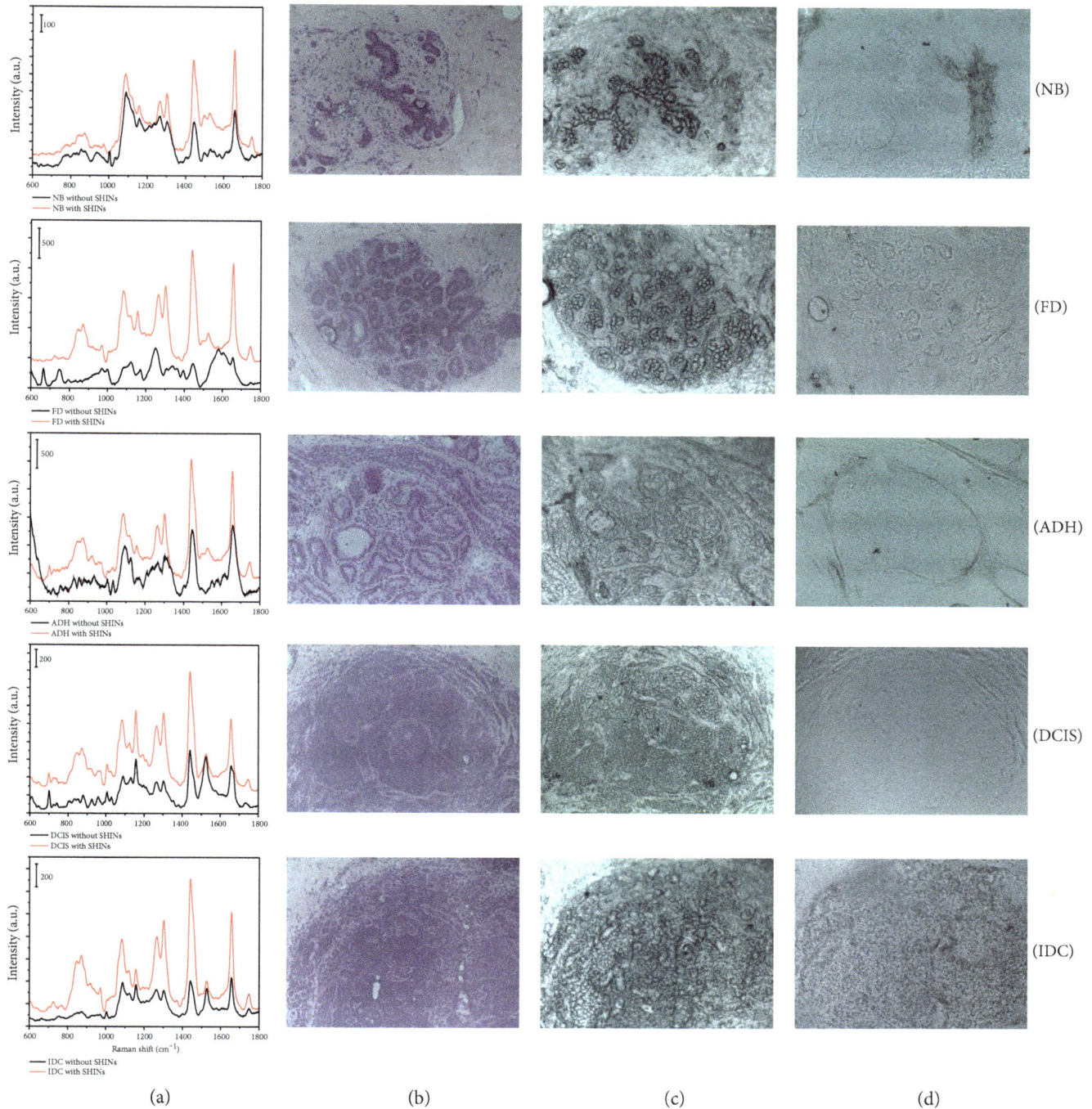

FIGURE 2: Diagnosis of breast cancer with Raman spectroscopy. (a) Mean Raman spectra and SHINERS spectra of breast tissues: normal breast (NB) tissues, fibroadenoma (FD), atypical ductal hyperplasia (ADH), ductal carcinoma in situ (DCIS), and invasive ductal carcinoma (IDC). (b) Corresponding images of frozen sections were stained with HE. (c) Corresponding images of frozen sections without shell-isolated nanoparticles (SHINs). (d) Corresponding images of frozen sections with SHINs. The images show an area of $0.8 \times 0.6 \, mm^2$ [24, 25].

cancer is successfully diagnosed, the running time of each section is 12–120 hours. Such a long running time is not suitable for real-time tissue diagnosis. To overcome this limitation, Horsnell et al. developed a method of recording only five spectra of each lymph node [39]. The spectra gained from 38 lymph nodes of 20 patients indicated that the fatty acid, collagen, and protein contents differ between metastatic and normal lymph nodes. Among the four standards in the study, if a node was classified based on the simple majority of the five random spectra, the sensitivity and specificity were higher than those in the other groups. Thereafter, to study the effect of the number of spectra, Horsnell et al. compared five- and 10-point probe analyses [40]. The assessment of the lymph node based on spectra from five points achieved 71%

sensitivity and 97% specificity, while the 10-point method achieved 81% sensitivity and 97% specificity. The sensitivity and specificity of 10-point spectra are comparable to those of other current methods of intraoperative assessment. The running time in these methods are 9 and 18 min, respectively, much shorter than the current methods, making Raman spectra a very promising technique in the diagnosis of lymph node metastasis and more suitable for clinical application than the currently used techniques.

3.2. Raman Spectroscopy in Cells

3.2.1. Raman Spectroscopy in Single Cells and the Cellular Microenvironment. Neugebauer et al. identified single cells in the circulation using Raman spectra [41]. Isolating leukocytes and erythrocytes from breast carcinoma-derived cells (MCF-7, BT-20) can achieve a sensitivity > 99.7% and a specificity > 99.5%. The team used SVMs with a linear kernel to develop a classification model. Moreover, using Raman-activated cell sorting, Dochow et al. successfully identified leucocytes extracted from blood, BT-20 and MCF-7 breast cancer cells, and OCI-AML3 leukemia cells [42, 43].

Many genetic and environmental factors contribute to the development of breast cancer. An enhanced understanding of these factors can provide more information on the pathogenesis of cancer and guide its treatment. Therefore, using various biomarkers and other testing methods to study tumor-associated molecules is a promising research direction. Raman is a good option because it can test unlabeled lesion as well as nucleic acid and surface receptor products using specific probes [6].

3.2.2. Efficacy of Chemotherapy in Breast Cancer with Raman Spectroscopy. Raman spectroscopy has been used to increase the efficacy of chemotherapy in breast cancer, to evaluate the resistance of chemotherapy, and to monitor the damage of cells due to chemotherapy. Hossain et al. first used SERS and a kind of biohybrid nanoparticles to increase the doxorubicin delivery. Meanwhile, this system also helped monitor the metabolism of drugs in breast cancer cells [44]. Lapatinib is another important oral anticancer drug with low absorption. Considering its high-binding ability with albumin, human serum albumin nanoparticles loaded with lapatinib obviously increased the efficacy of the drug [45]. Hu et al. designed a novel type of nanomedical platform with double-walled Au nanocage/SiO(2) nanorattle and improved biocompatibility as well as cell uptake ability, which in turn increased the efficacy of chemotherapy [46]. The human epidermal growth factor receptor 2 (HER2) proto-oncogene is amplified in 20–30% of human breast cancer cases. Studies have shown that patients with HER2-positive breast cancer are at increased risks of recurrence and metastasis [47]. HER2 proto-oncogene expression can be detected by Raman spectra, which have also been used to detect HER2 expression in the BT474 (HER2+), MCF-10A (HER2−), and MCF-10A (HER2+) cell lines [48]. In this study, the application of Raman spectra was first used to characterize the biochemical response of drug resistance and sensitivity to lapatinib, a targeted antitumor agent. After long time of chemotherapy,

cells would be damaged in vivo. Ilkhani et al. designed a novel biosensor to assess the damage of DNA after chemotherapy with the help of SERS [49].

3.3. Composition Analysis of Samples with Raman Spectroscopy. Raman spectra can exhibit spectral features including composition and structure changes such as peptide conformation, biomarkers, DNA sequences, base composition, microcalcifications, and lipids.

3.3.1. Raman Spectroscopy of Proteins. As a fast technique with high specificity, Raman spectra can also analyze proteins of cancer in a low-sample volume, especially in the detection of cell surface biomarkers. Current techniques such as enzyme-linked immunosorbent assays (ELISA) usually detect a sample with a concentration in the nanograms-to-grams per milliliter range. However, at the early cancer stage, the cancer proteins are at a low concentration that may be below the ELISA limit of detection. Dinish et al. developed a new hollow core photonic crystal fiber to detect low amounts of proteins [50]. It is well known that epidermal growth factor receptor (EGFR) is a very important biomarker in breast cancer. In that study, EGFR antibody is combined with SERS nanotags. Through this method, Raman spectra can detect proteins at a concentration in the range of ~100 pg in the volume of ~10 nL. Nilstad et al. also used SERS to detect epidermal growth factor receptor (EGFR) expression and investigate the therapeutic effects of cetuximab antibodies [51].

Besides EGFR, Lee et al. [52] used silica-encapsulated hollow gold nanospheres (SEHGNs) as a Raman nanotag that is conjugated with specific antibodies. This technique enables the assessment of the expression of epidermal growth factor (EGF), HER2, and insulin-like growth factor-1 receptors in the MDA-MB-468, KPL4, and SK-BR-3 human breast cancer cell lines. That study indicated that SEHGNs may possibly be used in the early diagnosis of breast cancer.

HER2 proto-oncogene encodes a protein that has extracellular, transmembrane, and intracellular domains that are consistent with the structures of a growth factor receptor. Raman spectra have shown a prominent presence of lipid in the MDA-MB-231 (HER2-negative), MDA-MB-435s (HER2-positive), and SK-BR-3 (HER2 overexpressed) cell lines [13]. Different from current tests, the application of SERS in tumor cell surface biomarkers such as HER2 is more effective than that of IHA and FISH.

Another successful application of SERS is the single-molecule detection of BRCA1 by Wabuyele et al. [53]. The probe has one DNA sequence complimentary to the target sequence and two arms, one that is connected with a SERS-active molecule and the other that is connected with a metallic nanoparticle. The metallic nanoparticle enhances the SERS signal. The probe looks like a hair loop without the target sequence and maintains a Raman label. Once the BRCA1 gene mutation occurs, the mutated target sequence combines with the probe, the Raman label separates from the metal nanoparticle, and the signal decreases.

A new simultaneous SERS method was developed by Li et al. [15] which consists of a novel Raman probe of branched DNA-gold nanoaggregates that can simultaneously monitor

(a)

(b)

(c)

(d)

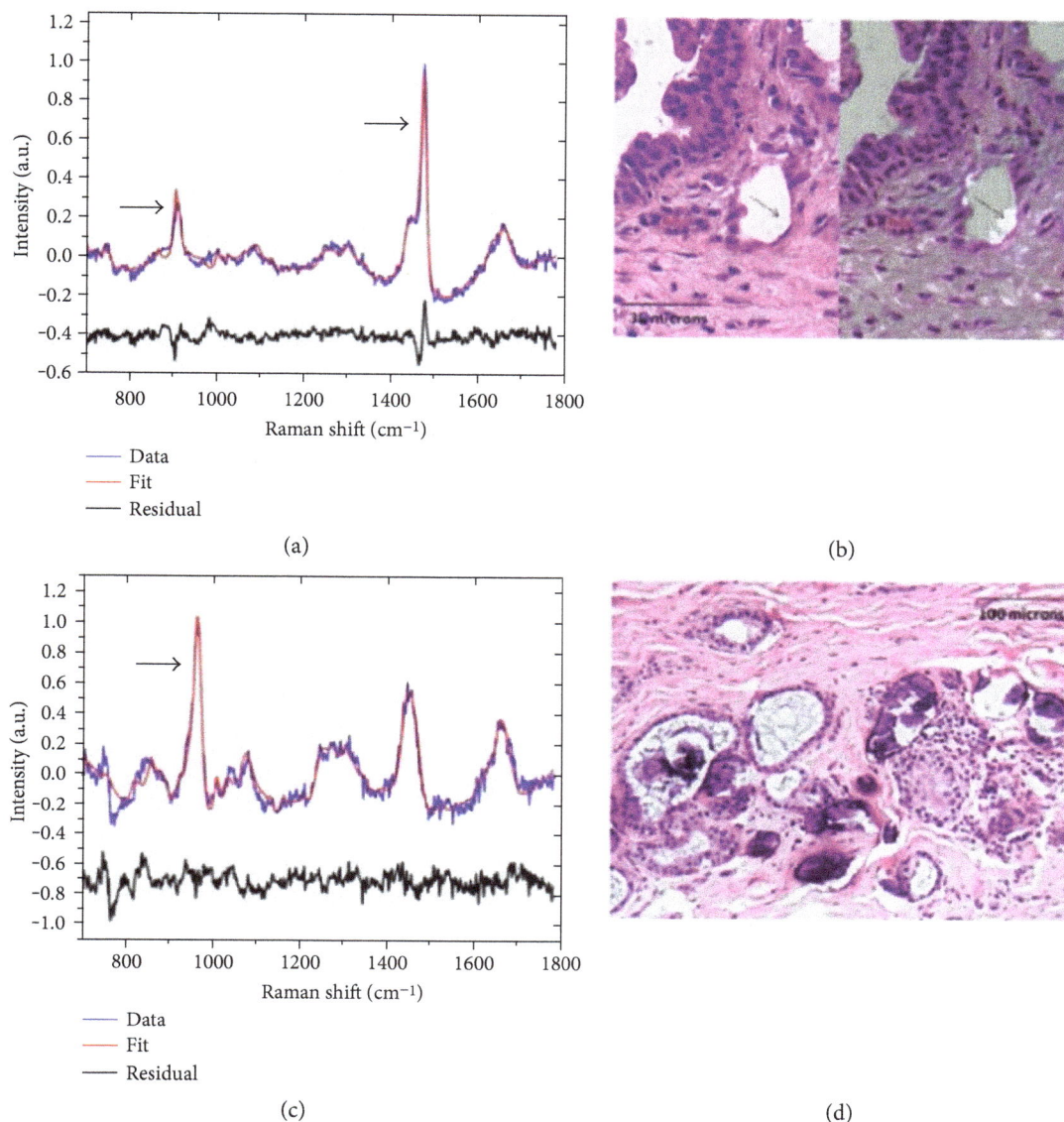

FIGURE 3: Typical Raman spectra and histopathology of breast lesions (fibrocystic change) with type I and II microcalcifications. The Raman spectrum of the breast lesion with type I microcalcifications in (a) shows prominent bands at $912 \, cm^{-1}$ and $1477 \, cm^{-1}$ (arrows) characteristic of calcium oxalate; the calcium oxalate crystals comprising the type I microcalcifications (b) do not bind H&E (left panel) and appear as colorless crystals (arrows) that are birefringent when viewed under polarized light (right panel). In contrast, the Raman spectrum of the breast lesion with type II microcalcifications in (c) shows a prominent band at $960 \, cm^{-1}$ (arrow) characteristic of calcium hydroxyapatite; the calcium hydroxyapatite rich type II microcalcifications appear as basophilic concretions on the H&E stain (d) and are nonbirefringent [59, 60].

two MCF-7 cell line markers. This development makes Raman spectroscopy more effective for clinical diagnosis. Using an optoelectrofluidic device and a liquid crystal display module, Hwang et al. successfully detected another human tumor marker, α-fetoprotein, in a sample volume of ~500 nL [54]. Combined with fluorescence dual-modal nanoprobes (DMNPs), Lee et al. also got the imaging of CD24 and CD44 expressed on the surface of breast cancer cells [55]. Kang et al. used SERS with channel-compressed spectrometry to detect EGFR, HER2, CD24, and CD44 in fresh tissue specimens with less errors and lower concentrations [56].

This simple, fast, and highly sensitive automated technique opens a new method of clinical diagnosis.

3.3.2. Raman Spectroscopy of Microcalcifications. Breast microcalcifications are key features for the diagnosis of malignant tumors and sometimes the only feature of mammography. In many clinical cases, calcification is a poor prognostic factor for patients with IDC. Mammography is currently the most effective technique to detect calcifications in clinical application despite its limitation of not distinguishing malignant from benign lesions. Mammographic images can show the morphological characteristics of different calcifications and help predict the tumor characteristics. Only after stereotactic vacuum-assisted breast biopsies and subsequent pathological diagnosis [57] can doctors make an accurate diagnosis of a lesion. However, this kind of invasive medical examination may fail to retrieve lesions with microcalcifications in about 15% of all cases. Follow-ups have shown that the cancer diagnosis is missed in 8% of patients [58]. Therefore, a new noninvasive technique that can gain information on microcalcifications in real time with higher sensitivity and specificity is urgently needed.

There are two types of microcalcifications in breast tumors (Figure 3) [59–61]. Type I microcalcifications, which consist of calcium oxalate, are mainly associated with benign breast disease, whereas type II microcalcifications, which consist of carbonated calcium hydroxyapatite, can be found in proliferative breast lesions, including invasive and in situ cancer. Raman spectroscopy can distinguish the two types of microcalcifications in breast diseases [62] and thus distinguish malignant from benign lesions. In type II microcalcifications, there is more carbonate in benign disease such as sclerosing adenosis than in malignant lesions, although malignant lesions usually contain greater amounts of proteins, amino acid residues, and carotenoids. Baker et al. observed a significant correlation between carbonate concentrations and carcinoma in situ subgrades [63]. In a recent study, an Au@SiO2 SHINERS technique was utilized in frozen sections of surgical resection or Mammotome biopsy. The SHINERS spectra can probe the presence of calcified deposits and distinguish among normal breast tissues, fibroadenoma, atypical ductal hyperplasia, ductal carcinoma in situ (DCIS), and invasive ductal carcinoma (IDC). By correlating the spectra with the corresponding histologic assessment, the study developed partial least squares discriminant analysis-derived decision algorithm that provides excellent diagnostic power in the fresh frozen sections (overall accuracy of 99.4% and 93.6% using SHINs for breast lesions with and without microcalcifications, resp.) [24, 64].

3.3.3. Raman Spectroscopy of Lipids and Other Metabolites. In addition to microcalcifications, there are significant differences in vibrational spectroscopy of proteins, lipids, and carotenoids between benign and malignant lesions, especially unsaturated fatty acids. It is well known that lipids are crucial in biological membranes. If abnormal metabolism occurs in the local enzymes responsible for lipid synthesis and degradation that results in alterations in the expression of genes encoding these enzymes, lipids may change and affect the membranes [26]. Another important factor of cell stability is carotenoid, which can neutralize reactive oxygen species and can reduce oxidative DNA damage and genetic mutations. Carotenoids can act as a protective factor against breast cancer [26, 65]. Raman spectroscopy can differentiate noncancerous from cancerous tissues through carotenoid and lipid levels [17]. Hartsuiker et al. reported variance in lipid levels among breast cancer cell lines with varying HER2 expression levels [13]. Bi et al. used Raman spectroscopy to detect lipids in cell lines [48]. In the cell lines treated with lapatinib, a drug-targeting HER2, the lipid levels were

increased in the drug-resistant cells. These results suggest a correlation between lipids and drug resistance.

Progestogen is a potent factor in the proliferation of breast carcinoma cells [66]. Potcoava et al. exposed T47D breast carcinoma cells to medroxyprogesterone acetate and the synthetic androgen R1881 [67]. Raman spectral imaging showed the process of hormone-mediated lipogenesis. Hormone-treated T47D cancer cells proliferate faster than untreated cell. Hedegaard et al. have first used Raman spectra to demonstrate the difference between metastatic cell lines and nonmetastatic cell lines through the content of unsaturated fatty acid [68]. Hartsuiker et al. evaluated lipid contents in two cell lines: SK-BR-3, with low metastatic potential, and MDA-MB, which are highly metastatic [13]. Raman imaging demonstrated that the lipid contents of MDA-MB cell lines are higher than those of SK-BR-3 cells, although HER2 expression is much higher in SK-BR-3 cells. These results indicate that lipid contents may be a more potent predictive factor of breast cancer than HER2 expression.

4. Conclusions

In this article, we have reviewed the clinical applications of Raman spectroscopy for breast cancer over the past few years. Raman spectroscopy has become a promising technique that can distinguish cancerous tissues from normal ones, combine with anticancer drugs for treatment, and detect components at the molecular level including proteins, lipids, and microcalcifications. The differences in components between cancerous and noncancerous tissues can provide information about cancer pathogenesis. While we are excited that Raman spectroscopy may become a new method of clinical diagnosis in breast cancer, we should note that some problems still need to be solved before it is widely used clinically, such as the stability of its sensitivity and predictability as well as how we can overcome the limitation of SERS to obtain better spectra. More importantly, it is necessary to construct a mathematical model using Raman spectrum data that are obtained from normal and malignant tissues based on various biochemical characteristics such as calcification, protein structure, and fat structure. This algorithm would allow us to identify early pathological changes, to distinguish the nature of breast tissues (normal, benign, or malignant), and to explore the inner link between breast cancer and premalignant lesions. The application of Raman spectroscopy can be an efficient way to reduce the need for unnecessary surgical biopsy and significantly save medical resources. Furthermore, it can create a beneficial economic advantage and decrease the psychological burden of patients.

Acknowledgments

This work was supported by the National Natural Science Foundation of China (Grant no. 81202078), the Seed Fund of the Second Hospital of Shandong University, and the National Construction of High Quality University Projects of Graduates from the China Scholarship Council (Grant no. 201406170141).

References

[1] L. A. Torre, F. Bray, R. L. Siegel, J. Ferlay, J. Lortet-Tieulent, and A. Jemal, "Global cancer statistics, 2012," *CA: A Cancer Journal for Clinicians*, vol. 65, no. 2, pp. 87–108, 2015.

[2] J. Depciuch, E. Kaznowska, I. Zawlik et al., "Application of Raman spectroscopy and infrared spectroscopy in the identification of breast cancer," *Applied Spectroscopy*, vol. 70, no. 2, pp. 251–263, 2016.

[3] Y. J. Choi, J. H. Kim, S. J. Nam, Y. H. Ko, and J. H. Yang, "Intraoperative identification of suspicious palpable lymph nodes as an integral part of sentinel node biopsy in patients with breast cancer," *Surgery Today*, vol. 38, no. 5, pp. 390–394, 2008.

[4] N. Y. Jung, R. Yoo Ie, B. J. Kang, S. H. Kim, B. J. Chae, and Y. Y. Seo, "Clinical significance of FDG-PET/CT at the postoperative surveillance in the breast cancer patients," *Breast Cancer (Tokyo, Japan)*, vol. 23, no. 1, pp. 141–148, 2016.

[5] U. Veronesi, G. Paganelli, V. Galimberti et al., "Sentinel-node biopsy to avoid axillary dissection in breast cancer with clinically negative lymph-nodes," *Lancet (London, England)*, vol. 349, no. 9069, pp. 1864–1867, 1997.

[6] H. Abramczyk and B. Brozek-Pluska, "Raman imaging in biochemical and biomedical applications. Diagnosis and treatment of breast cancer," *Chemical Reviews*, vol. 113, no. 8, pp. 5766–5781, 2013.

[7] C. Hu, J. Wang, C. Zheng et al., "Raman spectra exploring breast tissues: comparison of principal component analysis and support vector machine-recursive feature elimination," *Medical Physics*, vol. 40, no. 6, p. 063501, 2013.

[8] B. Bodanese, F. L. Silveira, R. A. Zangaro, M. T. Pacheco, C. A. Pasqualucci, and L. Silveira Jr., "Discrimination of basal cell carcinoma and melanoma from normal skin biopsies in vitro through Raman spectroscopy and principal component analysis," *Photomedicine and Laser Surgery*, vol. 30, no. 7, pp. 381–387, 2012.

[9] E. Vargas-Obieta, J. C. Martinez-Espinosa, B. E. Martinez-Zerega, L. F. Jave-Suarez, A. Aguilar-Lemarroy, and J. L. Gonzalez-Solis, "Breast cancer detection based on serum sample surface enhanced Raman spectroscopy," *Lasers in Medical Science*, vol. 31, no. 7, pp. 1317–1324, 2016.

[10] J. L. Gonzalez-Solis, J. C. Martinez-Espinosa, L. A. Torres-Gonzalez, A. Aguilar-Lemarroy, L. F. Jave-Suarez, and P. Palomares-Anda, "Cervical cancer detection based on serum sample Raman spectroscopy," *Lasers in Medical Science*, vol. 29, no. 3, pp. 979–985, 2014.

[11] X. Li, T. Yang, S. Li, D. Wang, and D. Guan, "Detecting esophageal cancer using surface-enhanced Raman spectroscopy (SERS) of serum coupled with hierarchical cluster analysis and principal component analysis," *Applied Spectroscopy*, vol. 69, no. 11, pp. 1334–1341, 2015.

[12] F. Draux, C. Gobinet, J. Sule-Suso et al., "Raman spectral imaging of single cancer cells: probing the impact of sample fixation methods," *Analytical and Bioanalytical Chemistry*, vol. 397, no. 7, pp. 2727–2737, 2010.

[13] L. Hartsuiker, N. J. Zeijen, L. W. Terstappen, and C. Otto, "A comparison of breast cancer tumor cells with varying expression of the Her2/neu receptor by Raman microspectroscopic imaging," *The Analyst*, vol. 135, no. 12, pp. 3220–3226, 2010.

[14] T. Tolstik, C. Marquardt, C. Matthaus et al., "Discrimination and classification of liver cancer cells and proliferation states by Raman spectroscopic imaging," *The Analyst*, vol. 139, no. 22, pp. 6036–6043, 2014.

[15] Y. Li, X. Qi, C. Lei, Q. Yue, and S. Zhang, "Simultaneous SERS detection and imaging of two biomarkers on the cancer cell surface by self-assembly of branched DNA-gold nanoaggregates," *Chemical Communications (Cambridge, England)*, vol. 50, no. 69, pp. 9907–9909, 2014a.

[16] D. Graham and K. Faulds, "Quantitative SERRS for DNA sequence analysis," *Chemical Society Reviews*, vol. 37, no. 5, pp. 1042–1051, 2008.

[17] H. Abramczyk, B. Brozek-Pluska, J. Surmacki, J. Jablonska-Gajewicz, and R. Kordek, "Raman 'optical biopsy' of human breast cancer," *Progress in Biophysics and Molecular Biology*, vol. 108, no. 1–2, pp. 74–81, 2012.

[18] H. N. Wang and T. Vo-Dinh, "Multiplex detection of breast cancer biomarkers using plasmonic molecular sentinel nanoprobes," *Nanotechnology*, vol. 20, no. 6, p. 065101, 2009.

[19] I. R. Ramos, A. Malkin, and F. M. Lyng, "Current advances in the application of Raman spectroscopy for molecular diagnosis of cervical cancer," *BioMed Research International*, vol. 2015, p. 561242, 2015.

[20] N. Kourkoumelis, I. Balatsoukas, V. Moulia, A. Elka, G. Gaitanis, and I. D. Bassukas, "Advances in the in vivo Raman spectroscopy of malignant skin tumors using portable instrumentation," *International Journal of Molecular Sciences*, vol. 16, no. 7, pp. 14554–14570, 2015.

[21] H. H. Kim, "Endoscopic Raman spectroscopy for molecular fingerprinting of gastric cancer: principle to implementation," *BioMed Research International*, vol. 2015, p. 670121, 2015.

[22] S. Li, G. Chen, Y. Zhang et al., "Identification and characterization of colorectal cancer using Raman spectroscopy and feature selection techniques," *Optics Express*, vol. 22, no. 21, pp. 25895–25908, 2014b.

[23] A. Saha, I. Barman, N. C. Dingari et al., "Precision of Raman spectroscopy measurements in detection of microcalcifications in breast needle biopsies," *Analytical Chemistry*, vol. 84, no. 15, pp. 6715–6722, 2012.

[24] C. Zheng, W. Shao, S. K. Paidi et al., "Pursuing shell-isolated nanoparticle-enhanced Raman spectroscopy (SHINERS) for concomitant detection of breast lesions and microcalcifications," *Nanoscale*, vol. 7, no. 40, pp. 16960–16968, 2015.

[25] C. Zheng, L. Liang, S. Xu et al., "The use of Au@SiO2 shell-isolated nanoparticle-enhanced Raman spectroscopy for human breast cancer detection," *Analytical and Bioanalytical Chemistry*, vol. 406, no. 22, pp. 5425–5432, 2014.

[26] B. Brozek-Pluska, J. Musial, R. Kordek, E. Bailo, T. Dieing, and H. Abramczyk, "Raman spectroscopy and imaging: applications in human breast cancer diagnosis," *The Analyst*, vol. 137, no. 16, pp. 3773–3780, 2012.

[27] . Emerman, S. J. Burwen, and D. R. Pitelka, "Substrate properties influencing ultrastructural differentiation of mammary epithelial cells in culture," *Tissue & Cell*, vol. 11, no. 1, pp. 109–119, 1979.

[28] M. J. Bissell, D. C. Radisky, A. Rizki, V. M. Weaver, and O. W. Petersen, "The organizing principle: microenvironmental influences in the normal and malignant breast," *Differentiation; Research in Biological Diversity*, vol. 70, no. 9–10, pp. 537–546, 2002.

[29] N. P. Damayanti, Y. Fang, M. R. Parikh, A. P. Craig, J. Kirshner, and J. Irudayaraj, "Differentiation of cancer cells in two-dimensional and three-dimensional breast cancer models by Raman spectroscopy," *Journal of Biomedical Optics*, vol. 18, no. 11, p. 117008, 2013.

[30] M. P. Shekhar, R. Pauley, and G. Heppner, "Host microenvironment in breast cancer development: extracellular matrix-stromal cell contribution to neoplastic phenotype of epithelial cells in the breast," *Breast Cancer Research: BCR*, vol. 5, no. 3, pp. 130–135, 2003.

[31] A. S. Haka, Z. Volynskaya, J. A. Gardecki et al., "In vivo margin assessment during partial mastectomy breast surgery using raman spectroscopy," *Cancer Research*, vol. 66, no. 6, pp. 3317–3322, 2006.

[32] M. D. Keller, S. K. Majumder, and A. Mahadevan-Jansen, "Spatially offset Raman spectroscopy of layered soft tissues," *Optics Letters*, vol. 34, no. 7, pp. 926–928, 2009.

[33] M. D. Keller, E. Vargis, N. de Matos Granja et al., "Development of a spatially offset Raman spectroscopy probe for breast tumor surgical margin evaluation," *Journal of Biomedical Optics*, vol. 16, no. 7, p. 077006, 2011.

[34] A. M. Mohs, M. C. Mancini, S. Singhal et al., "Hand-held spectroscopic device for in vivo and intraoperative tumor detection: contrast enhancement, detection sensitivity, and tissue penetration," *Analytical Chemistry*, vol. 82, no. 21, pp. 9058–9065, 2010.

[35] M. Gipponi, C. Bassetti, G. Canavese et al., "Sentinel lymph node as a new marker for therapeutic planning in breast cancer patients," *Journal of Surgical Oncology*, vol. 85, no. 3, pp. 102–111, 2004.

[36] J. Wong, W. S. Yong, A. A. Thike et al., "False negative rate for intraoperative sentinel lymph node frozen section in patients with breast cancer: a retrospective analysis of patients in a single Asian institution," *Journal of Clinical Pathology*, vol. 68, no. 7, pp. 536–540, 2015.

[37] M. Isabelle, N. Stone, H. Barr, M. Vipond, N. Shepherd, and K. Rogers, "Lymph node pathology using optical spectroscopy in cancer diagnostics," *Spectrometry International Journal*, vol. 22, no. 2–3, pp. 97–104, 2008.

[38] J. Smith, C. Kendall, A. Sammon, J. Christie-Brown, and N. Stone, "Raman spectral mapping in the assessment of axillary lymph nodes in breast cancer," *Technology in Cancer Research & Treatment*, vol. 2, no. 4, pp. 327–332, 2003.

[39] J. Horsnell, P. Stonelake, J. Christie-Brown et al., "Raman spectroscopy—a new method for the intra-operative assessment of axillary lymph nodes," *The Analyst*, vol. 135, no. 12, pp. 3042–3047, 2010.

[40] J. D. Horsnell, J. A. Smith, M. Sattlecker et al., "Raman spectroscopy—a potential new method for the intra-operative assessment of axillary lymph nodes," *The Surgeon: Journal of the Royal Colleges of Surgeons of Edinburgh and Ireland*, vol. 10, no. 3, pp. 123–127, 2012.

[41] U. Neugebauer, T. Bocklitz, J. H. Clement, C. Krafft, and J. Popp, "Towards detection and identification of circulating tumour cells using Raman spectroscopy," *The Analyst*, vol. 135, no. 12, pp. 3178–3182, 2010.

[42] S. Dochow, C. Krafft, U. Neugebauer et al., "Tumour cell identification by means of Raman spectroscopy in combination with optical traps and microfluidic environments," *Lab on a Chip*, vol. 11, no. 8, pp. 1484–1490, 2011.

[43] S. Dochow, C. Beleites, T. Henkel et al., "Quartz microfluidic chip for tumour cell identification by Raman spectroscopy in combination with optical traps," *Analytical and Bioanalytical Chemistry*, vol. 405, no. 8, pp. 2743–2746, 2013.

[44] M. K. Hossain, H. Y. Cho, K. J. Kim, and J. W. Choi, "In situ monitoring of doxorubicin release from biohybrid nanoparticles modified with antibody and cell-penetrating peptides in breast cancer cells using surface-enhanced Raman spectroscopy," *Biosensors & Bioelectronics*, vol. 71, pp. 300–305, 2015.

[45] X. Wan, X. Zheng, X. Pang, Z. Zhang, and Q. Zhang, "Incorporation of lapatinib into human serum albumin nanoparticles with enhanced anti-tumor effects in HER2-positive breast cancer," *Colloids and Surfaces. B, Biointerfaces*, vol. 136, pp. 817–827, 2015.

[46] F. Hu, Y. Zhang, G. Chen, C. Li, and Q. Wang, "Double-walled Au nanocage/SiO2 nanorattles: integrating SERS imaging, drug delivery and photothermal therapy," *Small*, vol. 11, pp. 985–993, 2015.

[47] J. A. Menendez, "Fine-tuning the lipogenic/lipolytic balance to optimize the metabolic requirements of cancer cell growth: molecular mechanisms and therapeutic perspectives," *Biochimica et Biophysica Acta*, vol. 1801, no. 3, pp. 381–391, 2010.

[48] X. Bi, B. Rexer, C. L. Arteaga, M. Guo, and A. Mahadevan-Jansen, "Evaluating HER2 amplification status and acquired drug resistance in breast cancer cells using Raman spectroscopy," *Journal of Biomedical Optics*, vol. 19, no. 2, p. 025001, 2014.

[49] H. Ilkhani, T. Hughes, J. Li, C. J. Zhong, and M. Hepel, "Nanostructured SERS-electrochemical biosensors for testing of anticancer drug interactions with DNA," *Biosensors & Bioelectronics*, vol. 80, pp. 257–264, 2016.

[50] U. S. Dinish, C. Y. Fu, K. S. Soh, B. Ramaswamy, A. Kumar, and M. Olivo, "Highly sensitive SERS detection of cancer proteins in low sample volume using hollow core photonic crystal fiber," *Biosensors & Bioelectronics*, vol. 33, no. 1, pp. 293–298, 2012.

[51] A. Nilstad, T. E. Andersen, E. Kristianslund et al., "Physiotherapists can identify female football players with high knee valgus angles during vertical drop jumps using real-time observational screening," *The Journal of Orthopaedic and Sports Physical Therapy*, vol. 44, no. 5, pp. 358–365, 2014.

[52] S. Lee, H. Chon, J. Lee et al., "Rapid and sensitive phenotypic marker detection on breast cancer cells using surface-enhanced Raman scattering (SERS) imaging," *Biosensors & Bioelectronics*, vol. 51, pp. 238–243, 2014.

[53] M. B. Wabuyele, F. Yan, and T. Vo-Dinh, "Plasmonics nanoprobes: detection of single-nucleotide polymorphisms in the breast cancer BRCA1 gene," *Analytical and Bioanalytical Chemistry*, vol. 398, no. 2, pp. 729–736, 2010.

[54] H. Hwang, H. Chon, J. Choo, and J. K. Park, "Optoelectrofluidic sandwich immunoassays for detection of human tumor marker using surface-enhanced Raman scattering," *Analytical Chemistry*, vol. 82, no. 18, pp. 7603–7610, 2010.

[55] S. Lee, H. Chon, S. Y. Yoon et al., "Fabrication of SERS-fluorescence dual modal nanoprobes and application to multiplex cancer cell imaging," *Nanoscale*, vol. 4, no. 1, pp. 124–129, 2012.

[56] S. Kang, Y. Wang, N. P. Reder, and J. T. Liu, "Multiplexed molecular imaging of biomarker-targeted SERS nanoparticles on fresh tissue specimens with channel-compressed spectrometry," *PloS One*, vol. 11, no. 9, Article ID e0163473, 2016.

[57] M. Tonegutti and V. Girardi, "Stereotactic vacuum-assisted

breast biopsy in 268 nonpalpable lesions," *La Radiologia Medica*, vol. 113, no. 1, pp. 65–75, 2008.

[58] R. J. Jackman and J. Rodriguez-Soto, "Breast microcalcifications: retrieval failure at prone stereotactic core and vacuum breast biopsy—frequency, causes, and outcome," *Radiology*, vol. 239, no. 1, pp. 61–70, 2006.

[59] I. Barman, N. C. Dingari, A. Saha et al., "Application of Raman spectroscopy to identify microcalcifications and underlying breast lesions at stereotactic core needle biopsy," *Cancer Research*, vol. 73, no. 11, pp. 3206–3215, 2013.

[60] A. Saha, I. Barman, N. C. Dingari et al., "Raman spectroscopy: a real-time tool for identifying microcalcifications during stereotactic breast core needle biopsies," *Biomedical Optics Express*, vol. 2, no. 10, pp. 2792–2803, 2011.

[61] M. P. Morgan, M. M. Cooke, and G. M. McCarthy, "Microcalcifications associated with breast cancer: an epiphenomenon or biologically significant feature of selected tumors?" *Journal of Mammary Gland Biology and Neoplasia*, vol. 10, no. 2, pp. 181–187, 2005.

[62] P. Matousek and N. Stone, "Emerging concepts in deep Raman spectroscopy of biological tissue," *The Analyst*, vol. 134, no. 6, pp. 1058–1066, 2009.

[63] R. Baker, K. D. Rogers, N. Shepherd, and N. Stone, "New relationships between breast microcalcifications and cancer," *British Journal of Cancer*, vol. 103, no. 7, pp. 1034–1039, 2010.

[64] L. Liang, C. Zheng, H. Zhang et al., "Exploring type II microcalcifications in benign and premalignant breast lesions by shell-isolated nanoparticle-enhanced Raman spectroscopy (SHINERS), Spectrochimica acta," *Part a, Molecular and Biomolecular Spectroscopy*, vol. 132, pp. 397–402, 2014.

[65] P. Toniolo, A. L. Van Kappel, A. Akhmedkhanov et al., "Serum carotenoids and breast cancer," *American Journal of Epidemiology*, vol. 153, no. 12, pp. 1142–1147, 2001.

[66] G. E. Dressing, T. P. Knutson, M. J. Schiewer et al., "Progesterone receptor-cyclin D1 complexes induce cell cycle-dependent transcriptional programs in breast cancer cells," *Molecular Endocrinology*, vol. 28, no. 4, pp. 442–457, 2014.

[67] M. C. Potcoava, G. L. Futia, J. Aughenbaugh, I. R. Schlaepfer, and E. A. Gibson, "Raman and coherent anti-Stokes Raman scattering microscopy studies of changes in lipid content and composition in hormone-treated breast and prostate cancer cells," *Journal of Biomedical Optics*, vol. 19, no. 11, p. 111605, 2014.

[68] M. Hedegaard, C. Krafft, H. J. Ditzel, L. E. Johansen, S. Hassing, and J. Popp, "Discriminating isogenic cancer cells and identifying altered unsaturated fatty acid content as associated with metastasis status, using k-means clustering and partial least squares-discriminant analysis of Raman maps," *Analytical Chemistry*, vol. 82, no. 7, pp. 2797–2802, 2010.

22

Spectral Quantitative Analysis Model with Combining Wavelength Selection and Topology Structure Optimization

Qian Wang,[1] Boyan Cai,[2] Yajie Yu,[2] and Hui Cao[2]

[1]*School of Automation and Information Engineering, Xi'an University of Technology, Xi'an, Shaanxi 710048, China*
[2]*Shaanxi Key Laboratory of Smart Grid and State Key Laboratory of Electrical Insulation and Power Equipment,*
 School of Electrical Engineering, Xi'an Jiaotong University, Xi'an, Shaanxi 710049, China

Correspondence should be addressed to Hui Cao; huicao@mail.xjtu.edu.cn

Academic Editor: K. S. V. Krishna Rao

Spectroscopy is an efficient and widely used quantitative analysis method. In this paper, a spectral quantitative analysis model with combining wavelength selection and topology structure optimization is proposed. For the proposed method, backpropagation neural network is adopted for building the component prediction model, and the simultaneousness optimization of the wavelength selection and the topology structure of neural network is realized by nonlinear adaptive evolutionary programming (NAEP). The hybrid chromosome in binary scheme of NAEP has three parts. The first part represents the topology structure of neural network, the second part represents the selection of wavelengths in the spectral data, and the third part represents the parameters of mutation of NAEP. Two real flue gas datasets are used in the experiments. In order to present the effectiveness of the methods, the partial least squares with full spectrum, the partial least squares combined with genetic algorithm, the uninformative variable elimination method, the backpropagation neural network with full spectrum, the backpropagation neural network combined with genetic algorithm, and the proposed method are performed for building the component prediction model. Experimental results verify that the proposed method has the ability to predict more accurately and robustly as a practical spectral analysis tool.

1. Introduction

Spectral quantitative analysis is a nondestructive and fast measurement technique and has been used in a variety of chemical fields [1–3]. The method measures the chemical composition dependent absorption of light that occurs at different wavelengths [4]. Based on the obtained wavelength signals, the spectral quantitative analysis model is built to predict the component concentrations by the regression algorithms [5].

Partial least squares (PLS) is a classical multivariate regression approach for spectroscopy quantitative analysis, and it could handle the multiple correlation among the input wavelength signals [6]. Nevertheless, PLS is a linear regression algorithm essentially [7], and the nonlinearity of wavelength signals may be generated by the instrument variation and the analyte characteristics [8]. To deal with the nonlinear factors, neural network is always adopted for spectral model. Neural network could approximate any function by some

simple interconnected processing units whose structure is inspired by animal brains [9, 10]. Backpropagation neural network (BPNN), as a popular neural network, uses the mean square error and the gradient descent for modifying the connection weights of the neurons [11]. The topology structure of BPNN is usually determined by the human experience [12] and may affect the model effectiveness. That may be one reason why three-layer BPNN is widely used [13–16].

Moreover, the spectral instrument usually records a large number of spectral wavelength signals and the regression model is generally performed based on the obtained wavelengths. However, not all of the obtained wavelengths have the useful information, and the wavelengths without any critical information would corrupt the prediction model [17, 18]. Therefore, wavelengths selection is a vital process for spectral quantitative analysis, and the goal of wavelengths selection is to determine a subset of the obtained spectral wavelengths that could generate the smallest possible errors of the regression models [19, 20]. Some statistical techniques have been adopted for the wavelength selection, and the importance of each wavelength could be estimated according to the statistical features of the prediction model [21, 22]. Uninformative variable elimination (UVE) is proposed to eliminate the wavelengths that do not contain much information for analyte prediction than random variables [23]. Although UVE is better than the statistical wavelength selection method [24, 25], the effectiveness of UVE would be affected by the quality of the random variables and the selection result is scattered throughout the spectrum [26].

BPNN could be optimized by the heuristic algorithm, and BPNN based on genetic algorithm (GA-BPNN) is proposed for determining the initial connection weights and the thresholds in a fixed topology structure [27, 28]. Furthermore, the wavelength selection could seem as a combinatorial problem; the genetic algorithm combined with PLS (GA-PLS) is presented, where GA finds the optimal subset of wavelengths associated with the PLS model [29, 30]. Because the model structure should be determined based on the number of the selected wavelengths, wavelength selection and the topology structure of BPNN would be optimized simultaneously, that is, a hybrid optimization problem. Evolutionary programming (EP) having no fixed structure outperforms with GA and is suitable for the hybrid optimization problem [31, 32]. Like GA, EP has the crossover operation and the mutation operation. However, the crossover operation of EP is limited by the chromosome form for the hybrid optimization problem, that may result in the side-effect, and EP without the crossover process would not reduce the search efficiency [33]. Furthermore, EP generally has the static mutation probability, and EP may fall into the local minima, which is similar to other searching algorithms [34].

In this paper, a spectral quantitative analysis model with combining wavelength selection and topology structure optimization is proposed. For the proposed method, BPNN is adopted for building the component prediction model, and the simultaneousness optimization of the wavelength selection and the topology structure of BPNN is realized by the nonlinear adaptive evolutionary programming (NAEP). The hybrid chromosome in binary scheme of NAEP has four fragments, which represent the number of the hidden layers of BPNN, the number of neurons in each hidden layer, the selection of spectral wavelengths, and two adaptive parameters of the mutation probability of NAEP, respectively. Hence, a chromosome represents an optimization plan. NAEP only has the mutation operation for the next generation, and the mutation probability of each chromosome is updated by a nonlinear equation with considering two adaptive parameters and the fitness values. For the initial generation of NAEP, each chromosome is encoded randomly. BPNN is performed on the calibration set based on different optimization plans represented by different chromosomes. The root-mean-squares error of cross-validation (RMSECV) is the fitness function; namely, the lower the RMSECV, the better the chromosome. The better parent chromosomes would be put into the next generation. The mutation probabilities of other chromosomes are updated according to the latest evaluation results, and the chromosomes are evolved only by the mutation operation.

The evolution process of NAEP terminates based on the stop condition. The chromosome with lowest fitness value is the final result; namely, the selected wavelength and the corresponding topology structure of BPNN are determined. Two real flue gas datasets are employed in the experiments. The effectiveness of PLS, BPNN, GA-BPNN, UVE, GA-PLS, and the proposed method is compared.

The remainder of this paper is organized as follows. In Section 2, The related methods are demonstrated. In Section 3, the proposed method is presented. In Section 4, the experimental results are discussed. Section 5 concludes the paper.

2. The Related Methods

2.1. PLS.

For PLS, X represents the input wavelength signals, and the component Y can be expressed by

$$Y = XM + \varepsilon, \tag{1}$$

where M is the matrix of regression coefficients and ε is the error vector.

It assumes that a small number of the latent variables are refined by linear combinations of the vectors of X. Then (1) can be transformed to

$$Y = TC + \varepsilon,$$
$$T = XQ\left(U^T Q\right)^{-1}, \tag{2}$$

where the matrix T is corresponding to the latent variables and C is the regression coefficients vector.

For T, X is the input matrix, Q is the matrix of weight loading representing the correlation between Y and X, and P is the matrix indicating the influence of X.

2.2. BPNN.

BPNN connects the input layer and the output layer by one or more hidden layers. For spectroscopy quantitative analysis, the wavelengths are the signals of the input layer and the component concentration is the signal of the output layer [35]. A neuron is an activation function which is described by the tansig function, and the transfer function of the output layer is a purelin function [36]. The training process BPNN has the information forward-propagation algorithm and the error backpropagation training algorithm [11]. For the information forward-propagation algorithm, the values of each layer are calculated based on the activation function and the values of the previous layer. For the error backpropagation training algorithm, the error is propagated from the output layer to the input layer, and the weights are regulated by feedback. The modification of the weights and the offset values makes the actual output be closer to the expected output.

2.3. GA-BPNN.

For GA-BPNN, the chromosome is encoded for the initial connection weights and the thresholds of a fixed topology structure. The individuals of the father generation are generated randomly. Then, BPNN is performed based on the information represented by each individual, and the fitness value of each individual is evaluated. Some individuals are reserved for the next generation during the selection operation and their fitness values have a great impact on the reserve probability. Some new individuals are obtained by the crossover operation and the mutation operation. The reserved individuals and the new individuals form the next generation, and the iteration procedure is running constantly until the program satisfies its requirements. After the initial weights are determined, the backpropagation training method is used to adjust the final weights of BPNN.

2.4. UVE. For UVE, an auxiliary matrix containing random noise is generated firstly and it has the same size as the input matrix. Then, the input matrix is combined with the auxiliary matrix to form the combination matrix, which has twice as many wavelength signals as the input matrix. PLS is performed on the combination matrix with the leave-one-out procedure. The criterion value of each column of the combination matrix is estimated by the average of its regression vector and its standard deviation. The original wavelength signal whose criterion value is not larger than a threshold is the uninformative wavelength and would be eliminated, where the threshold is set as the maximum value of the ratio of coefficient to the standard deviation of the auxiliary matrix region. Hence, UVE selects the wavelengths swiftly and practically.

2.5. GA-PLS. In the GA-PLS method, the chromosome is coded by a binary string, and the length of a chromosome is equal to the number of all the wavelengths. Each gene of the chromosome is 1 or 0, which indicates that the wavelength is selected or dropped. For GA-PLS, a random population including a number of chromosomes is initialized, and the PLS model is built for each chromosome, where each chromosome represents a solution of wavelength selection. The prediction precision of PLS model is adopted as the fitness values. A new population is generated by the selection, the crossover, and the mutation. The iteration process is repeated and terminates with reaching the condition, which is the number of iterations or a predefined fitness value. Then, the chromosome with the smallest fitness value is the final result of the wavelength selection.

3. The Proposed Method

For the proposed method, BPNN is adopted for building the component prediction model, and NAEP simultaneously optimizes the wavelength selection and the topology structure of BPNN.

For the new individuals of the next generation, NAEP has not the crossover operation and only has the mutation operation in evolving process. The mutation probability (P_m) is updated by

$$P_m = p_1 + \frac{1}{\sqrt{2\pi}\sigma} \exp\left(-\frac{\text{NTV}^2}{2\sigma^2}\right), \qquad (3)$$

where p_1 and σ are two adaptive parameters and NTV is the normalized fitness value.

The hybrid chromosome in binary scheme of NAEP has four fragments, which is shown in Figure 1. Fragment

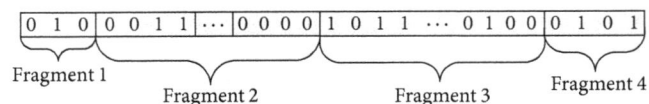

FIGURE 1: The structure of the chromosome.

1 represents the number of the hidden layers (N_{hl}). With considering the model complexity of BPNN, fragment 1 has three genes; namely, the maximum value of N_{hl} is 7. Fragment 2 has twenty-eight genes and every four genes are used for representing the number of neurons in the hidden layer; namely, the maximum value of neurons in each hidden layer is 15. If N_{hl} is the number of hidden layers determined by fragment 1, the values of the genes from the ($4N_{hl} + 1$) position to the end position of fragment 2 are all zero. Fragment 3 is used for the wavelength selection. The length of fragment 3 is equal to the number of all the wavelengths. Each gene of fragment 3 is 1 or 0, which represents that the corresponding wavelength is selected or dropped. Fragment 4 adopts two parts for representing p_1 and σ, respectively, and each part has two genes. The binary value of part 1 is 00, 01, 10, or 11, which represents that p_1 is 0.05, 0.1, 0.15, or 0.2, respectively. In the same way, the different binary values of part 2 represent that σ is 0.35, 0.45, 0.55, or 0.65.

The steps of the proposed method are described in the following.

Step 1 (initialization and evaluation of fitness values). Each chromosome is initialized randomly. BPNN is performed on the calibration set according to the information represented by each chromosome. For BPNN, the number of the hidden layers and the number of neurons in the hidden layer are determined by fragment 1 and fragment 2, respectively. The number of neurons in the input layer of BPNN is the number of selected wavelengths, which is represented by fragment 3. The number of neurons in output layer of BPNN is 1, which is the prediction value of a component concentration. The fitness function is RMSECV. The lower the RMSECV, the better the chromosome.

Step 2 (selection). The elitists strategy is adopted in the selection operation of the proposed method. The chromosomes in the parent generation are ranked based on the ascending order of the corresponding RMSECV values. The top 10% ranked chromosomes are retained for the next generation. For the elitists strategy, the quality information of population is preserved in the iteration process; namely, the search scope could be guided in the optimum direction and the convergence speed would be improved.

Step 3 (mutation). The mutation operation realizes the diversity of the population. The roulette wheel mechanism is used to perform the proportionally choice for the chromosomes being mutated. The mutation probability value of each chosen chromosome is calculated by (3), where p_1 and σ are determined by fragment 4.

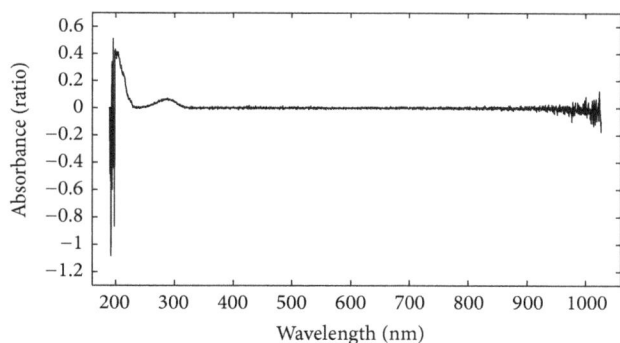

FIGURE 2: The spectrum of Dataset 1.

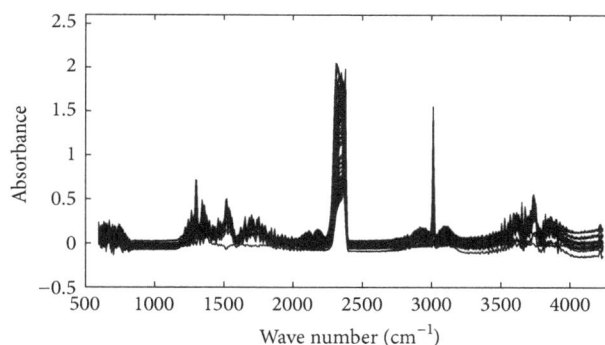

FIGURE 3: The spectrum of Dataset 2.

Step 4 (termination and output). When the number of iterations equals the predefined limit, the proposed algorithm is stopped and the chromosome with the smallest fitness value is output.

According to the best individual obtained by the proposed method, the topology structure of BPNN is determined and the wavelengths are selected. The spectral quantitative analysis model built by the optimized BPNN with the selected wavelengths would have higher accuracy. In the next section, the experiments will be performed to further verify the effectiveness of the proposed method.

4. Experimental Results

Two real flue gas datasets are employed in the experiments, and they are obtained during a combustion process.

4.1. Experimental Datasets

Dataset 1. The dataset is collected during the coal combustion process. It includes 98 samples and each sample consists of a spectrum for a mixture of sulfur dioxide (SO_2), nitrogen monoxide (NO), and nitrogen dioxide (NO_2). The concentration ranges are 0–1500 ppm, 0–3000 ppm, and 0–500 ppm for SO_2, NO, and NO_2, respectively. The absorbance spectra are measured by the USB2000t fiber optic spectrometer. The range of spectral number is from 187.87 nm to 1026.97 nm with the resolution of 0.35 nm. Each spectrum contains 2048 wavelengths. There is some noise in wavelengths less than 200 nm. To investigate the robustness of the proposed method, these noise wavelengths are still a part of input data. The spectrum of Dataset 1 is shown in Figure 2.

Dataset 2. The dataset is recorded during the nature gas combustion process by the GASMET DX4000 Fourier transform infrared (FTIR) gas analyzer and includes 106 samples. Each sample consists of different densities of methane (CH_4), carbon monoxide (CO), and carbon dioxide (CO_2). The wavelength range is from 549.44 cm^{-1} to 4238.28 cm^{-1} with an interval of 7.72 cm^{-1}; namely, each sample has 473 wavelength signals. The concentration ranges of CH_4, CO, and CO_2 are 0–0.0459 ppm, 0–0.4083 ppm, and 0–0.3818 ppm, respectively. The spectrum of Dataset 2 is shown in Figure 3.

4.2. Experimental Procedure.
In the experiments, PLS, BPNN, GA-BPNN, UVE, GA-PLS, and the proposed method are performed on the datasets. Each dataset is separated into a calibration set and a validation set with the shutter grouping strategy [11]. A fifth of the total samples would be put into the validation dataset and the rest of the samples are put into calibration dataset. The calibration set is used to build the prediction model, and the validation set is used for evaluating the effectiveness of the model.

For BPNN, three layers are used and they are the input layer, the hidden layer, and the output layer. The number of neurons in the input layer equals the number of all wavelengths. The number of neurons in the hidden layer is 15. The number of neurons in the output layer is 1; namely, the output is the component concentration. For PLS, UVE, and GA-PLS, the numbers of latent variables with the smallest RMSECV value are determined [37]. For GA-PLS, the latent variables for each individual in the population need to be redetermined at every iteration. For GA-BPNN, GA-PLS, and the proposed method, the population size is 40 and the fitness function is the RMSECV value. For GA-BPNN and GA-PLS, the crossover probability is 0.6, and the mutation probability is 0.01, which are empirically determined by experiences from the series of the GA-BPNN studies. In the experiments, 10-fold cross-validation is employed for the RMSECV value. Furthermore, the root mean-squared error of prediction (RMSEP), the squared cross-validation correlation coefficient (R_{cv}^2), the squared correlation coefficient of calibration (R_c^2), the squared correlation coefficient of prediction (R_p^2), and the compression ratio (CR) would be taken into account for comparing the predictive ability of different models. CR equals $(N_t - N_s)/N_t \times 100\%$, where N_t is the number of total wavelengths and N_s is the number of the selected wavelengths.

4.3. Results Analysis.
The experimental results of Dataset 1 for SO_2 are shown in Table 1. Although the CR value of the proposed method is smaller than that of UVE, the RMSEP value of the proposed method is smallest; namely, it is 40.57%, 69.74%, 34.59%, 21.66%, and 18.3% lower than that of PLS, BPNN, GA-BPNN, UVE, and GA-PLS, respectively. UVE may ignore the wavelengths with useful information. The CR value of the proposed method is larger than that of GA-PLS.

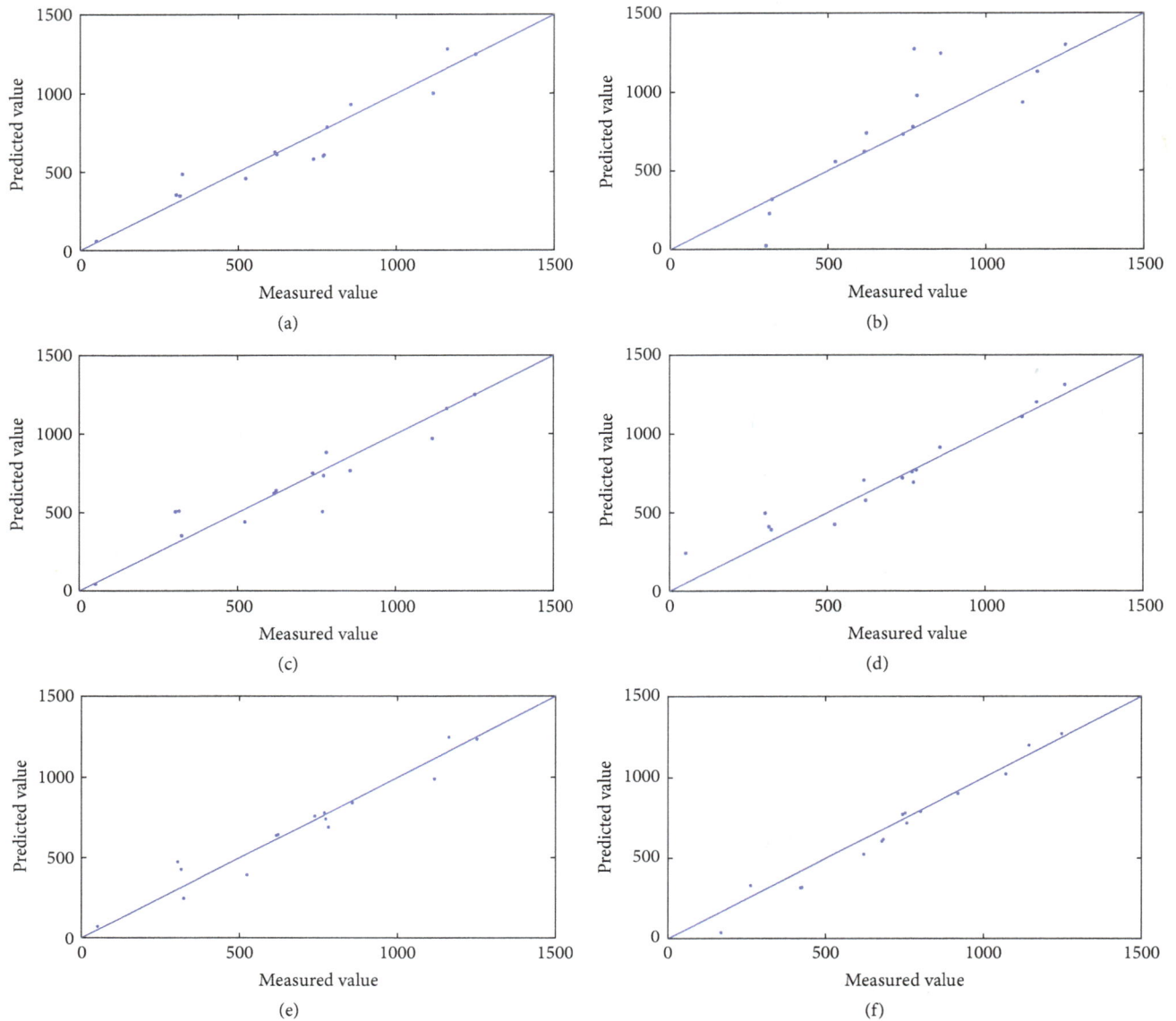

FIGURE 4: The predicted value versus the measured value scatter diagram of different methods for SO_2. (a) PLS. (b) BPNN. (c) GA-BPNN. (d) UVE. (e) GA-PLS. (f) The proposed method.

TABLE 1: The experimental results of Dataset 1 for SO_2.

SO_2	RMSECV	RMSEP	R^2_{cv}	R^2_p	CR
PLS	88.1731	67.2346	0.9600	0.9810	0
BPNN	219.4631	132.0324	0.9358	0.7939	0
GA-BPNN	66.3185	61.0886	0.9726	0.8700	0
UVE	57.9015	51.0014	0.9853	0.9860	0.8365
GA-PLS	55.1286	48.9536	0.9799	0.9855	0.5149
The proposed method	55.0993	39.9556	0.9834	0.9883	0.5509

The RMSECV value of the proposed method is 55.0993 and is also the smallest. Figure 4 shows the predicted value versus the measured value scatter diagram of different methods for

SO_2. PLS, UVE, and GA-PLS have better performance than BPNN. GA-BPNN is distributed as close to the diagonal line as PLS, UVE, and GA-PLS. The proposed method has the best result and is distributed more closed to the diagonal line on both sides. Therefore, the prediction ability of the proposed method is higher than the other methods for SO_2 of Dataset 1.

The experimental results of Dataset 1 for NO_2 are shown in Table 2. PLS has the worst performance and its RMSEP value is 406.1298. The RMSECV value of the proposed method is smallest, and the RMSEP value of the proposed method is 68.98%, 68.92%, 40.04%, 36.69%, and 52.08% lower than that of PLS, BPNN, GA-BPNN, UVE, and GA-PLS, respectively. Although the CR value of UVE-PLS is larger than that of the proposed method, other indicators

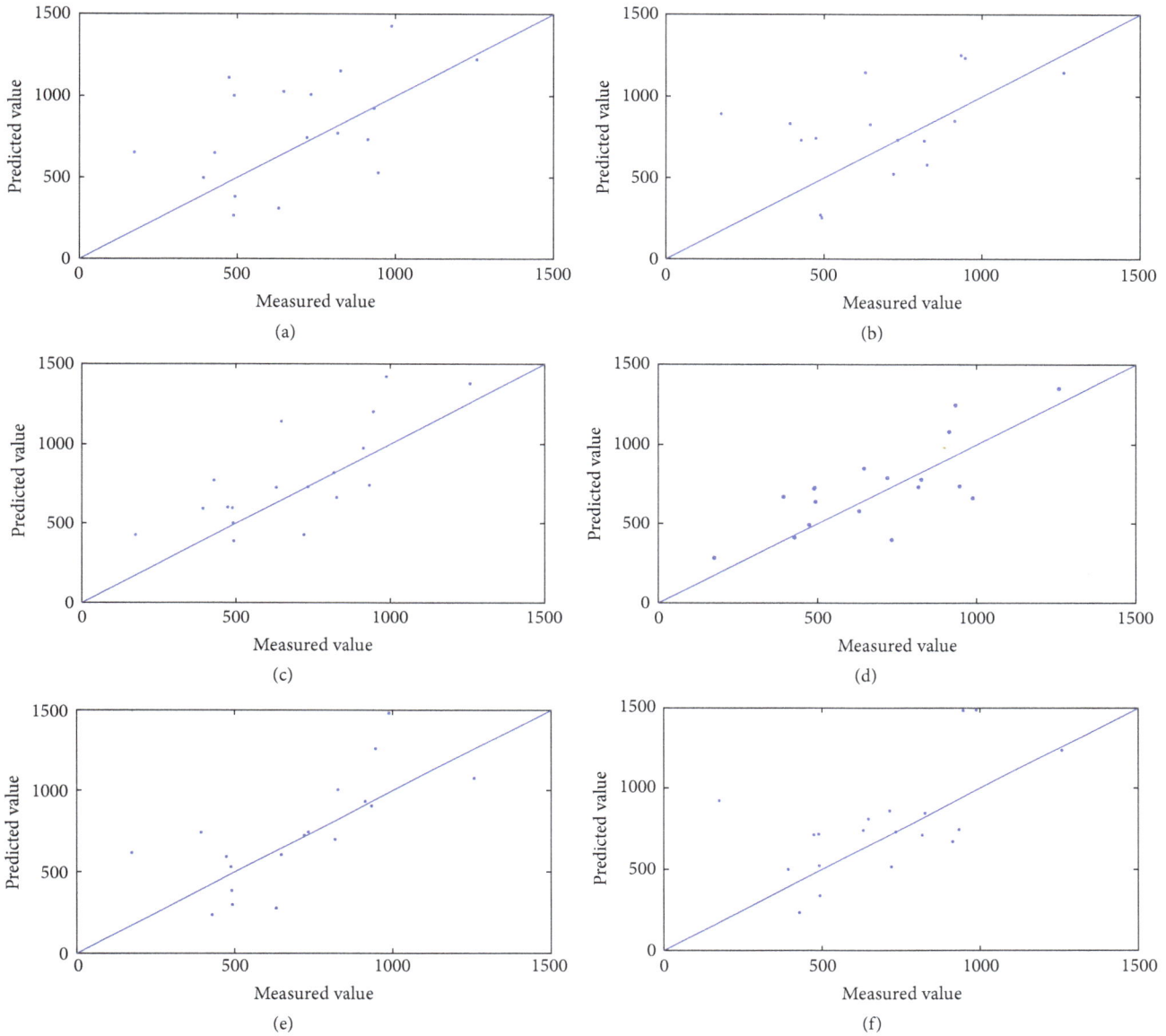

FIGURE 5: The predicted value versus the measured value scatter diagram of different methods for NO_2. (a) PLS. (b) BPNN. (c) GA-BPNN. (d) UVE. (e) GA-PLS. (f) The proposed method.

show the proposed method has better performance. Moreover, the CR value of the proposed method is higher than that of GA-PLS. Figure 5 shows the predicted value versus the measured value scatter diagram of different methods for NO_2. The points of PLS and BPNN spread on both sides of the diagonal line, while the points of GA-PLS, UVE, and GA-BPNN are more close to the line. The proposed method has the best performance as the points are closest to the line. Furthermore, the proposed method is the most robust of all as the points are plotted with roughly the same distance to the line. Hence, the proposed method has higher prediction precision for NO_2 of Dataset 1.

Table 3 demonstrates the experimental results of Dataset 1 for NO. The RMSECV value of the proposed method is smallest. The RMSEP value of BPNN is largest. The RMSEP

TABLE 2: The experimental results of Dataset 1 for NO_2.

NO_2	RMSECV	RMSEP	R_{cv}^2	R_p^2	CR
PLS	424.2092	406.1298	0.4910	0.2198	0
BPNN	488.1834	405.3482	0.6651	0.6515	0
GA-BPNN	213.2712	210.1223	0.8314	0.7028	0
UVE	183.9214	198.9831	0.9001	0.7713	0.9818
GA-PLS	249.2341	262.9014	0.8199	0.6310	0.5039
The proposed method	128.5434	125.9855	0.8813	0.8854	0.5218

value of the proposed method is 33.34%, 85.76%, 8.17%, 7.20%, and 41.93% smaller than that of PLS, BPNN, GA-BPNN, UVE, and GA-PLS, respectively. R_{cv}^2 and R_p^2 of the

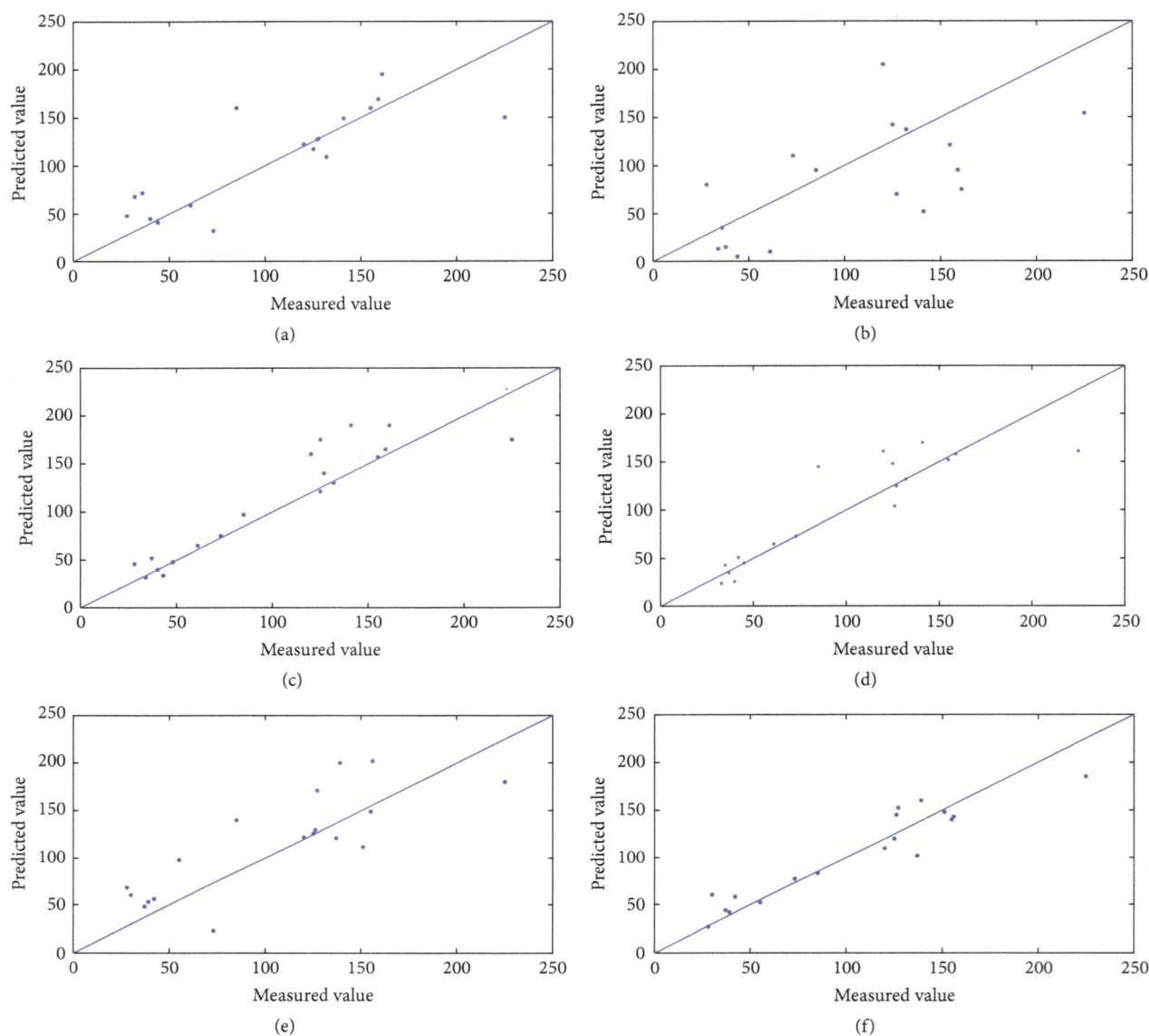

FIGURE 6: The predicted value versus the measured value scatter diagram of different methods for NO. (a) PLS. (b) BPNN. (c) GA-BPNN. (d) UVE. (e) GA-PLS. (f) The proposed method.

TABLE 3: Analytical results for NO.

NO	RMSECV	RMSEP	R_{cv}^2	R_p^2	CR
PLS	45.2984	33.1924	0.5319	0.6906	0
BPNN	145.7634	155.3482	0.6651	0.6165	0
GA-BPNN	31.8127	24.0923	0.8545	0.7314	0
UVE	25.6659	23.8409	0.8383	0.8790	0.8545
GA-PLS	24.7231	38.1008	0.8804	0.6172	0.5059
The proposed method	21.9856	22.1235	0.9882	0.9888	0.5926

proposed method are the largest. The predicted value versus the measured value scatter diagram of different methods for NO is shown in Figure 6. The PLS, BPNN, and GA-PLS do not obtain the good results as many points are far away from

the diagonal line. The points of GA-BPNN, UVE, and the proposed method are more close to the line. Some of points of the proposed method are on the line, and the average distance between the points and the line of the proposed method is smaller than those of GA-BPNN and UVE. Thus, the performance of the proposed method is best for the NO of Dataset 1.

In the same way, the analytical results for Dataset 2 are discussed in the following. The experimental results of Dataset 2 for CH_4 are shown in Table 4. Although the CR value of UVE is largest, the performance of UVE is worst because the RMSEP value of UVE is largest. The RMSEP value of the proposed method is 23.39%, 37.01%, 20.95%, 44.70%, and 21.68% smaller than that of PLS, BPNN, GA-BPNN, UVE, and GA-PLS, respectively. The RMSECV value

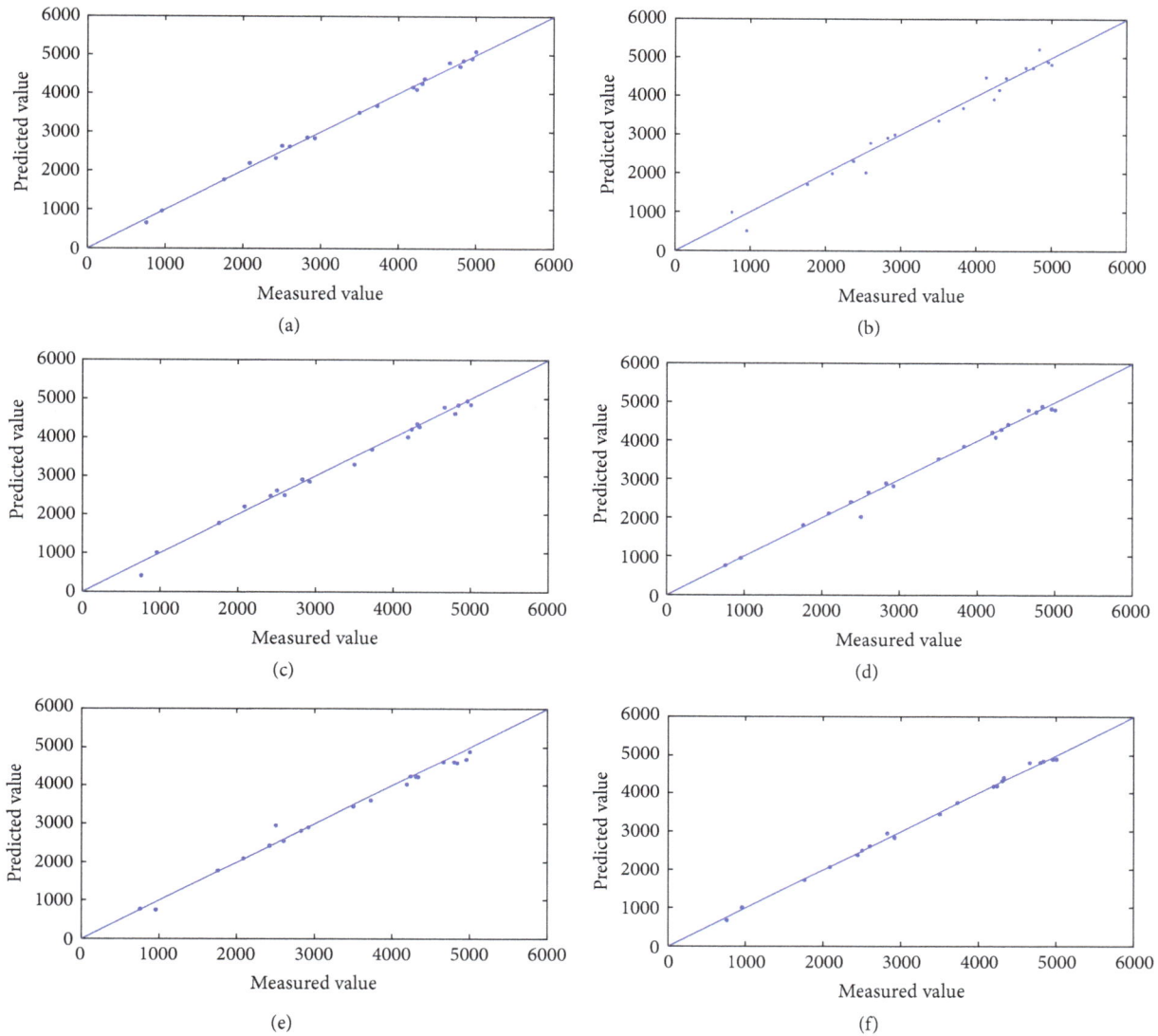

FIGURE 7: The predicted value versus the measured value scatter diagram of different methods for CH_4. (a) PLS. (b) BPNN. (c) GA-BPNN. (d) UVE. (e) GA-PLS. (f) The proposed method.

of the proposed method is also smallest. Figure 7 shows the predicted value versus the measured value scatter diagram of different methods for CH_4. PLS, BPNN, and GA-BPNN have less points which are right located on the diagonal line. There are one or two points of UVE and GA-PLS that are not close to the line. Most of points of the proposed method are on the diagonal line. Therefore, the accuracy and the robustness of the proposed method for the CH_4 of Dataset 2 are validated.

Table 5 lists the experimental results of Dataset 2 for CO. Although the RMSECV of UVE is smaller than that of the proposed method, the RMSEP of the proposed method is smallest. The RMSEP of the proposed method is 64.69%, 77.61%, 59.74%, 9.44%, and 38.89% smaller than that of PLS, BPNN, GA-BPNN, UVE, and GA-PLS, respectively. Moreover, the CR value of the proposed method is larger than that of GA-PLS. Figure 8 shows the predicted value versus

TABLE 4: Analytical results for CH_4.

CH_4	RMSECV	RMSEP	R^2_{cv}	R^2_p	CR
PLS	115.4345	99.5357	0.9939	0.9956	0
BPNN	101.0567	121.0567	0.9938	0.9752	0
GA-BPNN	97.2312	96.4683	0.9861	0.9844	0
UVE	119.4832	137.8875	0.9935	0.9914	0.8647
GA-PLS	81.8355	97.3635	0.9970	0.9919	0.5285
The proposed method	79.4571	76.2553	0.9978	0.9920	0.5327

the measured value scatter diagram of different methods for CO. The points of PLS, BPNN, and GA-BPNN spread on both sides of the diagonal line, while the points of UVE, GA-PLS, and the proposed method are more close to the diagonal

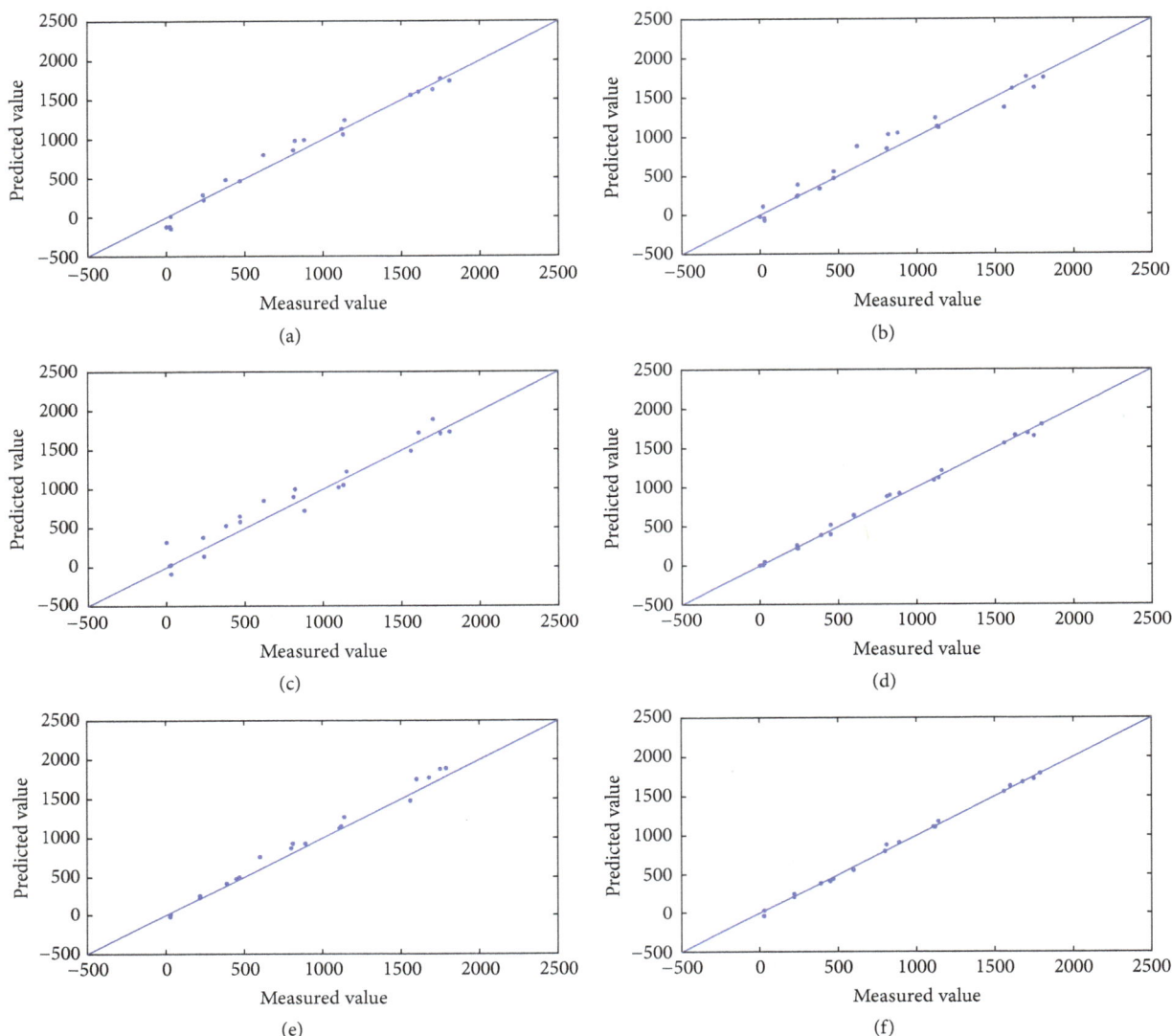

FIGURE 8: The predicted value versus the measured value scatter diagram of different methods for CO. (a) PLS. (b) BPNN. (c) GA-BPNN. (d) UVE. (e) GA-PLS. (f) The proposed method.

TABLE 5: Analytical results for CO.

CO	RMSECV	RMSEP	R_{cv}^2	R_p^2	CR
PLS	78.7763	77.12223	0.9842	0.9855	0
BPNN	150.2814	121.6186	0.9315	0.8482	0
GA-BPNN	82.9801	67.6445	0.9732	0.9581	0
UVE	29.0501	30.0698	0.9982	0.9976	0.8879
GA-PLS	41.4748	44.5643	0.9956	0.9950	0.5497
The proposed method	29.1491	27.2319	0.9993	0.9975	0.5530

line. Furthermore, most of points of the proposed method are settled on the diagonal line. Hence, the performance of the proposed method is the best for the CO of Dataset 2.

The experimental results of Dataset 2 for CO_2 are shown in Table 6. The RMSEP of BPNN equals 57.7472 which is largest. The RMSEP of the proposed method is 36.17%, 38.68%, 36.66%, 26.08%, and 17.32% smaller than that of PLS, BPNN, GA-BPNN, UVE, and GA-PLS, respectively. Figure 9 shows the predicted value versus the measured value scatter diagram of different methods for CO_2. The points of BPNN, GA-BPNN, and UVE are close to the diagonal line, and some points of PLS and GA-PLS are directly on the line. Almost all the points of the proposed method stay right on the line. Therefore, the prediction capability of the proposed method is the best for CO_2 of Dataset 2.

In summary, the experimental results verify that the proposed method is successfully employed for the spectral quantitative analysis of Dataset 1 and Dataset 2 with higher accuracy.

5. Conclusions

This paper proposes a spectral quantitative analysis model with combining wavelength selection and topology structure

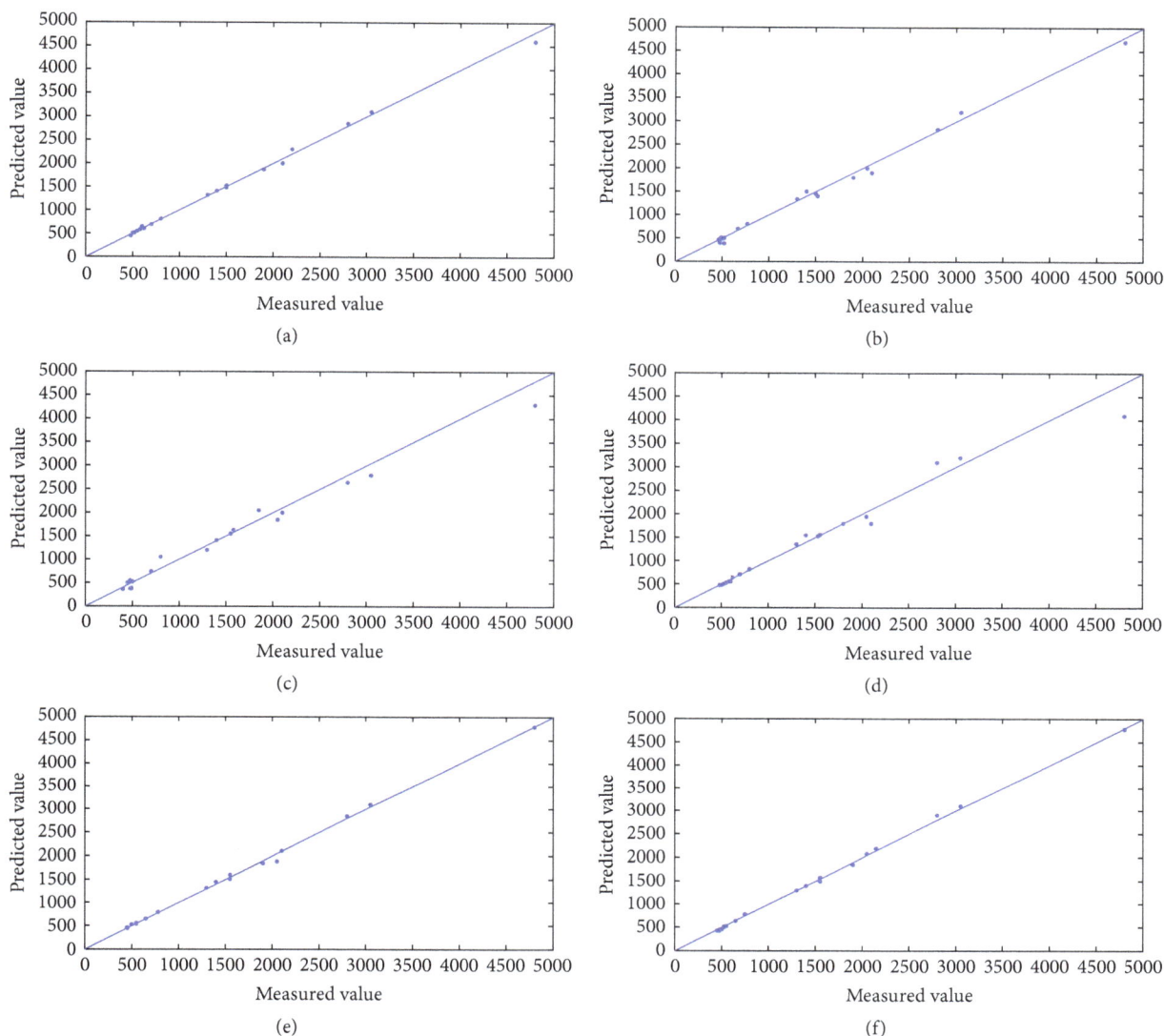

FIGURE 9: The predicted value versus the measured value scatter diagram of different methods for CO_2. (a) PLS. (b) BPNN. (c) GA-BPNN. (d) UVE. (e) GA-PLS. (f) The proposed method.

TABLE 6: Analytical results for CO_2.

CO_2	RMSECV	RMSEP	R^2_{cv}	R^2_p	CR
PLS	41.5959	55.4754	0.9971	0.9981	0
BPNN	55.8712	57.7472	0.9957	0.9895	0
GA-BPNN	51.1870	55.9048	0.9961	0.9901	0
UVE	45.4300	47.9081	0.9963	0.9910	0.9345
GA-PLS	43.5677	42.8292	0.9959	0.9982	0.5137
The proposed method	41.9192	35.4125	0.9980	0.9991	0.5370

optimization. The proposed method has some advantages as follows. First, the proposed method can be used for the spectral quantitative analysis. Second, the proposed method realizes the simultaneousness optimization of the wavelength selection and the topology structure. Third, the proposed method only has the mutation operation which can simplify the iteration procedure without decreasing the precision. The experiments results verify that the proposed method has higher predicative ability for spectral quantitative analysis and can be applied to different types of spectra.

Competing Interests

The authors declare no conflict of interests.

Acknowledgments

This work is supported by the National Natural Science Foundation of China (61375055), the Program for New Century Excellent Talents in University (NCET-12-0447), Natural Science Foundation of Shanxi Province of China (2014JQ8365), State Key Laboratory of Electrical Insulation and Power Equipment (EIPE16313), and the Fundamental

Research Funds for the Central University. The authors are grateful for the grammar modification provided by Yanxia Wang.

References

[1] L. Liang, J. Sun, Q. He, Z. Chen, S. Zhu, and L. Lin, "Semi-quantitative analysis on the content of berberine hydrochloride in compound berberine tablets with the fluorescence spectral imaging method," *Journal of Innovative Optical Health Sciences*, vol. 9, no. 2, Article ID 1650018, 7 pages, 2016.

[2] Y. Ma, M. Wei, X. Zhang, T. Zhao, X. Liu, and G. Zhou, "Spectral study of interaction between chondroitin sulfate and nanoparticles and its application in quantitative analysis," *Spectrochimica Acta—Part A: Molecular and Biomolecular Spectroscopy*, vol. 153, pp. 445–450, 2016.

[3] Z. Wu, L. B. Zeng, and Q. S. Wu, "Study of cervical exfoliated cell's DNA quantitative analysis based on multi-spectral imaging technology," *Spectroscopy and Spectral Analysis*, vol. 36, no. 2, pp. 496–501, 2016.

[4] V. Kopačková, "Using multiple spectral feature analysis for quantitative pH mapping in a mining environment," *International Journal of Applied Earth Observation & Geoinformation*, vol. 28, no. 1, pp. 28–42, 2014.

[5] C. Hua-zhou, S. Qi-qing, S. Kai, and J. Zhen, "Multidimensional scaling linear regression applied to FTIR spectral quantitative analysis of clinical parameters of human blood serum," *Spectroscopy and Spectral Analysis*, vol. 35, no. 4, pp. 914–918, 2015.

[6] Y. Li and J. Jing, "A consensus PLS method based on diverse wavelength variables models for analysis of near-infrared spectra," *Chemometrics and Intelligent Laboratory Systems*, vol. 130, pp. 45–49, 2014.

[7] C. Koch, A. E. Posch, H. C. Goicoechea, C. Herwig, and B. Lendl, "Multi-analyte quantification in bioprocesses by Fourier-transform-infrared spectroscopy by partial least squares regression and multivariate curve resolution," *Analytica Chimica Acta*, vol. 807, pp. 103–110, 2014.

[8] D.-S. Cao, Y.-Z. Liang, Q.-S. Xu, Q.-N. Hu, L.-X. Zhang, and G.-H. Fu, "Exploring nonlinear relationships in chemical data using kernel-based methods," *Chemometrics and Intelligent Laboratory Systems*, vol. 107, no. 1, pp. 106–115, 2011.

[9] V. N. Vapnik, "Statistical learning theory," *Encyclopedia of the Sciences of Learning*, vol. 41, no. 4, pp. 3185–3185, 2010.

[10] A. Rădulescu, "Neural network spectral robustness under perturbations of the underlying graph," *Neural Computation*, vol. 28, no. 1, pp. 1–44, 2016.

[11] J. Yuan and S. Yu, "Privacy preserving back-propagation neural network learning made practical with cloud computing," *IEEE Transactions on Parallel & Distributed Systems*, vol. 25, no. 1, pp. 212–221, 2014.

[12] J. A. K. Suykens, T. V. Gestel, J. D. Brabanter, B. D. Moor, and J. Van-Dewalle, "Least squares support vector machines," *Euphytica*, vol. 2, no. 2, pp. 1599–1604, 2002.

[13] L. Xing and L. Chen, "Quantitative analysis of near-infrared spectroscopy by combined stationary wavelet transform-support vector machine," *Spectroscopy Letters*, vol. 46, no. 1, pp. 47–53, 2013.

[14] J. C. L. Alves and R. J. Poppi, "Biodiesel content determination in diesel fuel blends using near infrared (NIR) spectroscopy and support vector machines (SVM)," *Talanta*, vol. 104, no. 2, pp. 155–161, 2013.

[15] Z.-Y. Deng, B. Zhang, W. Dong, and X.-P. Wang, "Research on prediction method of fatty acid content in edible oil based on raman spectroscopy and multi-output least squares support vector regression machine," *Spectroscopy and Spectral Analysis*, vol. 33, no. 11, pp. 2997–3001, 2013.

[16] R.-M. Luo, S.-M. Tan, Y.-P. Zhou et al., "Quantitative analysis of tea using ytterbium-based internal standard near-infrared spectroscopy coupled with boosting least-squares support vector regression," *Journal of Chemometrics*, vol. 27, no. 7-8, pp. 198–206, 2013.

[17] A.-X. Zhao, X.-J. Tang, Z.-H. Zhang, and J.-H. Liu, "The spectral characteristic wavelength selection and parameter optimization based on tikhonov regularization," *Spectroscopy and Spectral Analysis*, vol. 34, no. 7, pp. 1836–1839, 2014.

[18] L. Ling, L. I. Wei, Z. Rui-li, L. Rui-an, L. Gang, and W. Xiao-rong, "Wavelength selection of the oximetry based on test analysis of variance," *Spectroscopy and Spectral Analysis*, vol. 34, no. 7, pp. 2005–2009, 2014.

[19] X. Shao, G. Du, M. Jing, and W. Cai, "Application of latent projective graph in variable selection for near infrared spectral analysis," *Chemometrics & Intelligent Laboratory Systems*, vol. 114, pp. 44–49, 2012.

[20] M. Vohland, M. Ludwig, S. Thiele-Bruhn, and B. Ludwig, "Determination of soil properties with visible to near- and mid-infrared spectroscopy: effects of spectral variable selection," *Geoderma*, vol. 223–225, no. 1, pp. 88–96, 2014.

[21] L. Xu and I. Schechter, "Wavelength selection for simultaneous spectroscopic analysis. Experimental and theoretical study," *Analytical Chemistry*, vol. 68, no. 14, pp. 2392–2400, 1996.

[22] Y. Dong, B. Xiang, Y. Geng, and W. Yuan, "Rough set based wavelength selection in near-infrared spectral analysis," *Chemometrics & Intelligent Laboratory Systems*, vol. 126, pp. 21–29, 2013.

[23] V. Centner, D.-L. Massart, O. E. De Noord, S. De Jong, B. M. Vandeginste, and C. Sterna, "Elimination of uninformative variables for multivariate calibration," *Analytical Chemistry*, vol. 68, no. 21, pp. 3851–3858, 1996.

[24] Q.-J. Han, H.-L. Wu, C.-B. Cai, L. Xu, and R.-Q. Yu, "An ensemble of Monte Carlo uninformative variable elimination for wavelength selection," *Analytica Chimica Acta*, vol. 612, no. 2, pp. 121–125, 2008.

[25] Y. Hao, X. Sun, H. Zhang, and Y. Liu, "Application of effective wavelength selection methods to determine total acidity of navel orange," *Sensor Letters*, vol. 9, no. 3, pp. 1229–1234, 2011.

[26] J. Ghasemi, A. Niazi, and R. Leardi, "Genetic-algorithm-based wavelength selection in multicomponent spectrophotometric determination by PLS: application on copper and zinc mixture," *Talanta*, vol. 59, no. 2, pp. 311–317, 2003.

[27] R. Irani and R. Nasimi, "Evolving neural network using real coded genetic algorithm for permeability estimation of the reservoir," *Expert Systems with Applications*, vol. 38, no. 8, pp. 9862–9866, 2011.

[28] W. Sun and Y. Xu, "Financial security evaluation of the electric power industry in China based on a back propagation neural network optimized by genetic algorithm," *Energy*, vol. 101, pp. 366–379, 2016.

[29] D. Jouan-Rimbaud, D.-L. Massart, R. Leardi, and O. E. De Noord, "Genetic algorithms as a tool for wavelength selection in multivariate calibration," *Analytical Chemistry*, vol. 67, no. 23, pp. 4295–4301, 1995.

[30] M. Arakawa, Y. Yamashita, and K. Funatsu, "Genetic algorithm-based wavelength selection method for spectral calibration," *Journal of Chemometrics*, vol. 25, no. 1, pp. 10–19, 2011.

[31] R. Nordon, D. Lutz, R. Genzel et al., "The impact of evolving infrared spectral energy distributions of galaxies on star formation rate estimates," *The Astrophysical Journal*, vol. 745, no. 2, p. 167, 2011.

[32] H. Momm and G. Easson, "Evolving spectral transformations for multitemporal information extraction using evolutionary computation," *Journal of Applied Remote Sensing*, vol. 5, no. 1, Article ID 053564, 2011.

[33] Z. Michalewicz, *Genetic Algorithms + Data Structures = Evolution Programs*, Springer, Berlin, Germany, 1992.

[34] A. P. Engelbrecht, *Evolutionary Programming*, John Wiley & Sons, 2007.

[35] P. Sinha, A. Chandwani, and T. Sinha, "Algorithm of construction of optimum portfolio of stocks using genetic algorithm," *International Journal of Systems Assurance Engineering and Management*, vol. 6, no. 4, pp. 447–465, 2015.

[36] Y. Zhang, D. Guo, and Z. Li, "Common nature of learning between back-propagation and hopfield-type neural networks for generalized matrix inversion with simplified models," *IEEE Transactions on Neural Networks & Learning Systems*, vol. 24, no. 4, pp. 579–592, 2013.

[37] J. A. F. Pierna, B. Lecler, J. P. Conzen, A. Niemoeller, B. Baeten, and P. Dardenne, "Comparison of various chemometric approaches for large near infrared spectroscopic data of feed and feed products," *Analytica Chimica Acta*, vol. 705, no. 1-2, pp. 30–34, 2011.

23

Microanalysis of Organic Pigments in Ancient Textiles by Surface-Enhanced Raman Scattering on Agar Gel Matrices

Marilena Ricci,[1] Cristiana Lofrumento,[1] Emilio Castellucci,[1,2] and Maurizio Becucci[1,2]

[1]Department of Chemistry "Ugo Schiff", University of Florence, Via della Lastruccia 3-13, 50019 Sesto Fiorentino, Italy
[2]European Laboratory for Non-Linear Spectroscopy (LENS), Via N. Carrara 1, 50019 Sesto Fiorentino, Italy

Correspondence should be addressed to Maurizio Becucci; maurizio.becucci@unifi.it

Academic Editor: Christoph Krafft

We review some new methods based on surface-enhanced Raman scattering (SERS) for the nondestructive/minimally invasive identification of organic colorants in objects whose value or function precludes sampling, such as historic and archeological textiles, paintings, and drawing. We discuss in detail the methodology we developed for the selective extraction and identification of anthraquinones and indigoids in the typical concentration used in textiles by means of an ecocompatible homogeneous nanostructured agar matrix. The extraction system was modulated according to the chemical properties of the target analyte by choosing appropriate reagents for the extraction and optimizing the extraction time. The system has been found to be extremely stable, easy to use and produce, easy to store, and at the same time able to be analyzed even after long time intervals, maintaining its enhancement properties unaltered, without the detriment of the extracted compound. Highly structured SERS band intensities have been obtained from the extracted dyes adopting laser light excitations at 514.5 and 785 nm of a micro-Raman setup. This analytical method has been found to be extremely safe for the analyzed substrates, thus being a promising procedure for the selective analysis and detection of molecules at low concentration in the field of artworks conservation.

1. Introduction

Identification of dyes used in works of art, belonging to different molecular classes, is essential for dating, restoring, and conserving artwork and for studying art history in general. The importance of the cultural artifacts, such as archeological and ancient textiles, drawings, and paintings, requires techniques that can lead to the unambiguous identification of natural dyes from microscopic samples or directly from artworks. The analytical technique that is currently most common for the identification of historical dyes is high-performance liquid chromatography (HPLC) [1–3]. Although HPLC is a highly specific and very sensitive separation technique, laborious extraction and separation steps could induce molecular changes with a risk of sample loss in separation steps. In addition, its sample requirements (approximately 5 mm of a thread for textile samples) make it less than ideal for those cases where no or little sampling is allowed. Noninvasive and spectroscopic methods of analysis are usually limited to UV-visible absorption or fluorescence spectroscopy and carried out by means of fiber optics probes and they yield spectra that typically exhibit a poor specificity. Instead, vibrational spectroscopies can provide very specific information on the different materials. However, infrared absorption (or reflectance) is a vibrational spectroscopic technique not ideal for such type of applications because of matrix interference effects. Raman spectroscopy (RS) is another vibrational spectroscopy method. It is based on the observation of weak side-bands in the radiation scattered by the sample upon excitation with monochromatic light. Raman spectroscopy requires no sample preparation and offers the potential of noninvasive analysis, that is, a clear advantage when no sampling is allowed. However, only a limited number of Raman spectra of natural dyes are available in the literature [4–7]; most of the published body of work deals with inorganic pigments [8–11]. The two major factors that have hampered the widespread application of Raman spectroscopy for the identification of natural organic dyes are their very high tinting power (i.e., they are likely to be found in works of art in extremely low concentrations) and the strong fluorescence signal generated by excitation with visible light, even when red or near red emitting laser is used and that often obscures the inherently weak Raman scattering signal.

A phenomenon that can significantly enhance the Raman signal intensity is the surface-enhanced Raman scattering (SERS) effect. The Raman scattering efficiency of molecule can be enhanced when the molecule is effectively interacting with a nanostructured metal surface and it is spatially confined within the electromagnetic field of the localized surface plasmon resonance (LSPR) of the nanostructured system [12, 13]. The localized plasma resonance furnishes a mean to funnel electromagnetic energy in the proximity of the surface of the metallic nanoparticles (NPs), such that light scattering phenomena of species adsorbed on the metallic surface can benefit from a substantial increase in the incoming and outgoing (diffused) electromagnetic fields. SERS effect can give enhancements of Raman signals greater than 8 orders of magnitude. Also, SERS is a selective spectroscopic technique as only molecules effectively interacting with the NPs show enhanced signals. In our specific application on dyed textile fibers, the SERS effect is observed only for the dye molecules and not for the substrate. Finally, it must be mentioned that the presence of metal NPs provides a very efficient way to quench fluorescence from the excited molecules, thus removing a further serious limitation for the application of Raman spectroscopy to dye molecules [12, 13]. Therefore, SERS is also very effective for discrimination of molecules of different nature present in artistic samples, and the enhancement of the signal enables us to detect even trace quantities of materials.

Raman experiments can be carried out also at excitation frequencies that are close to (or even resonant) the electronic excitation frequency of the molecule under investigation. In that case, we are dealing with the so-called "resonance Raman" effect that produces much larger signals for specific vibrational modes [14]. Surface-enhanced [resonance] Raman scattering (SE[R]RS) is known to produce further enhancement of the Raman signal. When the LSPR of the enhancing substrate is also in the proper energy region, the SERRS enhancement factor is roughly the product of the enhancement factor for nonresonant SERS and the resonance Raman spectrum intensification factor of the molecule.

In order to improve the quality of the enhancement of the Raman scattering and the reproducibility of the spectra, a wide class of nanostructured substrates such as rough electrodes [15], colloids [16], nanoisland films [17, 18], nanostars [19, 20], nanorods [21], and other nanocomposite technological supports have been carried out. These advancements, together with solid phase microextraction techniques that use synthetic and natural polymeric materials [22], have allowed researchers in the field of cultural heritage to achieve conclusive identification of dyes from artworks with great molecular selectivity and specificity.

The first applications SERS methods for dyes identification on textile fibers were still not enough sensitive to allow nondestructive operations [23]. A first significant step for the increase of sensitivity of this method was represented by the work of Leona [24]. Very small (tens of micrometers) samples containing mordant dyes were treated with HF vapors in order to detach the dye from the mordant. Then, a droplet of colloidal solution containing silver NPs was added to the sample and the dye was free to move and to bind the NPs. Other methods were developed that use a dense dispersion of silver NPs directly on the fibers, leading to an irreversible contamination of the original sample [25, 26]. An innovative and effective method for dyes microextraction followed by SERS dyes identification was recently presented [27]. It was based on the use of common hydrogels based on hydroxymethacrylates (contact lens blanks), loaded with water, organic solvents, and chelating agents as extraction media. Another method was presented based on the application to the sample surface of a cellulose film loaded with silver NPs [28]. The film dries in minutes after application and SERS spectra of anthraquinonic dyes present on the surface were obtained with good sensitivity. The dimensional changes associated with the drying process possibly induces mechanical stress on the sample surface leading to the detachment of small fragments. The film self-detaches or is peeled off by the operator; its formulation was optimized in order to minimize the gel penetration in the samples, to obtain possibly its complete detachment, and to limit the detachment of small particles from the sample surface [29]. Comprehensive reviews on the application of SERS to forensic science and cultural heritage have been already published [30, 31].

This review will discuss the nondestructive microextraction procedure we developed for dye extraction from textile fibers, based on the use of an environmentally friendly nanocomposite hydrogel, to provide a new tool for those researchers who would like to embark on this kind of work using the SERS technique. We will present the results obtained on mordant dyes (anthraquinones) and vat dyes (indigoids) used both in mockups prepared in the laboratory and in ancient textiles.

2. Materials and Methods

2.1. Dyeing of Textiles. Dyes can be divided into three different categories (mordant, vat, and direct dyes), depending on the procedure according to which they are applied to fabrics. In what follows, we are providing information of the different classes of natural organic colorants used in textile dyeing in order to clearly outline the approach we are using.

Colorants named "mordant dyes" do not have a strong chemical affinity for the textile fibers, which therefore require being treated prior to the dyeing stage, involving a two-step chemical reaction. A solution of a "mordant" is first used to impregnate the fibers, allowing the metal ion to become complexed to appropriate functional groups in the structure of the textile. During the dyeing process, the colorant interacts with the mordant-fiber complex via ionic and coordinate covalent bonds to form insoluble brightly colored species. The complex thus formed within the fibers does not easily wash out of the textile and the resulting dyeing is therefore relatively fast. The mordant helped the dye bite onto the fiber so that it would hold fast during washing. A scheme of the binding mechanism for alizarin on cotton fibers is shown in Figure 1. Aluminum, iron, tin, chromium, or copper ions, as well as tannins, are examples of mordants; one of the substances most commonly employed in ancient times for this purpose was potassium alum ($KAl(SO_4)_2 \cdot 12H_2O$), but iron sulfate ($FeSO_4 \cdot 7H_2O$) and tin chloride ($SnCl_2$) were often used as well. Commonly employed in association with mordants are dye-assistants, such as cream of tartar or oxalic acid, which brighten the colors, protect the fibers, and help the absorption of mordants. Mordant dyes can be used with wool, silk, and protein yarns while cellulose fibers such as linen and cotton have to be chemically modified before dyeing. The vast majority of natural colorants belongs to this category and yields different colors when combined with different mordants: a typical example is represented by the madder lake, which can produce red, orange, or violet shades when associated with aluminum, tin, and iron mordants, respectively.

Colorants belonging to the "vat dyes" class are water-insoluble but, under reducing conditions, they can be converted into a *leuco* form, soluble in alkali. Nowadays the reducing reagent used is sodium dithionite, $Na_2S_2O_4$. Immersion of the textile into the dye solution allows the dissolved molecules of colorant to penetrate the fibers. After that, upon removal of the wet dyeing from the bath and exposure to atmospheric oxygen, these substances can be oxidized back to their colored forms which, thanks to their insolubility in water, are trapped on the surface of the fiber. The well-known indigo and woad blue colorants, as well as Tyrian purple, belong to this category. The relevant chemical reactions involved in the dyeing process with indigo are outlined in Figure 2.

Even if direct dyes are applied directly to the fiber without any special treatment, usually they are less wash- and lightfast than vat or mordant dyes. Examples of direct dyes include turmeric and saffron, which can be fixed to all fibrous materials in aqueous solution.

2.1.1. Mockups Dyed with Anthraquinones.

Three pieces of cotton fabric were prepared according to traditional dyeing methodologies, by using alizarin (Sigma-Aldrich), purpurin (Sigma-Aldrich), and carminic acid (Sigma-Aldrich). The textiles have been treated prior to the dyeing step with an alum (Zecchi, Firenze) mordant solution in distilled water (J. T. Baker HPLC Gradient Grade). After the dyeing step, they have been thoroughly washed with distilled water a couple of times and allowed to dry.

2.1.2. Mockups Dyed with Indigoids.

Pieces of cotton have been dyed according to the recipe reported by Schweppe [32]. The dyeing liquor was prepared by stirring 15 g of indigo powder with 75 mL of warm water in a beaker glass until it formed a paste. In a second vessel 30 g of sodium hydroxide, NaOH, was dissolved in 120 mL of warm water. A portion of this solution (60–70 mL) was poured over the indigo paste, stirring vigorously. Then, 30 g of sodium dithionite, $Na_2S_2O_4$, was added while keeping stirring and

1 L of warm water was added while stirring carefully. This mixture was heated to 55°C. Clean pieces of cotton were immersed in warm water until the fabric was thoroughly wet. The fabric was then put inside the dyeing liquor and kept in the dye-bath for some minutes. The pieces of cotton were then taken out of the vat squeezing the liquor out thoroughly. When the fabric comes out of the vat it has a typical green-yellow color, which turns blue when exposed to air. A special precaution was used with the Tyrian purple indigoid dye. The dyeing procedure adopted was similar to the previous one for indigo, using a colorant/NaOH/$Na_2S_2O_4$ ratio equal to 1 : 2 : 4. Since Tyrian purple is a mixture of brominated indigoids, 6-bromoindigo (MBI) and 6,6' dibromoindigo (DBI), it is subject to debromination because of the action of UV-light. For this reason, when the reagents were heated to 55°C, the light was turned off and the flask was wrapped in aluminum foil. A yellow-greenish homogeneous solution was obtained and the wet fabric was put into the dyeing bath for 15 minutes. Then it was removed from the flask and exposed to air while still protected from light. After one hour of air exposure, the fabric was rinsed in aqueous solution and allowed to dry.

2.2. Silver Nanoparticles for SERS.

We used the standard chemical synthesis of Ag nanoparticles (AgNPs) devised by Lee and Meisel [16]. In order to improve the production of suitable nanoparticles (NPs) with an average diameter of 40 nm, we took care of placing the reaction flask, after the reduction step, in an ice bath [33]. The average diameter of the particles was checked from their UV-Vis absorption spectrum, which showed a 100 nm broad peak centered at 418 nm, in good agreement with previous reports [16].

2.3. Agar Gel.

The agar gel usage as the extraction medium was suggested by its intrinsic safety for the operator and its widespread use for the cleaning of artistic materials [34]. It was prepared from agar-agar, a commonly gelling material used for food preparation. Agar-agar is a polymer containing mostly D-galactose units. Agar-agar was repeatedly washed in order to remove possible contaminants and chloride ions. The gel can be easily prepared by mixing 1 g of agar-agar with 50 mL of water, warming it for few seconds in a common microwave oven (300 W power) and cooling down the viscous solution in a convenient glassware (Petri dish). The agar gel becomes rapidly rigid and stable at room temperature. It is important to notice that the agar gel is a very good substrate for bacteria cultures so its practical lifetime could be very short and SERS experimental results can be altered by the presence of such kind of contamination. A convenient practice resulted to be using the colloidal solution containing AgNPs in place of water for the agar gel preparation. It provides a double advantage. On one side the AgNPs act as a bacteriostatic agent, thus preventing the agar gel biological contamination: we have found that this kind of material (Ag-agar), stored in closed containers, can be maintained at room temperatures in the dark for a long time (a few months). On the other side, the agar gel used as dyes extraction medium already carries the AgNPs needed

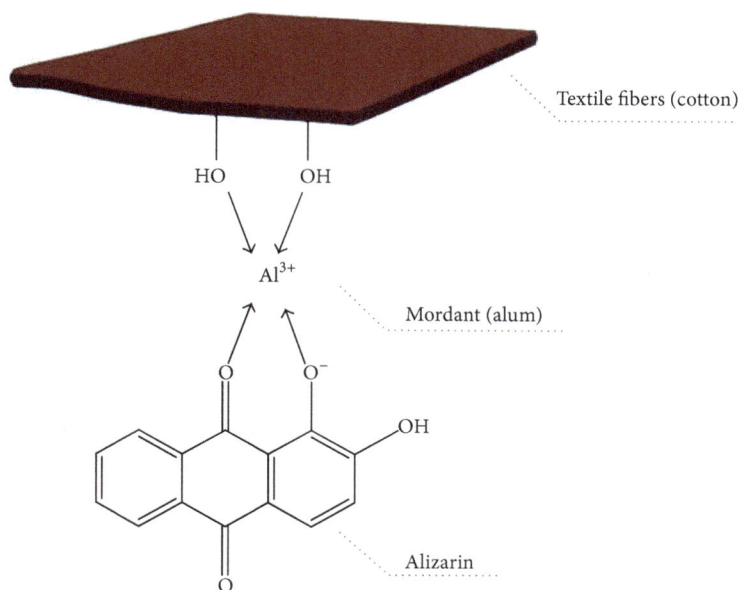

FIGURE 1: Chemical structure of alizarin anion and the bonding mechanism to the cotton fiber mediated by the presence of the Al^{3+} ion from the mordant.

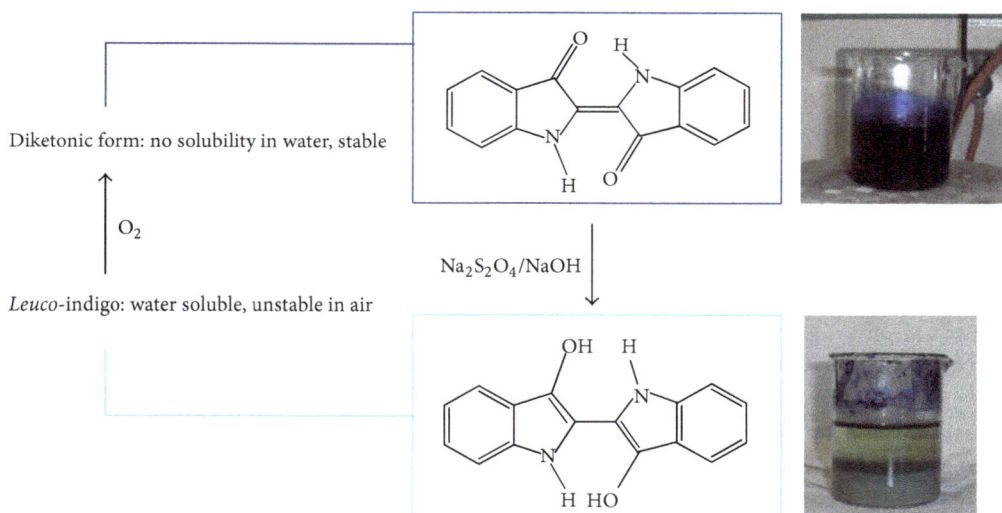

FIGURE 2: Chemical reactions involving indigo during the dyeing process. The indigo blue dye is the diketonic form.

for the SERS measurements. Anyway, droplets of the AgNPs colloidal solution can be poured in the top of the agar gel beads we use, after the extraction process, in order to further improve SERS measurements.

2.4. Sample Extraction. The agar gel can act as a suitable extraction substrate for dyes on textile fibers. Clearly, water already present in the agar gel is not, *per se*, a suitable solvent for the dyes extraction. Different extraction strategies can be used for different dyes classes. Here we are reporting

examples of studies on anthraquinonic [35, 36] and indigoid dyes [37].

Anthraquinonic dyes, like alizarin and carminic acid, are bound to textile fibers by the use of a mordant like alum that introduces Al^{3+} ions. These ions act as a bridge between the fiber and the dye molecule (Figure 1). Both molecules can be tentatively transferred from the textile fiber to the agar gel either by using a solvent in which they are highly soluble (e.g., an alcohol) or by detaching them from the fiber by specific chemical processes. Therefore, in the first approach we added a few ethyl alcohol droplets to the Ag-agar and then we place it in contact with the fiber for a suitable amount of time (the extraction time was 30 minutes). In the second approach, we tried to detach only the dye from the fiber. Even though the chemical etching with HF method was already well established [23], we did try to avoid it as it requires the use of highly toxic chemicals in sealed chamber and the removal of fibers from the original artifact. We used a polydentate chelating agent, ethylenediaminetetraacetic (EDTA) acid, as the reactant is able to detach the dye by selective removal of the Al^{3+} ions from the system. We added 0.5 g of EDTA (or equivalent amount of its disodium salt) to the mixture described above, during the agar gel preparation. The use of EDTA leads to an acidic system (pH $\approx$ 3) while the use of its disodium salt leads to solutions at pH values almost neutral. A bead (a small cube of 5 mm side) of Ag-agar containing also EDTA was then placed in contact with the textile for some time (the extraction time ranges from 15 to 30–50 minutes depending on the textiles being analyzed).

Indigo and related dyes (like Tyrian purple) are water-insoluble dyes that are dissolved in the textile fibers in their reduced *leuco* form. Once dispersed within the fiber, the natural oxidation process, due to the exposure of atmospheric oxygen, reverts the dye to its colored form which then sticks to the fiber. We added to the Ag-agar gel the chemical reactants used for the chemical reduction of the indigoid dye. Therefore, during the contact of the bead, we tried to convert indigo into its *leuco* form, which is water soluble, thus possibly transferable to the agar gel. Specifically, we added to the bead a droplet of the $NaOH/Na_2S_2O_4$ 1:2 w/w water solution described above. The bead was then applied to the textile fibers, covered with a glass to reduce evaporation, and let in contact for 5–15 minutes

In both cases, after the chosen contact time with the fiber, the bead (still wet) was removed from the textile and let dry in air. In about 30' the bead loses most of its water content (we measure a loss of about 92% in weight) and reduces considerably its dimensions (from a 5 mm size cube it goes down to a 1 or 2 mm side film, about 0.2 mm thick). The dried bead was then moved to the Raman microscope for the measurement of the SERS spectrum.

2.5. SERS Measurements. We used a micro-Raman spectrometer RM2000 from Renishaw operating with laser excitation either at 514 or 785 nm and equipped with a CCD detector thermoelectrically cooled. It operates in backscattering geometry and uses a 50x microscope objective from Leica. The typical dimension of the active area of the sample is about 2 μm and the used laser power is in the order of 20–200 μW at the sample.

3. Results and Discussion

The Ag-agar gel shows an open, and porous structure with channels of 50–200 nm diameter. The AgNPs can be found in the support either as isolated particles or as aggregates; their sizes range between 50 and 450 nm, in accordance with the UV-Vis absorption spectrum. A high-resolution image of this nanostructured material (Figure 3) was obtained by

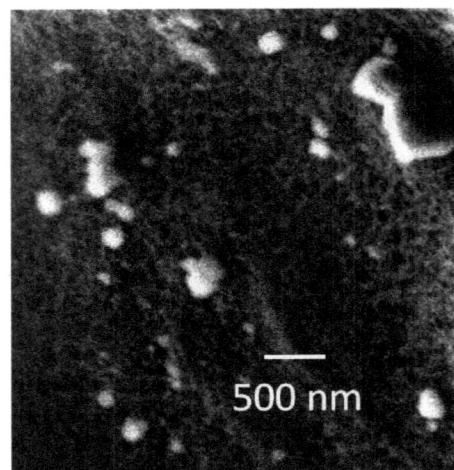

FIGURE 3: Helium ion microscope image of the native agar gel loaded with Ag nanoparticles.

using a Helium ion scattering apparatus (Orion Plus by Carl Zeiss), a system well suitable for the characterization of soft materials without any preliminary metallization treatment. The experimental details for sample preparation and the measurement settings are described in our previous report [37].

Upon a convenient contact time with the fibers, the Ag-agar bead was removed from the sample and allowed to dry in air, thus decreasing in volume. It is a factor that positively affects the Raman signal as it corresponds to an effective sample concentration process and also by decreasing the distance between the metal NPs, it favors the presence of stronger local electric fields [13]. Also, if the dehydration process is not completed the dimension of the bead is not constant and the focusing on the sample in the microscope is troublesome. As a final remark, all our tests demonstrated that the agar gel can be conveniently used as SERS substrate for microextraction as it shows an intrinsic negligible Raman/SERS signal that makes the measurements practically background-free.

The first tests for our microextraction/concentration/ SERS determination approach were carried out on the mock-ups prepared with the alizarin dye [35]. These tests were further extended to include different kinds of fibers and substrates [36]. The preliminary tests were mostly oriented to verify the extraction efficiency and the safety with respect to the textile fiber. We found that both the use of ethanol and EDTA solution as an extraction-promoting reagent on the Ag-agar beads made it possible to observe the SERS signal from alizarin already after 10-minute extraction time. In difficult cases, the extraction time can be as long as 50 minutes. In both cases, during extraction, the bead must be covered in order to prevent fast evaporation of the solvents.

We report in Figure 4 the SERS spectra obtained from the dried beads after extraction of the red dye from a pre-Columbian textile sample with Ag-agar gel spiked with either ethanol or disodium-EDTA solution. Both spectra correspond to the well-known Raman spectrum of the

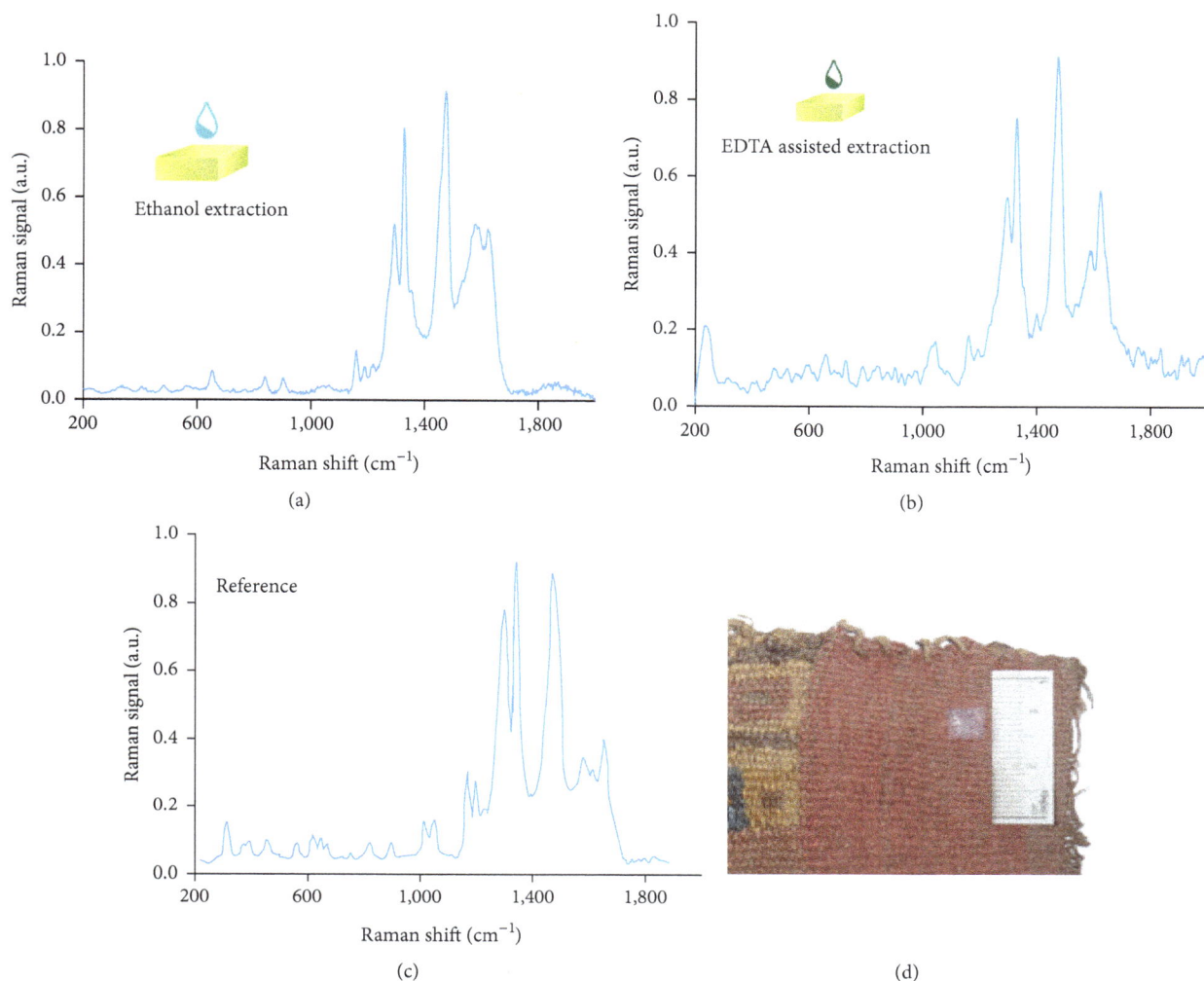

FIGURE 4: SERS spectra (514 nm excitation) obtained from the dried beads after extraction of the red dye from a pre-Columbian textile sample with Ag-agar gel spiked either with ethanol or disodium-EDTA solution ((a) and (b), resp.) or from reference alizarin solutions in ethanol (c). (d) The textile sample is shown with a 25 mm ruler on top; the 5 mm dimension agar gel bead is placed close to the ruler.

monoanionic species of the alizarin dye [38]. However some attention is needed when the experimental data are evaluated. Alizarin is an acid and it is well known that different species exist in solution at different pH values. Alizarin exists as a neutral molecule in acidic solutions while in neutral solutions the most abundant species is the monoanionic one while at pH above 10 the dianionic species is dominant. Alizarin is also an acidochromic molecule that changes its color with pH: the neutral molecule is red, the monoanion is yellow and the dianion is blue. Thus, while performing Raman experiments on alizarin two key factors must be taken into account [38]. The first is the pH as it controls the effective species present in solution, the second is the excitation wavelength used for the Raman experiment. The chemical nature of the dye changes with pH; then both its color and the possible resonance Raman effects change as well. All of these phenomena will reflect directly on the shape and intensity of the different vibrational bands observed in the Raman spectrum. Therefore, a very good knowledge of the chemical

system is needed and a clear feeling of the relevance of the different experimental parameters is also mandatory. In this context, the prediction of Raman spectra of alizarin and other dyes by density functional theory has provided a systematic information on these systems and assisted the interpretation of the experimental data collected by SERS [38, 39].

Additional tests were carried out concerning the alizarin extraction from textile substrates made with different fibers: wood and silk. We added the EDTA reagent to the Ag-agar bead for both the considered textile substrates, and we left the agar beads in contact with the fibers for 50 min and 15 min, respectively. We succeeded in acquiring excellent SERS spectra of alizarin in both cases, thus confirming the capability of the procedure to extract alizarin from different textile fibers [36].

The safety of the process with respect to the textile fiber was verified by the determination of the color of the fiber in the sampling area before and after the contact with the Ag-agar bead. Colorimetric measurements on the textile

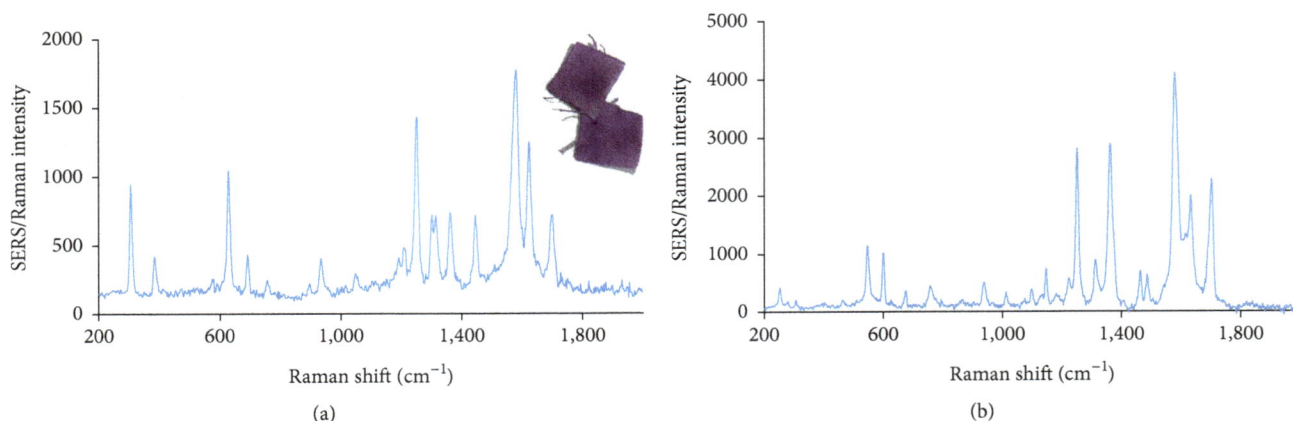

FIGURE 5: SERS spectrum (514 nm excitation) of Tyrian purple extracted from a cotton fiber mockup by using an agar gel bead containing the *leuco*-reduction chemicals (a) and Raman spectrum indigo crystals (b).

fiber were made by using a CM-2600d Konica-Minolta portable spectrophotometer equipped with the integrative sphere and a Xenon lamp. The measurement aperture is 3 mm diameter; the light that reaches the detector is reflected by the illuminated surface with an angle of 8°. Color coordinates are based on CIE $L^*a^*b^*$ system using an illuminant D65 with an observer angle of 10°. The value of $\Delta E^* = 2.3$ corresponds to a just noticeable difference in color [40]. The results reported in Table 1 demonstrate that the fibers are virtually nonaltered by our extraction process.

After extraction, traces of the chemical components of the bead were searched on the textile surface by spectroscopic methods (ATR-FTIR and X-ray fluorescence methods). X-ray fluorescence analysis was performed with a Bruker TRACeR III-SD spectrometer and the spectra were acquired in 50 seconds at 40 KV/15 μA on the rhodium X-ray tube. No evidence for the presence of silver on the fiber was detected. ATR-FTIR did not reveal any surface contamination from the agar gel.

The first test for extraction of indigoids by water solubilization of the dye in its reduced *leuco* form was carried out on a mockup textile purposely prepared of Tyrian purple [37]. The test is important in order to assess the extraction efficiency and the invasiveness of the process with respect to the textile sample. Even though the chemical reagents involved are quite strong (water solution of NaOH and $Na_2S_2O_4$) we consider them rather safe for the textile fiber as they are used also in the original dyeing process. Also, they are applied only in very small amounts and in a limited area. Finally, in the case of relevant contamination, being water soluble, they can be washed out easily.

The efficiency of the extraction process was measured by looking at the SERS spectrum that was possibly obtained from the dye extracted on the bead. The SERS spectrum obtained from the dye extracted on the bead and measured on the dried bead with the micro-Raman spectrometer upon 514 nm excitation is reported in Figure 5. The excitation conditions we used do not allow for the best sensitivity in the indigo detection by Raman measurements as it is well known

that electronic resonance effects strongly enhance the indigo Raman signal when a red excitation is used [41]. However, a very good signal to noise ratio is obtained in the observed spectrum. It is quite relevant to appreciate that the extraction process does not lead to the decomposition of the Tyrian purple ($6,6'$ dibromoindigotin) dye. It is well known that this dye can decompose by the breaking of the C-Br bonds. The bands corresponding to the C-Br stretching mode at 305 cm^{-1} are clearly observed in the SERS spectrum.

The method was then applied to a small piece of blue wire in wool, silk, silver, and gilded silver obtained from the Medici's XVIth century tapestry (Joseph escaping from Potiphar's wife, designed by Bronzino). The Ag-agar bead containing the *leuco*-reagents was applied for 10 minutes to the fiber. No sensible discoloration was appreciated on the fiber after the bead removal. The SERS spectrum obtained from the dried bead is reported in Figure 6 and it clearly shows the characteristic pattern of bands associated with indigo (shown in Figure 5(b)).

The same tests for the safety of the extraction procedure described above were repeated also in the case of this series of measurements. As reported in Table 2, the changes of chromatic properties of the mockup textile fiber upon extraction are negligible, even for extraction times much longer than those needed for obtaining a sample for SERS measurements (10 minutes).

X-ray fluorescence analysis revealed a single detectable signal upon application of the extraction bead. It was related to Sulphur and it was observed only after 30' contact time. ATR-FTIR did not show any evidence of surface contamination.

Finally, we verified if the methods we propose were specific for the different dyes. It is quite clear that extraction promoted by the presence of a solvent, for example, ethanol, is not specific. All the dyes that are not water soluble but have some affinity with organic solvents can be extracted, even if with different efficiency. We did try to use the EDTA method developed for mordant dye with the mockup dyed with indigo and no SERS signal from indigo was detectable. The use of the

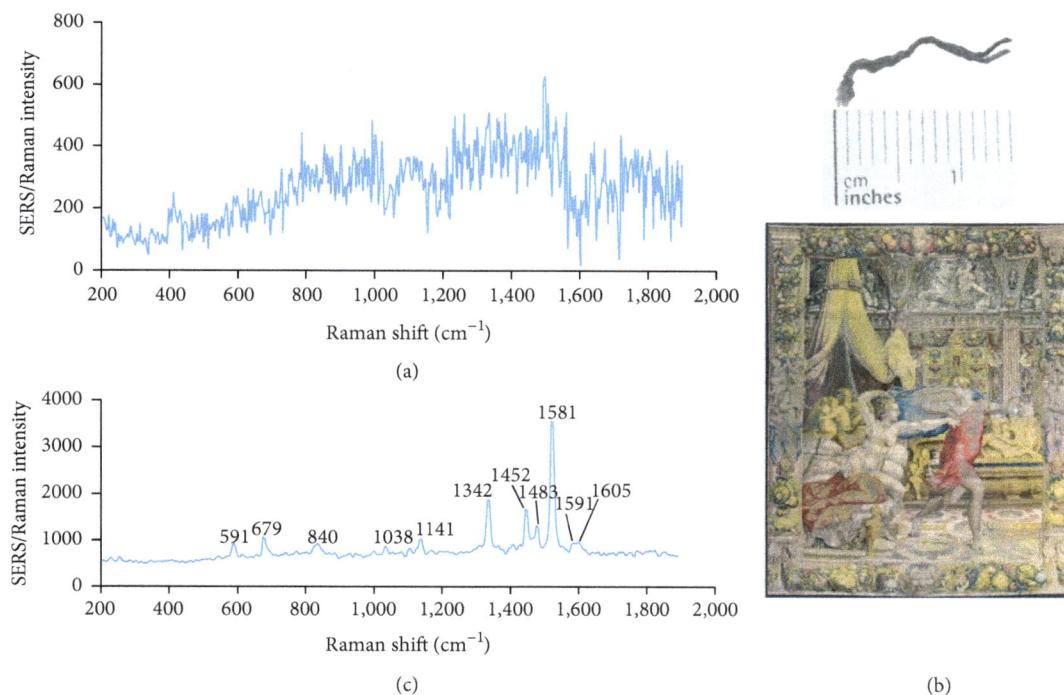

(a)

(c)

(b)

FIGURE 6: Upper trace: Raman spectrum from the blue thread of an XVIth century tapestry (Italian artwork) shown in (b). Lower trace: SERS spectrum (514 nm excitation) of indigo extracted from the same sample by using an agar gel bead containing the *leuco*-reduction chemicals. Adapted from [37].

TABLE 1: Change in color (CIE $L^*a^*b^*$ system) of the cotton mockup dyed with alizarin upon extraction with the Ag-agar gel added with either ethanol or disodium-EDTA solution.

	Before			After			
	L^*	a^*	b^*	ΔL^*	Δa^*	Δb^*	ΔE^*
Ethanol 15 minutes	44.39	45.73	21.02	0.06	−0.04	−0.03	0.08
Disodium-EDTA 30 minutes	45.08	45.9	20.8	0.08	−0.03	−0.04	0.09

TABLE 2: Change in color (CIE $L^*a^*b^*$ system) of the cotton mockup dyed with indigo upon extraction with the Ag-agar gel added with the *leuco*-reagents.

	Before			After			
	L^*	a^*	b^*	ΔL^*	Δa^*	Δb^*	ΔE^*
Point 1: 60′ contact time	36.36	−1.59	−22.51	0.00	0.11	−0.42	0.43
Point 2: 120′ contact time	37.29	−1.77	−22.30	−0.16	−0.03	0.18	0.24

leuco-reagents on the alizarin mockup have made the fiber to change temporarily color from red to blue (the high pH value of the *leuco*-reagents solution induced deprotonation of alizarin) but no extraction of the dye was observed.

4. Conclusions

The paper summarizes the development of a SERS substrate (synthesized by mixing a hydrogel, agar-agar, with a colloidal dispersion of silver nanoparticles) used to extract minimal amounts of dye molecules from textiles. The intent is to provide a simple, efficient, and safe method to extract colorants from textiles fibers of historical, artistic, and archeological interest, exploiting then the high sensitivity of SERS spectroscopy for their identification. The method was found to be very efficient and has advantages with respect to other possible methodologies [23–29]. It is a nondestructive method that is possibly applied directly on the original fibers, without sampling. The safety of the methodology was assessed by means of X-ray fluorescence, ATR-FTIR, and colorimetric analyses that confirmed that no residuals or discoloring effects have been observed on the textile surface after the microextraction, once the extraction process has been optimized with tests on mockups. The simplicity of the procedure to perform SERS measurements, the stability of the dry nanocomposite gel, and the strong enhancement of the observed SERS signal are factors that assure the detection and the recognition of dyes. The high level of enhancement achieved is due also to the shrinkage of Ag-agar gel structure upon drying, which concentrates the extracted molecules and possibly favors the interaction of silver nanoparticles that create high plasmon density sites. Moreover, using a gel as a medium for the solvent mixture allows for a more localized and controlled microextraction essay, which involves the analysis of extremely small areas of the object under study.

Furthermore, the possibility of selecting different solvents and components within the gel structure for the improvement of the extractive performance of the nanocomposite matrix offers the possibility of tailoring the gel for the extraction of specific molecules and for the potential discrimination and recognition of single components of a mixture.

In particular, the efficacy of the proposed methodology has been widely demonstrated by its application to several textiles of artistic relevance, validating the suitability of the proposed method on the investigated samples. The high efficiency of this methodology confirms that the tailored gel extraction is a promising nondestructive technique for SERS identification of dyes.

Further studies are under way considering other substrates for extraction, in order to extend the application of the method to other classes of materials of artistic and historical interest, such as illuminated manuscripts, paintings, and sculptures. As a possible development of this method, we foresee its integration with current cleaning methods based on the use of agar gel. This would allow for simultaneous cleaning and sampling actions to be followed by instrumental characterization of the removed materials.

Competing Interests

The authors declare that there are no competing interests regarding the publication of this paper.

Acknowledgments

This work was supported by Italian MIUR (PRIN, Grant 2010329WPF_007) and by Ente Cassa di Risparmio di Firenze (Grant no. 2014.0405A2202.8044). Opificio delle Pietre Dure (Florence, I) and the Museo del Tessuto (Prato, I) made available samples of ancient textile artworks for our research. Their support is kindly acknowledged.

References

[1] I. Degano, E. Ribechini, F. Modugno, and M. P. Colombini, "Analytical methods for the characterization of organic dyes in artworks and in historical textiles," *Applied Spectroscopy Reviews*, vol. 44, no. 5, pp. 363–410, 2009.

[2] M. R. van Bommel, I. V. Berghe, A. M. Wallert, R. Boitelle, and J. Wouters, "High-performance liquid chromatography and non-destructive three-dimensional fluorescence analysis of early synthetic dyes," *Journal of Chromatography A*, vol. 1157, no. 1-2, pp. 260–272, 2007.

[3] Z. C. Koren, "HPLC analysis of the natural scale insect madder and indigoid dyes," *Journal of the Society of Dyers & Colourists*, vol. 110, no. 9, pp. 273–277, 1994.

[4] L. Burgio and R. J. H. Clark, "Library of FT-raman spectra of pigments, minerals, pigment media and varnishes," *Spectrochimica Acta Part A: Molecular and Biomolecular Spectroscopy*, vol. 57, no. 7, pp. 1491–1521, 2001.

[5] F. Casadio, M. Leona, J. R. Lombardi, and R. Van Duyne, "Identification of organic colorants in fibers, paints, and glazes by surface enhanced Raman spectroscopy," *Accounts of Chemical Research*, vol. 43, no. 6, pp. 782–791, 2010.

[6] S. Bruni, V. Guglielmi, and F. Pozzi, "Historical organic dyes: a surface-enhanced Raman scattering (SERS) spectral database on Ag Lee-Meisel colloids aggregated by $NaClO_4$," *Journal of Raman Spectroscopy*, vol. 42, no. 6, pp. 1267–1281, 2011.

[7] P. Colomban and D. Mancini, "Lacquerware pigment identification with fixed and mobile raman microspectrometers: a potential technique to differentiate original/fake artworks," *Arts*, vol. 2, no. 3, pp. 111–123, 2013.

[8] A. Zoppi, C. Lofrumento, M. Ricci, E. Cantisani, T. Fratini, and E. M. Castellucci, "A novel piece of Minoan art in Italy: the first spectroscopic study of the wall paintings from Phaistos," *Journal of Raman Spectroscopy*, vol. 43, no. 11, pp. 1663–1670, 2012.

[9] L. Burgio and R. J. H. Clark, "Comparative pigment analysis of six modern Egyptian papyri and an authentic one of the 13th century BC by Raman microscopy other techniques," *Journal of Raman Spectroscopy*, vol. 31, no. 5, pp. 395–401, 2000.

[10] D. Bersani, P. P. Lottici, F. Vignali, and G. Zanichelli, "A study of medieval illuminated manuscripts by means of portable Raman equipments," *Journal of Raman Spectroscopy*, vol. 37, no. 10, pp. 1012–1018, 2006.

[11] A. Zoppi, C. Lofrumento, N. F. C. Mendes, and E. M. Castellucci, "Metal oxalates in paints: a Raman investigation on the relative reactivities of different pigments to oxalic acid solutions," *Analytical and Bioanalytical Chemistry*, vol. 397, no. 2, pp. 841–849, 2010.

[12] R. Aroca, *Surface-Enhanced Vibrational Spectroscopy*, John Wiley & Sons, Chichester, UK, 2006.

[13] E. C. Le Ru and P. G. Etchegoin, *Principles of Surface-Enhanced Raman Spectroscopy*, Elsevier, New York, NY, USA, 2009.

[14] A. C. Albrecht, "On the theory of raman intensities," *The Journal of Chemical Physics*, vol. 34, no. 5, p. 1476, 1961.

[15] M. Fleischmann, P. J. Hendra, and A. J. McQuillan, "Raman spectra of pyridine adsorbed at a silver electrode," *Chemical Physics Letters*, vol. 26, no. 2, pp. 163–166, 1974.

[16] P. C. Lee and D. Meisel, "Adsorption and surface-enhanced Raman of dyes on silver and gold sols," *Journal of Physical Chemistry*, vol. 86, no. 17, pp. 3391–3395, 1982.

[17] E. Vogel, W. Kiefer, V. Deckert, and D. Zeisel, "Laser-deposited silver island films: an investigation of their structure, optical properties and SERS activity," *Journal of Raman Spectroscopy*, vol. 29, no. 8, pp. 693–702, 1998.

[18] A. V. Whitney, R. P. Van Duyne, and F. Casadio, "Silver island films as substrate for Surface-Enhanced Raman Spectroscopy

(SERS): a methodological study on their application to Artists' red dyestuffs," in *Advanced Environmental, Chemical, and Biological Sensing Technologies III*, vol. 5993 of *Proceedings of SPIE*, Boston, Mass, USA, October 2005.

[19] E. Nalbant Esenturk and A. R. Hight Walker, "Surface-enhanced Raman scattering spectroscopy via gold nanostars," *Journal of Raman Spectroscopy*, vol. 40, no. 1, pp. 86–91, 2009.

[20] A. Guerrero-Martínez, S. Barbosa, I. Pastoriza-Santos, and L. M. Liz-Marzán, "Nanostars shine bright for you: colloidal synthesis, properties and applications of branched metallic nanoparticles," *Current Opinion in Colloid & Interface Science*, vol. 16, no. 2, pp. 118–127, 2011.

[21] J.-Q. Hu, Q. Chen, Z.-X. Xie et al., "A Simple and effective route for the synthesis of crystalline silver nanorods and nanowires," *Advanced Functional Materials*, vol. 14, no. 2, pp. 183–189, 2004.

[22] S. E. J. Bell and S. J. Spence, "Disposable, stable media for reproducible surface-enhanced Raman spectroscopy," *Analyst*, vol. 126, no. 1, pp. 1–3, 2001.

[23] M. Leona, J. Stenger, and E. Ferloni, "Application of surface-enhanced Raman scattering techniques to the ultrasensitive identification of natural dyes in works of art," *Journal of Raman Spectroscopy*, vol. 37, no. 10, pp. 981–992, 2006.

[24] M. Leona, "Microanalysis of organic pigments and glazes in polychrome works of art by surface-enhanced resonance Raman scattering," *Proceedings of the National Academy of Sciences of the United States of America*, vol. 106, no. 35, pp. 14757–14762, 2009.

[25] A. Idone, M. Gulmini, A.-I. Henry et al., "Silver colloidal pastes for dye analysis of reference and historical textile fibers using direct, extractionless, non-hydrolysis surface-enhanced Raman spectroscopy," *Analyst*, vol. 138, no. 20, pp. 5895–5903, 2013.

[26] Z. Jurasekova, E. del Puerto, G. Bruno, J. V. García-Ramos, S. Sanchez-Cortes, and C. Domingo, "Extractionless non-hydrolysis surface-enhanced Raman spectroscopicdetection of historical mordant dyes on textile fibers," *Journal of Raman Spectroscopy*, vol. 41, no. 11, pp. 1455–1461, 2010.

[27] M. Leona, P. Decuzzi, T. A. Kubic, G. Gates, and J. R. Lombardi, "Nondestructive identification of natural and synthetic organic colorants in works of art by surface enhanced raman scattering," *Analytical Chemistry*, vol. 83, no. 11, pp. 3990–3993, 2011.

[28] B. Doherty, B. G. Brunetti, A. Sgamellotti, and C. Miliani, "A detachable SERS active cellulose film: a minimally invasive approach to the study of painting lakes," *Journal of Raman Spectroscopy*, vol. 42, no. 11, pp. 1932–1938, 2011.

[29] B. Doherty, F. Presciutti, A. Sgamellotti, B. G. Brunetti, and C. Miliani, "Monitoring of optimized SERS active gel substrates for painting and paper substrates by unilateral NMR profilometry," *Journal of Raman Spectroscopy*, vol. 45, no. 11-12, pp. 1153–1159, 2014.

[30] F. Pozzi and M. Leona, "Surface-enhanced raman spectroscopy in art and archaeology," *Journal of Raman Spectroscopy*, vol. 47, no. 1, pp. 67–77, 2016.

[31] C. Muehlethaler, M. Leona, and J. R. Lombardi, "Review of surface enhanced Raman scattering applications in forensic science," *Analytical Chemistry*, vol. 88, no. 1, pp. 152–169, 2015.

[32] H. Schweppe, "Indigo and woad," in *Artists' Pigments: A Handbook of Their History and Characteristics*, R. L. Feller, E. West FitzHugh, and A. Roy, Eds., vol. 3, pp. 81–107, National Gallery of Art, Washington, DC, USA, 1997.

[33] Z. Zhou, G. G. Huang, T. Kato, and Y. Ozaki, "Experimental parameters for the SERS of nitrate ion for label-free semi-quantitative detection of proteins and mechanism for proteins

to form SERS hot sites: a SERS study," *Journal of Raman Spectroscopy*, vol. 42, no. 9, pp. 1713–1721, 2011.

[34] D. Stulik, D. Miller, H. Khanjian et al., *Solvent Gels for the Cleaning of Works of Art: The Residue Question*, edited by V. Dorge, Getty Publications, 2004.

[35] C. Lofrumento, M. Ricci, E. Platania, M. Becucci, and E. Castellucci, "SERS detection of red organic dyes in Ag-agar gel," *Journal of Raman Spectroscopy*, vol. 44, no. 1, pp. 47–54, 2013.

[36] E. Platania, J. R. Lombardi, M. Leona et al., "Suitability of Ag-agar gel for the microextraction of organic dyes on different substrates: the case study of wool, silk, printed cotton and a panel painting mock-up," *Journal of Raman Spectroscopy*, vol. 45, no. 11-12, pp. 1133–1139, 2014.

[37] E. Platania, C. Lofrumento, E. Lottini, E. Azzaro, M. Ricci, and M. Becucci, "Tailored micro-extraction method for Raman/SERS detection of indigoids in ancient textiles," *Analytical and Bioanalytical Chemistry*, vol. 407, no. 21, pp. 6505–6514, 2015.

[38] C. Lofrumento, E. Platania, M. Ricci, C. Mulana, M. Becucci, and E. M. Castellucci, "The SERS spectra of alizarin and its ionized species: the contribution of the molecular resonance to the spectral enhancement," *Journal of Molecular Structure*, vol. 1090, pp. 98–106, 2015.

[39] C. Lofrumento, F. Arci, S. Carlesi, M. Ricci, E. Castellucci, and M. Becucci, "Safranin-O dye in the ground state. A study by density functional theory, Raman, SERS and infrared spectroscopy," *Spectrochimica Acta Part A: Molecular and Biomolecular Spectroscopy*, vol. 137, pp. 677–684, 2015.

[40] S. Gaurav and B. Raja, *Digital Color Imaging Handbook*, Edited by G. Sharma, CRC Press, New York, NY, USA, 2002.

[41] I. T. Shadi, B. Z. Chowdhry, M. J. Snowden, and R. Withnall, "Semi-quantitative analysis of indigo by surface enhanced resonance Raman spectroscopy (SERRS) using silver colloids," *Spectrochimica Acta Part A: Molecular and Biomolecular Spectroscopy*, vol. 59, no. 10, pp. 2213–2220, 2003.

24

Quantitative Estimation of Organic Matter Content in Arid Soil Using Vis-NIR Spectroscopy Preprocessed by Fractional Derivative

Jingzhe Wang,[1,2] **Tashpolat Tiyip,**[1,2] **Jianli Ding,**[1,2] **Dong Zhang,**[1,2] **Wei Liu,**[1,2] **and Fei Wang**[1,2]

[1]*College of Resources and Environment Science, Xinjiang University, Urumqi 830046, China*
[2]*Key Laboratory of Oasis Ecology, Xinjiang University, Urumqi 830046, China*

Correspondence should be addressed to Tashpolat Tiyip; tash_xju@163.com

Academic Editor: Tino Hofmann

Soil organic matter (SOM) content is an important index to measure the level of soil function and soil quality. However, conventional studies on estimation of SOM content concerned about the classic integer derivative of spectral data, while the fractional derivative information was ignored. In this research, a total of 103 soil samples were collected in the Ebinur Lake basin, Xinjiang Uighur Autonomous Region, China. After measuring the Vis-NIR (visible and near-infrared) spectroscopy and SOM content indoor, the raw reflectance and absorbance were treated by fractional derivative from 0 to 2nd order (order interval 0.2). Partial least squares regression (PLSR) was applied for model calibration, and five commonly used precision indices were used to assess the performance of these 22 models. The results showed that with the rise of order, these parameters showed the increasing or decreasing trends with vibration and reached the optimal values at the fractional order. A most robust model was calibrated based on 1.8 order derivative of R, with the lowest RMSEC ($3.35\,g\,kg^{-1}$) and RMSEP ($2.70\,g\,kg^{-1}$) and highest R_c^2 (0.92), R_p^2 (0.91), and RPD (3.42 > 3.0). This model had excellent predictive performance of estimating SOM content in the study area.

1. Introduction

Soil organic matter content (SOM) is an important index to measure the level of soil function and soil quality, and detection of SOM content is an important approach to understand the local soil fertility [1, 2]. As known to all, the correlation between SOM and soil organic carbon (SOC) is significant. Soil represents the largest carbon sinks on earth and plays a major role in the global carbon cycle [3, 4]. More seriously, along with global warming, intensified human activities, and other factors, the loss of SOC is more severe especially in the arid and semiarid region [5, 6]. The capacities of SOC and carbon sequestration through SOM management have attracted considerable attention in recent years.

Traditionally, SOM is determined by capacity and combustion methods in laboratory. These methods are generally laborious. Besides, there are possible risks of air contamination during the operational procedure. Because of high efficiency, low-cost, large-scale, nondestruction, and rapid data acquisition, remote sensing technology has been proved to be a promising tool to strengthen or perfect traditional methods [7, 8]. And it provides a fresh approach for quantitative research of SOM content. Due to the lofty high spectral resolution, convenience, and controllability, the analysis on the laboratory Vis-NIR spectroscopy of soil is fashionable, especially. It is precisely the significant quantitative relationship between SOM and SOC; the estimation of SOM content by remote sensing has been proved as a feasible approach to grasp the condition of local SOC storage. Many studies have shown that SOM has unique spectral response in the Vis-NIR (visible and near-infrared) bands [9–12]. To some extent, the soil spectral reflectance could reflect the content of SOM. Fast detection of SOM content could be conducted by using raw spectral reflectance data (R) and through its mathematical transformations.

The pretreatment of Vis-NIR spectroscopy is very necessary and effective to improve the accuracy of the spectral estimation model. To remove the effect of soil moisture from the spectra, Minasny et al. preprocessed the measured raw spectral data using EPO (external parameter orthogonalisation) algorithm, and it improved the stability and accuracy of SOC predicting model in southern New South Wales, Australia [13]. Using spectral reflectance pretreated by Savitzky-Golay (SG) smoothing, first derivative with SG smoothing (FD), and other mathematical methods, the predicting performance of the support vector machine regression model (SVMR) was perfected [14]. These spectral inversion models of SOM are mainly based on the R and some commonly appropriate pretreatment methods, that is, inversion ($1/R$), logarithm ($\lg R$), logarithm-inversion ($1/\lg R$), and root mean square ($\sqrt{R}$) and their first or second order derivative. As the high-dimensional data source with massive information, the raw spectra were pretreated by conventional integer order derivative generally, but it might influence the effective information detection and cause the loss of spectral information to some extent. Furthermore, the accuracy of modeling will also be constrained. Fractional order derivative broadened the concept of the classic integer derivative [15, 16]. Due to the better accuracy and higher efficiency, it has been widely used in system control and diagnosis, digital filtering, signal and image processing, and other related fields [17–20]. For spectrum analysis, this algorithm had been introduced in the spectral reflectance pretreatment of saline soil in the Ebinur Lake basin [21]. The research demonstrated that it was desirable to extract potential spectral information of soil Vis-NIR spectroscopy using fractional derivative algorithm in arid desert region.

Desert soil is a typical soil in the Ebinur Lake basin of Northwest China. The Ebinur Lake basin is a typical lake wetland in arid areas. There are few studies on fractional derivative applied in Vis-NIR spectroscopy of desert soil, and to this regard and motivated by the previous research, the objective of this study was to utilize laboratory Vis-NIR spectroscopy treated by fractional derivative algorithm combining with SOM content data to establish a predicting model with better accuracy and stability than existing models.

2. Materials and Methods

2.1. Study Area. The study area is located in the Ebinur Lake basin ($82°36'–83°10'E$, $44°30'–45°09'N$) in the southwest of Junggar Basin, Xinjiang Uighur Autonomous Region, China. The basin is surrounded by the mountains on 3 sides, north, west, and south side, separately [22, 23]. This region is a major function area to prevent dust of ecological protection system of the northern slope of Tianshan Mountain. Due to the arid desert climate of the study area, the annual precipitation in Ebinur Lake basin is approximately 102 mm, whereas the potential evaporation can reach 1447 mm. The annual average temperature ranges from 6.6 to 7.8°C. Strong

winds are typical in this region as well [24]. The main geomorphic types are stone desert, gravel desert, salt desert, swamp, and so on. The soil types are mainly *Piedmont psephitic* and *Gypsum desert soil* [25, 26]. The Ebinur Lake basin is a normal closed oasis system in the inland arid areas and also an integrated region which is composed of wetland, hydrology, and human activities.

2.2. Soil Sample Collection and Chemical Analysis. Considering the typical landscape features of the study area, such as oasis, desert, the soil condition, and site accessibility, we set up 103 sites (30×30 m square area, 5 samples per site). In every measuring unit, the corresponding coordinate of each sample point was recorded by GPS (Figure 1). Each soil sample (about 0.5 kg) was put into a water-tight bag, sealed, numbered, and then brought back to the laboratory. A total of 103 topsoil (depth 0~20 cm) samples were obtained from the Ebinur Lake basin of Xinjiang Uighur Autonomous Region, China, from 18 to 29 May 2015. In order to reduce the effect of water content, all samples were air dried sufficiently, after then, these soil samples were crushed and sieved through the 2 mm screen to remove the stones, plant residue, and other impurities. Every sample was divided into 2 equal parts for soil chemical analysis and spectral reflectance measurement in laboratory, respectively. The potassium dichromate method was used for the determination of SOM.

2.3. Laboratory Reflectance Measurement. All of air-dried soil samples were individually put into wide round containers with a diameter of 12 cm and a depth of 1.8 cm (1.5 cm is considered optically infinitely thick for soil). To avoid the contamination in the period of the measurement, these containers had been painted black previously [27]. And the surfaces had to be scraped with a plastic ruler to ensure the same flat measuring surface, as pressing can affect the porosity of the soil and result in false measurements [28, 29]. For the controlled light conditions, the reflectance spectra of all soil samples measurement were conducted in a dark laboratory with an ASD FieldSpec®3 portable spectrometer (Analytical Spectral Device, Boulder, CO, USA). The sampling intervals of this spectrometer are 1.4 nm (350~1000 nm) and 2 nm (1000~2500 nm), while the resampling interval is 1 nm. A 50 W halogen lamp served as the light source for the laboratory reflectance measurement, shining 8° from vertical and being placed 50 cm above each soil sample surface. The optical sensor was installed with a distance of 15 cm from the flat of each soil sample with a 30° zenith angle. Each reflectance measurement was calibrated by a standardized plate with 100% reflectance to ensure the accuracy [30]. For each soil sample, twenty spectra curves were collected, and the mean value of the twenty spectra was taken as the final reflectance.

2.4. Spectral Processing and Data Analysis. The real sample information was inevitably contaminated by the instrument noise [31, 32]. In order to reduce the noise, the ViewSpecPro software version 6.0 was applied to correct and eliminate the breakpoints and remove the marginal wavebands with large noise (350~400 nm and 2401~2500 nm). The SG

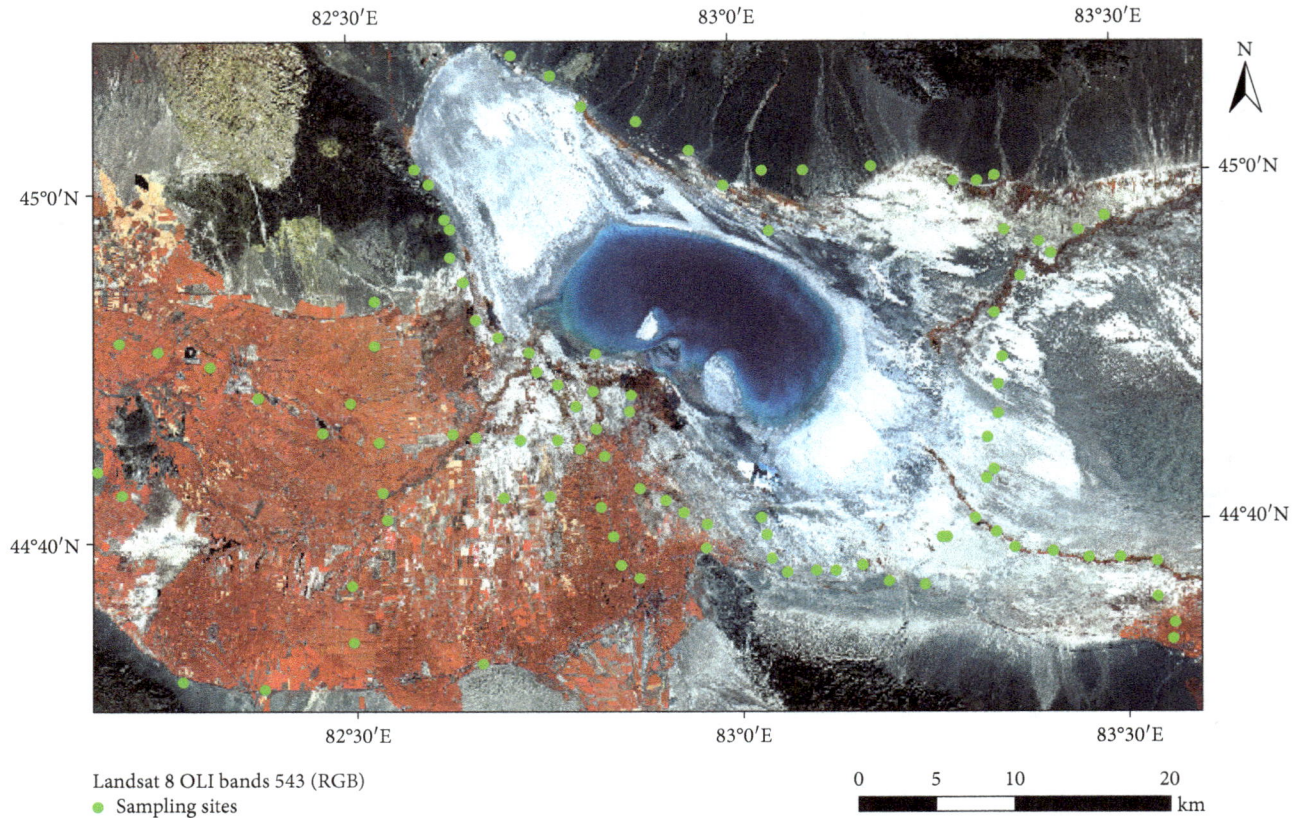

Figure 1: Distribution of the all sampling sites and the location of the study area.

smoothing method (polynomial order of 2 and frame size of 5) was employed for the smoothness of 103 spectral curves with OriginPro version 9.0.0. The processed spectra constituted the final data for further analysis. The processed spectral reflectance of all soil samples is shown in Figure 2.

Fractional calculus is a theory branch of mathematics and generalizes the classic integer derivative to arbitrary (noninteger) order [31, 32]. Fractional derivative has different definitions, that is, Grümwald-Letnikov (G-L), Riemann-Liouville (R-L), and Capotu. For less computational cost, G-L definition was applied in this research. Thereinto, v means the order, and zeroth order means the data are not processed by the algorithm.

$$\frac{d^v f(x)}{dx^v} \approx f(x) + (-v)f(x-1) + \frac{(-v)(-v+1)}{2}f(x-2)$$
$$+ \cdots + \frac{\Gamma(-v+1)}{n!\Gamma(-v+n+1)}f(x-n). \tag{1}$$

Commonly, $\lg(1/R)$ spectra was used because it represented the absorbance, in spectrum analysis. For more modeling results and the improvement of nonliner relations, the smoothed and predenoised reflectance data were transformed by the absorbance ($\lg(1/R)$). According to (1), R and its absorbance 0~2nd fractional derivatives (order interval 0.2) were computed under the platform Eclipse.

2.5. Data Modeling and Validating. Due to its advantage of dimension reduction, synthesis, and solving colinearity problems among independent variables, partial least squares regression (PLSR) has been proved as a robust and reliable approach in spectral quantitative research [7, 33–35]. For modeling, the benefit of PLSR is that it uses significance test wavelengths in the range selected to arrive at a prediction equation that uses wavelengths highly correlated to the analyte and gives little weight to the nonpredictive wavelengths. In order to take full advantage of the spectral reflectance, all wavelengths ranging from 401 to 2400 nm were applied in modeling calibration by PLSR. Ranking based the principle from the highest to lowest. The calibration set ($n = 69$) and the validation set ($n = 34$) were selected at equal interval for the calibration and precision test.

The capacity of estimation models were tested by five performance indices: ratio of performance to deviation (RPD), the determinant coefficients of calibration (R_c^2), root mean square errors of calibration (RMSEC), and accordingly in prediction (R_p^2, RMSEP). The optimal models are represented by high values of R_c^2, R_p^2, and RPD, but low RMSEC and RMSEP. Generally, if $1.5 < \text{RPD} \leq 2.0$, it indicates that the model only estimates high and low level of SOM poorly. If $2.0 < \text{RPD} \leq 2.5$, it indicates a better predictive ability, while if $2.5 < \text{RPD} \leq 3.0$, a very good predictive ability, and if $\text{RPD} > 3.0$, the model has excellent predictive performance [12, 36]. All of the above indicators were calculated by MATLAB software version R2012a (MathWorks, Natick,

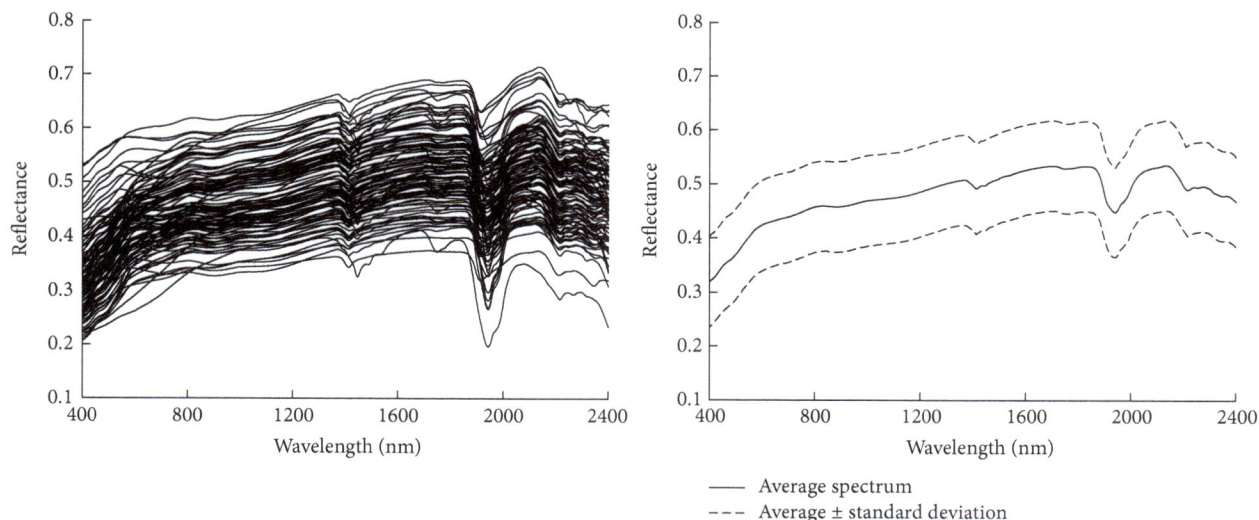

FIGURE 2: Spectral reflectance of soil samples after processing.

TABLE 1: The statistical characteristics of organic matter content of soil samples ($g\,kg^{-1}$).

Type of samples	Observations	Min	Max	Mean	Standard error	Standard deviation	Coefficient of variation
Whole set	103	0.68	78.39	21.43	1.07	10.81	50.46%
Calibration set	69	0.68	78.39	21.69	1.39	11.56	53.32%
Validation set	34	4.79	39.16	20.90	1.59	9.24	44.22%

MA, USA). The final results were used to assess the performance of the models.

3. Results

3.1. Statistical Analysis of SOM Content. The descriptive statistical characteristics for organic matter of soil samples of the whole dataset, the calibration set, and the validation set were presented in Table 1. Compared with the range of SOM content ($0.68–78.39\,g\,kg^{-1}$) for both the whole dataset and the calibration set, the validation set had a narrower range with $4.79–39.16\,g\,kg^{-1}$, because of the deficient soil samples. The average SOM content and coefficient of variation of whole set were $21.43\,g\,kg^{-1}$ and 50.46% between the range of the values of calibration and validation set, respectively, while the descriptive statistical characteristics of SOM content in the calibration and validation set were similar to the six parameters of the whole set. Thus, the SOM content of the calibration and validation set could represent those of the whole dataset sufficiently.

3.2. Reflectance of Different Soil Organic Matter Content. In the visible region, absorption bands related to soil color are because of electron excitations, which assist the measurement of SOM, the content of SOM, and the spectral reflectance are correlative [12, 37]. For researching the relationship between SOM content and spectral reflectance of the corresponding soil sample, five representative soil samples with different contents were selected for the curve plotting. The diagram

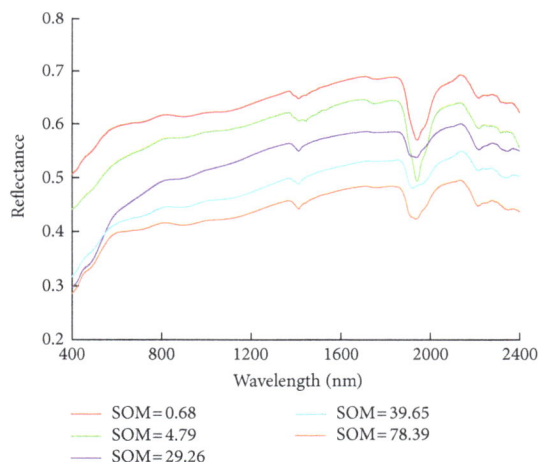

FIGURE 3: Spectral reflectance of soils with different organic matter content in the Ebinur Lake basin ($g\,kg^{-1}$).

showed that SOM content of $0.68\,g\,kg^{-1}$ and $78.39\,g\,kg^{-1}$ corresponded to the highest and lowest reflectance, separately. Spectral curves of soil samples with different organic matter content had similar reflectance and curve slopes, and there were three main obvious absorption features located near 1400, 1900, and 2200 nm, respectively (Figure 3). The absorption peak at 1400 nm is a typical absorption band for water which is associated with the bending and stretching of the O–H bonds of free water. The regions near 1900 and

TABLE 2: Performance statistics of R model for calibration set and validation set based on PLSR.

Order	Principal components	Calibration set		Validation set		
		R_c^2	RMSEC ($g\,kg^{-1}$)	R_p^2	RMSEP ($g\,kg^{-1}$)	RPD
0	2	0.12	10.87	0.16	8.50	1.09
0.2	2	0.28	9.84	0.23	8.13	1.14
0.4	2	0.36	9.24	0.28	7.84	1.18
0.6	2	0.38	9.11	0.28	7.87	1.17
0.8	3	0.41	8.91	0.25	8.01	1.15
1	3	0.38	9.09	0.23	8.09	1.14
1.2	4	0.42	8.79	0.36	7.42	1.24
1.4	4	0.38	9.07	0.47	6.70	1.38
1.6	4	0.42	8.84	0.51	6.47	1.43
1.8	5	0.92	3.35	0.91	2.70	3.42
2	5	0.90	3.64	0.91	2.77	3.34

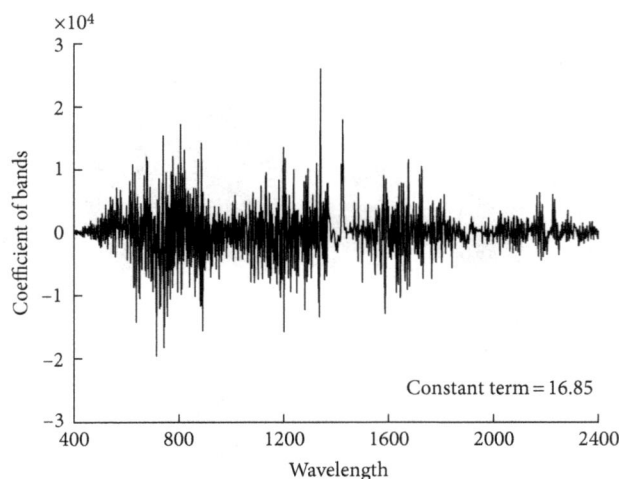

FIGURE 4: The coefficients of all bands and the constant term.

2200 nm in the combination range are due to the bending and stretching vibration of Al–OH and Mg–OH, respectively [38–40]. From 401 to 760 nm, reflectance increased sharply with increasing wavelength. Reflectance gradually decreased and tended to flat between 760 and 1900 nm. The second spectral absorption peak was measured around 1900 nm. Reflectance had acute changes with wavelength increasing from 1900 to 2400 nm. There was a negative correlation relationship between soil spectral reflectance and SOM content in the range of 401–2400 nm; that soil reflectance increased with SOM content decreasing and vice versa. It was easy to distinguish the soil reflectance with different SOM contents through the entire spectrum range, although spectral curves of five soil samples had some overlap sections but could be discriminated approximately from 400 to 600 nm and from 1900 to 2000 nm. The results were consistent with conventional researches [30, 41, 42].

3.3. Model Calibration and Validation.
Model calibration with all wavelengths could take advantage of the whole spectral information of the reflectance. The derivative pretreatment could effectively eliminate the effect of background noise on the target spectrum and highlight the spectral characteristics of analyte. In this research, all raw spectral reflectance and according absorbance data pretreated by fractional derivative algorithm were applied in the process of model calibration. As the order interval set to 0.2, all 22 inversion models were built by PLSR. The five performance indices of calibration and validation were summarized in Table 2 and Table 3. For R, during the range from 0 to 1st order, the preference of models did not increase significantly, the highest R_c^2, R_p^2, and RPD were only 0.41, 0.28, and 1.18, respectively. And the parameters did not reach the maximum at the same order. Five performance indices had a slight improvement with increasing order from 1st to 1.6 order. The RMSEC and RMSEP of model based on 1.6 order derivative reached 8.84 and 6.47 $g\,kg^{-1}$, separately. When the order reached 1.8, the performance of this model had

significant promotion with the lowest RMSEC (3.35 $g\,kg^{-1}$) and RMSEP (2.70 $g\,kg^{-1}$) and highest R_c^2 (0.92), R_p^2 (0.91), and RPD (3.42 > 3.0). With the order increasing to 2, the capability of model decreased slightly.

The variation trend of absorbance model built-up by PLSR was similar with R model from 0 to 1st order. The RMSEC and RMSEP of 6 models were kept in the high level of values, that is, significant error. When the order is greater than 1, R_c^2 and R_p^2 increased sharply and reached highest at 1.8 order. The stability and accuracy of this model were perfected with the lowest RMSEC (3.06 $g\,kg^{-1}$) and RMSEP (3.06 $g\,kg^{-1}$). The sensitivity of the spectrum to SOM, the stability, and accuracy of models were enhanced. Both the models based on 1.8 order derivative of R and absorbance had the best predicting accuracy.

After repeated siftings for excellent predictive performance, there were 2 models having acceptable results with RPD > 3, R, and its absorbance model based on 1.8 order derivative, respectively. And among these 22 models, there was only one best model which was built-up based on 1.8 order derivative of R, represented the high values of R_c^2, R_p^2, and RPD, but low RMSEC and RMSEP, relatively. The coefficients of all bands and the constant term were demonstrated in Figure 4. The scatter plot of measured and predicted SOM content of the optimal model is shown in Figure 5. R^2 of measured and predicted values in calibration and validation set both reached 0.91, and the whole performance indices meant the model based on Vis-NIR spectroscopy treated by fractional derivative could be used to predict the SOM content in the Ebinur Lake basin.

4. Discussion

Due to the massive information, continuous bands, and high resolution of the spectral reflectance, the measured spectra are easily effected by individual differences (the particle size of samples, the angle of light source, the condition of analyte, etc.), and substantial noises [43, 44]. Therefore, the necessary

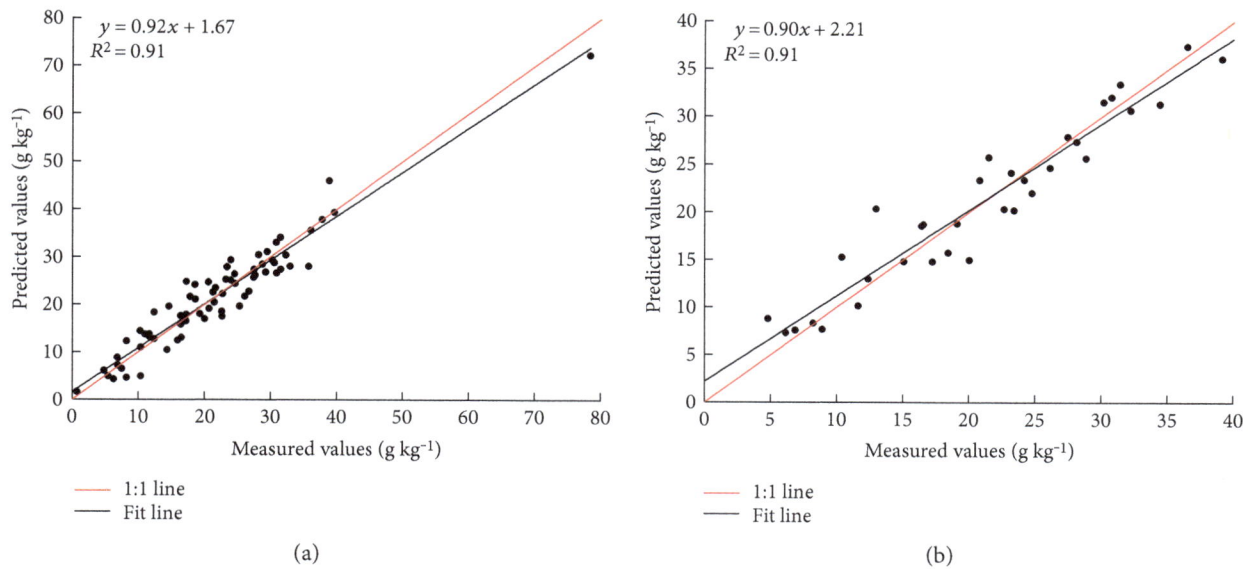

(a)   (b)

FIGURE 5: Comparison of measured SOM content and estimated values of modeling sample (a) and testing sample (b) through R 1.8-order derivative.

TABLE 3: Performance statistics of absorbance model for calibration set and validation set based on PLSR.

Order	Principal components	Calibration set		Validation set		
		R_c^2	RMSEC (g kg^{-1})	R_p^2	RMSEP (g kg^{-1})	RPD
0	2	0.11	10.91	0.12	8.69	1.06
0.2	2	0.20	10.35	0.19	8.31	1.11
0.4	2	0.35	9.34	0.24	8.04	1.15
0.6	2	0.38	9.13	0.26	7.98	1.16
0.8	3	0.36	9.27	0.20	9.27	1.12
1	3	0.36	9.27	0.21	8.21	1.13
1.2	3	0.38	9.11	0.32	7.60	1.22
1.4	4	0.52	8.03	0.58	5.99	1.54
1.6	4	0.44	8.63	0.50	6.55	1.41
1.8	5	0.93	3.06	0.89	3.06	3.02
2	5	0.92	3.32	0.88	3.15	2.94

pretreatment should be applied to minimize the irrelevant and useless information of the spectra and increase the correlation between the spectra and measured values. The usual pretreatment methods of soil spectrum mainly include smoothing, denoising, normalization, derivative processing, and multiple scatter correction [45]. For derivative processing, the applications of first and second derivatives are popular [46]. The first derivative could reduce the effect caused by the background noise of partially linear or near linear. Through the second derivative, signal wander of spectra could be weakened.

The 1st and 2nd derivatives mean the slope and curvature of spectral curves, respectively. Although the explicit spectral meaning of fractional derivative has not been clarified yet, the nonlocal and genetic characteristics of fractional derivative are widely recognized. But it suggests

that between 0 and 2nd order of fractional derivative could be identified as the sensitivity to the slope and curvature of spectral curves. The derivative value becomes more sensitive to the slope and less sensitive to reflectance with the order increasing from 0 to 1st, and from 1st to 2nd order, the derivative value become more sensitive to the curvature and less sensitive to the slope [15]. In the case of R model in this research, RPD and other parameters of regression models did not increase or decrease monotonously as the order is increasing. The process of change was undulant. They achieved optimal values at fractional order (1.8 order). These indices did not continue to improve at 2nd order as expected; the capability of model decreased slightly. The sensitivity of the spectrum to SOM was enhanced by pretreatment. RPD, R_c^2, R_p^2, RMSEC, and RMSEP all revealed the sensitivity. For conventional researches based on integer order derivative, these process details were ignored, which might cause the concealment of better models.

The soil spectral reflectance differ due to the influence of the parent material and soil type [47]. Shi et al. compared the correlations between the reflectance and SOM in different types, just like limestone soils and red soils. The results manifested that the reflectance in the wavelength from 580 to 820 nm could be used to predict the SOM content [12]. Liu et al. confirmed that the reflectance in the range of 620–810 nm was relevant to SOM, and the maximum correlation coefficient was discovered at 710 nm [48]. With SOM content of 2% as a boundary, that is, when SOM content exceeded 2%, the SOM played a principal role in masking out the spectral features, while the SOM content was less than 2%, it became less effective [30, 37, 49]. Though, it is hard to estimate SOM content of desert soil precisely when it is less than 2%. In this research, the spectral reflectance displayed the higher correlation with the SOM content in the range of 600–900 nm (Figure 4). Our results are consistent with the finding of above studies.

For higher SOM content, prediction based on Vis-NIR spectroscopy was widely researched in the black-soil region [7, 30, 50]. However, this kind of application is relatively few in the arid and semiarid desert soils. Yang et al. discovered an optimal model to estimate SOM content in brown calcic soil region of Xinjiang and model with $R^2 = 0.89$ and RMSE = 0.32 [50]. Nawar et al. used multivariate adaptive regression splines with first derivative reflectance data to predict SOM in El-Tina Plain, Egypt; the values of R_p^2 and RPD reached 0.76 and 1.98, respectively [51]. Comparing our results with previous research, in this study, not only considering the single band reflectance, we excavated more potential spectral information by using fractional derivative. It reduced the loss of information, detailed the variation trend of 5 accuracy indexes based on R and absorbance models of 11 order derivatives.

For abundant spectral information of all wavelengths, models based on feature bands only utilize part of all wavelengths, and the number of bands is very limited. The significance test at the level of 0.01 is used in current selection methods for choosing feature bands. This method might miss some suboptimal bands and lead to the loss of some important spectral information. For PLSR model with all wavelengths, spectral parameters of each wavelength in the whole spectral region are considered. Because of the advantages of PLSR, some problems just like fewer samples, more independent variables, and multiple correlations between variables could be solved effectively. In addition, due to the introduction of fractional order algorithm, the related fractional order spectral information of SOM is released, which has been ignored previously. Thus, the performance of estimating model is increased to some extent. The Ebinur Lake basin is the typical arid and semiarid region. Our research could enrich the SOM Vis-NIR spectroscopy studies and provide a new perspective to estimate SOM content in the special areas, where the organic matter content of desert soil with mass fraction is less than 2%.

The characteristics of soil reflectance spectra are not only directly relevant to SOM and water content but also obey the obvious regional differentiation rules. The Ebinur Lake basin is also the representative area with severe salinization. For the predicting model of SOM content, the salt content and texture may have a certain impact on the accuracy to some degree. For a better precision, the next step for further research is to distinguish the features of salt and SOM from spectral reflectance curves.

5. Conclusion

The pretreatment of Vis-NIR spectroscopy is very necessary and effective to improve the accuracy of the spectral estimation model. In this research, the fractional derivative algorithm was employed for pretreatment to determine the most accurate model for SOM content in the Ebinur Lake basin. We found that the whole 5 five performance indices,

that is, R_c^2, R_p^2, RMSEC, RMSEP, and RPD did not increase or decrease monotonously with the increasing order. With the rise of order, these parameters showed the increasing or decreasing trends with vibration and reached the optimal values at the fractional order. Through the comparison of the 22 models, a most robust model was calibrated based on 1.8 order derivative of R, with the lowest RMSEC $(3.35\,\mathrm{g\,kg^{-1}})$ and RMSEP $(2.70\,\mathrm{g\,kg^{-1}})$ and highest R_c^2 (0.92), R_p^2 (0.91), and RPD $(3.42 > 3.0)$. This model had excellent predictive performance of estimating SOM content in the study area.

Acknowledgments

This study was supported by the National Natural Science Foundation of China (41130531, 41561089, and 41661046) and the National Plan on Key Technology Research and Development Program (2014BAC15B01). The authors would like to thank Master Huiyun Yuan (School of Politics, Philosophy, Language and Communication Studies, University of East Anglia) for the linguistic assistance.

References

[1] M. H. Beare, P. F. Hendrix, M. L. Cabrera, and D. C. Coleman, "Aggregate-protected and unprotected organic matter pools in conventional- and no-tillage soils," *Soil Science Society of America Journal*, vol. 58, no. 3, pp. 787–795, 1994.

[2] A. J. Franzluebbers, "Soil organic matter stratification ratio as an indicator of soil quality," *Soil and Tillage Research*, vol. 66, no. 2, pp. 95–106, 2002.

[3] G. J. Canadell, A. H. Mooney, D. D. Baldocchi et al., "Commentary: carbon metabolism of the terrestrial biosphere: a multitechnique approach for improved understanding," *Ecosystems*, vol. 3, no. 2, pp. 115–130, 2000.

[4] Z. Li, F. Han, Y. Su et al., "Assessment of soil organic and carbonate carbon storage in China," *Geoderma*, vol. 138, no. 1-2, pp. 119–126, 2007.

[5] J. Ardö and L. Olsson, "Assessment of soil organic carbon in semi-arid Sudan using GIS and the CENTURY model," *Journal of Arid Environments*, vol. 54, no. 4, pp. 633–651, 2003.

[6] J. Albaladejo, R. Ortiz, N. Garcia-Franco et al., "Land use and climate change impacts on soil organic carbon stocks in semi-arid Spain," *Journal of Soils and Sediments*, vol. 13, no. 2, pp. 265–277, 2013.

[7] X. Yu, Q. Liu, Y. Wang, X. Liu, and X. Liu, "Evaluation of MLSR and PLSR for estimating soil element contents using visible/near-infrared spectroscopy in apple orchards on the Jiaodong peninsula," *Catena*, vol. 137, pp. 340–349, 2016.

[8] H. Liu, X. Zhang, W. Yu, B. Zhang, K. Song, and J. Blackwell, "Simulating models for Phaeozem hyperspectral reflectance," *International Journal of Remote Sensing*, vol. 32, no. 13, pp. 3819–3834, 2011.

[9] D. J. Brown, K. D. Shepherd, M. G. Walsh, M. Dewayne Mays, and T. G. Reinsch, "Global soil characterization with VNIR diffuse reflectance spectroscopy," *Geoderma*, vol. 132, no. 3-4, pp. 273–290, 2006.

[10] V. Bellon-Maurel and A. McBratney, "Near-infrared (NIR) and mid-infrared (MIR) spectroscopic techniques for assessing the amount of carbon stock in soils – critical review and research perspectives," *Soil Biology and Biochemistry*, vol. 43, no. 7, pp. 1398–1410, 2011.

[11] L. Chen, J. Gong, B. Fu, Z. Huang, Y. Huang, and L. Gui, "Effect of land use conversion on soil organic carbon sequestration in the loess hilly area, loess plateau of China," *Ecological Research*, vol. 22, no. 4, pp. 641–648, 2007.

[12] Z. Shi, Q. Wang, J. Peng et al., "Development of a national VNIR soil-spectral library for soil classification and prediction of organic matter concentrations," *Science China Earth Sciences*, vol. 57, no. 7, pp. 1671–1680, 2014.

[13] B. Minasny, A. B. McBratney, V. Bellon-Maurel et al., "Removing the effect of soil moisture from NIR diffuse reflectance spectra for the prediction of soil organic carbon," *Geoderma*, vol. 167-168, pp. 118–124, 2011.

[14] X. Peng, T. Shi, A. Song, Y. Chen, and W. Gao, "Estimating soil organic carbon using Vis/NIR spectroscopy with SVMR and SPA methods," *Remote Sensing*, vol. 6, no. 4, pp. 2699–2717, 2014.

[15] J. M. Schmitt, "Fractional derivative analysis of diffuse reflectance spectra," *Applied Spectroscopy*, vol. 52, no. 6, pp. 840–846, 1998.

[16] J. T. Machado, V. Kiryakova, and F. Mainardi, "Recent history of fractional calculus," *Communications in Nonlinear Science and Numerical Simulation*, vol. 16, no. 3, pp. 1140–1153, 2011.

[17] B. Kuldeep, V. K. Singh, A. Kumar, and G. K. Singh, "Design of two-channel filter bank using nature inspired optimization based fractional derivative constraints," *ISA Transactions*, vol. 54, pp. 101–116, 2015.

[18] C. C. Tseng and S. L. Lee, "Design of linear phase FIR filters using fractional derivative constraints," *Signal Processing*, vol. 92, no. 5, pp. 1317–1327, 2012.

[19] A. Arikoglu, "A new fractional derivative model for linearly viscoelastic materials and parameter identification via genetic algorithms," *Rheologica Acta*, vol. 53, no. 3, pp. 219–233, 2014.

[20] J. Zhang and K. Chen, "Variational image registration by a total fractional-order variation model," *Journal of Computational Physics*, vol. 293, pp. 442–461, 2015.

[21] D. Zhang, T. Tiyip, J. Ding et al., "Quantitative estimating salt content of saline soil using laboratory hyperspectral data treated by fractional derivative," *Journal of Spectroscopy*, vol. 2016, pp. 1–11, 2016.

[22] F. Zhang, T. Tashpolat, H.-t. Kung, and J. Ding, "The change of land use/cover and characteristics of landscape pattern in arid areas oasis: an application in Jinghe, Xinjiang," *Geo-Spatial Information Science*, vol. 13, no. 3, pp. 174–185, 2010.

[23] A. Jilili, Z. Zhang, and F. Jiang, "Evaluation of the pollution and human health risks posed by heavy metals in the atmospheric dust in Ebinur Basin in Northwest China," *Environmental Science and Pollution Research International*, vol. 22, no. 18, pp. 14018–14031, 2015.

[24] F. Zhang, T. Tashpolat, V. C. Johnson et al., "The influence of natural and human factors in the shrinking of the Ebinur Lake,

Xinjiang, China, during the 1972–2013 period," *Environmental Monitoring and Assessment*, vol. 187, no. 1, pp. 1–14, 2014.

[25] L. Ma, J. Wu, A. Jilili, and W. Liu, "Geochemical responses to anthropogenic and natural influences in Ebinur Lake sediments of arid Northwest China," *PloS One*, vol. 11, no. 5, article e0155819, 2016.

[26] J. Yao, Q. Zhao, and Z. Liu, "Effect of climate variability and human activities on runoff in the Jinghe River Basin, Northwest China," *Journal of Mountain Science*, vol. 12, no. 2, pp. 358–367, 2015.

[27] W. Liu, F. Baret, X. Gu, Q. Tong, L. Zheng, and B. Zhang, "Relating soil surface moisture to reflectance," *Remote Sensing of Environment*, vol. 81, no. 2-3, pp. 238–246, 2002.

[28] G. Zheng, D. Ryu, C. Jiao, and C. Hong, "Estimation of organic matter content in coastal soil using reflectance spectroscopy," *Pedosphere*, vol. 26, no. 1, pp. 130–136, 2016.

[29] H. Liu, Y. Zhang, X. Zhang et al., "Quantitative analysis of moisture effect on black soil reflectance," *Pedosphere*, vol. 19, no. 4, pp. 532–540, 2009.

[30] X. Jin, J. Du, H. Liu, Z. Wang, and K. Song, "Remote estimation of soil organic matter content in the Sanjiang Plain, Northest China: the optimal band algorithm versus the GRA-ANN model," *Agricultural and Forest Meteorology*, vol. 218-219, pp. 250–260, 2016.

[31] B. Li and W. Xie, "Adaptive fractional differential approach and its application to medical image enhancement," *Computers & Electrical Engineering*, vol. 45, pp. 324–335, 2015.

[32] V. E. Tarasov, "On chain rule for fractional derivatives," *Communications in Nonlinear Science and Numerical Simulation*, vol. 30, no. 1–3, pp. 1–4, 2016.

[33] A. Volkan Bilgili, H. M. van Es, F. Akbas, A. Durak, and W. D. Hively, "Visible-near infrared reflectance spectroscopy for assessment of soil properties in a semi-arid area of Turkey," *Journal of Arid Environments*, vol. 74, no. 2, pp. 229–238, 2010.

[34] J. Farifteh, F. Van der Meer, C. Atzberger, and E. J. M. Carranza, "Quantitative analysis of salt-affected soil reflectance spectra: a comparison of two adaptive methods (PLSR and ANN)," *Remote Sensing of Environment*, vol. 110, no. 1, pp. 59–78, 2007.

[35] H. Martens and M. Martens, "Modified Jack-knife estimation of parameter uncertainty in bilinear modelling by partial least squares regression (PLSR)," *Food Quality and Preference*, vol. 11, no. 1-2, pp. 5–16, 2000.

[36] W. Niu, X. Liu, G. Huang, L. Chen, and L. Han, "Physicochemical composition and energy property changes of wheat straw cultivars with advancing growth days at maturity," *Energy & Fuels*, vol. 27, no. 10, pp. 5940–5947, 2013.

[37] B. Stenberg, R. A. Viscarra Rossel, A. M. Mouazen, and J. Wetterlind, "Chapter five-visible and near infrared spectroscopy in soil science," *Advances in Agronomy*, vol. 107, pp. 163–215, 2010.

[38] L. S. Galvão, M. A. Pizarro, and J. C. N. Epiphanio, "Variations in reflectance of tropical soils: spectral-chemical composition relationships from AVIRIS data," *Remote Sensing of Environment*, vol. 75, no. 2, pp. 245–255, 2001.

[39] J. T. Bushong, R. J. Norman, and N. A. Slaton, "Near-infrared reflectance spectroscopy as a method for determining organic carbon concentrations in soil," *Communications in Soil Science and Plant Analysis*, vol. 46, no. 14, pp. 1791–1801, 2015.

[40] P. K. Mutuo, K. D. Shepherd, A. Albrecht, and G. Cadisch, "Prediction of carbon mineralization rates from different

soil physical fractions using diffuse reflectance spectroscopy," *Soil Biology and Biochemistry*, vol. 38, no. 7, pp. 1658–1664, 2006.

[41] T. Udelhoven, C. Emmerling, and T. Jarmer, "Quantitative analysis of soil chemical properties with diffuse reflectance spectrometry and partial least-square regression: a feasibility study," *Plant and Soil*, vol. 251, no. 2, pp. 319–329, 2003.

[42] E. Ben-Dor, Y. Inbar, and Y. Chen, "The reflectance spectra of organic matter in the visible near-infrared and short wave infrared region (400–2500 nm) during a controlled decomposition process," *Remote Sensing of Environment*, vol. 61, no. 1, pp. 1–15, 1997.

[43] C. T. Chiou, S. E. McGroddy, and D. E. Kile, "Partition characteristics of polycyclic aromatic hydrocarbons on soils and sediments," *Environmental Science & Technology*, vol. 32, no. 2, pp. 264–269, 1998.

[44] X. Li, Y. Zhang, Y. Bao et al., "Exploring the best hyperspectral features for LAI estimation using partial least squares regression," *Remote Sensing*, vol. 6, no. 7, pp. 6221–6241, 2014.

[45] T. Shi, Y. Chen, Y. Liu, and G. Wu, "Visible and near-infrared reflectance spectroscopy—an alternative for monitoring soil contamination by heavy metals," *Journal of Hazardous Materials*, vol. 265, pp. 166–176, 2014.

[46] T. Shi, L. Cui, J. Wang, T. Fei, Y. Chen, and G. Wu, "Comparison of multivariate methods for estimating soil total nitrogen with visible/near-infrared spectroscopy," *Plant and Soil*, vol. 366, no. 1, pp. 363–375, 2013.

[47] C. Lin, S. Zhou, and S. Wu, "Using hyperspectral reflectance to detect different soil erosion status in the subtropical hilly region of southern China: a case study of Changting, Fujian Province," *Environmental Earth Sciences*, vol. 70, no. 4, pp. 1661–1670, 2013.

[48] H. Liu, Y. Zhang, and B. Zhang, "Novel hyperspectral reflectance models for estimating black-soil organic matter in Northeast China," *Environmental Monitoring and Assessment*, vol. 154, no. 1–4, pp. 147–154, 2009.

[49] S. D. L. Gregory, J. D. Lauzon, I. P. O'Halloran, and R. J. Heck, "Predicting soil organic matter content in southwestern Ontario fields using imagery from high-resolution digital cameras," *Canadian Journal of Soil Science*, vol. 86, no. 3, pp. 573–584, 2006.

[50] H. Yang and J. Li, "Predictions of soil organic carbon using laboratory-based hyperspectral data in the northern Tianshan mountains, China," *Environmental Monitoring and Assessment*, vol. 185, no. 5, pp. 3897–3908, 2013.

[51] S. Nawar, H. Buddenbaum, J. Hill, J. Kozak, and A. M. Mouazen, "Estimating the soil clay content and organic matter by means of different calibration methods of vis-NIR diffuse reflectance spectroscopy," *Soil and Tillage Research*, vol. 155, pp. 510–522, 2016.

25

Rancidity Estimation of Perilla Seed Oil by Using Near-Infrared Spectroscopy and Multivariate Analysis Techniques

Suk-Ju Hong,[1] Shin-Joung Rho,[1] Ah-Yeong Lee,[1] Heesoo Park,[1] Jinshi Cui,[1] Jongmin Park,[2] Soon-Jung Hong,[3] Yong-Ro Kim,[1] and Ghiseok Kim[1,4]

[1]*Department of Biosystems and Biomaterials Science and Engineering, Seoul National University, 1 Gwanak-ro, Gwanak-gu, Seoul 08826, Republic of Korea*
[2]*Department of Bio-Industrial Machinery Engineering, Pusan National University, 1268-50 Samnangjin-ro, Cheonghak-ri, Samnangjin-eup, Miryang-si 50463, Republic of Korea*
[3]*Rural Human Resource Development Center, Rural Development Administration, 420 Nongsaengmyeong-ro, Wansan-gu, Jeonju-si 54874, Republic of Korea*
[4]*Research Institute of Agriculture and Life Sciences, Seoul National University, 1 Gwanak-ro, Gwanak-gu, Seoul 08826, Republic of Korea*

Correspondence should be addressed to Ghiseok Kim; ghiseok@snu.ac.kr

Academic Editor: Jose S. Camara

Near-infrared spectroscopy and multivariate analysis techniques were employed to nondestructively evaluate the rancidity of perilla seed oil by developing prediction models for the acid and peroxide values. The acid, peroxide value, and transmittance spectra of perilla seed oil stored in two different environments for 96 and 144 h were obtained and used to develop prediction models for different storage conditions and time periods. Preprocessing methods were applied to the transmittance spectra of perilla seed oil, and multivariate analysis techniques, such as principal component regression (PCR), partial least squares regression (PLSR), and artificial neural network (ANN) modeling, were employed to develop the models. Titration analysis shows that the free fatty acids in an oil oxidation process were more affected by relative humidity than temperature, whereas peroxides in an oil oxidation process were more significantly affected by temperature than relative humidity for the two different environments in this study. Also, the prediction results of ANN models for both acid and peroxide values were the highest among the developed models. These results suggest that the proposed near-infrared spectroscopy technique with multivariate analysis can be used for the nondestructive evaluation of the rancidity of perilla seed oil, especially the acid and peroxide values.

1. Introduction

Recently, westernization and gentrification of food has increased the variety of processed food, and the expanding consumption pattern aimed at pursuing wellbeing and health is increasing the interest and demand in healthy high-functioning vegetable oils. Also, with the advent of problems, such as obesity and adult diseases, caused by trans fats and cholesterol from vegetable oils like soybean oil and corn oil, which were previously extensively used, the demand for high-quality vegetable oils, such as olive oil and grape seed oil, is increasing; recently, the interest in premium vegetable oils, for example, canola oil, green tea oil, and brown rice oil, has rapidly increased. Oils are known to be not only a high-energy source, as one of the three major nutrient groups of carbohydrates, proteins, and oils, but also an important and useful component of the human body; it is present in cell membranes as a fat-soluble carrier and also protects hypodermic tissues and organs [1]. In particular, perilla seed oil, a traditional Korean vegetable oil, contains 60% linolenic acid, one of the essential fatty acids, which is known to be useful in brain and nervous tissue development and in adult disease prevention. Generally, perilla (*Perilla frutescens Britton* var. *japonica Hara*) is an annual plant belonging to the family Labiatae. Its native habitat is Southeast Asia and it is commonly found in Korea, China, India, and Japan.

Records indicate that it was raised approximately 1300 years ago in Korea along with sesame, and it grows easily in every area of a mountain or roadside swamp to be raised on a national scale [2–4].

Note that the smell, fragrance, and taste of vegetable oils used in most homes, restaurants, and the food processing industry tends to change because of the various chemical and microbial factors during storage and processing. This decreases its nutritive value, which in turn deteriorates quality as well as that of the processed foods, and sometimes, toxic agents that are harmful to humans are also generated. This oil deterioration is called rancidity [5, 6]. Oil deterioration is known to occur by various causes, including auto-oxidation if the oil component absorbs oxygen. Auto-oxidation of vegetable oil increases its weight and causes the formation of aldehydes and ketones. Vegetable oil oxidation caused by heating at high temperatures increases the content of free fatty acids, lowers the smoke point, and leads to bubbling; acrolein, which is harmful to humans, may also form. In particular, vegetable oils containing a large amount of unsaturated fatty acids and, if heated for a long time and repetitively, may generate the toxic chemical compound 4-hydroxy-*trans*-2-nonenal, which can cause cardiovascular disorders [7–10].

As an existing method to measure oil rancidity, physicochemical titration is widely used to measure the acid value, peroxide value, and so forth. However, physicochemical titration analysis may generate errors resulting from the experimenter's skills and is expensive and time-consuming and hence is not suitable for repetitive experiments. Therefore, the development of a technique to nondestructively analyze vegetable oil rancidity in real time is attracting much interest. Recently, a study has been conducted into spectroscopy to nondestructively analyze and predict variation in the components and quality of agricultural products or food. Near-infrared (NIR) spectroscopy can obtain a signal of relatively high energy, relative to far-infrared radiation and microwaves, and can detect a particular spectrum of an intrinsic component of a test object during wavelength bandwidth measurement. Its relatively simple device composition of sensor and light source makes it easy to implement; therefore, NIR spectroscopy is being widely used in researches into the nondestructive quality analysis of agricultural products and food [11, 12].

Generally, near-infrared rays have a high energy level relative to mid- and far-infrared rays, and hence, they have low optical absorbance on test samples and excellent penetrability; this means that the technique is little influenced by test sample thickness and there is no need for the preprocessing of the test sample. Also, the components of near-infrared ray light splitters, such as fiber-optic cables, monochromators, and detectors, are reported to be easy to set up and operate relative to mid-infrared ray light splitters and have excellent measurement repeatability. In particular, fiber-optic communications have been developed based on near-infrared rays; consequently, their use makes it possible to measure and analyze spectra remotely in real time and allows the development of telemetering device. Therefore, near-infrared rays can be used for the composition of real-time

analysis devices for material process control and quality analysis. In spite of the many merits of NIR spectroscopy, the measurement environment and instruments may cause noise in the NIR spectra, which may lead to the degradation of spectrum analysis. Therefore, it is very important to perform appropriate preprocessing methods to remove these noise components and to develop a suitable optimal model to analyze and predict sample components by using spectra data. Until now, NIR spectroscopy research on agricultural products has been applied to the internal quality analysis of fruits and vegetables by using multivariate analysis methods, such as PLSR (partial least square regression), PCR (principal component regression), and MLR (multiple linear regression) [13–18]. In addition, during the 2000s, artificial neural network (ANN) techniques have been steadily applied to the internal quality analysis of food and agricultural products with visible/NIR spectrum analysis, and various kinds of research have been conducted on production date range distinction of milk products and component analysis and quality prediction for fruit juice, wine, and so forth [19–22].

This study employed NIR spectroscopy and multivariate analysis techniques to nondestructively discern the rancidity of perilla seed oil, in conjunction with the storage conditions, and to evaluate the effectiveness of the analysis methods. For this purpose, perilla seed oil was stored in specific environments for a certain period of time, and the acid value, peroxide value, and transmittance spectra of the same perilla seed oil were measured by physicochemical titration analysis and NIR spectroscopy. We used PCR, PLSR, and ANN analysis methods to quantitatively predict the rancidity extent of perilla seed oil from the collected data, and then we evaluated the performance of each model in conformity with the applied preprocessing methods.

2. Material and Methods

2.1. Sample Materials. Perilla (*Perilla frutescens Britton* var. *japonica Hara*) seeds harvested in October 2016 were acquired twice from a local producer (Miryang, South Korea) with a one-week interval to prepare two a little different sample groups, and they were cleaned with water and air-dried to remove moisture. After drying, the perilla seeds were roasted for 60 min at 200°C, and the heated perilla seeds were squeezed for 15 min with a pressure of 600 kgf/cm^2 by using an oil press to extract the perilla seed oil. The extracted perilla seed oil was kept in an air-tight opaque container under refrigeration at 5°C before the experiments. For our test, perilla seed oil samples were separated into two groups and stored in two different environments to artificially accelerate rancidity. The first group was stored at a controlled temperature of 60°C with a relative humidity of 40% for 4 d (96 h), and the second sample group was stored for 6 d (144 h) at a higher temperature of 80°C with a lower relative humidity of 10%.

2.2. Determination of Acid Value. The acid value (AV) of the samples was determined by the method of the American Oil Chemists' Society (1977) with some modifications [23]. One gram of each sample was dissolved in 20 mL of ether : ethanol

(2 : 1, *v/v*) solution, and then, 1% (*v/v*) phenolphthalein solution was added as an indicator. The mixture was agitated and titrated with 0.1 N KOH solution until the appearance of a pink color. Titration analysis was also performed for a blank sample under the same conditions. The AV was expressed as milligrams of KOH per gram of oil (mg/g) and determined according to the following equation.

$$AV = \frac{5.611 \times A \times F}{W}, \qquad (1)$$

where AV is the acid value, A denotes the volume of KOH (mL), 5.611 is the constant value equivalence of mass of 0.1 N KOH, W is the weight of sample, and F is the normality of the standard KOH solution.

2.3. Determination of Peroxide Value.

The peroxide value (PV) is a measure of the peroxides contained in the oil during storage. The PV of the oil was measured by iodine released from potassium iodide (KI) according to the method of the Association of Analytical Communities (1995) with slight modifications [24]. One gram of each sample was mixed with 25 mL of chloroform : acetic acid (2 : 3, *v/v*) solution and 1 mL of saturated KI solution with stirring and kept in a dark place for 5 min. Subsequently, the solution was mixed with 30 mL of distilled water and 1% (*w/v*) starch solution as an indicator. The mixture was titrated with 0.01 N $Na_2S_2O_3$ until the color of the reactant completely disappeared. The same preparation without an oil sample served as a blank experiment. The PV was represented as milliequivalents of oxygen per kilogram of oil (meq/kg) and was calculated by the following equation.

$$PV = \frac{(A - B) \times F \times 100}{W}, \qquad (2)$$

where PV is the peroxide value, A and B denote the volume of $Na_2S_2O_3$ (mL) in the sample and blank, respectively, W is the weight of sample, and F is the normality of the standard $Na_2S_2O_3$ solution.

2.4. Measurement of NIR Transmittance Spectra.

As illustrated in Figure 1, the transmittance spectra of perilla seed oil samples were measured by using a NIR spectrometer (CDI-NIR128, Control Development Inc., USA), with a wavelength range from 900 to 1700 nm, and a tungsten-halogen lamp (LS-1, Ocean optics, USA) was used as the light source.

Perilla seed oil samples were separated into two groups and stored in two different environments. The first sample group was stored at a controlled temperature of 60°C with a relative humidity of 40% for 4 d (96 h) to accelerate rancidity. Samples were taken at time zero, and then 10 g of 12 oil samples were collected every 24 h in order to obtain transmittance spectra after different periods. Therefore, we acquired 60 spectra for the first perilla seed oil sample group in 4 d (96 h). The environmental conditions for the second sample group were similar but employed higher temperature and lower relative humidity, that is, the temperature and relative humidity of the storage chamber were controlled as 80°C, 10% for 6 d (144 h), respectively. Samples were again taken

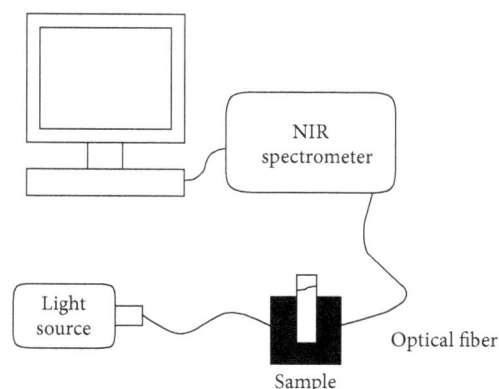

FIGURE 1: Schematic representation of the NIR spectra measuring instrument.

at time zero, and then in this case, 10 g of nine perilla seed oil samples were collected every 24 h to obtain transmittance spectra after different periods. Therefore, we acquired 63 spectra for the second perilla seed oil sample group in 6 d (144 h). At the end, we had prepared 123 spectra for the two different groups of perilla seed oil samples. All spectra were prepared with ten replicates at each measuring point with 50 ms of measuring exposure time.

The transmittance of measured spectra was calibrated by using both white-referenced and dark-referenced spectra as shown in (3). The white-referenced spectrum was measured from the transmittance through the blank cuvette used in the test, and the dark-referenced spectrum measured with the fiber optic cable was completely blocked.

$$\text{Transmittance (\%)} = \frac{S_i - D}{R - D} \times 100, \qquad (3)$$

where S_i is the raw transmittance of the samples and D and R represent the raw transmittance of the dark and white referenced spectra, respectively.

2.5. Model Development by Using Multivariate Analysis.

Light scattering, optical path changes, and noise can be induced in the obtained transmittance spectra from the measuring environment and instruments; therefore, preprocessing methods should be applied to the measured spectra in order to remove noise components and correct the spectra. In this study, we employed scatter correction techniques, such as MSC (multiplicative scatter correction) and SNV (standard normal variate), and normalization techniques, including maximum normalization, mean normalization, and range normalization, as preprocessing methods.

We used the PCR and PLSR methods that are defined as (4) to predict the acid value and peroxide value, which can be used to estimate the extent of oil rancidity, according to storage conditions and periods. The performances of the developed models are compared and analyzed according to the applied preprocessing methods. Among the obtained 123 spectra, calibration and validation modeling operations used 82 and 41 spectra, respectively. Algorithms for preprocessing and regression modeling were developed by

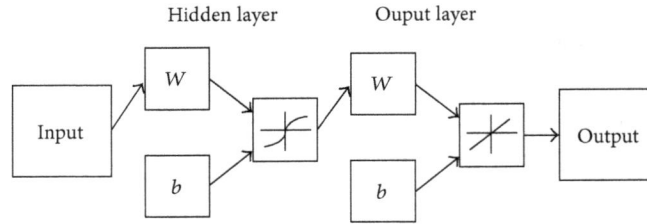

FIGURE 2: Schematic structure of the ANN model.

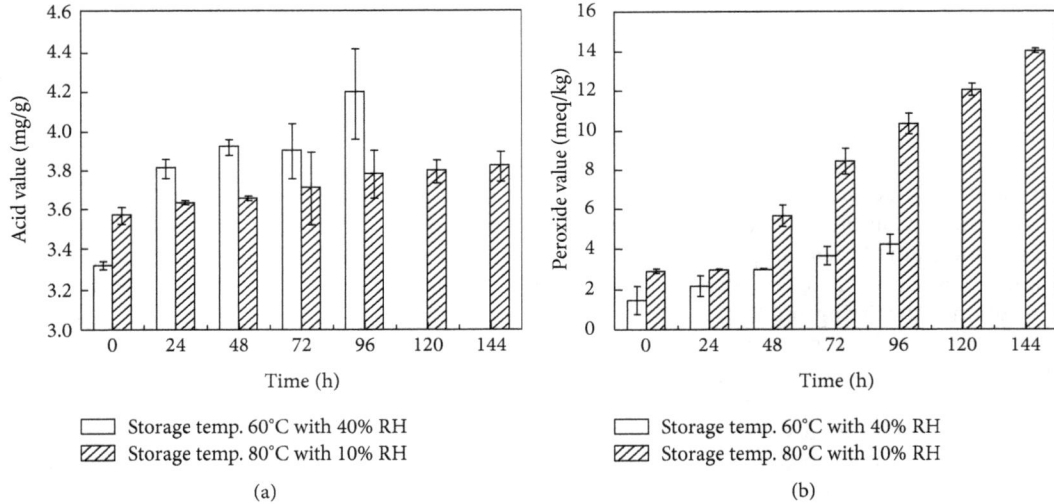

FIGURE 3: Acid values and peroxide values of perilla seed oil for various storage temperatures, relative humidity, and time periods. (a) Acid value. (b) Peroxide value.

TABLE 1: Statistics of acid values and peroxide values of perilla seed oil according to storage time.

Measurements	Storage conditions	Time (h)						
		0	24	48	72	96	120	144
Acid value (mg/g)	60°C, 40% RH	3.32 ± 0.02	3.81 ± 0.05	3.92 ± 0.04	3.9 ± 0.14	4.19 ± 0.23	—	—
	80°C, 10% RH	3.57 ± 0.04	3.63 ± 0.01	3.65 ± 0.01	3.71 ± 0.19	3.78 ± 0.13	3.80 ± 0.06	3.82 ± 0.08
Peroxide value (meq/kg)	60°C, 40% RH	1.48 ± 0.68	2.18 ± 0.50	2.99 ± 0.30	3.65 ± 0.45	4.22 ± 0.45	—	—
	80°C, 10% RH	2.91 ± 0.10	2.98 ± 0.38	5.64 ± 0.54	8.45 ± 0.66	10.33 ± 0.50	12.03 ± 0.29	14.00 ± 0.11

using commercial software (Matlab ver. R2016, MathWorks, Natick, MA, USA).

$$X = TP^T + E,$$
$$Y = TQ^T + F, \tag{4}$$

where X is the n by m matrix of predictors, Y is the n by p matrix of responses, T is the n by 1 matrix of the score matrix, E and F are error terms, and P and Q are the m by 1 and p by 1 loading matrices, respectively.

In the development process of both PCR and PLSR models, the selection of the number of PCs (principal components) should be significantly considered. If the number of PCs is too small, the regression model becomes inaccurate because the complete measurement information cannot be

sufficiently reflected in the developing model. By contrast, overfitting of the regression model, which may reduce its performance, can be induced if a large number of PCs are selected. The performance of the PCR and PLSR models that were developed to estimate the extent of rancidity was assessed with the RMSE (root-mean-square error), as defined in (5). The regression model with a small RMSE and large coefficient of determination was selected as the appropriate model.

$$\text{RMSE} = \sqrt{\frac{\sum_{i=1}^{m} \left(y_{\text{obs}} - y_{\text{pre}} \right)^2}{m}}, \tag{5}$$

where RMSE represents the sample standard deviation of the differences between predicted values and observed values.

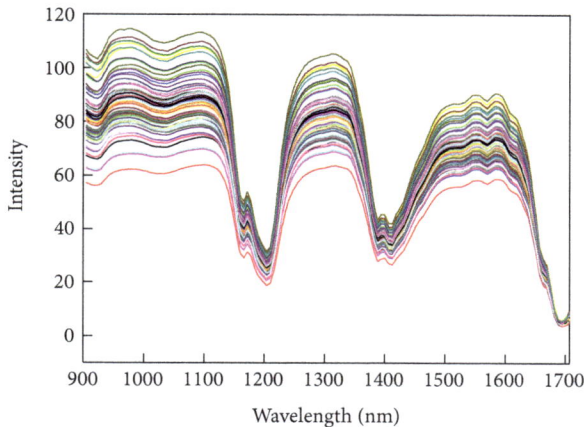

FIGURE 4: Transmittance spectra of perilla seed oil samples according to storage time.

TABLE 2: Locations of function groups and corresponding vibration modes in the near-infrared region.

Wavelength (nm)	Vibration mode	Structure
900	C–H str. third overtone	CH_3
1195	C–H str. second overtone	CH_3
1215	C–H str. second overtone	CH_3
1395	2C–H str. + 2C–H def.	CH_3
1410	O–H str. first overtone	ROH

y_{obs} and y_{pre} are the observed and predicted values, respectively. m is the number of data.

We also developed ANN-based regression models that can predict the acid and peroxide values by using NIR transmittance spectra obtained from perilla seed oil samples. As shown in Figure 2, the models fundamentally consisted of three parts, namely, input, hidden, and output layers, and both supervised-learning method and Levenberg-Marquardt back-propagation algorithms were used in this study. The operating system, CPU, and RAM capacity that were used for the development of the ANN-based regression models were OS X El Capitan, 2.6 GHz Intel Core i5, and 8 GB, respectively. The algorithms needed for the ANN process were created by using commercial software (Matlab ver. R2016, MathWorks, Natick, MA, USA). For the development of the ANN-based models, the NIR transmittance spectra of perilla seed oil samples were set to input (dependent variable) and both acid and peroxide values were set to output (independent variable). We designed four kinds of hidden layers by selecting different node numbers as 50, 70, 90, and 110. In addition, a sigmoid function was used as the activation function, and the iteration was limited to a maximum of 1000 to prevent overflow during calculations. Among 123 spectra (100%), the ANN operations of training, validation, and testing used 73 (60%), 25 (20%), and 25 spectra (20%), respectively.

3. Results and Discussion

3.1. Titration Analysis of Perilla Seed Oil. Oil oxidation, the main cause of rancidity, is the process by which the double bonds of unsaturated fatty acids in the oil combine with oxygen in the air to form oxidation products, with accompanying off-flavors and smells. The AVs and PVs measured in this study are widely used as key indicators of the extent to which rancidity reactions have occurred during the initial and intermediate stages of oil oxidation. They could be used as an indication of the quality and stability of oils. Generally, the more the oil rancidity has progressed, the more these

values have increased [25, 26]. The extent of rancidity of the perilla seed oil according to the storage environment and period was measured as the AV and PV, and the results are summarized in Figure 3 and Table 1. The oils were stored at two different conditions of temperature and relative humidity (condition 1: 60°C with 40% RH and condition 2: 80°C with 10% RH). There were significant rises in the AVs and PVs under all conditions during the storage period. The AVs of oil samples stored for 4 d (96 h) under conditions 1 and 2 were 4.19 ± 0.23 mg/g and 3.78 ± 0.13 mg/g, respectively, and the PVs were 4.22 ± 0.45 meq/kg and 10.33 ± 0.5 meq/kg, respectively.

From the incremental results of AVs and PVs that were obtained during the initial and intermediate processes of oil oxidation under the two storage conditions, it can be observed that the AV of the oil sample stored under condition 1 (60°C, 40% RH) was higher than that of the oil sample stored under condition 2 (80°C, 10% RH). That is, the AV of the oil sample stored at the lower relative humidity of 10% was generally stabilized during the six days at under 3.82 mg/g, whereas the observed AV at the higher relative humidity of 40% increased continuously during the four days until it reached 4.19 mg/g. In contrast to the AV, the PV of the oil sample stored under condition 2 (80°C, 10% RH) increased rapidly above the PV of the oil sample stored under condition 1 (60°C, 40% RH). That is, the PV was generally stabilized during the four days at less than 4.2 meq/kg at a 60°C, whereas the observed PV of the oil sample stored at the higher temperature of 80°C increased rapidly until it reached 10.3 meq/kg on the same fourth day. Moreover, the PV of the oil sample stored at the higher temperature of 80°C condition was measured as 14 meq/kg on the sixth day. Based on a previous study, which reported that rancid tastes and odors are clearly noticeable when the PV exceeds 20 meq/kg [27], we assume that the perilla seed oil sample can be significantly degraded within six days by high-temperature conditions 80°C, as suggested in this study.

These results indicate that the free fatty acids that were produced by the oil oxidation process can be more easily affected by relative humidity than temperature, whereas the peroxides that were formed by iodine released from potassium iodide by an oil oxidation process were more significantly affected by temperature than relative humidity during the initial and intermediate processes of the oil oxidation under the storage environments in this study. Similarly to our study results, previous studies showed that the oxidation process of oil stored at around 60°C was significantly affected by both the moisture content of the

TABLE 3: Results of PCR calibration and validation for acid and peroxide values of perilla seed oil according to preprocessing methods.

Preprocessing methods	Acid value				Peroxide value			
	Calibration		Validation		Calibration		Validation	
	R^2	RMSE	R^2	RMSE	R^2	RMSE	R^2	RMSE
Raw data	0.8979	0.0699	0.6501	0.1293	0.8240	1.6833	0.7748	1.9136
MSC	0.8645	0.0805	0.7091	0.1179	0.8787	1.4042	0.7328	2.0841
SNV	0.8373	0.0882	0.6859	0.1225	0.8120	1.7484	0.7552	1.9951
Normalization								
Max	0.8390	0.0877	0.5344	0.1492	0.8403	1.6113	0.7654	1.9529
Mean	0.8785	0.0762	0.6985	0.1201	0.8575	1.5156	0.7833	1.8771
Range	0.8105	0.0952	0.6454	0.1302	0.8157	1.7308	0.7474	2.0265

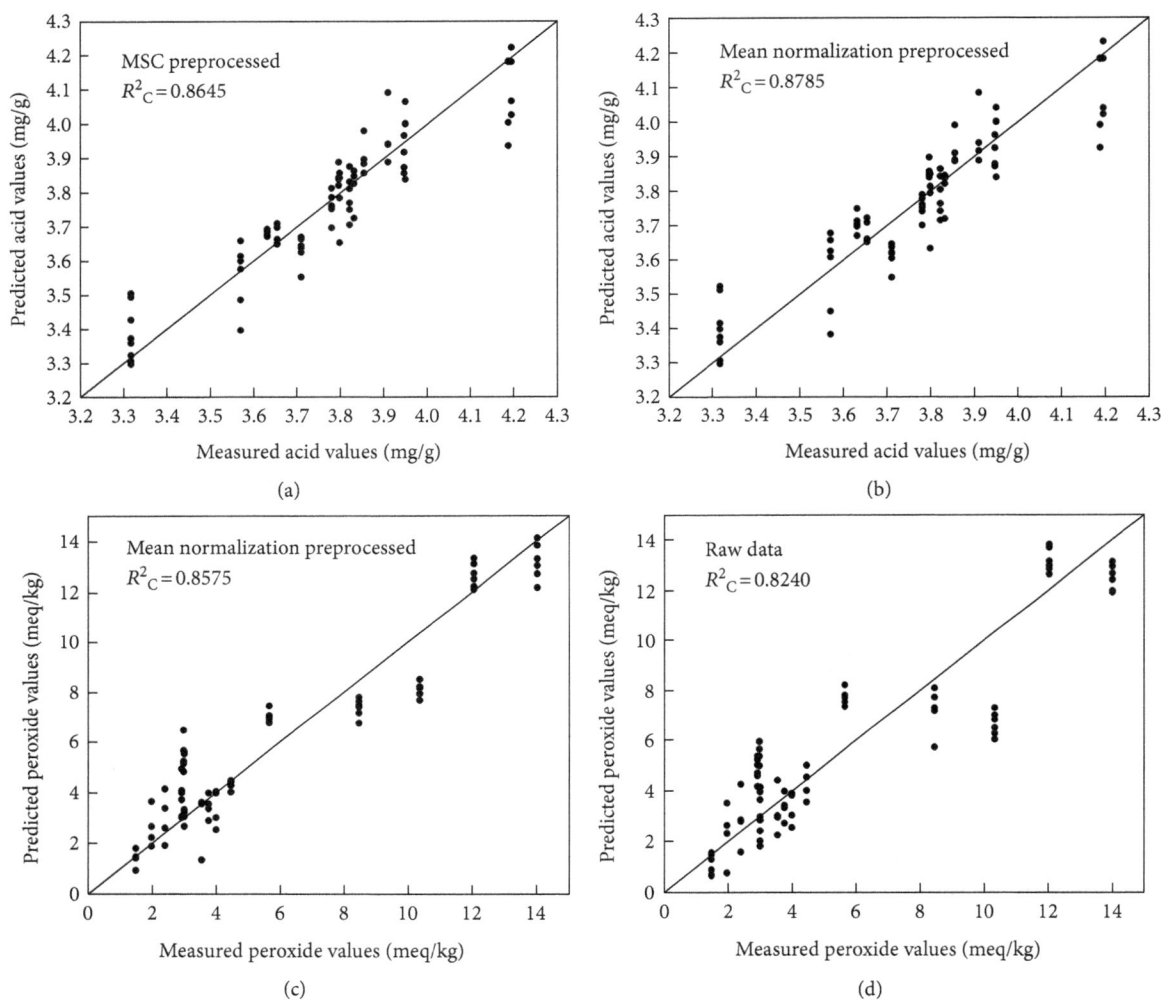

(a)

(b)

(c)

(d)

FIGURE 5: Prediction results of PCR models for acid and peroxide values from the highest R^2 values. (a) Acid value with MSC preprocessed. (b) Acid value with mean normalization. (c) Peroxide value with mean normalization. (d) Peroxide value with raw data.

oil itself and the relative humidity of the storage environment [28–31].

3.2. Transmittance Spectra of Perilla Seed Oil. Figure 4 shows the measured transmittance spectra of the oil samples over the storage period. As the storage period increases, the spectral absorption in the near-infrared band gradually increases

on the whole, and large absorption peak can be observed at around wavelengths of 1200 and 1400 nm. Generally, fatty acids in vegetable oils like perilla seed oil contain about 60% α-linolenic acid, which is a highly unsaturated fatty acid known as omega 3 and which is prone to induce rancidity due to the storage environment (temperature and moisture). It is known that hydroperoxide and carbonyl compounds,

TABLE 4: Results of PLSR calibration and validation for acid and peroxide values of perilla seed oil according to preprocessing methods.

Preprocessing methods	Acid value				Peroxide value			
	Calibration		Validation		Calibration		Validation	
	R^2	RMSE	R^2	RMSE	R^2	RMSE	R^2	RMSE
Raw data	0.8257	0.0913	0.6887	0.1220	0.8209	1.7061	0.8036	1.7869
MSC	0.8643	0.0806	0.7369	0.1122	0.8255	1.6842	0.7858	1.8660
SNV	0.8703	0.0787	0.7272	0.1142	0.8268	1.6781	0.7905	1.8455
Normalization								
Max	0.8741	0.0776	0.7225	0.1152	0.8340	1.6428	0.7914	1.8416
Mean	0.8328	0.0894	0.6681	0.1260	0.8037	1.7863	0.7816	1.8844
Range	0.9163	0.0633	0.7413	0.1112	0.8460	1.5821	0.7829	1.8787

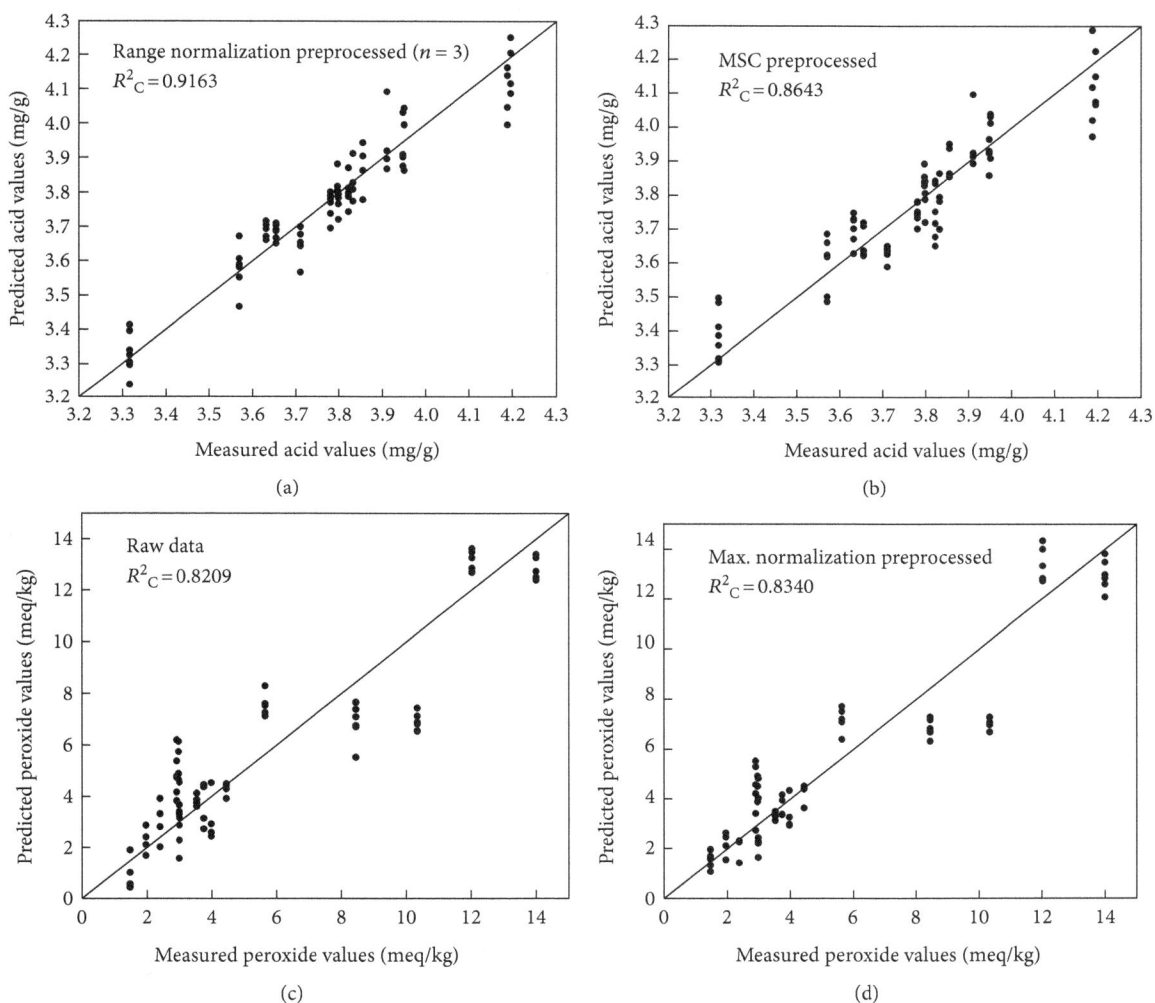

FIGURE 6: Prediction results of PLSR models for acid and peroxide values from the highest R^2 values. (a) Acid value with range normalization ($n = 3$). (b) Acid value with MSC preprocessed. (c) Peroxide value with raw data. (d) Peroxide value with maximum normalization.

which are not found in fresh vegetable oil (perilla seed oil), can be formed and cause an increase in the –ROOH, –OH, =O, and –CHO groups. Table 2 shows the commonly known wavelength locations of function groups and corresponding vibration modes which can be activated in the near-infrared region, and this information demonstrates the results of large spectral absorption at around the wavelength of 1200 and 1400 nm when the rancidity of vegetable oil induces the increasing acid value and peroxide value as shown in Figure 4.

3.3. *Results of Principal Component Regression.* The performances of PCR models for determining the AVs and PVs of the oil samples are summarized in Table 3 by showing

TABLE 5: Performance of ANN models for acid value predictions.

Nodes	Training			Validation			Test		
	R^2	RMSE	Elapsed time (sec)	R^2	RMSE	Elapsed time (sec)	R^2	RMSE	Elapsed time (sec)
50	0.8440	0.0847	1	0.6001	0.1586	1	0.6764	0.1177	1
70	0.8942	0.0745	1	0.7728	0.1008	1	0.6572	0.1181	1
90	0.9037	0.0694	1	0.8175	0.0963	1	0.8555	0.1112	1
110	0.9374	0.0611	2	0.8393	0.0779	2	0.6777	0.0935	2

TABLE 6: Performance of ANN models for peroxide value predictions.

Nodes	Training			Validation			Test		
	R^2	RMSE	Elapsed time (sec)	R^2	RMSE	Elapsed time (sec)	R^2	RMSE	Elapsed time (sec)
50	0.8439	1.5760	1	0.8266	1.9282	1	0.6458	2.1710	1
70	0.9388	0.9862	1	0.8304	1.6262	1	0.8140	2.0883	1
90	0.9210	1.1986	1	0.9341	0.9806	1	0.8286	1.5867	1
110	0.9195	1.1220	2	0.8224	1.8588	2	0.8159	1.7379	2

the coefficients of determination and RMSE values. Based on the validation results of the regression models for the AV, the MSC model shows the highest performance (R^2_C: 0.8645, RMSE$_C$: 0.0805; R^2_V: 0.7091, RMSE$_V$: 0.1179), and the performance of the model with mean normalization preprocessing was estimated as the second best (R^2_C: 0.8785, RMSE$_C$: 0.0762; R^2_V: 0.6985, RMSE$_V$: 0.1201). As for the validation results of the regression models for the PV, the performance of the model with mean normalization preprocessing was analyzed as the highest (R^2_C: 0.8575, RMSE$_C$: 1.5156; R^2_V: 0.7833, RMSE$_V$: 1.8771), and the model with raw data shows the second best performance (R^2_C: 0.8240, RMSE$_C$: 1.6833; R^2_V: 0.7748, RMSE$_V$: 1.9136). The loading matrix P, score matrix T, and beta matrix which are calculated during the PCR are summarized in Tables S1 available online at https://doi.org/10.1155/2017/1082612, S2, and S3, respectively. Figure 5 shows the prediction results of the two most highly correlated PCR models for acid and peroxide values.

3.4. *Results of Partial Least Squares Regression (PLSR).* Table 4 shows the performance of PLSR models for determining the acid and peroxide values of the oil samples according to preprocessing methods. The performance of each PLSR model was compared and evaluated by using the coefficient of determination and the RMSE value. On the basis of the validation results for the developed models, prediction model of AV with the range normalization preprocessing method shows the best performance (R^2_C: 0.9163, RMSE$_C$: 0.0633; R^2_V: 0.7413, RMSE$_V$: 0.1122), and the performance of the MSC model was estimated as the second best (R^2_C: 0.8643, RMSE$_C$: 0.0806; R^2_V: 0.7369, RMSE$_V$: 0.1122). As for the validation results of regression models for the PV, the performance of the model with raw data was analyzed as the best (R^2_C: 0.8209, RMSE$_C$: 1.7061; R^2_V: 0.8036, RMSE$_V$: 1.7869), and the model using maximum normalization shows the second best performance (R^2_C: 0.8340, RMSE$_C$: 1.6428, R^2_V: 0.7914, RMSE$_V$: 1.8416). In this PLSR

process, the number of principle components was decided as five and each spectrum has 804 wavelengths. The loading matrices P and Q, score matrix T, and beta matrix are summarized in Tables S4, S5, S6, and S7, respectively. Figure 6 shows the prediction result of two most highly correlated PLSR models for acid and peroxide values.

3.5. *Results of ANN Models.* Tables 5 and 6 represent the prediction performances of ANN models for the acid and peroxide values of perilla seed oils. Four different ANN models were developed by applying four different node numbers to the models, and the performance of each model was evaluated by using the coefficient of determination and RMSE value. Based on the test results of the ANN models, the prediction performance for acid values (R^2_{tra}: 0.9037, RMSE$_{tra}$: 0.0694; R^2_{val}: 0.8175, RMSE$_{val}$: 0.0963; R^2_{test}: 0.8555, RMSE$_{test}$: 0.1112) was evaluated to be slightly better than that for peroxide values (R^2_{tra}: 0.9210, RMSE$_{tra}$: 1.1986; R^2_{val}: 0.9341, RMSE$_{val}$: 0.9806; R^2_{test}: 0.8286, RMSE$_{test}$: 1.5867). Figure 7 shows the prediction results of most highly correlated ANN models for acid and peroxide values; the number of hidden nodes was set to 90.

4. Conclusions

In this study, we developed prediction models and estimated the acid and peroxide values of perilla seed oil to nondestructively estimate the rancidity in conjunction with the storage conditions by using NIR spectroscopy and multivariate analysis methods. These methods have some merits in reducing time-consuming repetitive experiments and supporting reliable results by minimizing the errors that may originated from the operator's skills. It is generally known that the NIR spectroscopy technique we used in the study can precisely and quantitatively estimate the optical response of functional groups for a target molecular structure through multivariate analysis methods. We employed several multivariate analysis methods, such as PCR, PLSR, and ANN,

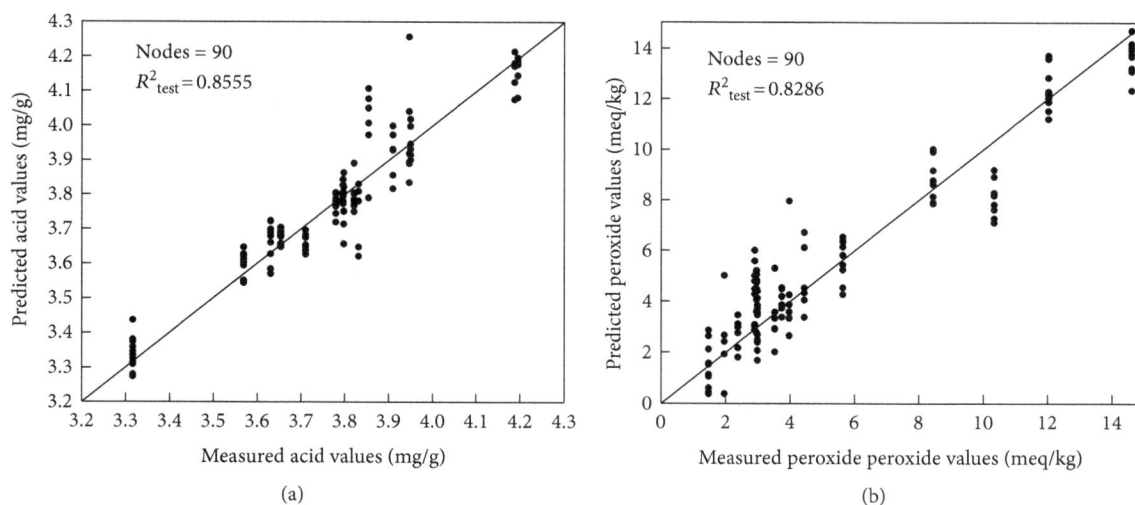

FIGURE 7: Graphs of the best results for the prediction of acid and peroxide values (90 nodes). (a) Acid value. (b) Peroxide value.

and it was determined that the prediction results for ANN models for both the acid value (R^2_{tra}: 0.9037, R^2_{val}: 0.8175, and R^2_{test}: 0.8555) and peroxide value (R^2_{tra}: 0.9210, R^2_{val}: 0.9341, and R^2_{test}: 0.8286) were the highest among the developed models.

The NIR spectroscopy technique used in this study has been commonly used in both qualitative and quantitative analyses of target samples by considering the vibrational energies of molecules, and the low optical absorbance characteristics of near-infrared rays facilitate deeper penetration than those of mid-infrared rays. In spite of its merits, NIR spectroscopy has limited ability to resolve noise, which is induced from measurement environments and the overlap tendencies of NIR spectra. A great deal of research is being performed in order to employ possible preprocessing methods and develop proper multivariate analysis models. We expect that our application of the PCR, PLSR, and ANN multivariate analysis methods for the nondestructive evaluation of the rancidity of perilla seed oil by using NIR spectroscopy techniques has great potential for use in the component analysis of agricultural products and foods, as well as many other scientific areas, such as life sciences and biomaterial researches.

Acknowledgments

This work was supported by the National Research Foundation of Korea (Grant 2017R1D1A1A02019090).

References

[1] W. A. May, R. J. Peterson, and S. S. Chang, "Chemical reactions involved in the deep-fat frying of foods: IX. Identification of the volatile decomposition products of triolein," *Journal of the American Oil Chemists' Society*, vol. 60, no. 5, pp. 990–995, 1983.

[2] I. H. Kim, S. Y. Jung, J. S. Jo, and Y. E. Kim, "Changes in components and sensory attribute of the oil extracted from perilla seed roasted at different roasting conditions," *Agricultural Chemistry and Biotechnology*, vol. 39, no. 2, pp. 118–122, 1996.

[3] Y. S. Cho, B. K. Kim, J. K. Park, J. W. Jeong, S. W. Jeong, and J. H. Lim, "Influence of thermal treatment on chemical changes in cold-pressed perilla seed oil," *Korean Journal of Food Preservervation*, vol. 16, no. 6, pp. 884–892, 2009.

[4] C. U. Choi, "History and science of perilla seed oil and sesame oil," *Proceedings of the Korean Society of Food and Cookery Science Conference*, vol. 37, no. 1, pp. 443–452, 1998.

[5] H. K. Chung, C. S. Choe, J. H. Lee, M. J. Chang, and M. H. Kang, "Oxidative stability of the pine needle extracted oils and sensory evaluation of savored laver made by extracted oils," *Journal of the Korean Society of Food Culture*, vol. 18, no. 2, pp. 89–95, 2003.

[6] F. W. Summerfield and A. L. Tappel, "Detection and measurement by high-performance liquid chromatography of malondialdehyde crosslinks in DNA," *Analytical Biochemistry*, vol. 143, no. 2, pp. 265–271, 1984.

[7] D. S. Kim, B. S. Koo, and M. S. Ahn, "A study on the formation of trans fatty acids with heating and storage of fats and oils (I) - the change of physicochemical characteristics and total trans fatty acids content," *Korean Journal of Food and Cookery Science*, vol. 6, no. 2, pp. 37–50, 1990.

[8] N. R. Artman, "The chemical and biological properties of heated and oxidized fats," *Advances in Lipid Research*, vol. 7, pp. 245–330, 1969.

[9] S. Husain, G. S. R. Sastry, and N. P. Raju, "Molecular weight averages as criteria for quality assessment of heated oils and fats," *Journal of the American Oil Chemists Society*, vol. 68, no. 11, pp. 822–826, 1991.

[10] S. L. Taylor, C. M. Berg, N. H. Shoptaugh, and E. Traisman, "Mutagen formation in deep-fat fried foods as a function of frying conditions," *Journal of the American Oil Chemists' Society*, vol. 60, no. 3, pp. 576–580, 1983.

[11] E. Ben-Dor, J. R. Irons, and G. F. Epema, *Remote Sensing for the Earth Sciences: Manual of Remote Sensing*, John Wiley & Sons, New York, Third edition, 1999.

[12] E. Choe, F. van der Meer, F. van Ruitenbeek, H. van der Werff, B. de Smeth, and K. W. Kim, "Mapping of heavy metal pollution in stream sediments using combined geochemistry, field spectroscopy, and hyperspectral remote sensing: a case study of the Rodalquilar mining area, SE Spain," *Remote Sensing of Environment*, vol. 112, no. 7, pp. 3222–3233, 2008.

[13] C. H. Choi, K. J. Lee, and B. S. Park, "Prediction of soluble solid and firmness in apple by visible/near-infrared spectroscopy," *Journal of the Korean Society for Agricultural Machinery*, vol. 22, no. 2, pp. 256–265, 1997.

[14] R. F. Lu, D. E. Guyer, and R. M. Beaudry, "Determination of firmness and sugar content of apples using nearinfrared diffuse reflectance," *Journal of Texture Studies*, vol. 31, no. 6, pp. 615–630, 2000.

[15] Y. T. Kim and S. R. Suh, "Comparison of performance of models to predict hardness of tomato using spectroscopic data of reflectance and transmittance," *Journal of Biosystems Engineering*, vol. 33, no. 1, pp. 63–68, 2008.

[16] S. Kawano, H. Watanabe, and M. Iwamoto, "Determination of sugar content in intact peaches by near infrared spectroscopy," *Journal of Japanese Society for Horticultural Science*, vol. 61, no. 2, pp. 445–451, 1992.

[17] G. G. Dull, R. G. Leffler, G. S. Birth, and D. A. Smittle, "Instrument for non-destructive measurement of soluble solids in honeydew melons," *Transactions of the American Society of Agricultural Engineers*, vol. 35, no. 2, pp. 735–737, 1992.

[18] H. I. Chung and H. J. Kim, "Near-infrared spectroscopy: principles," *Analytical Science & Technology*, vol. 13, no. 1, pp. 138–151, 2000.

[19] M. J. Martelo-Vidal and M. Vázquez, "Application of artificial neural networks coupled to UV–VIS–NIR spectroscopy for the rapid quantification of wine compounds in aqueous mixtures," *CyTA - Journal of Food*, vol. 13, no. 1, pp. 32–39, 2015.

[20] C. Cimpoiu, V. Cristea, A. Hosu, M. Sandru, and L. Seserman, "Antioxidant activity prediction and classification of some teas using artificial neural networks," *Food Chemistry*, vol. 127, no. 3, pp. 1323–1328, 2011.

[21] M. Gestal, M. P. Gomez-Carracedo, J. M. Andrade et al., "Classification of apple beverages using artificial neural networks with previous variable selection," *Analytica Chimica Acta*, vol. 524, no. 1-2, pp. 225–234, 2004.

[22] A. Gori, C. Cevoli, A. Fabbri, M. F. Caboni, and G. Losi, "A rapid method to discriminate season of production and feeding regimen of butters based on infrared spectroscopy and artificial neural networks," *Journal of Food Engineering*, vol. 109, no. 3, pp. 525–530, 2012.

[23] AOCS, *Official Methods and Recommended Practices of the American Oil Chemists' Society*, method Te 1a-64, American Oil Chemists' Society, Champaign, IL, USA, 1977.

[24] AOAC, *Official Methods of Analysis, Association of Official Agricultural Chemists*, AOAC Press, Washington DC, 16th edition, 1995.

[25] H. Lawson, *Common Chemical Reactions in Food Oils and Fats, Food Oils and Fats*, CBS Publishers and Distributors, New Delhi, India, 1997.

[26] F. Shahidi and U. N. Wanasundara, *Methods for Measuring Oxidative Rancidity in Fats and Oils. Food Lipids: Chemistry, Nutrition and Biotechnology*, CRC Press, New York, Second edition, 2002.

[27] R. S. Kirk and R. Sawyer, *Pearson's Composition and Analysis of Foods*, Addison Wesley Longman, Harlow, UK, 9th edition, 1991.

[28] J. W. Park, J. Y. Kim, M. J. Kim, and J. H. Lee, "Evaluation of oxygen-limitation on lipid oxidation and moisture content in corn oil at elevated temperature," *Journal of the American Oil Chemists' Society*, vol. 91, no. 3, pp. 439–444, 2014.

[29] W. Chaiyasit, R. J. Elias, D. J. McClements, and E. A. Decker, "Role of physical structures in bulk oils on lipid oxidation," *Critical Reviews in Food Science and Nutrition*, vol. 47, no. 3, pp. 299–317, 2007.

[30] D. J. Mcclements and E. A. Decker, "Lipid oxidation in oil-in-water emulsions: impact of molecular environment on chemical reactions in heterogeneous food systems," *Journal of Food Science*, vol. 65, no. 8, pp. 1270–1282, 2000.

[31] K. Schwarz, S. W. Huang, J. B. German, B. Tiersch, J. Hartmann, and E. N. Frankel, "Activities of antioxidants are affected by colloidal properties of oil-in-water and water-in-oil emulsions and bulk oils," *Journal of Agricultural and Food Chemistry*, vol. 48, no. 10, pp. 4874–4882, 2000.

Permissions

List of Contributors

Dong Zhang, Tashpolat Tiyip, Jianli Ding, Fei Zhang, Ilyas Nurmemet, Ardak Kelimu and Jingzhe Wang
College of Resources and Environment Science, Xinjiang University, Urumqi 830046, China
Key Laboratory of Oasis Ecology, Xinjiang University, Urumqi 830046, China

H. Jull and R. Künnemeyer
School of Engineering, University of Waikato, Hamilton 3240, New Zealand
The Dodd-Walls Centre for Photonic and Quantum Technologies, Hamilton 3240, New Zealand

P. Ewart
School of Engineering, University of Waikato, Hamilton 3240, New Zealand
Waikato Institute of Technology, Hamilton 3200, New Zealand

P. Schaare
The New Zealand Institute for Plant & Food Research Ltd, Hamilton 3240, New Zealand

Jie Yu, Xiaoyan Zhang, Dibo Hou, Fang Chen, Tingting Mao, Pingjie Huang and Guangxin Zhang
State Key Laboratory of Industrial Control Technology, College of Control Science and Engineering, Zhejiang University, Hangzhou 310027, China

Jair C. C. Freitas, Daniel F. Cipriano, Carlos G. Zucolotto, Alfredo G. Cunha and Francisco G. Emmerich
Laboratory of Carbon and Ceramic Materials, Department of Physics, Federal University of Espírito Santo, Av. Fernando Ferrari 514, 29075-910 Vitória, ES, Brazil

Zhihong Xu, Shanshan Wu, Maowen Chen, Xiaosong Ge and Xueliang Lin
Key Laboratory of Optoelectronic Science and Technology for Medicine, Ministry of Education, Fujian Normal University, Fuzhou 350007, China

Wei Huang
Key Laboratory of Optoelectronic Science and Technology for Medicine, Ministry of Education, Fujian Normal University, Fuzhou 350007, China
Fujian Metrology Institute, Fuzhou 350003, China

Duo Lin
Key Laboratory of Optoelectronic Science and Technology for Medicine, Ministry of Education, Fujian Normal University, Fuzhou 350007, China
College of Integrated Traditional Chinese andWestern Medicine, Fujian University of Traditional Chinese Medicine, Fuzhou, Fujian 350122, China

Liqing Sun
Affiliated Fuzhou First Hospital of Fujian Medical University, Fuzhou 350009, China

Wenjing Liu
State Key Laboratory of Environmental Chemistry and Ecotoxicology, Research Center for Eco-Environmental Sciences, Chinese Academy of Sciences, Beijing 100085, China
University of Chinese Academy of Sciences, Beijing 100049, China

Zhaotian Sun and Chuanbo Jing
No. 4 Hospital, Jinan 250031, China

Jinyu Chen
National Center for Mathematics and Interdisciplinary Sciences, Academy of Mathematics and Systems Science, Chinese Academy of Sciences, Beijing 100190, China

M. Z. Khan, I. Rehan and R.Muhammad
Department of Applied Physics, Federal Urdu University of Arts, Science and Technology, Islamabad 44,000, Pakistan

A. Ali
Department of Applied Physics, Federal Urdu University of Arts, Science and Technology, Islamabad 44,000, Pakistan
National Center for Physics, Quaid-i-Azam University Campus, Islamabad 44,000, Pakistan

K. Rehan
Department of Applied Physics, Federal Urdu University of Arts, Science and Technology, Islamabad 44,000, Pakistan
International College, UCAS, Beijing 100190, China

Jinhui Zhao, Haichao Yuan, Yijie Peng, Qian Hong and Muhua Liu
Optics-Electrics Application of Biomaterials Lab, College of Engineering, Jiangxi AgriculturalUniversity, Nanchang 330045, China

Ping Liu
College of Animal Science and Technology, Jiangxi Agricultural University, Nanchang 330045, China

Yuxiang Zheng
Shanghai Ultra-Precision Optical Manufacturing Engineering Center, Department of Optical Science and Engineering, Fudan University, Shanghai 200433, China

Ramzan Ullah
Shanghai Ultra-Precision Optical Manufacturing Engineering Center, Department of Optical Science and Engineering, Fudan University, Shanghai 200433, China
Department of Physics, COMSATS Institute of Information Technology, Islamabad 45550, Pakistan

Ishaq Ahmad
Department of Physics, COMSATS Institute of Information Technology, Islamabad 45550, Pakistan

Patrick Kilcullen, Mark Shegelski and Matthew Reid
Department of Physics, University of Northern British Columbia, Prince George, BC, Canada V2N 4Z9

MengXing Na, David Purschke and Frank Hegmann
Department of Physics, University of Alberta, Edmonton, AB, Canada T6G 2E1

Giuseppe Pesce
Department of Physics "E. Pancini", University of Naples Federico II, Via Cintia, 80126 Naples, Italy

Giulia Rusciano and Antonio Sasso
Department of Physics "E. Pancini", University of Naples Federico II, Via Cintia, 80126 Naples, Italy
National Institute of Optics (INO), National Research Council (CNR), Via Campi Flegrei 34, 80078 Pozzuoli, Italy

Gianluigi Zito
Institute of Protein Biochemistry (IBP), National Research Council (CNR), Via Pietro Castellino 111, 80131 Napoli, Italy

A. M. Dunaev, V. B. Motalov, L. S. Kudin and M. F. Butman
Research Institute of Thermodynamics and Kinetics, Ivanovo State University of Chemistry and Technology, Ivanovo 153000, Russia

K.W. Krämer
Department of Chemistry and Biochemistry, University of Bern, 3012 Bern, Switzerland

Samuel P. Hernández-Rivera
ALERT DHS Center of Excellence for Explosives Research, Department of Chemistry, University of Puerto Rico-Mayagüez, Mayagüez, PR 00681, USA

John R. Castro-Suarez
ALERT DHS Center of Excellence for Explosives Research, Department of Chemistry, University of Puerto Rico-Mayagüez, Mayagüez, PR 00681, USA
Molecular Spectroscopy Research Group, Antonio de Arevalo Technological Foundation, TECNAR, Cartagena, Colombia

Leonardo C. Pacheco-Londoño
ALERT DHS Center of Excellence for Explosives Research, Department of Chemistry, University of Puerto Rico-Mayagüez, Mayagüez, PR 00681, USA
Environmental Engineering Program, Vice-Rectory for Research, Universidad ECCI, Bogota, Colombia

Joaquín Aparicio-Bolaño
Department of Physics, University of Puerto Rico, Ponce, PR 00732, USA

Klara Dégardin, Aurélie Guillemain and Yves Roggo
F. Hoffmann-La Roche Ltd., Bldg 250 Room 3.504.01, Wurmisweg, 4303 Kaiseraugst, Switzerland

Melissa A. Kerr and Fei Yan
Department of Chemistry, North Carolina Central University, Durham, NC 27707, USA

T. I. Milenov
"E. Djakov" Institute of Electronics, Bulgarian Academy of Sciences, 72 Tzarigradsko Chaussee Blvd., 1784 Sofia, Bulgaria

E. Valcheva and V. N. Popov
Faculty of Physics, University of Sofia, 5 James Bourchier Blvd., 1164 Sofia, Bulgaria

Myeongsu Seong, Songhyun Lee and Hyeryun Jeong
Department of Biomedical Science and Engineering, Institute of Integrated Technology, Gwangju Institute of Science and Technology, 123 Cheomdan Gwagiro, Bukgu, Gwangju 61005, Republic of Korea

Jae Gwan Kim
Department of Biomedical Science and Engineering, Institute of Integrated Technology, Gwangju Institute of Science and Technology, 123 Cheomdan Gwagiro, Bukgu, Gwangju 61005, Republic of Korea

School of Electrical Engineering and Computer Science, Gwangju Institute of Science and Technology, 123 Cheomdan Gwagiro, Bukgu, Gwangju 61005, Republic of Korea

NoSoung Myoung and Sang-Youp Yim
Advanced Photonics Research Institute, Gwangju Institute of Science and Technology, 123 Cheomdan Gwagiro, Bukgu, Gwangju 61005, Republic of Korea

Bang-Cheng Tang, Zeng-Yan Zhou and Wei Shi
Research Institute of Applied Chemistry, College of Material and Chemical Engineering, Tongren University, Tongren, Guizhou 554300, China

Lu Xu
Research Institute of Applied Chemistry, College of Material and Chemical Engineering, Tongren University, Tongren, Guizhou 554300, China
The Modernization Engineering Technology Research Center of Ethnic Minority Medicine of Hubei Province, College of Pharmacy, South-Central University for Nationalities,Wuhan 430074, China

Hai-Yan Fu and Qiao-Bo Yin
The Modernization Engineering Technology Research Center of Ethnic Minority Medicine of Hubei Province, College of Pharmacy, South-Central University for Nationalities,Wuhan 430074, China

Yuan-Bin She
College of Chemical Engineering, Zhejiang University of Technology, Hangzhou 310014, China

J. J. Roberts, J. Ch. Motin, D. Swain and D. Cozzolino
School of Health, Medical and Applied Sciences, Central Queensland Innovation and Research Precinct (CQIRP), Central Queensland University (CQU), Bruce Highway, North Rockhampton, QLD 4701, Australia

Hamid M. Ghaithan, Mamduh J. Aljaafreh, Abdullah Ahmed and Reem al Thagafi
Department of Physics and Astronomy, College of Science, King Saud University, P.O. Box 2455, Riyadh 11451, Saudi Arabia

Vadivel Masilamani, Saradh Prasad and Mohamad S. Alsalhi
Department of Physics and Astronomy, College of Science, King Saud University, P.O. Box 2455, Riyadh 11451, Saudi Arabia
Research Chair on Laser Diagnosis of Cancers, Department of Physics and Astronomy, College of Science, King Saud University, Riyadh 11451, Saudi Arabia

Pin Gao, Bing Han, Ye Du, Gang Zhao and Zhimin Fan
Department of Breast Surgery, The First Hospital of Jilin University, Changchun 130021, China

Zhigang Yu
Department of Breast Surgery, The Second Hospital of Shandong University, Jinan 250033, China

Chao Zheng
Department of Breast Surgery, The Second Hospital of Shandong University, Jinan 250033, China
Department of Mechanical Engineering, Johns Hopkins University, Baltimore, MD 21218, USA

Weiqing Xu
State Key Laboratory for Supramolecular Structure and Materials, Jilin University, Changchun 130012, China

Qian Wang
School of Automation and Information Engineering, Xi'an University of Technology, Xi'an, Shaanxi 710048, China

Boyan Cai, Yajie Yu and Hui Cao
Shaanxi Key Laboratory of Smart Grid and State Key Laboratory of Electrical Insulation and Power Equipment, School of Electrical Engineering, Xi'an Jiaotong University, Xi'an, Shaanxi 710049, China

Marilena Ricci and Cristiana Lofrumento
Department of Chemistry "Ugo Schiff", University of Florence, Via della Lastruccia 3-13, 50019 Sesto Fiorentino, Italy

Emilio Castellucci and Maurizio Becucci
Department of Chemistry "Ugo Schiff", University of Florence, Via della Lastruccia 3-13, 50019 Sesto Fiorentino, Italy
European Laboratory for Non-Linear Spectroscopy (LENS), Via N. Carrara 1, 50019 Sesto Fiorentino, Italy

Jingzhe Wang, Tashpolat Tiyip, Jianli Ding, Dong Zhang, Wei Liu and Fei Wang
College of Resources and Environment Science, Xinjiang University, Urumqi 830046, China
Key Laboratory of Oasis Ecology, Xinjiang University, Urumqi 830046, China

Suk-Ju Hong, Shin-Joung Rho, Ah-Yeong Lee, Heesoo Park, Jinshi Cui and Yong-Ro Kim
Department of Biosystems and Biomaterials Science and Engineering, Seoul National University,1 Gwanak-ro, Gwanak-gu, Seoul 08826, Republic of Korea

Ghiseok Kim
Department of Biosystems and Biomaterials Science and Engineering, Seoul National University, 1 Gwanak-ro, Gwanak-gu, Seoul 08826, Republic of Korea
Research Institute of Agriculture and Life Sciences, Seoul National University, 1 Gwanak-ro, Gwanak-gu, Seoul 08826, Republic of Korea

Jongmin Park
Department of Bio-Industrial Machinery Engineering, Pusan National University, 1268-50 Samnangjin-ro, Cheonghak-ri, Samnangjin-eup, Miryang-si 50463, Republic of Korea

Soon-Jung Hong
Rural Human Resource Development Center, Rural Development Administration, 420 Nongsaengmyeong-ro, Wansan-gu, Jeonju-si 54874, Republic of Korea

Index